Handbook of Food and Beverage Stability

Chemical, Biochemical, Microbiological, and Nutritional Aspects

FOOD SCIENCE AND TECHNOLOGY

A SERIES OF MONOGRAPHS

A list of books in this series is available from the publisher on request.

Handbook of Food and Beverage Stability

Chemical, Biochemical, Microbiological, and Nutritional Aspects

Edited by

GEORGE CHARALAMBOUS

St. Louis, Missouri

1986 

ACADEMIC PRESS, INC.

Harcourt Brace Jovanovich, Publishers

Orlando San Diego New York Austin
London Montreal Sydney Tokyo Toronto

ACADEMIC PRESS, INC.
Orlando, Florida 32887

United Kingdom Edition published by
ACADEMIC PRESS INC. (LONDON) LTD.
24–28 Oval Road, London NW1 7DX

Library of Congress Cataloging in Publication Data

Handbook of food and beverage stability.

(Food science and technology)
Includes index.
1. Food–Analysis–Handbooks, manuals, etc. 2. Food spoilage–Handbooks, manuals, etc. 3. Food–Shelf-life dating–Handbooks, manuals, etc. I. Charalambous, George, Date . II. Series.
TX535.H34 1986 664'.028 85-43102
ISBN 0–12–169070–9 (alk. paper)

PRINTED IN THE UNITED STATES OF AMERICA

86 87 88 89 9 8 7 6 5 4 3 2 1

Contents

Contributors

Numbers in parentheses indicate the pages on which the authors' contributions begin.

MILTON E. BAILEY (75), Department of Food Science and Nutrition, University of Missouri, Columbia, Missouri 65211

UMBERTO BRACCO (391), Nestlé Research Laboratories, CH-1800 Vevey, Switzerland

RONALD J. CLARKE (685), Donnington, Chichester, Sussex P020 7PW, England

LEOPOLDO G. ENRIQUEZ (113), Department of Food Science and Technology, Virginia Polytechnic Institute and State University, Blacksburg, Virginia 24061

GEORGE J. FLICK, JR. (113), Department of Food Science and Technology, Virginia Polytechnic Institute and State University, Blacksburg, Virginia 24061

THOMSEN J. HANSEN (423), Department of Nutrition and Food Sciences, Drexel University, Philadelphia, Pennsylvania 19104

IAN HORMAN (391), Nestlé Research Laboratories, CH-1800 Vevey, Switzerland

JANIS B. HUBBARD (113), Department of Food Science and Technology, Virginia Polytechnic Institute and State University, Blacksburg, Virginia 24061

KAREL KULP (1), American Institute of Baking, Manhattan, Kansas 66502

DAVID C. LEWIS (353), Department of Environmental Toxicology, University of California, Davis, Davis, California 95616

WILLIAM W. MENZ (621), Winston–Salem, North Carolina 27104

ROBERT R. MOD (489), Southern Regional Research Center, United States Department of Agriculture, New Orleans, Louisiana 70179

STEVEN NAGY (719), Scientific Research Department, State of Florida Department of Citrus, Lake Alfred, Florida 33850

JOHN H. NELSON (33), Quality Assurance/Regulatory Compliance, Kraft, Inc., Glenview, Illinois 60025

TOSHITERU OHBA (773), National Research Institute of Brewing, Tokyo 114, Japan

ROBERT L. ORY (489), Southern Regional Research Center, United States Department of Agriculture, New Orleans, Louisiana 70179

THOMAS M. RADKE (467), Food Science Research Center, Chapman College, Orange, California 92666

PASCAL RIBÉREAU-GAYON (745), Institut d'OEnologie, Université de Bordeaux II, F-33405 Talence, France

LOUIS B. ROCKLAND (467), Food Science Research Center, Chapman College, Orange, California 92666

RUSSELL L. ROUSEFF (719), Scientific Research Department, State of Florida Department of Citrus, Lake Alfred, Florida 33850
MAKOTO SATO (773), National Research Institute of Brewing, Tokyo 114, Japan
TAKAYUKI SHIBAMOTO (353), Department of Environmental Toxicology, University of California, Davis, Davis, California 95616
JAMES S. SWAN (801), Pentlands Scotch Whisky Research Ltd., Edinburgh EH11 1QU, Scotland
JAMES VETTER (1), American Institute of Baking, Manhattan, Kansas 66502
TEI YAMANISHI (665), Ochanomizu University, Tokyo 167, Japan
TAMOTSU YOKOTSUKA (517), Kikkoman Corporation, 399 Noda-shi, Chiba-ken 278, Japan

Preface

A recently compiled list of world needs amenable to solution through chemistry was submitted to leaders in the world chemical community for comment and discussion. The application of chemistry to alleviate hunger was allotted high priority by almost everyone. One way of achieving this, as the population of the world expands and the migration to urban centers where food is not grown continues, is through an improvement in the stability of foods and beverages. The prevention of spoilage and thus waste in the face of dwindling resources in the food supply has long been an objective. In many ways, however, chemistry and agriculture, also related endeavors, have developed along parallel or independent paths.

Fortunately, chemistry—the root of all life processes—is becoming better understood and more accessible. A strong synergism between the chemical, agricultural, and related sciences is highly desirable. This handbook attempts to provide in easily accessible detail up-to-date information relevant to the stability of foods and beverages. Highly qualified scientists have compiled an extraordinary amount of data on the chemical, biochemical, and microbiological stability, along with sensory aspects, of selected foods and beverages. These data have been distilled and are presented mostly in tabular form, with a minimum of commentary whenever possible.

A total of 17 chapters (10 on food, 7 on beverages) by renowned experts in their particular fields from the United States, Europe, and Japan present a wealth of food and beverage stability information in handbook format. In particular, the chapters on fish and shellfish, cheese, and meat are remarkable in presenting data not readily available in an easily digestible form.

This handbook, encompassing as it does aging, shelf life, and stability—in short, the knowledge necessary to ensure preservation of our food supply—should help to bring about the above-mentioned synergism between chemical, agricultural, and related sciences. It is expected to fill a need, especially through the convenience of its tabular presentations.

The editor wishes to thank his far-flung authors for their considerable efforts in compiling up-to-date and not always readily available information, compressing it in tables for handbook format. He also expresses his appreciation of the publisher's advice and assistance.

CHAPTER 1

EFFECT OF AGING ON FRESHNESS OF WHITE PAN BREAD

KAREL KULP
JAMES VETTER
American Institute of Baking
Manhattan, Kansas

Handbook of Food and Beverage Stability: Chemical, Biochemical, Microbiological, and Nutritional Aspects

TABLE I

Definitions of Changes Affecting Freshness of Bread During Storage[a]

Term Describing Loss of Freshness	Definition	Characteristics	Cause
Staling	A series of changes that cause a decrease in consumer acceptance other than that resulting from the action of spoilage microorganisms.	Crust staling: loss of crispness.	Moisture migration from crumb to crust.
		Crumb staling: firming, development of crumbliness, loss of flavor, and emergence of stale flavor.	Retrogradation, complexation of flavorants with amylose and oxidative changes.
Microbial Spoilage	Growth of microorganisms in crust or crumb during storage.	Molds (various Aspergilli, Penicillia) wild yeasts, spore formers.	Contaminated ingredients, air, and equipment during processing.
Flavor Deterioration	Part of staling (see Staling above).	Loss of fresh bread flavor and aroma. Development of stale bread flavor.	Complexation of flavorants with amylose.
			Oxidative changes in flavorants, migration of flavorants from crust to crumb.

[a]Kulp and Ponte (1981).

TABLE II

Methods of Determination of Freshness of Breads

Method	Principle	Reference/Use
	I. PHYSICAL METHODS	
Compressibility	A uniform square of crumb is compressed to constant deformation using Baker's Compressimeter or Instron. The required compression force increases with bread age.	AACC Method 74-10 (AACC, 1983). Most common research & control method.
Capacitance/Conductance	Both properties increase with the age of bread.	Kay and Willhoft (1972), research method.
Cell-Wall Firmness Measurements	Determines compressibility of a bread crumb pellet; eliminates effect of loaf volume.	Guy and Wren (1968), research method.
Differential Thermal Analysis	Emergence of an endothermic peak indicative of starch crystallization.	Axford and Collwell (1967), Russell (1983).
X-Ray Diffraction	Change of X-ray diffraction pattern from A to B/V.	Zobel (1973), research. NOTE: not usable in bread with bacterial α-amylase.

(table continues)

TABLE II (Continued)

Method	Principle	Reference/Use
Nuclear Magnetic Resonance	Measures decreases of water mobility which decreases with age of bread.	Leung *et al.* (1983).
Rate of Starch Crystallization	Avrami Equation $[\Theta = (E_L - E_t)/(E_L - E_O) = \exp(-kt^n)]$ is used to estimate the rate constant k; 1/k = time constant of crystallization of the system is generally reported (the higher the value, the slower the rate); Θ is noncrystalline portion, E_L is limiting modulus and E_t, E_O moduli at times t and O, respectively.	Comford *et al.* (1964). widely used in research.
	II. CHEMICAL METHODS	
Iodine Absorption	This value decreases with bread aging; it may be determined colorimetrically, iodometrically, or potentiometrically.	Pelshenke and Hampel (1962), research.
	III. SENSORY METHODS	
Panel Test Evaluation of Staling	Bread samples rating by panel for degree of staleness.	AACC 77-30 (AACC, 1983), widely used in control and research work.

TABLE III

Theories of Bread Staling

Theory According to:	Basis
Schoch and French (1947)	Retrogradation of starch polymers. Amylose crystallizes rapidly and produces initial firmness. Firming during storage is attributed to crystallization of amylopectin (Fig. 2).
Lineback (1984)	Essentially same as Schoch's except it emphasizes intergranular interaction (Fig. 4).
Knyaginichev (1965)	Formation of structured gel, consisting of starch, protein, and water.
Erlander and Erlander (1969)	Interaction of gliadin and glutenin with starch chains.
Willhoft (1971)	Implicates gluten in addition to starch (Fig. 5).

TABLE IV

Effect of Protein Content of Flour on Avrami Exponent (n) and Time Constant of Bread Stored at 21°C

Flour	Bread	
Protein Content[a]	Avrami Exponent	Time Constant
10.6	0.94	3.75
11.0	0.92	3.74
13.9	0.92	5.44
21.6	1.04	11.25

[a]From Kim and D'Appolonia (1977b).

TABLE V

Effect of Soluble and Insoluble Flour Pentosans on Staling of Starch Gel and Bread Stored at 21°C[a]

Gels	Avrami Exponent	Overall Time Constant	Time Constant During the First Day of Storage
	Starch		
Starch (S)	0.98	3.80	3.70
S-Soluble Pentosans	0.70	5.33	3.29
S-Insoluble Pentosans	0.83	7.41	5.75
	Bread		
Control (C)	0.92	5.44	4.80
C-1% Soluble Pentosans	0.73	6.53	4.23
C-1% Insoluble Pentosans	0.77	8.54	5.88

[a]From Kim and D'Appolonia (1977a).

TABLE VI

Effect of Flour α-Amylase on Firmness Values (g/cm) of Breads

Falling Number[a] of Flour (Sec.)	Day				
	0	1	2	3	5
SPONGE/DOUGH BREAD PROCESS					
411	11.5	24.3	31.6	37.8	43.2
353	8.5	18.0	23.3	25.0	25.3
247	7.3	16.5	17.5	23.1	23.3
148	6.7	11.6	18.2	17.3	24.8
STRAIGHT DOUGH PROCESS					
411	5.2	11.2	17.8	27.8	27.0
353	6.3	11.2	17.2	25.0	31.5
247	5.5	11.3	14.0	24.4	29.3
148	6.5	13.4	14.9	24.9	32.5

[a]Falling number indicates α-amylase activity (the higher the number, the higher the activity).
(From D'Appolonia, 1984).

TABLE VII

Effect of Formulation of White Pan Bread on Freshness[a]

Formula Ingredient	Crust Freshness[b]	Crumb Freshness[b]
Flour Protein	+	+
Sugars	+	+
Oligosaccharides	+	+
Dextrins	+	+
Milk Ingredients	+	-
Milk Replacers	+	±
Salt	±	±
Shortening	-	+
Water Absorption:		
High	-	-
Optimum	+	+
Low	-	-
Enzymes:		
Malt	+	+
Fungal Amylases	+	+
Bacterial Amylase	+	++
Surfactants	+	++

[a]From Kulp (1979).

[b]+ = Improves freshness retention.
± = No effect on freshness retention.
- = Reduces freshness retention.

TABLE VIII

FDA Regulations
Dough Strengtheners/Crumb Softeners Standardized Products[a]

Product	Limitation	Reference	Function
Calcium Stearoyl-2-Lactylate	0.5% max.[b]	CFR21, 136.110 (c)(15)	Dough strengthener (excellent) Crumb softener (good)
Sodium Stearoyl-2-Lactylate	0.5% max.[b]	CFR21, 136.110 (c)(15)	Dough strengthener (excellent) Crumb softener (very good)
DATA Esters	No limit	CFR21, 136.110 (c)(6)(ii)	Dough strengthener (excellent) Crumb softener (fair)
Mono- and diglycerides	No limit	CFR21, 136.110 (c)(5)(ii)	Dough strengthener (none) Crumb softener (excellent)
Succinylated monoglycerides	0.5% max.[b]	CFR21, 136.110 (c)(15)	Dough strengthener (good) Crumb softener (good)
Polysorbate 60	0.5% max.[b]	CFR21, 136.110 (c)(15)	Dough strengthener (fair) Crumb softener (good)
Ethoxylated monoglycerides	0.5% max.[b]	CFR 21, 136.110 (c)(15)	Dough strengthener (very good) Crumb softener (poor)
Sucrose Esters	GMP[c]	CFR 170.3 (n)(1)	Dough strengthener (very good) Crumb softener (good)

[a]From Dubois (1979).
[b]Total alone or in combination cannot exceed 0.5% based on flour.
[c]Use in accordance with good manufacturing practices.

TABLE IX

Effect of Surfactants on X-Ray Pattern During Aging of Bread[a]

Bread No.	Description of Additives	Compression Data[b]	Intensity of Diffraction Lines[c,d]				Complexing Index[e]
			Structure		B	V	
1	Control	13.8	B	V	10	7	-
2	Hydrated mono-diglycerides, 20%	12.1	V	B	6	10	28
6	Succinylated monoglycerides	11.8	V	B	6	9	63
4	60% Mono-di, 40% ethoxylated monoglycerides	11.6	V	B	5	9	30
3	Sodium stearoyl-2-lactylate	10.6	V	B	4	10	72
5	75% Hydrated mono-diglycerides, 25% Polyoxethylene Sorbitan 20	10.5	V	B	4	10	29

[a] For mechanism, see Figure 3.

[b] After 3 days aging.

[c] After 6 days aging.

[d] Scale, 1-10 with 10 being the most intense; B, retrograded starch structure; V, amylose complex structure.

[e] Data from Krog (1971).

TABLE X

Processing Variables Affecting Staling Rate[a]

Operational Steps	Crust Freshness[b]	Crumb Freshness[b]
Dough Mixing:		
Overmixing	-	-
Optimum	+	+
Undermixing	-	-
Fermentation Time:		
Short	-	-
Normal	+	+
Long	-	+
Baking Rate:		
Slow	-	-
Fast	-	+

[a]From Kulp (1979).

[b]+ = Improves freshness retention.
 - = Reduces freshness retention.

TABLE XI

Effect of Storage Temperature of Bread Firming During Storage[a]

Temperature, ^{o}F	Chorleywood Bread Process, Time Constant	Bulk Fermented, Time Constant
30	1.44	1.39
50	1.84	1.89
70	3.28	3.68
90	5.02	5.51
110	9.0	--
130	13.5	--
150	23.3	--

[a]Cornford *et al.* (1964).

TABLE XII

The Effect of Bread Storage on Flavor Score, Carbonyl Content, and GLC Headspace Area[a]

Bread Storage, Days	Average Panel Flavor Score	Total Carbonyl Compounds, ppm	Total GLC Headspace area, cm^2
0	1.95	224	64.2
1		136	61.9
2	3.55	176	63.7
3		280	66.6
4	4.50	280	67.8
5		328	69.5

[a]From Lorenz and Maga (1972).

TABLE XIII

Bread Carbonyl Composition, Changes During Aging[a]

	Freshly Baked Bread		5-Day Old Bread	
	GLC Area, cm^2	%	GLC Area, cm^2	%
Formaldehyde	1.2	2.5	0.1	0.2
Acetaldehyde	2.0	4.0	0.4	0.9
Acetone	2.2	4.5	0.8	1.7
Propanal	6.7	13.8	0.6	1.3
Butanal	11.1	22.7	0.3	0.8
2-Butanone	5.4	11.0	6.4	14.7
2-Hexanone	4.0	8.2	29.4	67.4
Hexanal	4.3	8.9	1.0	2.4
2-Heptanone	1.7	3.4	0.5	1.1
Heptanal	0.6	1.3		
Nonanal	6.9	14.2	4.1	9.5
Unknown	2.7	5.5		
	48.6	100.0	43.6	100.0

[a]From Lorenz and Maga (1972).

TABLE XIV

Migration of Moisture from Crust to Crumb[a]

Storage Bread[b]	Time (Days)	Storage Temperature 2°C Without Crust %	2°C With Crust %	2°C Moisture Migration[c] %	30°C Without Crust %	30°C With Crust %	30°C Moisture Migration[c] %
Control	0	45.42	-	-	45.28	-	-
	1	45.16	44.83	0.34	44.95	43.56	1.62
	4	44.94	43.73	1.44	45.31	40.70	4.48
		45.17			45.18		
With SSL	0	45.36	-	-	45.36	-	-
	1	45.06	44.64	0.47	44.81	42.85	2.32
	4	44.92	42.84	2.27	45.34	39.84	5.33
		45.11			45.17		
With Tandem 8	0	45.31	-	-	45.33	-	-
	1	45.01	44.46	0.61	45.02	43.54	1.69
	4	44.90	42.73	2.34	45.22	38.66	6.57
		45.07			45.19		
With Atmul 500	0	45.28	-	-	45.22	-	-
	1	44.96	44.66	0.41	44.82	43.54	1.69
	4	44.98	43.12	1.95	45.65	38.66	6.57
		45.07			45.23		

[a]From Pisesookbunterng and D'Appolonia (1983).
[b]Bread moisture values at zero day storage are those of breads 2 hrs. after baking.
[c]Values reported are percentage point change between moisture content of bread stored with crust intact and bread stored without crust.

TABLE XV

Refreshening of Bread: Firmness Values (in g/cm)[a]

Storage Time (days)	Refreshening Process	Bread					
		Control		With SSL		With Atmul 500	
		2°C	30°C	2°C	30°C	2°C	30°C
0	--	82	84	96	82	75	75
2	Before 1st refreshening	393	233	331	176	264	137
	After 1st refreshening	99	146	97	125	83	117
4	No refreshening	459	277	393	216	339	234
	Refreshened once/before 2nd refreshening	432	285	294	234	269	187
	After 2nd refreshening	119	180	100	165	74	140
6	No refreshening	516	345	411	263	395	306
	Refreshened twice/before 3rd refreshening	391	315	312	245	289	226
	After 3rd refreshening	124	248	120	198	104	180

[a]From Pisesookbunterng *et al.* (1983).

TABLE XVI

Effect of Production Methods on Bread Softness[a]

Production Method	Crust Freshness[b]	Crumb Freshness[b]
Continuous Mix	1	1
Sponge Dough	2	2
Liquid Ferment, 0% Flour	2.5	2.5
" " 20% Flour	2	2
" " 50% Flour	2	2
Straight Dough	3	3
No-Time Dough	4	4

[a]From Kulp (1979).

[b]Lower number, softer.

TABLE XVII

Product Variables Affecting Staling Rate of White Pan Bread[a]

Bread	Crust Freshness[b]	Crumb Freshness[b]
Specific Volume (Higher)	+	+
Moisture Content (Higher)	+	+
Crust Thickness (Higher)	+	-

[a]Kulp (1979).

[b]+ = Improves.
- = Reduces.

TABLE XVIII

United States Regulatory Status of Antimicrobial Agents[a]

Agent	CFR[b]	Restrictions on Use
Acetic Acid	182.1005	Generally recognized as safe as a multipurpose food substance when used in accordance with good manufacturing practices.
Propionic Acid	184.1081	Affirmed generally recognized as safe direct food substance when used as an antimicrobial agent and a flavoring agent at levels not to exceed good manufacturing practices in baked goods; cheeses; confections; and frostings; gelatins; puddings; and fillings; and jams and jellies.
Calcium Propionate	184.1221	Affirmed generally recognized as safe direct food substance when used as an antimicrobial agent and a flavoring agent at levels not to exceed good manufacturing practices in baked goods; cheeses; confections and frostings; gelatins; puddings and fillings; and jams and jellies.
Sodium Propionate	184.1784	Affirmed generally recognized as safe direct food substance when used as an antimicrobial agent and a forming agent at levels not to exceed good manufacturing practices in baked goods; nonalcoholic beverages; cheeses; confections and frostings; gelatins, puddings, and fillings; jams and jellies; meat products; and soft candy.

(table continues)

TABLE XVIII (Continued)

Agent	CFR[b]	Restrictions on Use
Sorbic Acid	182.3089	Generally recognized as safe as a chemical preservative when used in accordance with good manufacturing practices.
Calcium Sorbate	182.3225	Generally recognized as safe as a chemical preservative when used in accordance with good manufacturing practices.
Potassium Sorbate	182.3640	Generally recognized as safe as a chemical preservative when used in accordance with good manufacturing practices.
Sodium Sorbate	182.3795	Generally recognized as safe as a chemical preservative when used in accordance with good manufacturing practices.
Methyl Paraben	184.1490	Affirmed generally recognized as safe direct food substance when used as an antimicrobial agent at levels not to exceed 0.1 percent in food.
Propyl Paragen	184.167	Affirmed generally recognized as safe direct food substance when used as an antimicrobial agent at levels not to exceed 0.1 percent in food.
Benzoic Acid	184.1021	Affirmed generally recognized as safe direct food substance when used as an antimicrobial agent and as a flavoring agent and adjuvant at a level not to exceed 0.1 percent in food.

(table continues)

TABLE XVIII (Continued)

Agent	CFR[b]	Restrictions on Use
Sodium Benzoate	184.1733	Affirmed generally recognized as safe direct food substance when used as an antimicrobial agent and as a flavoring agent and adjuvant at a level not to exceed 0.1 percent in food.
Ethyl Alcohol	184.1293	Affirmed generally recognized as safe direct food substance when used as an antimicrobial agent on pizza crusts prior to final baking at levels not to exceed 2.0 percent by product weight.

[a]Regulatory status may change. Information presented is current as of date of publication.

[b]U.S. Code of Federal Regulations: 21 CFR, Food and Drugs.

TABLE XIX

U.S. Regulatory Status of Antioxidants Which May Be Used in Bakery Products[a]

Antioxidant	CFR[b]	Limitations
Butylated hydroxyanisole	182.3169	Generally recognized as safe for use in food at a total antioxidant level not to exceed 0.02 percent of the fat or oil, including essential (volatile) oil content of the food.
Butylated hydroxytoluene	182.3173	Generally recognized as safe for use in food at a total antioxidant level not to exceed 0.02 percent of the fat or oil, including essential (volatile) oil content of the food.
Propyl gallate	184.1660	Affirmed generally recognized as safe for use in food at a total antioxidant level not to exceed 0.02 percent of the fat or oil, including essential (volatile) oil content of the food.
Tertiary butylhydroquinone	172.185	Food additive which may be used alone or in combination with BHA and/or BHT (not propyl gallate) at a total antioxidant level not to exceed 0.02 percent of the fat or oil, including essential (volatile) oil content of the food.

[a] Regulatory status may change. Information presented is current as of date of publication.

[b] U.S. Code of Federal Regulations: 21 CFR, Food and Drugs.

TABLE XX

Spectrum of Effective Action of Antimicrobial Agents Commonly Used in Bakery Foods[a,b,c]

Agent	pH	Effective Against:
Benzoates	4.5 or below	Yeasts, molds, many bacteria.
Propionates	5.5 or below	Mold; limited antibacterial potency, but effective against "rope" (B. subtilis); essentially no effect on yeasts.
Sorbates	6.5 or below	Yeasts; molds; many bacteria, including B. subtilis, but generally not lactic acid bacteria.

[a]Barrett (1970). [b]Brachfeld (1969). [c]King (1981).

TABLE XXI

Recommended Levels of Calcium or Sodium Propionates in Baked Goods[a]

Product	oz. Propionate/100 lb. flour
White bread, buns and rolls	2.5-3.0 under normal conditions; up to 5.0 under severe conditions
Dark breads, whole or cracked wheat, rye bread, rolls or buns	3.0-4.0 under normal conditions; up to 6.0 under severe conditions
Angel Food Cake	1.5-3.5
Cheesecake	2.0-4.0
Chocolate or Devil's Food Cake	5.0-7.0
Fruitcake	2.0-6.0
Pound, white or yellow cake	4.0-6.0
Pie Crust	2.0-5.0
Pie Fillings	2.0-4.0

[a]Furia (1972).

TABLE XXII

Recommended Levels of Sorbates in Bakery Goods[a]

Product	% Batter Wt.	Method of Application
Angel Food Cake	0.03-0.06	Dry blend with flour.
Cheesecake	0.1-0.3	Dry blend with sugar and milk powder.
Chocolate Cake	0.1-0.3	Dry blend with flour or add during creaming.
Devil's Food Cake	0.3	Dry blend with flour or add during creaming.
Pound, white, yellow cake	0.075-0.15	Dry blend with flour or add during creaming
Cake Mixes	0.05-0.1	Dry blend with flour and other dry ingredients.
Fruitcake	0.1-0.4	Dry blend with flour. For added protection, soak fruit in 1.0% potassium sorbate solution.
Fillings, Icings, Fudges, Toppings	0.05-0.1	Add after heating when temperature has dropped below 160°F.
Pie Crust Dough	0.05-0.1	Dry blend with flour or dough.
Pie Fillings	0.05-0.1	Add after heating when temperature has dropped below 160°F.
Doughnut Mixes	0.05-0.1	Dry blend with flour and other dry ingredients.

[a]Monsanto Company (1978).

TABLE XXIII

Effect of Sorbates and Propionates on Mold-Free Shelf Life of Baking Foods[a]

Additive	% Flour Basis in Dough	% Product Weight on Surface	Average Days Without Mold
Variety Bread:			
None	0	0	3
Ca Propionate	0.16	0	5
K Sorbate	0	0.016	9
K Sorbate	0.02	0.016	11+
Hamburger Buns:			
None	0	0	7
Ca Propionate	0.3	0	16
K Sorbate	0	0.02	23
K Sorbate	0.02	0.02	28+
English Muffins:			
None	0	0	5
Ca Propionate	1.0	0.1[b]	9
K Sorbate	0.12	0.1	12
K Sorbate	0.12	0.2	26+
Brown & Serve Rolls:			
None	0	0	4
Ca Propionate	0.4	0	9
K Sorbate	0.05	0.06	14
K Sorbate	0.05	0.12	27+
Flour Tortillas:			
None	0	0	2
Ca Propionate	0.25	0	2
K Sorbate	0.05	0.1	9
K Sorbate	0.05	0.2	36+

[a]Monsanto Company (1977).

[b]In dusting flour.

TABLE XXIV

Effect of Antimicrobial Agents on Mold-Free Shelf Life of Bread[a]

Antimicrobial Agent	Usage oz./CWT	Mold-Free Shelf Life, Days[b]	
		Sponge/ Dough	No-Time Dough
Ca Propionate	6	7	5
	5	6	4
	4	6	4
	3	4	4
	0	3	3
Sodium Diacetate	6	7	5
	5	6	4
	4	6	4
	3	4	4
	0	3	3
200 Grain Vinegar	16	5	5
	8	3	3
	6	3	3
	4	3	3
	0	3	3
Potassium Sorbate	2.5	3	4
	1.5	3	3
	0	3	3
Ca Propionate/Na Diacetate	6/0	6	5
	6/1	5	4
	4.5/1.5	5	4
	3/3	4	3
	0/0	3	3
Ca Propionate/200 Grain Vinegar	6/4	8	5
	5/4	6	5
	4/4	6	5
	3/4	6	4
	0/4	3	3
Ca Propionate/Mono-calcium Phosphate	6/4	8	5
	5/4	8	5
	4/4	7	5
	4/5	6	5
	0/4	4	4

[a]Briscoe (1978).

[b]Slices of bread inoculated with mold spores.

TABLE XXV

Effect of Antimicrobial Agents on Rope-Free Shelf Life in Bread[a]

Antimicrobial Agent	%, Flour Basis	Days of Rope-Free Shelf Life
Acetic Acid	0.1	3-6
	0.2	>21
Calcium Acetate	0.3	3-6
	0.4	6-10
	0.5	>21
Sodium Diacetate	0.2	3-6
	0.3	17-21
Propionic Acid	0.1	3-6
	0.2	>21
Calcium Propionate	0.18	3-6
	0.23	6-10
	0.35	>21
Sodium Propionate	0.2	3-6
None - Control	0.0	0-3

[a]Ingram *et al.* (1956).

TABLE XXVI

Effect of pH and Equilibrium Relative Humidity (ERH) on Mold-Free Shelf Life of Cakes With and Without 0.1% Sorbic Acid[a]

ERH %	Sorbic Acid %	Mold-Free Shelf Life, Days, at pH: 5.5	6.0	6.5	7.0
92	0.0	4	4	4	4
	0.1	8	6	5	4
90	0.0	4	4	4	4
	0.1	11	8	6	5
88	0.0	5	5	5	5
	0.1	15	11	8	7
86	0.0	7	7	7	7
	0.1	22	16	12	9
84	0.0	10	10	10	10
	0.1	>30	24	17	13

[a]Seiler (1962).

TABLE XXVII

Food Ingredients Suggested for Antimicrobial Activity in Bread

Product	Approximate Use Level, Percent, Flour Basis
200 Grain Vinegar	1.0
Raisin Juice Concentrate	5.0-10.0
Cultured Dairy Product	1.0- 3.0
Cultured Wheat Product	0.5

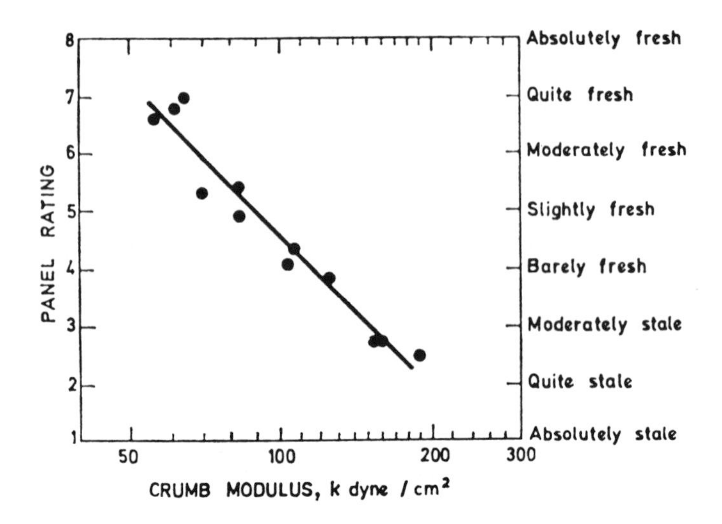

Fig. 1. Relationship between panel ratings and crumb firmness values. From Axford *et al.* (1968).

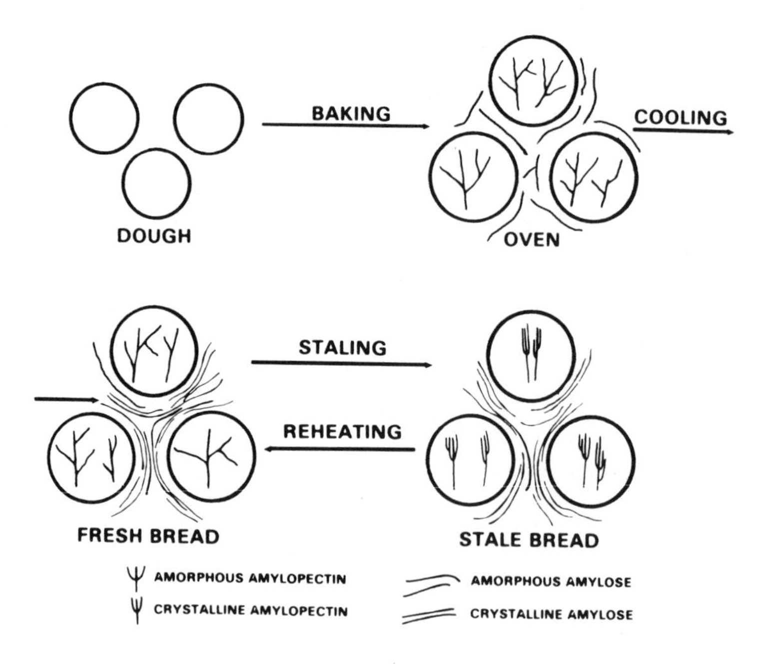

Fig. 2. Schoch's mechanism of bread staling. From Schoch and French (1947).

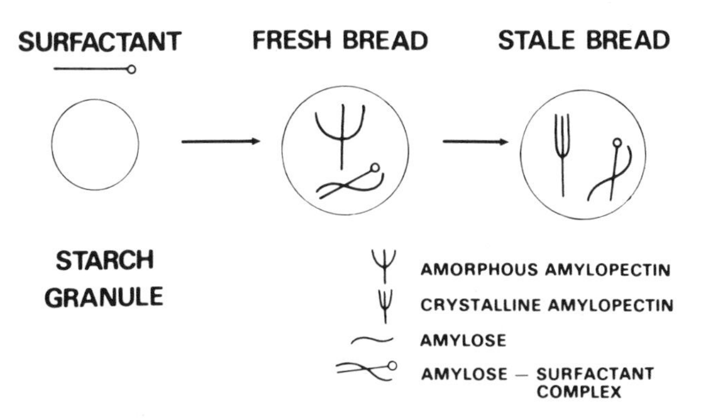

Fig. 3. Schoch's mechanism of action of surfactants (Schoch, 1965).

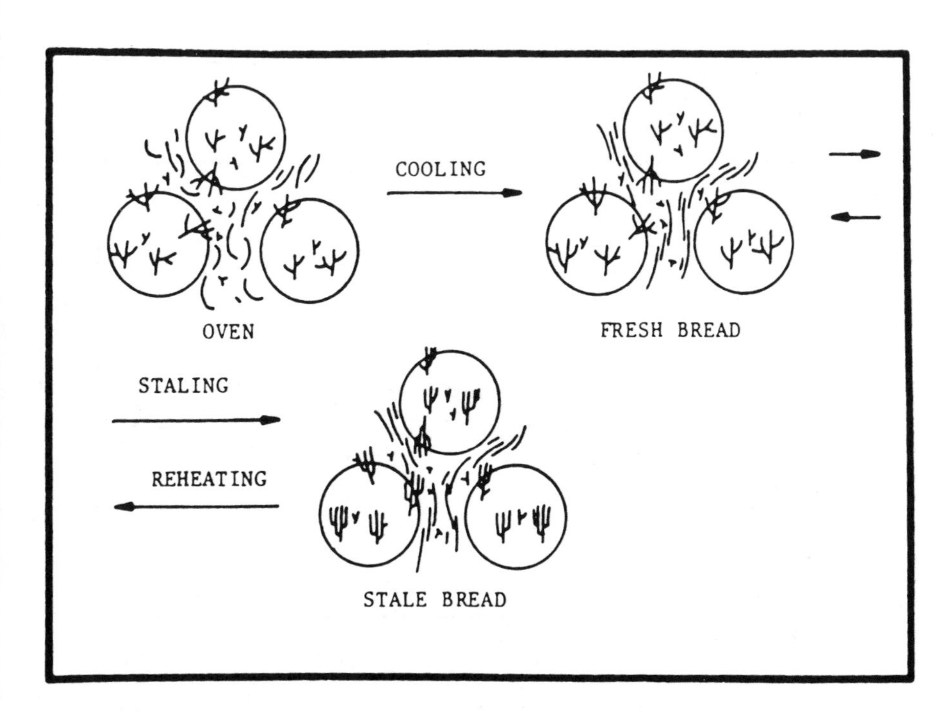

Fig. 4. Modification of Schoch's mechanism as proposed by Lineback. From Lineback (1984).

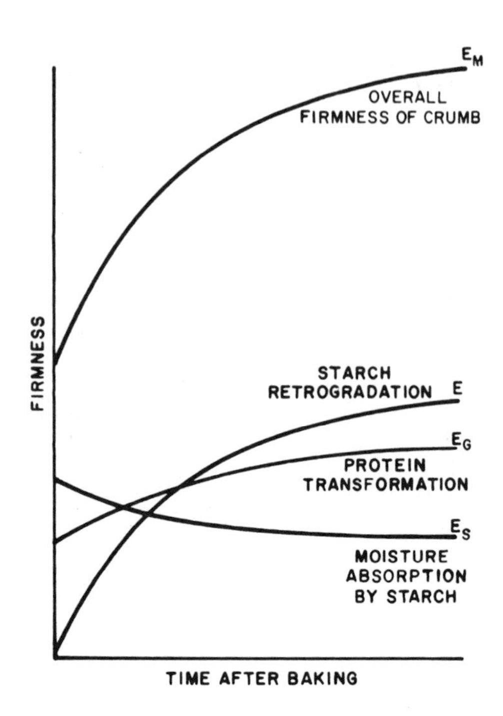

Fig. 5. Graphic representation of Willhoft's equation. From Willhoft (1971).

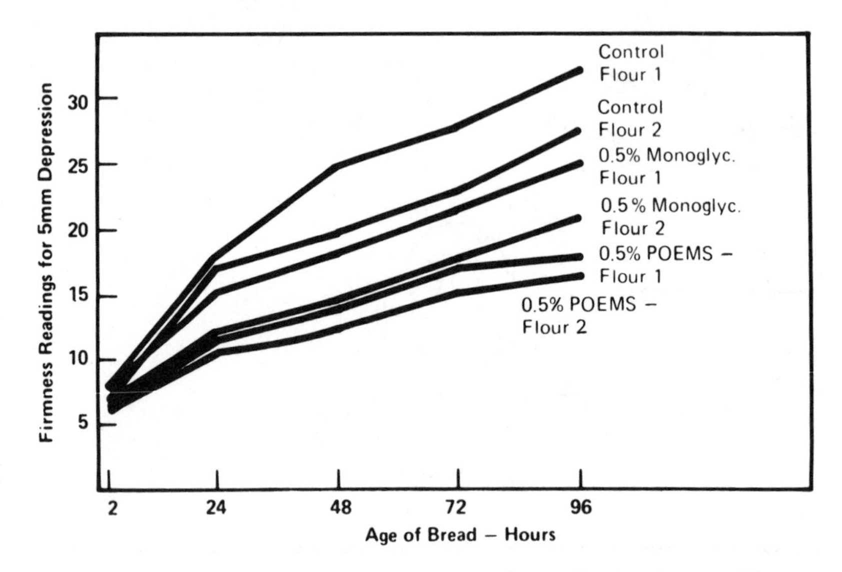

Fig. 6. Effect of bread storage and surfactants on firmness. Control doughs have 3% shortening, others 2%. From Skovholt and Dowdle, 1950.

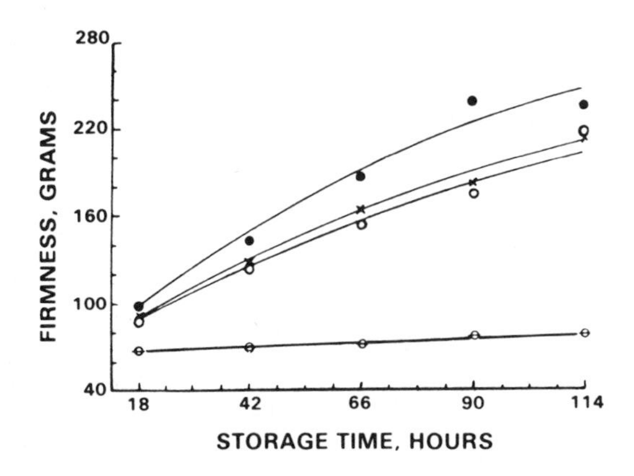

Fig. 7. Effects of equivalent levels of malted flour (×), fungal (o), and bacterial α-amylase (⊖) on bread firmness. Control (●). From Miller *et al.*, (1953).

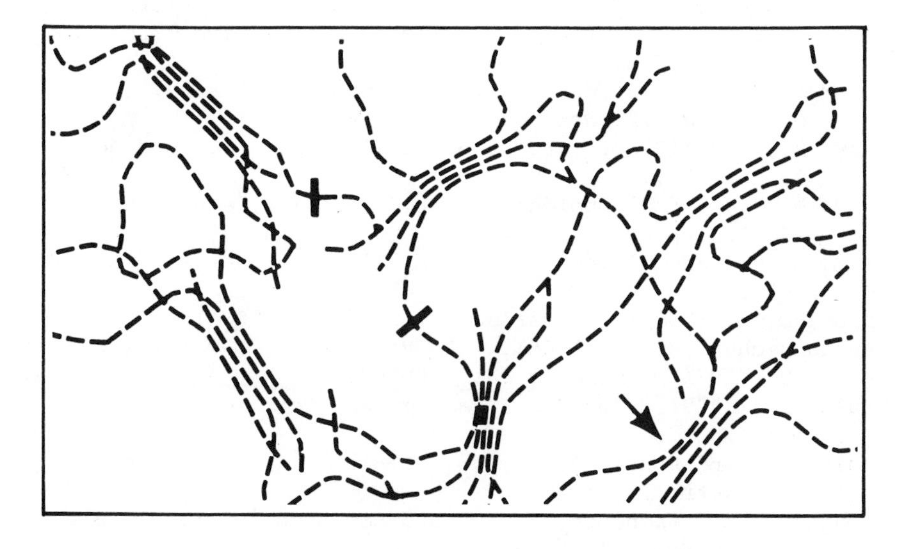

Fig. 8. Network structure of starch gel showing crystalline regions (the arrow points to one) and possible points in amorphous regions at which bacterial α-amylase cleaves the structure during bread storage. From Senti and Dimler, 1960.

REFERENCES

American Association of Cereal Chemists (AACC) (1983). "Approved Methods of the AACC." AAAC, St. Paul, Minnesota.
Axford, D. W. E., and Colwell, K. H. (1967). *Chem. Ind. (London)*, pp. 467-468.
Axford, D. W. E., Colwell, K. H., Cornford, S. J., and Elton, G. A. H. (1968). *J. Sci. Food Agric. 19*, 95-101.
Barrett, F. (1970). *Baker's Dig. 44* (4), 48-49, 67.
Brachfeld, B. A. (1969). *Baker's Dig. 43* (5), 60-62, 65.
Briscoe, R. (1978). *Baker's J. 39* (2), 12-13, 31-32.
Comford, S. J., Axford, D. W. E., and Elton, G. A. H. (1964). *Cereal Chem. 41*, 216-229.
D'Appolonia, B. L. (1984). *In* "International Symposium on Advances in Baking Science and Technology," pp. T1-T18. Department of Grain Science, Kansas State University, Manhattan.
Dubois, D. (1979). *Am. Inst. Baking Res. Dep. Tech. Bull. 1* (4).
Erlander, S. R., and Erlander, L. G. (1969). *Staerke 21*, 305-315.
Furia, T. E., ed. (1972). "Handbook of Food Additives." CRC Press, Cleveland, Ohio (Now Boca Raton, Florida).
Guy, R. C. E., and Wren, J. J. (1968). *Chem. Ind. (London)*, pp. 1727-1728.
Ingram, M., Ottaway, F. L. M., and Coppock, J. B. M. (1956). *Chem. Ind. (London)*, pp. 1154-1163.
Kay, M., and Willhoft, E. M. A. (1972). *J. Sci. Food Agric. 23*, 321-331.
Kim, S. K., and D'Appolonia, B. L. (1977a). *Cereal Chem. 54*, 150-160.
Kim, S. K., and D'Appolonia, B. L. (1977b). *Cereal Chem. 54*, 207-215.
King, B. D. (1981). *Baker's Dig. 55* (5), 8-10, 12.
Knyaginichev, M. I. (1965). *Zh. Vses. Khim. O-va. 10*, 277.
Krog, N. (1971). *Staerke 23*, 206-209.
Kulp, K. (1979). *Am. Inst. Baking Res. Dep. Tech. Bull. 1* (8).
Kulp, K., and Ponte, J. G., Jr. (1981). *CRC Crit. Rev. Food Sci. Nutr. 15*, 1-48.
Leung, H. K., Magnuson, J. A., and Bruinsma, B. L. (1983). *J. Food Sci. 41*, 297-300.
Lineback, D. R. (1984). *In* "International Symposium on Advances in Baking Science and Technology," pp. 51-59. Department of Grain Science, Kansas State University, Manhattan.

Lorenz, K., and Maga, J. (1972). *J. Agric. Food Chem. 20,* 211-218.
Miller, B. S., Johnson, J. A., and Balmer, D. L. (1953). *Food Technol. 7,* 38-42.
Monsanto Company (1977). "Potassium Sorbate Surface Treatment for Yeast-raised Bakery Products," p. 3. Monsanto Co., St. Louis.
Monsanto Company (1978). "Sorbic Acid and Potassium Sorbate for Preserving Food Freshness and Market Quality," p. 15. Monsanto Co., St. Louis.
Pelshenke, P. F., and Hampel, G. (1962). *Baker's Dig. 36* (3), 48-50, 52-54, 56-57.
Pisesookbunterng, W., and D'Appolonia, B. L. (1983). *Cereal Chem. 60,* 298-300.
Pisesookbunterng, W., D'Appolonia, B. L., and Kulp, K. (1983). *Cereal Chem. 60,* 301-305.
Russell, P. L. (1983). *J. Cereal Sci. 1,* 297-303.
Schoch, T. J. (1965). *Baker's Dig. 39* (2), 48-52, 54-57.
Schoch, T. J., and French, D. (1947). *Cereal Chem. 24,* 231-249.
Seiler, D. A. L. (1962). *In* "Microbial Inhibitors in Foods" (M. Molin, ed.), pp. 211-220. Almqvist & Wiksell, Stockholm.
Senti, F. R., and Dimler, R. J. (1960). *Baker's Dig. 34* (1), 28-32, 70-71.
Skovholt, O., and Dowdle, R. L. (1950). *Cereal Chem. 27,* 26-30.
Willhoft, E. M. A. (1971). *J. Sci. Food Agric. 22,* 176-180.
Zobel, H. F. (1973). *Baker's Dig. 47* (4), 52-53, 56-61.

CHAPTER 2

CHARACTERISTICS, COMPOSITION, AND SHELF-LIFE OF CHEESE

JOHN H. NELSON
Quality Assurance/Regulatory Compliance
Kraft, Inc.
Glenview, Illinois

Handbook of Food and Beverage
Stability: Chemical, Biochemical,
Microbiological, and Nutritional Aspects

I. DEFINITIONS OF CHEESE

Webster's unabridged dictionary defines cheese as "curd that has been separated from whey, consolidated by molding for soft cheese or subjected to pressure for hard cheese, and ripened for use as a food."

II. NATURAL CHEESE

A. Definitions

There is no single definition for natural cheese in U.S. Standards of Identity promulgated by the Food and Drug Administration; instead, each variety is defined in a separate standard.

The Codex Alimentarius General Standard for Cheese (Codex Standard No. A-6) defines natural cheese as follows:

> Natural cheese is the fresh or ripened curd obtained:
> a. by coagulating milk, skim milk, cream or buttermilk or any combination of these, through the action of lactic acid producing bacteria, the action of rennet or other suitable coagulating agents or by a combination of the two, and by partially draining the whey resulting from such coagulation; or
> b. by techniques applied to milk and/or materials derived from milk which give an end-product with essentially the same physical and organoleptic characteristics as the product defined under a.

B. Major Natural Cheese Groups

1. Extra Hard (Grating)
   Ripened by bacteria (e.g., Parmesan)

2. Hard
   a. Ripened by bacteria, without eyes (e.g., Cheddar)
   b. Ripened by bacteria, with eyes (e.g., Swiss)

3. Semisoft and Semihard
    a. Ripened by bacteria (e.g., Gouda)
    b. Ripened by bacteria and surface microorganisms (e.g., brick)
    c. Ripened principally by mold in the interior (e.g., blue)

4. Soft
    a. Ripened by bacteria (e.g., feta)
    b. Ripened by bacteria and surface microorganisms (e.g., Liederkranz)
    c. Ripened principally by white mold on the surface (e.g., Camembert)
    d. Unripened (e.g., cottage)

Table I describes the more natural cheeses, including not only major American varieties but also major continental varieties.

Table II categorizes natural cheese by hardness, fat content, and ripening method. Two sets of standards are references: U.S. Code of Federal Regulations, Chapter 21, and the Codex Alimentarius General Standard for Cheese A-6.

C. Natural Cheese Ingredients

Cow's milk is by far the predominant type of milk used for cheese making. The milk of cows is defined as the whole, fresh lacteal secretion obtained by the complete milking of one or more healthy cows, excluding that obtained within 15 days before and 5 days after calving, or such longer period as may be necessary to render the milk practically colostrum-free; it contains not less than 8.25% milk solids not fat and not less than 3.25% milk fat.

Most cheese is made from milk, but the other derivatives of milk, such as whey, buttermilk, and ultrafiltration retentate, are used in some varieties of cheese.

Table III lists the ingredients, other than milk and milk-derived ingredients, permitted in natural cheese, including the limits of addition, if any, and the function(s) of each ingredient.

D. Natural Cheese Manufacture

Figure 1 portrays schematically the principal steps in the manufacture of natural cheese. There are numerous modifications of and alternative steps to this basic process, depending upon the choice of equipment and procedures, the variety of cheese manufactured and the final utilization of the cheese (sold as natural cuts, converted into process cheese, or used as an ingredient in other foods).

III. PROCESS CHEESE AND RELATED PRODUCTS

A. Definition

Process cheese is defined in the U.S. Standards of Identity as "the food prepared by comminuting and mixing, with the aid of heat, one or more cheeses of the same or two or more varieties ... with an emulsifying agent ... into a homogeneous plastic mass."

B. Related Products

Products resembling process cheese are described in the U.S. Standards including (1) process cheese food and process cheese spreads in which a minor part of the cheese ingredient is replaced by other milk-derived ingredients, and (2) process _blended_ cheese, cheese food, and cheese spread, which are not permitted to contain emulsifying salts.

Cold-pack or club cheese is the food prepared by comminuting and mixing, without the aid of heat, one or more types of cheese into a homogeneous, plastic mass.

Grated cheese is the food prepared by grinding, grating, shredding, or otherwise comminuting extra hard or hard natural cheese. Grated cheese is usually partially dried to prevent molding and to extend shelf-life.

Table IV contains the composition and description of process cheese and related products defined by U.S. Standards.

C. Process Cheese Ingredients

Table V lists the ingredients permitted in process cheese and related products by U.S. Standards, including limits of addition, if any, and the function of each ingredient.

D. Process Cheese

Figure 2 portrays schematically the basic steps for the manufacture of process cheese and related products. A variety of equipment and techniques are used in process cheese manufacture, particularly after the cheese has been cooked. Hot, molten cheese may be packaged into jars or shaped into loaves, or formed into slices.

IV. NUTRITIONAL QUALITIES OF CHEESE

Cheese is a significant source of protein and calcium.
Tables VI, VII, and VIII contain nutritional profiles of major cheese varieties, including calories, major constituents, minerals, and vitamins.

V. SHELF STABILITY OF CHEESE

The shelf stability of cheese is dependent on a number of variables, including moisture content, packaging, storage temperatures, adequacy of manufacturing controls, and most significantly, cheese type--natural cheese versus pasteurized process cheeses and related products.

Most natural cheeses are biochemically active, incorporating an ongoing, multifaceted fermentation. This fermentation is a prime determinant of typical, desirable organoleptic characteristics--flavor, aroma, and texture. Although self-limiting to some degree, most natural cheese fermentations can continue beyond the optimum point of flavor development. The resulting intense, often unbalanced, flavors are not appealing to most consumers and thus make the cheese unpalatable.

Stored refrigerated (4°C), soft, high-moisture natural cheese varieties commonly have shelf-lives of a few weeks. Hard natural cheeses have shelf-lives measured in months. Some very hard cheeses, including grating cheeses, have a shelf-life measured in years if stored at less than 10°C.

Pasteurized process cheeses and related products are biochemically inert as a consequence of controlled heat treatment during production. Provided Good Manufacturing Practices have been observed, such cheeses and related products do not exhibit microbial degradation. Instead, after prolonged storage of several months, such cheeses may exhibit the results of oxidative processes such as rancidity and nonenzymatic browning.

VI. CHEESE SPOILAGE

The most common cause of cheese spoilage is mold. Cheese is a favorable medium for mold growth. Meticulous enforcement of Good Manufacturing Practices are required to exclude mold contamination and prevent mold growth. Control of the processing environment, adequate packaging, and use of mold inhibitors are the primary deterrents to mold spoilage. Cheese will support the growth of some toxigenic molds under laboratory conditions. Although there have been reports of toxin production in cheese at ambient temperatures or above, refrigeration inhibits toxin production in cheese.

Controlled mold growth is, of course, typical and desirable in some cheese types (refer to Table I).

Natural cheeses, like most foods, are at risk for contamination by pathogenic microorganisms, such as *Escherichia coli*, *Salmonella*, and *Staphylococcus* spp. Milk can be readily contaminated by these species. Good Manufacturing Practices, particularly adequate heat treatment of milk, coupled with uncompromising control of the cheese fermentation process will prevent proliferation of these microorganisms. The use of active lactic cultures to induce vigorous but controlled acid production during the make procedure is critical.

Pasteurized process cheese and related products are usually processed to temperatures of 175 to 180°F and packaged while hot. These practices virtually eliminate the risk of microbial spoilage or contamination by heat-labile pathogens.

Some pasteurized process cheese-type products packaged in hermetically sealed containers fall within the definition for "low-acid canned foods" promulgated by the U.S. Food and Drug Administration (21 CFR, Part 113--pH greater than 4.6 and a water activity a_w greater than 0.85). Properly formulated cheese products exhibiting pH and water activity values above these limits have been demonstrated to inhibit botulinum outgrowth and toxin production. Research has substantiated that, in addition to moisture and pH, phosphate, salt, and lactate are significant in preventing outgrowth and toxin production in cheese products by *Clostridium botulinum*. The exact compositional boundaries for cheese products that preclude the risk of botulism are in the process of being defined.

VII. OUT-OF-REFRIGERATION DISPLAY OF CHEESES

Out-of-refrigeration display of cheeses has become an increasingly popular way to promote sales. Soft cheeses should always be refrigerated, but hard cheese varieties, pasteurized process cheese, and related products may safely be displayed out of refrigeration provided the following conditions are maintained.

1. The cheese is in the original, sealed package.
2. Ambient temperatures do not exceed 78°F (25.5°C).
3. Display at ambient temperature does not exceed 2 weeks.

Table IX summarizes refrigeration guidelines for major U.S. cheese varieties.

ACKNOWLEDGMENT

The major, skilled assistance of Dr. R. W. Kunkel, Kraft, Inc., is gratefully acknowledged.

Table I - Description of the More Common Natural Cheeses

Major Group	Variety	Description When Ready for Consumption		
		Color	Texture	Flavor
Extra hard-ripened by bacteria	Asiago	White to light yellow	Brittle, flaky, granular	Aged 12 months; sharp, tangy
	Parmesan	White to light yellow	Brittle, flaky granular	Sharp, piquant, distinctive
	Romano	White to light yellow	Brittle, granular	Sharp, piquant, slightly rancid
	Sap Sago	Pale green	Dry	Sharp, aroma herb-like from added clover
Hard-ripened by bacteria - without eyes	Caciocavallo-Siciliano	White to light yellow	Smooth to granular	Young: mild, tangy. Aged: spicy and salty
	Cheddar	White to deep yellow	Smooth, waxy, supple	Mild to extra sharp, nutty
	Cheshire	Light yellow	Open, slightly granular	Mild, acid, tangy
	Colby	Light yellow to yellow	More open than Cheddar	Mild, sweet
	Provolone	White to light yellow	Smooth, supple to granular, varies with age	Mild, to robust varies with age. May be smoked.
Hard-ripened by bacteria - with eyes	Gruyere	Light yellow to amber	Smooth, supple	Sharp, nutty

Table I - Description of the More Common Natural Cheeses (Cont.)

Major Group	Variety	Description When Ready for Consumption		
		Color	Texture	Flavor
Hard-ripened by bacteria - with eyes (Cont'd)	Samsoe	White to yellow	Smooth, supple	Mild, sweet, buttery, nutty
	Swiss/ Emmantaler	White to light yellow	Smooth, supple	Mild to medium, sweet, nutty
Semi-soft - ripened by bacteria	Edam	Light yellow to yellow	Smooth, supple	Mild, acid, buttery to sharp, salty, full
	Fontina	Light yellow	Smooth, supple	Mild, sweet, nutty, distinctive
	Fynbo	White to yellow	Smooth, supple	Mild to sharp, buttery, aromatic
	Gouda	Light yellow to yellow	Smooth, supple	Mild to medium, creamy, buttery, distinctive
	Monterey	White to light yellow	Smooth, supple, resilient	Mild, buttery
	Muenster	Light yellow	Slight open, supple to pasty	Mild to medium, tangy, spicy, aromatic

Table I - Description of the More Common Natural Cheeses (Cont'd)

Major Group	Variety	Description When Ready for Consumption		
		Color	Texture	Flavor
Semi-soft - ripened by bacteria plus surface micro-organisms	Brick	White to yellow	Slight open, sliceable	Mild to full flavored, sweet to pungent, to highly aromatic
	Havarti	White to yellow	Open, sliceable	Full, smooth piquant to pungent with age
	Limburger	White to light yellow	Smooth, creamy	Strong, tangy, pungent, gamey, powerful aroma
	Port Salut	Light yellow to yellow	Smooth, resilient	Mild to full, buttery creamy to piquant with age
	Saint Paulin	Light yellow to yellow	Smooth, resilient	Mild to full, buttery, tangy, distinctive
	Tilsiter	White to light yellow	Open, sliceable	Full, tangy, piquant, strong aroma with age
Semi-soft - ripened principally by mold in the interior	Blue	White to light yellow with veins of blue	Open, crumbly, buttery	Strong, tangy, pungent
	Gorgonzola	White to light yellow with veins of greenish-blue mold	Open, crumbly, creamy	Strong, rich, creamy, pungent

Table I - Description of the More Common Natural Cheeses (Cont'd)

Major Group	Variety	Description When Ready for Consumption		
		Color	Texture	Flavor
Semi-soft - ripened principally by mold in the interior (Cont'd)	Roque-fort	White veins of blue mold	Open, crumbly, buttery	Sharp, tangy, rich, pungent, salty, lingering aftertaste
Soft-ripened by bacteria	Feta	White	Open, crumbly, moist	Sharp, salty
	Mozza-rella	White	Smooth, plastic	Mild, creamy, slightly sweet
Soft-ripened by bacteria and surface micro-organisms	Lieder-kranz	Light yellow	Smooth, very creamy	Full, pungent, rich, distinctive
Soft-ripened principally by white mold on surface	Brie	Light yellow	Smooth, flowing	Delicate, rich, creamy, tangy
	Camem-bert	Light yellow	Smooth, flowing	Delicate, rich, creamy, tangy
Soft-unripened	Cottage	White	Soft curd particles in creamy mixture	Bland, mildly acid
	Cream	White	Smooth, creamy	Rich creamy, mildly acid
	Neuf-chatel	White	Smooth, creamy	Creamy, mildly acid
	Ricotta	White	Crumbly, moist	Bland, sweet

Table II - Natural Cheese Categorized by Hardness, Fat Content and Ripening Method

Variety	U.S. 21CFR	Codex A6	Hardness Ready for Consumption	Moisture (%)	Fat (%)*	Ripening Method
Amsterdam		C28	Semi-hard to soft	max. 47	min. 48	By bacteria
Asiago - soft	133.102		Hard	max. 45	min. 50	By bacteria
Asiago - medium	133.103		Extra Hard	max. 35	min. 45	By bacteria
Asiago -	133.104		Extra	max. 32	min. 42	By bacteria
Blue	133.106		Semi-soft	max. 46	min. 50	Principally by blue mold in the interior
Brick	133.108		Semi-soft	max. 44	min. 50	By bacteria and surface micro-organisms
Brie		C34	Soft	max. 56	min. 40	Principally by white mold on surface
Butterkase		C17	Semi-soft to soft	max. 52	min. 45	By bacteria
Caciocavallo-Siciliano	133.111		Hard	max. 40	min. 42	By bacteria, without eyes

*Fat content-percent on dry basis, unless otherwise indicated

Table II - Natural Cheese Categorized by Hardness, Fat Content and Ripening Method (Cont'd)

Variety	U.S. 21CFR	Codex A6	Hardness Ready for Consumption	Moisture (%)	Fat (%)*	Ripening Method
Camembert		C33	Soft	max. 56	min. 45 and min. 30 - label to show level	Principally by white mold on surface
Cheddar	133.113		Hard	max. 39	min. 50	By bacteria, without eyes
Cheshire		C8	Hard	max. 44	min. 48	By bacteria, without eyes
Colby	133.118		Hard	max. 40	min. 50	By bacteria, without eyes
Cook (Koch kaese)	133.127		Soft	max. 80	no minimum, lowfat, from skim milk	By bacteria
Cottage	133.128		Soft	max. 80	min. 4 on as is basis	Unripened

*Fat content-percent on dry basis, unless otherwise indicated

Table II - Natural Cheese Categorized by Hardness, Fat Content and Ripening Method (Cont'd)

Variety	U.S. 21CFR	Codex A6	Hardness Ready for Consumption	Moisture (%)	Fat (%)*	Ripening Method
Dry curd cottage	133.129		Soft	max. 80	min. 0.5 on as is basis	Unripened
Lowfat cottage	133.131		Soft	max. 82.5	range 0.5 to 2 on as is basis	Unripened
Coulommiers		C18	Soft	max. 56	min. 40	Principally by white mold on surface
Cream	133.133		Soft	max. 55	min. 33 on as is basis	Unripened
Washed/soaked curd	133.136		Hard	max. 42	min. 50	By bacteria, without eyes
Danbo		C3	Semi-hard	max. 46	min. 45	By bacteria, with eyes
Edam	133.138		Semi-hard	max. 45	min. 40	By bacteria, with eyes
Esrom		C26	Semi-hard	max. 50	min. 45	By bacteria and surface micro-organism

*Fat content-percent on dry basis, unless otherwise indicated

Table II - Natural Cheese Categorized by Hardness, Fat Content and Ripening Method (Cont'd)

Variety	U.S. 21CFR	Codex A6	Hardness Ready for Consumption	Moisture (%)	Fat (%)*	Ripening Method
Feta	(Non-standardized)		Soft	typical 55	typical 21 on as is basis	By bacteria
Fontina	(Non-standardized)		Semi-hard	typical 38	typical 31 on as is basis	By bacteria
Friese (Friesian)		C30	Hard	max. 41	min. 40	By bacteria, without eyes
Fynbo		C25	Semi-hard to hard	max. 46	min. 45	By bacteria, with eyes
Gammelost	133.140		Semi-hard	max. 52	no minimum, lowfat, from skim milk	By bacteria
Gjetost	(Non-standardized)		Hard	typical 13	typical 29.5 on as is basis	Unripened
Gorgonzola	133.141		Semi-soft	max. 42	min. 50	Principally by blue mold in interior

*Fat content-percent on dry basis, unless otherwise indicated

Table II - Natural Cheese Categorized by Hardness, Fat Content and Ripening Method (Cont'd)

Variety	U.S. 21CFR	Codex A6	Hardness Ready for Consumption	Moisture (%)	Fat (%)*	Ripening Method
Gouda	133.142		Semi-hard	max. 45	min. 46	By bacteria
Granular and stirred curd	133.144		Hard	max. 39	min. 50	By bacteria, without eyes
Gruyere	133.149		Hard	max. 39	min. 45	By bacteria, with eyes
Hard grating	133.148		Hard	max. 34	min. 32	By bacteria
Harzer-kase		C20	Soft	max. 68	Range 0-10	By bacteria and surface micro-organisms
Havarti		C6	Semi-hard	max. 50	min. 45	By bacteria and surface micro-organisms
Herrgard-sost		C21	Hard	max. 41	min. 45	By bacteria, with eyes
Hushell-sost		C22	Semi-hard	max. 46	min. 45	By bacteria, with eyes
Leidse (Leyden)		C29	Hard	max. 41	min. 40	By bacteria, without eyes

*Fat content-percent on dry basis, unless otherwise indicated

Table II - Natural Cheese Categorized by Hardness, Fat Content and Ripening Method (Cont'd)

Variety	U.S. 21CFR	Codex A6	Hardness Ready for Consumption	Moisture (%)	Fat (%)*	Ripening Method
Limburger	133.152		Semi-soft	max. 50	min. 50	By bacteria and surface micro-organisms
Liederkranz	(Non-standardized)		Soft	typical 50	typical 24 on as is basis	By bacteria and surface micro-organisms
Maribo		C24	Semi-hard to hard	max. 43	min. 45	By bacteria and surface micro-organisms
Monterey and Monterey Jack	133.153		Semi-soft	max. 44	min. 50	By bacteria
High Moisture Jack	133.154		Semi-soft	min. 44	min. 50	By bacteria
Mozzarella & Scamorze	133.155		Soft	min. 52 max. 60	min. 45	Unripened or ripened by bacteria

*Fat content-percent on dry basis, unless otherwise indicated

Table II - Natural Cheese Categorized by Hardness, Fat Content and Ripening Method (Cont'd)

Variety	U.S. 21CFR	Codex A6	Hardness Ready for Consumption	Moisture (%)	Fat (%)*	Ripening Method
Low moisture part skim mozzarella and scamorze	133.157		Soft	min. 45 max. 52	min. 30 max. 45	Unripened or ripened by bacteria
Muenster	133.160		Semi-soft	max. 46	min. 50	By bacteria
Neufchatel	133.162		Soft	max. 65	min. 20 max. 33 on as is basis	Unripened
Norvegia		C23	Semi-hard	max. 44	min. 45	By bacteria, with eyes
Nuworld	133.164		Semi-soft	max. 46	min. 50	Principally by white mold in interior
Parmesan and Reggiano	133.165		Extra-hard	max. 32	min. 32	By bacteria
Port Salut	(Non-standardized)		Semi-hard	typical 45.5	typical 28 on as is basis	By bacteria and surface micro-organisms

*Fat content-percent on dry basis, unless otherwise indicated

Table II - Natural Cheese Categorized by Hardness, Fat Content and Ripening Method (Cont'd)

Variety	U.S. 21CFR	Codex A6	Hardness Ready for Consumption	Moisture (%)	Fat (%)*	Ripening Method
Provolone and pasta filata	133.181		Hard	max. 45	min. 45	By bacteria, without eyes
Ricotta	(Non-standardized)		Soft	typical 72 to 82	typical 0.5 to 13 on as is basis	Unripened
Romadur		C27	Soft	max. 65	min. 20	By bacteria and surface micro-organisms
Romano	133.183		Extra-hard	max. 34	min. 38	By bacteria
Roquefort	133.184		Semi-soft	max. 45	min. 50	Principally by blue mold in interior
Saint Paulin		C13	Semi-hard	max. 56	min. 40	By bacteria and surface micro-organisms
Samsoe	133.185		Hard	max. 41	min. 45	By bacteria, with eyes

*Fat content-percent on dry basis, unless otherwise indicated

Table II - Natural Cheese Categorized by Hardness, Fat Content and Ripening Method (Cont'd)

Variety	U.S. 21CFR	Codex A6	Hardness Ready for Consumption	Moisture (%)	Fat (%)*	Ripening Method
Sap Sago	133.186		Extra-hard	max. 38	no minimum, lowfat, from skim milk	By bacteria
Spiced (e.g. Caraway)	133.190		Hard to semi-	No limits	min. 50	By bacteria
Spiced, flavored	133.193		Conforms to standard of identity for the specific natural cheese which is spiced or flavored			
e.g. Monterey Jack with jalapeno pepper			Semi-soft	max. 44	min. 50	By bacteria
Svecia		C14	Semi-hard to hard	max. 41	min. 45	By bacteria
Swiss and Emmentaler	133.195					
Tilsiter		C11	Semi-hard	max. 47	min. 45	By bacteria and surface micro-organism

*Fat content-percent on dry basis, unless otherwise indicated

Table III - Ingredients Permitted in Natural Cheese

INGREDIENT	LIMITS	PURPOSE
Microorganisms - bacteria Lactobacillus bulgaricus Lactobacillus helveticus Lactobacillus lactis Lactobacillus casei Leuconostoc lactis Leuconostoc cremoris Micrococcus species Propioni bacterium shermani Streptococcus lactis Streptococcus lactis subspecies diacetylactis Streptococcus cremoris Streptococcus thermophilus	none	Added to milk to produce lactic acid and assist in the ripening process.
Brevi bacterium linens		Added to cheese surface to ripen cheese
Microorganisms - mold Penicillium roqueforti	none	Added to curd to ripen blue mold veined cheeses (Blue and Roquefort)
Penicillium roqueforti - white mutant	none	Added to curd to ripen Nuworld cheese
Penicillium camemberti	none	White mold added to cheese surface to ripen cheese
Microorganisms - mold Penicillium glaucum	none	Added to curd to ripen greenish-blue veined cheese (Gorgonzola)

Table III - Ingredients Permitted in Natural Cheese (Cont'd)

INGREDIENT	LIMITS	PURPOSE
Microorganisms - yeast Candida	none	Grow naturally on surface of blue mold veined cheeses, they utilize lactic acid and enhance mold growth
Enzymes		
Rennet or other safe and suitable milk clotting enzymes including:		Added to milk to catalyze curd formation and assist in the cheese ripening process
A. Rennets		
1. Animal source a. veal rennet b. bovine rennet	Used at levels not to exceed current good manufacturing practice	
2. Fermentation derived a. Mucor mehei b. Mucor pusillus c. Endothia parasitica d. Bacillus cereus	Used in an amount not in excess of the minimum required to produce the intended effect.	
B. Swine pepsin		
C. Blends of two or more of the above		
Enzymes used in curing or flavor development including: A. Pregastric esterase from calf, lamb or kid goat source. B. Microbial esterase 1. Mucor mehei 2. Aspergillus species	Some standards limit addition to 0.1% by weight of milk. Addition may be during the cheese making operation. Recent revisions of standards impose no limits.	Assist in cheese ripening process

Table III - Ingredients Permitted in Natural Cheese (Cont'd)

INGREDIENT	LIMITS	PURPOSE
Enzymes used in curing or flavor development (Cont'd)		
C. Pancreatin		
D. Microbial protease 1. Aspergillus oryzae Var.		
Catalase		
A. Animal source 1. Bovine liver	Maximum 20 ppm by weight of the milk. To eliminate hydrogen peroxide when it is used.	To eliminate residual peroxide used to destroy milk spoilage organisms.
B. Microbial source 1. Aspergillus niger 2. Micrococcus lysodeikticus		
Salts		
1. Sodium chloride	Standards impose no limits except use is prohibited in low sodium cheddar and colby.	Influence flavor, moisture content, texture, rind formation and controls ripening
2. Potassium chloride	Standards impose no limits.	Substitute for sodium chloride in low sodium varieties.
3. Calcium chloride	Added to milk at maximum of 0.2% by weight of the milk.	Aids curd forming and firming.

Table III - Ingredients Permitted in Natural Cheese (Cont'd)

INGREDIENT	LIMITS	PURPOSE
Salts (Cont'd)		
4. Sodium and potassium nitrate	Not permitted by U.S. standards. Permitted in some varieties by Codex at a maximum of 0.2% by weight of milk.	To control excessive gas formation by bacteria.
5. Sodium dihydrogen phosphate and disodium hydrogen	Not permitted by U.S. standards. Codex allows a maximum of 0.2% by weight of the milk in Scandinavian cheese varieties.	
6. Calcium carbonate and sodium hydrogen carbonate.	Not permitted by U.S. standards. Codex allows a maximum of 3.0% by weight of acid curd in Harzerkase.	Adjust pH of curd
Colors		
1. Annatto 2. Beta carotene	Codex limits addition to 0.06% by weight of the milk when used singly or in combination. U.S. standards impose no limits.	To impart uniform yellow color to certain varieties of cheese.
3. FD&C Blue No. 1 and 2 and FD&C Green No. 3	U.S. Standards impose no limit.	To neutralize the yellow color in blue mold veined cheeses and certain Italian cheese varieties.

Table III - Ingredients Permitted in Natural Cheese (Cont'd)

INGREDIENT	LIMITS	PURPOSE
Colors (Cont'd)		
4. Benzoyl Peroxide (not a color per se, used as a bleach).	Maximum 0.002% of the milk. When used, Vitamin A must be added to the curd to restore that destroyed in the bleaching process.	To bleach the yellow color in blue mold veined cheeses and certain Italian cheese varieties.
Antimycotics		
1. Sorbic acid, potassium sorbate, sodium sorbate, or any combination.	Maximum of 0.3% by weight as sorbic acid on the surface of some varieties of consumer size cuts and slices and some varieties of Italian cheese.	Retard mold growth.
2. Natamycin (Pimaricin)	Aqueous solutions containing 200 to 300 ppm natamycin may be applied to the surface of some varieties of consumer size cuts and slices in some varieties of Italian cheese.	Retard mold growth.
Acidifying agents		
1. Food grade phosphoric acid, citric acid or hydrochloric acid.	Added to milk to reach a pH of 4.5 to 4.7 in the manufacture of dry curd cottage cheese.	An alternate make procedure which substitutes lactic acid producing bacteria and requires the label statement "Directly set".

Table III - Ingredients Permitted in Natural Cheese (Cont'd)

INGREDIENT	LIMITS	PURPOSE
Acidifying Agents (Cont'd)		
2. D-Glucono-delta-lactone	Added to milk to reach a maximum pH of 4.7 in the manufacture of dry curd cottage cheese.	An alternate make procedure substitutes for lactic acid producing bacteria and requires the label statement "Directly set".
Other		
1. Hydrogen peroxide	U.S. standards limit use to certain hard varieties of cheese including cheddar, colby, and Swiss. Maximum 0.05% of the milk followed by catalase to eliminate the hydrogen peroxide. Weight of catalase shall not exceed 20 ppm of the weight of the milk.	Destroy milk spoilage organisms.
2. Spice and spice oils	Spice minimum 0.015 oz./1 lb. cheese and may contain spice oils.	To impart flavor.
3. Smoke or smoke flavor	none	To impart to hard cheese varieties such as Provolone and cheddar.

Table IV - Composition and Description of Process Cheese and Related Products Defined by U. S. Standards

Standard 21CFR	Type	DESCRIPTION Moisture	Fat	Body/Texture	Flavor
133.169	Pasteurized process cheese	Generally 1% greater than max. for natural cheese(s) used.	Min. on solids basis equal to natural cheese(s) used.	Firm, smooth, sliceable; melts smoothly	Mild to sharp, typical of varieties used, e.g., cheddar, Swiss; may have spice or flavor added.
133.170	Pasteurized process cheese with fruits, vegetables, or meats	May be 1% greater than 133.169	May be 1% less than 133.169	Firm, quite smooth, sliceable, will melt.	Typical of cheese varieties plus added food, may have spice or flavor added.
133.171	Pasteurized process pimento cheese.	Max. 41%.	Min. on solids basis 49%.	Firm, quite smooth, sliceable, will melt.	Mild to sharp, typical of cheddar plus pimento.
133.173	Pasteurized process cheese food	Max. 44%.	Min. 23% on as is basis.	Semi-firm, smooth, sliceable, melts smoothly.	Mild, typical of cheese varieties used. May have spice or flavor added.
133.174	Pasteurized process cheese food with fruits, vegetables or meats.	Max. 44%	Min. 22% on as is basis	Semi-firm, quite smooth, sliceable, will melt.	Mild, typical of cheese varieties plus added food. May have spice or flavor added.

Table IV - Composition and Description of Process Cheese and Related Products Defined by U. S. Standards (Cont'd)

Standard 21CFR	Type	DESCRIPTION Moisture	Fat	Body/Texture	Flavor
133.179	Pasteurized process cheese spread.	Min. 44%, max. 60%	Min. 20% on as is basis	Soft, smooth, may slice or spread, melts smoothly	Mild, cheesy, slight sweet. May have spice or flavor added.
133.180	Pasteurized process cheese spread with fruits, vegetables or meats.	Min. 44%, max. 60%	Min. 20% on as is basis	Soft, quite smooth, may slice or spread, will melt.	Mild, cheesy, slight sweet plus flavor of added food. May have spice or flavor added.
133.175	Pasteurized process cheese spread.	Min. 44%, max. 60%	Min. 20% on as is basis	Soft, usually spreadable, variable melt.	Mild, cheesy, slight sweet. May have spice or flavor added.
133.176	Pasteurized cheese spread with fruits, vegetables or meats.	Min. 44%, max. 60%	Min. 20% on as is basis	Soft, usually spreadable, variable melt.	Mild, cheesy, slight sweet plus flavor of added food. May have spice or flavor added.

Table IV - Composition and Description of Process Cheese and Related Products Defined by U. S. Standards (Cont'd)

Standard 21CFR	Type	DESCRIPTION Moisture	Fat	Body/Texture	Flavor
133.178	Pasteurized neufchatel cheese spread with other foods	Max. 65%	Min. 20% on as is basis	Soft, spreadable.	The amount of added food must be sufficient to so differentiate the blend that it does not simulate neufchatel cheese
133.167	Pasteurized blended cheese.	Generally 1% greater than max. for natural cheese(s) used.	Min. on solids basis, equal to natural cheese(s) used.	Variable from firm to semifirm and sliceable to spreadable.	Mild to sharp, typical of cheese varieties used. May be spiced or flavored.
133.123	Cold pack and club cheese.	Generally the same as for natural cheese(s)	Generally the same as for natural cheese(s)	Soft, may be quite smooth, spreadable.	Mild to sharp, typical of cheese varieties used. May be spiced or flavored.
133.124	Cold pack cheese food.	Max. 44%.	Min. 23% on as is basis	Soft, may be quite smooth, spreadable.	Mild to sharp, typical of cheese variety used. May be spiced or flavored.

Table IV - Composition and Description of Process Cheese and Related Products Defined by U. S. Standards (Cont'd)

Standard 21CFR	Type	DESCRIPTION Moisture	Fat	Body/Texture	Flavor
133.125	Cold pack cheese food with fruits, vegetables or meats.	Max. 44%	Min. 22% on as is basis.	Soft, spreadable.	Mild to sharp, typical of cheese variety used plus added food. May be spiced or flavored.
133.146	Grated cheeses.	May be reduced below that of the natural cheese used.	Min. on solids basis not more than 1% below the natural cheese used.	Discrete particles of cheese.	Mild to sharp, typical of cheese variety used.
133.147	Grated American cheese food.	May be reduced below that of the natural cheese used.	Min 23% on as is basis.	Uniform blend in powder or granular form.	Mild to medium, typical of cheddar and related varieties, e.g. Colby.

Table V - Ingredients Permitted in Process Cheese and Related Products by U.S. Standards

INGREDIENT	LIMITS	PURPOSE
Cheeses		
Natural cheeses other than cream, neufchatel, cottage, cook, skim milk and part skim cheese and hard grating cheese.	Minimum of 51% cheese in pasteurized process cheese food, pasteurized process cheese spread and pasteurized cheese spread.	Provide the desired sensory characteristics and composition.
Optional dairy ingredients are:		
cream, milk, skim milk, buttermilk, cheese whey, any of the foregoing from which water has been removed, anhydrous milk fat, dehydrated cream, albumin from cheese whey, whey protein concentrate and skim milk cheese for manufacturing.	Not permitted in pasteurized process cheese. Permitted in pasteurized process cheese food, pasteurized process cheese spread and pasteurized cheese spread. All except buttermilk are permitted in pasteurized neufchatel cheese with other foods.	Provide the desired sensory characteristics and composition.
Emulsifying agents include one or any mixture of two or more of the following: monosodium phosphate, disodium phosphate, dipotassium phosphate, trisodium phosphate, sodium metaphosphate, sodium acid pyrophosphate, tetrasodium pyrophosphate, sodium aluminum phosphate, sodium citrate, calcium citrate, sodium tartrate and sodium potassium tartrate.	Maximum 3% by weight of the pasteurized process cheese, pasteurized process cheese food or pasteurized process cheese spread and pasteurized cheese spread.	Prevent fat separation and for desirable body and texture.

Table V - Ingredients Permitted in Process Cheese and Related Products by U.S. Standards (Cont'd)

INGREDIENT	LIMITS	PURPOSE
Acidifying agents		
One or any mixture of two or more of the following: vinegar, lactic acid, citric acid, acetic acid and phosphoric acid.	In pasteurized process cheese the quantity used shall not reduce pH below 5.3. In pasteurized process cheese food, pH shall not be below 5.0. In pasteurized process cheese spread and pasteurized cheese spread pH shall not be below 4.0. No limit in pasteurized neufchatel cheese spread with other foods and grated American cheese food.	Microbial control and flavor.
Sweetening agents		
One or any mixture of two or more of the following: sugar, dextrose, corn sugar, corn syrup, glucose, maltose, malt syrup and hydrolyzed lactose.	Use is limited to pasteurized process cheese spread, pasteurized cheese spread and pasteurized neufchatel cheese spread. No limit in pasteurized neufchatel cheese spread with other foods.	Flavor.
Stabilizer gums		
One or any mixture of two or more of the following: carob bean gum, gum karaya, gum tragacanth, guar gum, gelatin, carboxymethyl cellulose, carrageenan, oat gum, sodium alginate or xanthan gum.	Use is limited to 0.8% by weight of pasteurized process cheese spread, pasteurized cheese spread and pasteurized neufchatel cheese spread with other foods.	Prevent water separation.

Table V - Ingredients Permitted in Process Cheese and Related Products by U.S. Standards (Cont'd)

INGREDIENT	LIMITS	PURPOSE
Stabilizer gums (Cont'd)		
Guar gum, xanthan gum or both.	Use is limited to 0.3% by weight of cold pack cheese food.	Prevent water separation.
Antimycotics		
1. Sorbic acid, potassium sorbate, sodium sorbate or any combination.	Maximum 0.2% by weight of pasteurized process cheese, pasteurized process cheese food or pasteurized process cheese spread in consumer size cuts and slices.	Retard mold growth.
	Maximum of 0.3% by weight as sorbic acid in grated cheeses, cold pack or chub cheese and cold pack cheese food.	Retard mold growth.
2. Sodium propionate, calcium propionate or a combination thereof.	Maximum of 0.3% by weight of pasteurized process cheese, pasteurized process cheese food or pasteurized process cheese spread in consumer size cuts and slices, cold pack or chub cheese and cold pack cheese food.	Retard mold growth.
Anticaking agents		
Silicon dioxide, sodium silicoaluminate, microcrystalline cellulose or any combination of two or more of these.	Maximum 2% by weight of grated cheese.	Retard caking.

Table V - Ingredients Permitted in Process Cheese and Related Products by U.S. Standards (Cont'd)

INGREDIENT	LIMITS	PURPOSE
Other		
Harmless artificial color	Permitted in all types except grated cheese.	Provide desired character.
Spices and flavorings including smoke	Permitted in all types except grated American cheese food and pasteurized neufchatel cheese spread with other foods.	Flavor.
Salt (sodium chloride)	Permitted in all types except grated cheese and pasteurized neufchatel cheese spread with other foods.	Microbial control and flavor.
Enzyme modified cheese	Permitted in pasteurized process cheese, pasteurized process cheese food, pasteurized process cheese spread and pasteurized cheese spread.	Flavor.
Lecithin	Maximum 0.03% by weight in pasteurized process cheese, pasteurized process cheese food, pasteurized process cheese spread and pasteurized cheese spread in consumer size packages.	Optional anti-sticking agent.
Dioctyl sodium sulfosuccinate	Maximum of 0.5% by weight of stabilizer gums when they are used.	As a solubilizing agent on gums.

Table VI - Nutritional Profiles of Major Cheese Varieties - Proximate

Average Amount in a Serving (1 oz/28g) of Cheese, Except Cottage (4 oz/113g)

Cheese Variety	Cal-ories	Pro-tein (g)	Fat (g)	Car-bohy-drate (g)	Choles-terol (g)	Ash (g)
Blue	100	6.07	8.15	.66	21	1.45
Brick	105	6.59	8.41	.79	27	.90
Camembert	85	5.61	6.88	.13	20	1.04
Cheddar	114	7.06	9.40	.36	30	1.11
Colby	112	6.74	9.10	.73	27	.95
Creamed Cottage	117	14.11	5.10	3.03	17	1.54
Dry Curd Cottage	96	19.52	.48	2.09	8	.78
Cream	99	2.14	9.89	.75	31	.33
Edam	101	7.08	7.88	.40	25	1.20
Feta	75	4.03	6.03	1.16	25	1.47
Gouda	101	7.07	7.78	.63	32	1.12
Limburger	93	5.68	7.72	.14	26	1.07
Monterey	106	6.94	8.58	.19	--*	1.01
Mozzarella	80	5.51	6.12	.63	22	.74
Mozzarella Low Moist. Part skim	79	7.79	4.85	.89	15	1.05
Muenster	104	6.64	8.52	.32	27	1.04
Parmesan, grated	129	11.78	8.51	1.06	22	1.99
Provolone	100	7.25	7.55	.61	20	1.34
Romano	110	9.02	7.64	1.03	29	1.90
Swiss	107	8.06	7.78	.96	26	1.00
Tilset	96	6.92	7.36	.53	29	1.38
Past. Process Cheese, American	106	6.28	8.86	.45	27	1.66
Past. Process Cheese Food	93	5.56	6.97	2.07	18	1.52
Past. Process Cheese Spread	82	4.65	6.02	2.48	16	1.70

*Dashes denote lack of reliable data on a constituent believed to be present in measureable amounts

Table VII - Nutritional Profiles of Major Cheese Varieties - Minerals

	Average Amount in a Serving (1 oz/28g) of Cheese, Except Cottage (4 oz/113g)						
Cheese Variety	Calcium (mg)	Iron (mg)	Magnesium (mg)	Phosphorus (mg)	Potassium (mg)	Sodium (mg)	Zinc (mg)
Blue	950	.09	7	110	73	396	.75
Brick	191	.12	7	128	38	159	.74
Camembert	110	.09	6	98	53	239	.68
Cheddar	204	.19	8	145	28	176	.88
Colby	194	.22	7	129	36	171	.87
Creamed cottage	68	.16	6	149	95	457	.42
Dry curd cottage	36	.26	4	118	37	14	.53
Cream	23	.34	2	30	34	84	.15
Edam	207	.12	8	152	53	274	1.06
Feta	140	.18	5	96	18	316	.82
Gouda	198	.07	8	155	34	232	1.11
Limburger	141	.04	6	111	36	227	.60
Monterey	212	.20	8	126	23	152	.85
Mozzarella	147	.05	5	105	19	106	.63
Mozzarella Low Moisture Part Skim	207	.07	7	149	27	150	.89
Meunster	203	.12	8	133	38	178	.80
Parmesan, grated	390	.27	14	229	30	528	.90
Provolone	214	.15	8	141	39	248	.92
Romano	302	--*	--*	215	--*	340	--*
Swiss	272	.05	10	171	31	74	1.11
Tilset	198	.06	4	142	18	213	.99
Past. Process Cheese, American	174	.11	6	211	46	406	.85
Past. Process Cheese Food	163	.24	9	130	79	337	.85
Past. Process Cheese Spread	159	.09	8	202	69	381	.73

*Dashes denote lack of reliable data for a constituent believed to be present in measureable amount

Table VIII - Nutritional Profiles of Major Cheese Varieties - Vitamins

	Average Amount in a Serving (1 oz./28g) of Cheese, Except Cottage (4 oz./113g)								
	Ascorbic acid (mg)	Thiamine (mg)	Ribo. (mg)	Niacin (mg)	Pantothenic Acid (mg)	Vit. B_6 (mg)	Folacin (mcg)	Vit. B_{12} (mcg)	Vit. A IU
Blue	0	.008	.108	.288	.490	.047	10	.345	204
Brick	0	.004	.100	.033	.082	.018	6	.356	307
Camembert	0	.008	.138	.179	.387	.064	18	.367	262
Cheddar	0	.008	.106	.023	.117	.021	5	.234	300
Colby	0	.004	.106	.026	.060	.022	-	.234	293
Creamed cottage	trace	.024	.184	.142	.241	.076	14	.704	184
Dry curd cottage	0	.028	.160	.175	.184	.093	17	.932	34
Cream	0	.005	.056	.029	.077	.013	4	.120	405
Edam	0	.010	.110	.023	.080	.022	5	.435	260
Feta	0	-*	-	-	-	-	-	-	-
Gouda	0	.009	.095	.018	.096	.023	6	-	183
Limburger	0	.023	.143	.045	.334	.024	16	.295	363
Monterey	0	-	.111	-	-	-	-	-	269
Mozzarella	0	.004	.069	.024	.018	.016	2	.185	225
Mozzarella Low Moist. Part Skim	0	.006	.097	.034	.026	.022	3	.262	178

Table VIII - Nutritional Profiles of Major Cheese Varieties - Vitamins (Cont'd)

Average Amount in a Serving (1 oz./28g) of Cheese, Except Cottage (4 oz./113g)

	Ascorbic acid (mg)	Thiamine (mg)	Ribo. (mg)	Niacin (mg)	Pantothenic Acid (mg)	Vit. B_6 (mg)	Folacin (mcg)	Vit. B_{12} (mcg)	Vit. A IU
Muenster	0	.004	.091	.029	.054	.016	3	.418	318
Parmesan, grated	0	.013	.109	.089	.149	.030	2	-	199
Provolone	0	.005	.091	.044	.135	.021	3	.415	231
Romano	0	-	.105	.022	-	-	2	-	162
Swiss	0	.006	.103	.026	.122	.024	2	.475	240
Tilset	0	.017	.102	.058	.098	-	-	.595	296
Past. Process Cheese American	0	.008	.100	.020	.137	.020	2	.191	343
Past. Process Cheese Food	0	.008	.125	.040	.158	-	-	.317	259
Past. Process Cheese Spread	0	.014	.122	.037	.194	.033	2	.113	223

*Dashes denote lack of reliable data for a constituent believed to be present in measureable amount

Table IX - Guidelines for Out-of-Refrigeration Display of Major U.S. Cheese Varieties

Variety	Maximum Moisture %	
SOFT CHEESES - Refrigeration Essential		
Cottage	80	
Bakers	(80)	
Neufchatel	65	
Cook (Koch)	80	
Cream	55	
High Moisture Jack	50	(44 minimum)
Low Moisture Mozzarella, Scamorze	52	(45 miminum)
Mozzarella, Scamorze	60	(45 minimum)
Part Skim Mozzarella, Scamorze	60	(52 miminum)
Ricotta	(70)	
SEMI-SOFT CHEESES - Refrigeration Desirable		
Surface Ripened		
Brie	(50)	
Camembert	(50)	
Brick	44	
Limburger	50	
Mold Ripened		
Blue	46	
Gorganzola	46	
Roquefort	45	
Other Varieties		
Edam, Gouda	45	
Monterey, Monterey Jack	44	
Muenster, Munster	46	
HARD CHEESES - Refrigeration Optional		
Cheddar	39	
Colby	40	
Swiss	41	
HARD GRATING CHEESES, GRATED CHEESES - Refrigeration Unnecessary		
Parmesan	32	
Romano	34	
Grated	(18)	
PROCESS CHEESES AND RELATED PRODUCTS - Refrigeration Optional		
Pasteurize Process:		
Cheese	40-42	
Cheese Food	44	
Cheese Spread	60	(44 minimum)

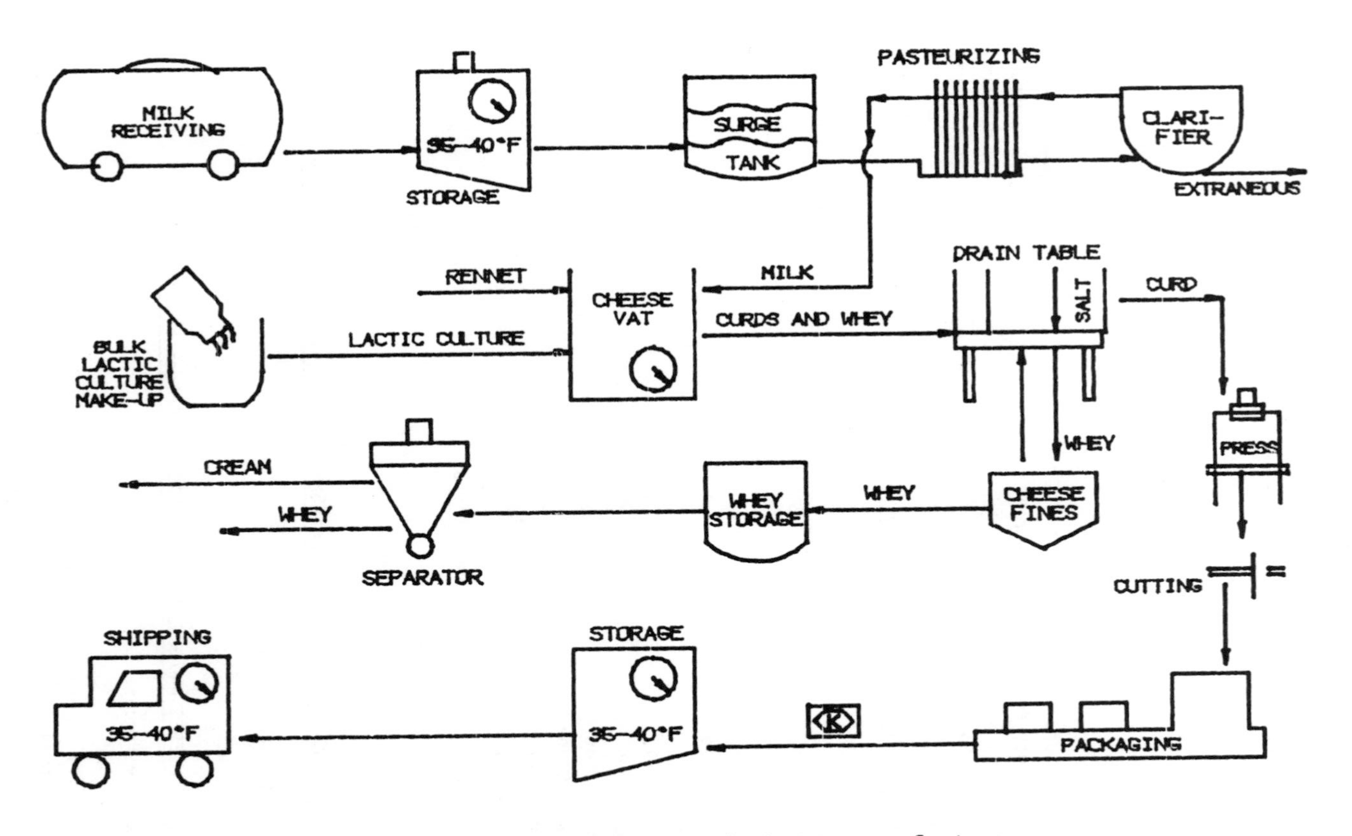

Fig. 1. Schematic for natural cheese manufacture.

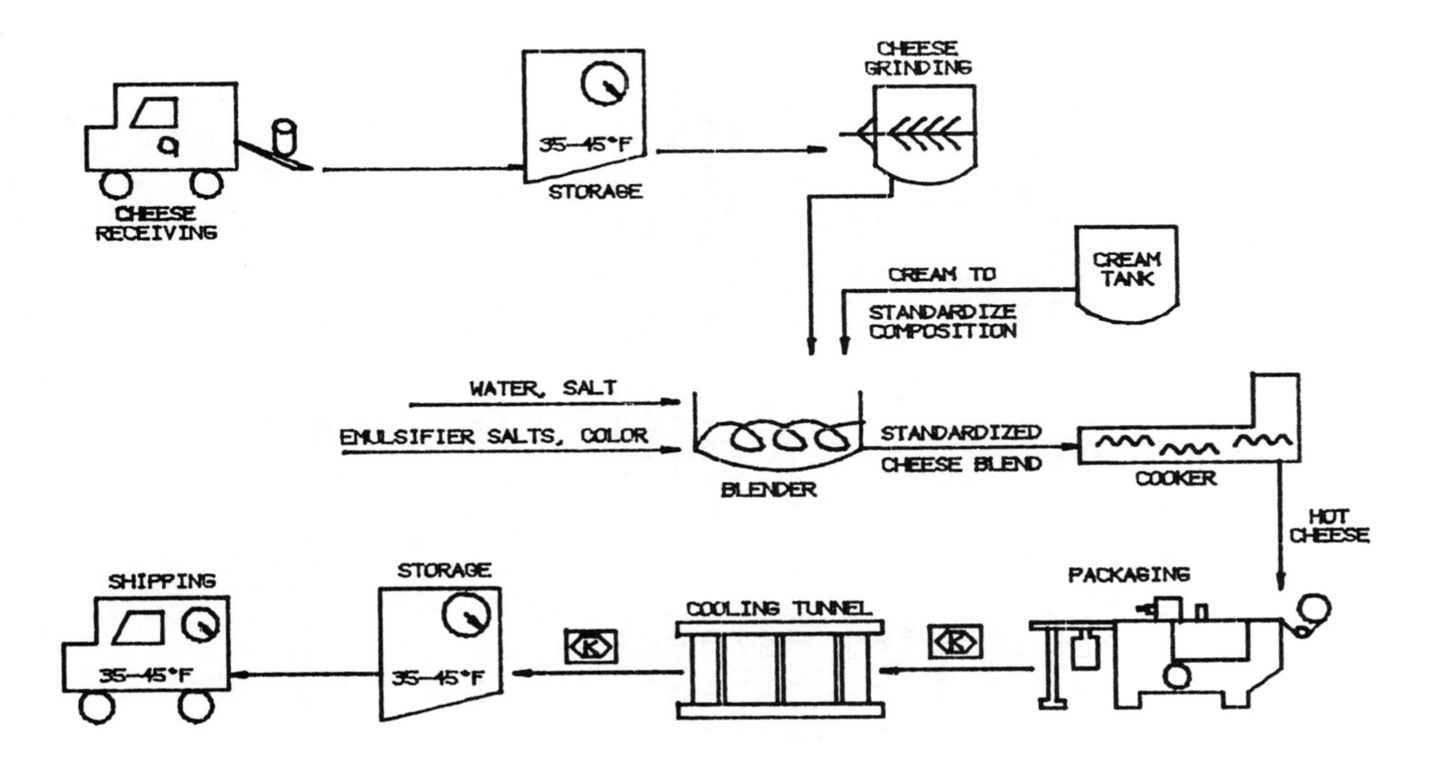

Fig. 2. Schematic for process cheese manufacture.

REFERENCES

Agricultural Handbook (1976). "Composition of Foods, Dairy and Egg Products," Agric. Handb. No. 8-1, Revised November 1976. U.S. Gov. Printing Office, Washington, D.C.

Code of Federal Regulations (1984). Chapter 21, Parts 133, Revised April 1, 1984. U.S. Gov. Printing Office, Washington, D.C.

Federal Register (1984). Docket No. 80N-0373 (includes Codex Standard No. A-6). *Fed. Regist. 49*, 17020.

Kosikowski, F. (1977). "Cheese and Fermented Milk Foods," 2nd ed., Edwards Brothers, Ann Arbor, Michigan.

Nelson, J. H. (1983). Should cheese be refrigerated? *Dairy Food Sanitation 3*, 372.

Price, W. V., and Bush, M. G. (1974a). The process cheese industry in the United States: A review. I. Industrial growth and problems. *J. Milk Food Technol. 37*, 135.

Price, W. V., and Bush, M. G. (1974b). The process cheese industry in the United States: A review. II. Research and development. *J. Milk Food Technol. 37*, 179.

Tanaka, N. (1982). Challenge of pasteurized process cheese spreads with *Clostridium botulinum* using in-process and post-process inoculation. *J. Food Prot. 45*, 1044.

Tanaka, N. *et al.* (1979). A challenge of pasteurized process cheese spread with *Clostridium botulinum* spores. *J. Food Prot. 42*, 747.

Van Slyke, L. L., and Price, W. V. (1979). "Cheese," 2nd ed., Ridgeview Publ. Co., Reseda, California.

Widcombe, R. (1983). "The Cheese Book." Omega Books Limited, Ware Hertfordshire, England.

CHAPTER 3

CHANGES IN QUALITY OF MEAT DURING AGING AND STORAGE

MILTON E. BAILEY
Department of Food Science and Nutrition
University of Missouri
Columbia, Missouri

I. INTRODUCTION

Because of the subjective nature of food preferences, the term quality as it is related to red meat by different people is difficult to define to everyone's complete satisfaction. Certain attributes evaluated during purchase, cooking, and consumption are important to most consumers, including those related to sensory acceptance as color, odor, flavor, tenderness, juiciness (water binding), nutritional quality, and safety.

Handbook of Food and Beverage
Stability: Chemical, Biochemical,
Microbiological, and Nutritional Aspects

Several volumes the size of this book would be required to review adequately published data on this subject. Retrieval information from one abstract source published since 1968 enumerated over 30,000 pertinent references, and 17,000 of these references are concerned with aging and storage of meat. This information has not been satisfactorily compiled or integrated.

There is an urgent need for comprehensive review and tabulation of these data into useful form, but this chapter is not intended to serve that purpose. It is rather a limited discussion of some of the changes in parameters occurring during storage of red meat (beef, pork, and lamb) that undoubtedly influence its acceptance and utilization by the consumer.

Meat quality is influenced by many factors, including those contributed by animal genetics, sex, and maintenance (antemortem factors) and those associated with slaughter, aging, processing, storage, and distribution (postmortem factors). This is a discussion of only changes occurring during aging and storage.

Aging of meat usually consists of refrigerating it at 4°C for several days or longer to improve tenderness, color, water-binding capacity, and flavor.

Changes that occur during storage include discoloration, spoilage by microorganisms, oxidation, dehydration, and changes caused by natural enzyme catalysis. These changes are retarded by different processing and preservation techniques, including curing, smoking, dehydration, heat processing, irradiation, packaging, refrigeration, and freezing. The latter three are the most important procedures utilized for the preservation of fresh meat.

II. POSTMORTEM CHANGES

The chemical and ultrastructural changes occurring in muscle postmortem have a tremendous influence on meat quality during subsequent storage. Much of this change is dependent on antemortem factors such as animal genetics, nutritional regimen, and animal management, and is beyond the scope of this chapter. Important variables that regulate carcass grade are fat content, degree of marbling, color, tenderness, and severity of muscle contraction.

An important physiological feature that predominates despite these variables is susceptibility and degree of stress on the meat animal several days or shortly before slaughter. Other processing procedures during the first 2 weeks of storage that influence subsequent quality are the slaughter procedures employed and conditions utilized for aging.

A. Influence of Stress

The major physiological conditions that might influence quality compared to that of the normal animal are the availability of muscle glycogen, rate of glycolysis, availability of ATP residues, and ultimate pH during chilling and freezing and their influence on tenderness, flavor, and water binding. Many aspects of postmortem changes were reviewed by Asghar and Pearson (1980).

The normal muscle of meat animal carcasses has high glycogen reserve that degrades slowly during chilling, so that the muscle pH is about 5.5 at rigor mortis (Wismer-Pedersen, 1976). The ultimate pH of stressed animals is higher or lower than normal depending on the duration of stress prior to slaughter. High-pH (6.0-6.8) muscle is called "dry, firm and dark" (DFD) or "dark cutting" because of the dark appearance of muscle upon cutting. The cause of "dark-cutting" beef is a severe and continuous form of stress several hours or days before slaughter. The stress results in rapid glycolysis and increase in glucose in urine and blood. The lack of muscle glycogen at slaughter results in high pH (Hedrick, 1981).

Undesirable aspects of DFD meat include less acceptable flavor, dark color, sticky texture, and greater susceptibility to microbiological growth during storage. According to Dransfield (1981), high-pH beef has less flavor and is less acceptable compared to meat with normal pH.

As part of the European Economic Community beef coordination program, quality assessments were compared among five institutes for flavor and juiciness relative to pH. Juiciness of the steaks was not significantly related to pH, but two of the stations reported significant negative correlations between flavor and pH as shown in Table I.

Flavor of roast *M. longissimus dorsi* was also found to be disliked more as pH increased in steer beef, although there were considerable individual preferences of nine panelists. Eight disliked the flavor of DFD meat, but only four had significant preferences. Dransfield (1981) also reported results from consumer assessment of DFD meat and concluded that DFD beef had poorer flavor.

Lack of desirable flavor at high pH might be a result of the closed structure and greater adsorption of flavor precursors as suggested by Lawrie (1979), or it may be due to the lack of sugar and sugar phosphate precursors necessary for formation of browned products responsible for desirable flavor. Volatiles from lamb and beef are different at pH 6.0 compared to pH 5.5 (Park and Murray, 1975), and this difference may be the formation of heterocyclic compounds needed for desirable flavor.

The dark red color of DFD meat causes it to be less attractive to consumers, who may relate the color with that from older animals or with meat that has been poorly stored (Sornay *et al.*, 1981). Green discoloration may occur in DFD meat because of bacterial metabolism and other abnormal characteristics that can occur at higher pH.

Nichol *et al.* (1970) stored beef at 1 to 2°C under low O_2 tension and found that *Aspergillus putrefaciens* produced H_2S when O_2 tension was low, and that pH of the meat above 6.0 resulted in production of sulfomyoglobin and greening of meat. Bem *et al.* (1976) also found that high-pH meat had poor keeping quality and was unsuitable for vacuum packaging.

Patterson and Gibbs (1977) detected off-odor due to gram-negative organisms such as *Proteus, Hafnia, Serratia, Pseudomonas, Aeromonas,* and *Alcaligenes* during storage of vacuum-packaged beef (pH 6.6) at 0 to 2°C for 8 weeks. Patterson and Bolton (1981) reported a large increase in sulfur compounds after 100 hr storage of high-pH meat at 15°C. Similar odor components were identified in both high- and low-pH meat, but they appeared much sooner in samples with higher pH.

Petaja *et al.* (1981) reported data on the storage quality of ground beef and vacuum-packaged fillets. Ground beef made of DFD (pH 6.0-6.4) meat maintained its red color longer than normal (pH 5.2) beef during storage at 6°C. Vacuum-packaged fillet steaks turned green after 2 weeks' storage, while no discoloration was observed for normal fillet steaks. Ground beef with normal pH had putrefactive off-odors after 3 days' storage at 6°C, while DFD meat produced undesirable odors after 1 day of storage. Total bacterial counts in various packaged DFD fillet steaks was 10^7/g after storage for 2 weeks at 6°C, whereas below pH 6.0 the value of 10^7/g was not exceeded until after 4 weeks of storage.

Another physiological condition that results in reduced shelf-life of meat during storage is pale-soft and exudative (PSE) muscle resulting from stress immediately prior to slaughter. This type of stress causes a very rapid drop in pH (5.3) postmortem before the carcass has been cooled to 4°C. The combination of low pH and high temperature disrupts muscle structure and denatures sarcoplasmic proteins (Bendall and Wismer-Pedersen, 1962; Briskey, 1963).

Topel *et al*. (1976) found that lack of color of PSE pork was responsible for poor consumer preferences. This kind of pork has low water-holding capacity, a soft texture, is not desirable for emulsion-type products, and is not desirable for processed meat production (Briskey, 1964).

Weeping or dripping becomes excessive during refrigeration or frozen storage of PSE pork (Lawrie, 1979). This can result in drying under normal conditions of cookery.

A similar condition is sometimes found in beef. Stringer's data (1963) revealed that normal firm beef loin steaks maintained desirability longer than soft steaks. The color of firm steaks remained desirable for 6 days at 32°F, while soft steaks were undesirable after 4 days of storage at this temperature. This difference was undoubtedly a pH effect, since the firm steaks had higher bacterial counts than soft steaks initially and after 7 days storage at 32°F. There was a high correlation (r = .88) between total bacterial counts and color.

B. Changes Occurring during Aging

After the animal's death, its skeletal muscles live on for a short period, but because of the lack of ATP and ATP precursors the oxygen-starved muscles enter rigor mortis. These muscles lose their rubberlike extensibility and become quite rigid.

The extent of muscle contraction at rigor is likely responsible for the degree of tenderness and is dependent on muscle temperature and pH. Muscle that is cooled prerigor to temperatures below 20°C begins to shorten, and shortens to a maximum at about 15°C. Cold shortening is very slow and develops only about 5% of the power of physiological contraction (Davey *et al*., 1967). Prerigor muscle also goes through thaw rigor if frozen and thawed while ATP is available for contraction. Shortening in this muscle can be as great as 80% (Jungk *et al*., 1967).

Toughness can be avoided if prerigor muscle is allowed to go into rigor before chilling or if the carcass is maintained briefly at 35 to 45°C before chilling. Shortening can also be prevented by electrical stimulation of carcasses to induce muscle rigor by depleting ATP and its precursors (Chrystall and Devine, 1978).

Muscle in various forms of rigor is very rigid but becomes more flaccid during storage at 2 to 4°C. The phrase "resolution of rigor" has been used to describe this phenomenon of meat aging, but there is little evidence that rigor is resolved. Rather, the bulk of the data indicate that myofibrillar and supportive proteins are degraded during this aging period. An aging period of 6 days has been shown to produce satisfactory tenderness (Martin *et al.*, 1970), and little tenderness is gained after aging for 11 days or more (Culp *et al.*, 1973). High-temperature (15-44°C) aging helps reduce storage time and cold shortening, but bacterial growth and excessive shrinkage are major problems (Sleeth and Naumann, 1960).

Some measurable changes that occur during postmortem aging are lengthening of rigor-shortened sarcomeres, increase in ATPase activity of myofibrils, increase in protein solubility, degradation of Z-lines, and weakening of actin-myosin bridges (Venugopal, 1970; Kim, 1974). These and other changes were discussed by Davey (1983).

Z-disk disappearance was the first identifiable structural change reported to occur during aging (Davey and Gilbert, 1967), and according to Chin-Shing and Parrish (1978), bonding of α-actinin to the Z-disk is apparently weakened during aging. Z-disk degradation results from the activity of calcium activation factor (Dayton *et al.*, 1981) and other muscle proteases (Cho, 1976), as well as that of blood leukocyte lysosomal proteinases (Venugopal, 1970; Cho, 1976).

Perhaps the greatest changes occurring in proteins during aging are those associated with the cytoskeletal muscle. Three recently discovered proteins (desmin, titin, and nebulin) have been shown to degrade appreciably during postmortem aging (Robson and Huiatt, 1983). Desmin filaments are believed to encircle the Z-line and radiate perpendicularly to the myofibrillar axis and connect to adjacent myofibrils. Titin consists of thin, longitudinally running elastic filaments that apparently help maintain myosin filaments in longitudinal register. Nebulin may be a structural component of the N_2-line in the I-band and may be attached to titin in the sarcomere.

Desmin progressively disappears during postmortem aging of muscle and is presumably destroyed by proteolysis. There is also a possible relationship between titin degradation and aging, particularly at higher temperatures (Locker, 1982).

Nebulin is especially sensitive to proteolytic degradation (Wang, 1981), and N_2-lines are readily removed by treatment with CAF (Yamaguchi *et al.*, 1983).

III. QUALITY CHANGES DURING STORAGE

A. Refrigerated Storage

The retail storage of meat is severely limited by the growth of microorganisms. The shelf-life at the normal temperature of storage is usually ~3 days. It is possible through careful sanitary procedures to extend this period to 5 to 7 days, although this is usually not attained under normal commercial conditions.

1. Microbiological Spoilage

Microorganisms are the most important cause of spoilage of refrigerated meats. Bacterial spoilage is primarily due to the activities of psychrotrophiles, which produce undesirable flavors and odors and discolor meat by catalyzing oxidation of myoglobin. Important factors outlined by Grau and Macfarlane (1980) necessary for preventing spoilage by bacteria are (1) use of good sanitation practices during slaughter and processing to limit initial contamination, (2) destroying or removing contaminating microorganisms, (3) reducing the rate of growth of contaminating organisms by maintaining low temperature during processing, transportation, and storage, and (4) knowing time-temperature limitations for maintaining quality and turning product over within these limitations.

One American packing company with a good reputation for maintaining good quality of both fresh and processed meats has a general product motto relative to keeping meat clean, keeping it cool, and keeping it moving through their processing and sales divisions.

Many aspects of microbiological contamination and control have been reviewed by Lawrie (1979) and by Grau and Macfarlane (1980). They pointed out that the animal itself and slaughterhouse tactics are the primary sources of microbiological contamination. It is particularly important to avoid dirt from animal sources, such as feet, hides, and fleece (Bryce-Jones, 1969), and to prevent exposure of meat surfaces to gastrointestinal contents, feces, and other parts of the animal carcass.

Improvements in slaughterhouse practice have been used to bring about a 20-fold reduction in contamination with psychrotrophic microorganisms (Empey and Scott, 1939).

An important contribution to reducing carcass surface microorganisms has been the innovative procedure of Anderson *et al.* (1980), who used an automatic spraying system for washing and sanitizing beef carcasses. The sanitizing solution was 3% acetic acid, which could be used to reduce the microbial population on the meat surface by an initial 1.49 logs. The difference between washed and sanitized carcasses after 1 week (168 hr) was 0.92 log. Reduction of bacterial counts due to sanitizing increased considerably with storage time. These sanitation procedures were designed to automate spraying techniques of Naumann and Balasundaram (1975), who found that sanitizing with acetic acid markedly reduced the surface bacterial counts of beef short loins and steaks.

Similar use (Eustace *et al.*, 1979) of dilute acetic acid for sanitizing lamb carcasses extended storage life of vacuum-packaged lamb by at least 4 weeks.

Good hygiene is essential in all carcass and meat processing areas. Contamination by contact with unhygienic surfaces by personnel and by airborne organisms is a possibility in all processing operations including chilling, cutting, packaging, refrigeration, freezing, transport, and sales.

The most efficient procedure for retarding microbiological growth is decreasing the temperature, and obviously lower temperatures are more effective than higher temperatures. Short-term storage is normally at 1 to 4°C and longer-term storage at freezing temperatures. Growth of microorganisms during meat storage at refrigerated temperature can result in changes in flavor, color, and texture.

Temperature of storage is very important in maintaining color quality. Stringer (1963) found that shelf-life of beef steak could be extended to 7 days if temperature was kept at 0°C instead of 3.3°C. After 7 days of storage, bacteria of steaks stored at 0°C were 1×10^7/g, while those stored at 3.3°C were 1×10^8/g.

Pseudomonas fragi organisms are present during storage of pork at 2°C, and some proteolysis of myofibrils occurs but not before spoilage odors develop (Dainty *et al.*, 1975). Alford and Pierce (1961) found that lipolytic activity of microorganisms also occurs at low or intermediate temperatures.

Discolorations are due to changes in myoglobin to brown metmyoglobin or combining with H_2S to form green sulfmyoglobin (Jensen, 1949) or oxidation to yellow or green pigments by hydrogen peroxide produced by microorganisms (Watts, 1954).

Light also contributes to discoloration of fresh meat at refrigeration temperatures. Marriott *et al.* (1967) reported that beef stored at 1°C under direct illumination deteriorated faster than similar samples stored in the dark, although the surface temperature of the samples stored under light might have been somewhat higher, since light enhanced bacterial growth. The light effect on discoloration, however, has been demonstrated by other workers (Lentz, 1971).

Higher energy light (shorter wavelengths) is more detrimental to color than low-energy light (Sester *et al.*, 1973). A more extensive review of the effects of retail display conditions on meat color and other environmental variables was published by Kropf (1980), and other factors were discussed by Seideman *et al.* (1983).

Another typical index of the end point of spoiled product during refrigeration is odor. This may be the most critical characteristic of judging quality, since color can be undesirable on some meat samples that still have desirable flavor. Amino acids and proteins are readily degraded to form undesirable odors. *Micrococci* can decarboxylate amino acids to form amines. *Pseudomonas* can deaminate amino acids and proteins, and putrefactive organisms can produce H_2S from cysteine. Other metabolic products produced by microorganisms have undesirable odors (Gill and Newton, 1978).

Scott (1937) found the shelf-life of beef could be extended by 50% by reducing temperature to $-1^{\circ}C$. Bacteria grow very little at the freezing point of meat, and there is essentially no growth of microorganisms at $-18^{\circ}C$.

Some other data cited by the International Commission on Microbiological Specifications for Food (1980) on the influence of temperature on shelf-life of meat at different initial bacterial counts are given in Table II.

2. Packaging of Fresh Meat

The key to good quality protection of meat during low-temperature storage is use of appropriate packaging. A large percentage of fresh meat is sold through the supermarket, and packaging is one of the most important factors relative to the shelf-life. Three methods were discussed by Hermansen (1983) for packaging of fresh meat:

1. Ordinary retail packaging, where the product is packaged in a food container and wrapped in a thin film that is permeable to oxygen
2. Packaging in modified atmospheres (MA), where the product is packaged in altered concentrations of gases such as 80% O_2 plus 20% CO_2 or inert atmospheres such as N_2
3. Vacuum packaging, where the product is sealed in a thick oxygen-impermeable film and evacuated.

All three types of packages lower evaporation losses and protect the product, but the MA and vacuum packaging have greater potential for prolonging the storage life of centralized packaged meat.

Naumann and Balasundaram (1975) demonstrated that storage of beef in high-CO_2 atmospheres inhibited growth of microorganisms, and normal logarithmic growth commenced only after ∿14 days of storage at 4°C.

These results were confirmed by Partmann *et al.* (1975), who reported that the highest scores during storage for 6 weeks at 1°C for color, visual freshness, and the lowest bacterial counts were obtained for samples packaged in 20% CO_2-8% N_2 compared to samples packaged in O_2 or air. Similar results were reported by Clark and Lentz (1973) and by Adams and Huffman (1972).

A summary of the storage characteristics found by Hermansen (1983) for the two types of packaging materials is given in Table III. Somewhat better results were obtained by Clark and Lentz (1973) for CO_2 MA packaging as shown in Table IV.

Modified atmospheres have only few advantages compared to vacuum packaging in maintaining quality during storage. Smith and Carpenter (1973) demonstrated that steaks from vacuum-packaged cuts had longer shelf-life than those from PVC-wrapped cuts and were preferred to those from CO_2-chilled cuts. Vacuum-packaged rounds were more desirable, had lower bacterial counts, and required less trimming than CO_2-chilled rounds. These same workers (Seidemann *et al.*, 1976) reported that total desirability ratings for primal cuts packaged with high vacuum were higher than those for prime cuts packaged with low vacuum and obtained up to 28 days' storage at 1 to 3°C.

Vacuum packaging has also been used to extend the storage life of pork cuts (Smith *et al.*, 1974). Vacuum-packaged pork loins refrigerated longer than 1 week sustained less shrinkage, and had less surface discoloration and higher consumer acceptance ratings than loins wrapped in PVC film or parchment paper. Chops from these loins also received higher ratings.

The appearance of MA-packaged meat might be superior to that of vacuum-packaged meat if O_2 is one of the gases. The color of vacuum-packaged meat is usually darker because of the lack of oxygen to oxygenate myoglobin. Packaging in some modified atmospheres reduces the shelf-life and requires more space for storage and is unsuitable for freezing. Vacuum-packaged meats have longer shelf-life, reduced drip loss, require less space, and can be stored frozen (Hermansen, 1983).

The predominant bacteria on the surface of fresh meat at refrigeration temperatures are *Achromobacter* and *Pseudomonas*. These organisms have low survival in the presence of CO_2;

thus CO_2 is frequently used in gaseous flushes for meat packaging. Microbiological growth is inhibited with increasing concentrations of CO_2 (Niedzielski and Krala, 1981). Hess (1980) showed that pseudomonads and enterobacteria were completely inhibited during 56 days storage at 1°C in an atmosphere of pure CO_2. Total aerobic bacteria and lactobacilli grew slowly after a log phase of 12 days.

CO_2 is a good bacteriostatic agent because it can penetrate bacterial membranes and reduce pH (Turin and Wamer, 1977). The main disadvantages of using CO_2 is that it results in pigment and fat oxidation.

Vacuum packaging is largely replacing MA packaging in North America. Its use can reduce shrinkage, reduce storage costs, and inhibit psychrotolerant bacteria. Storage life can be extended appreciably compared to meat packaged in gas-permeable films as shown by Roth and Clark (1972).

These workers also found that the shelf-life of fresh beef stored at 5°C was also increased in evacuated gas-permable packages, because pseudomonads were inhibited and non-odor-producing lactobacilli flourished.

The type of flora developed on vacuum-packaged meat is determined by the pH of the meat and the oxygen permeability of the packaging film (Campbell *et al.*, 1979). When normal low-pH (5.5) meat is packaged in films of low oxygen permeability, enterobacteria, along with pseudomonads as discussed above, are inhibited. Growth of these strains increase with increase in oxygen permeability and increase in pH ($\geq$ 6.0).

CO_2 formed in the package appears to be a controlling factor, since it has been found to accumulate and also inhibit growth of aerobic organisms (Roth and Clark, 1972).

An important type of biochemical change in storage of fresh meat for long periods of time is that due to the innate enzymes remaining in muscle following animal death. Undesirable flavors can develop even when the growth of microorganisms is slow (Grau and Macfarlane, 1980). Certain ninhydrin-positive compounds in beef increased considerably during storage at 0 to 2°C in vacuum packages, probably peptides and amino acids, and acidic odors observed following storage for 3 or 4 weeks were attributed to the presence of short-chain fatty acids, mainly acetic acid (Sutherland *et al.*, 1976).

3. Predicted Shelf-Life

Labuza (1982) summarized data concerning shelf-life of beef, lamb, and pork stored at 5°C. Selected data are given in Table V. Odor, color, flavor, or bacterial counts were used to determined shelf-life by different workers. The results were quite variable and undoubtedly depended on factors other than temperature. Labuza (1982) used the data from different authors to plot log of shelf-life against temperature (in degrees Celsius) and calculated activation energy values similar to those used to define the activity of enzymes at various temperatures. The variability of the activation energies also showed that several factors other than temperature were responsible for quality differences. All factors should be considered when attempting to predict shelf-life from previously published or new data.

4. Warmed-over Flavor

Cooked meat exposed to oxygen develops undesirable oxidized flavors within a very short period following cooking, and these flavors are more apparent following reheating. Warmed-over flavor (WOF) has been used to describe this quality attribute of meat. Results of past experiments indicate that it is caused by iron catalysis of oxidative degradation of cellular lipids. The iron becomes more active as a catalyst following denaturation of the protein in myoglobin by heating, which also disrupts membranes and exposes phospholipids to environments containing oxygen. Physical procedures such as grinding or chopping of raw meat can also cause disruption of membrane systems, exposing bound lipids that are oxidized to produce WOF in raw meat. This undesirable flavor attribute can also occur when cured meat is irradiated with UV light, which results in destruction of the cured meat pigment and activates iron. WOF was scientifically recongized as a flavor problem in cooked meat products by Tims and Watts (1958), and Watts and her student studied the problem for many years. There are a number of review articles describing conditions necessary for causing WOF, the mechanism of catalysis of tissue lipids by iron, and methodologies useful for studying the problem (Watts, 1961, 1962; Sato and Hegarty, 1971; Love, 1972; Pearson *et al.*, 1977; Pearson and Gray, 1983).

WOF results when phospholipid fatty acids of muscle cell membranes and other cellular lipids are exposed to an oxidative environment that also contains iron from denatured myoglobin or iron and/or other metal ions from other sources. Unsaturated fatty acids are rapidly oxidized and degraded to low molecular weight constituents that contribute to undesirable flavor.

WOF develops in muscle from different species at different rates perhaps as a result of differences in degree of unsaturation of fatty acids subjected to oxidation. Phospholipid fatty acids are more highly unsaturated and are very unstable in oxidative environments. They are in imminent contact with cellular ingredients that contain iron in various forms as well as other catalytic metals.

In cooked meat, fish, and poultry, WOF develops very rapidly and flavor deterioration is very imminent following protein denaturation by heating. M. E. Bailey, S. Y. Lee, G. Joseph, and L. Fernando (unpublished data, 1985) obtained TBA and volatile aldehyde data, presented in Table VI, indicating that WOF development begins immediately after cooking and cooling of roast beef and that saturated and unsaturated aldehydes, alcohols, 2,3-octanedione, and 2-pentyl furan may be good marker compounds during these changes. Data in references cited above and specifically that of Younathan and Watts (1960), Witte *et al.* (1970), and Pearson *et al.* (1977) reveal that WOF develops very rapidly after cooking. The refrigerated shelf-life of unprotected cooked meat is very short indeed.

5. Summary of Quality Changes during Refrigerated Storage

In summary of information concerning storage of meat at refrigeration (-1 to 5°C) temperatures, the most extensive shelf-life will be attained when the meat producer becomes a meat hygienist at all levels of slaughter, processing, storage, and distribution. Preventative measures for retarding growth of microorganisms should begin at the slaughterhouse and continue through processing. Animals should be thoroughly cleaned before evisceration, and the carcasses should be washed and sanitized before evisceration. Storage facilities and processing areas and equipment should be kept immaculately clean and essentially sterile.

Meat prepared under these conditions, packaged in oxygen-permeable film, and stored in display cases under light at 0 to 1°C should have a shelf-life of at least 7 days. Extended shelf-life can be attained by MA packaging with CO_2 or by vacuum packaging.

The most accurate indicators of shelf-life or hygienic status of fresh meat are the numbers of psychrotrophic obligate aerobic pseudomonads and enterobacteria. These organisms can be reduced to low levels by packaging with CO_2 or by vacuum packaging. Shelf-life can be extended for several weeks by appropriate packaging and storage at low, nonfluctuating temperatures 0 ± 1°C, and can be predicted by establishing relationships between shelf-life and storage temperature similar to activation energies of enzyme reactions.

Storage life of cooked meat at refrigeration temperatures is very short unless the meat is treated to prevent oxidation and warmed-over flavor. Since this undesirable quality is apparently catalyzed by free iron, it begins immediately after cooking and is difficult to prevent and preserve the flavor of fresh meat. Nonprotected meat develops undesirable flavor in less than 1 day storage at 4°C.

B. Frozen Storage of Meat

1. Causes of Quality Changes during Frozen Storage

It is generally believed that shelf-life of frozen meat depends only on storage temperature and that it can be lengthened by lowering the temperature. Quality of meat during any kind of storage is no better than its initial quality, which can differ depending on animal species, breed, management, maturity, and stress, as well as certain product, processing, and packaging aspects. The latter include the color, pH, moisture content, and microbiological contamination, the degree of aging prior to freezing, the rate of freezing and thawing, and the packaging environment.

There are voluminous data in the literature relating meat quality to time and temperature tolerances that did not include consideration of the influence of other variables; yet many of these data are pertinent to various aspects of commercial and consumer use and must be considered as one phase of continued growth and improvement in the frozen-food industry (Diehl, 1969).

Van Arsdel (1969) distinguished three different ways of determining rates of quality change in frozen foods: (1) objective measurements of a characteristic related to food quality, (2) sensory analysis of "just-noticeable difference" (JND) of a quality attribute, and (3) consensus of opinion concerning the useful "storage life" expected of the product. All of these methods are useful for judging the quality of meat during frozen storage.

Chrystall (1972) reviewed the literature concerning the influence of storage temperature on the quality of frozen meat and discussed some of the changes that occur in meat during frozen storage, including weight loss, changes in structure, loss of nutrient content, protein denaturation, and changes in sensory attributes such as color and flavor. Similar studies were discussed by Berry (1983).

The greatest effect on weight loss is change in structural integrity, as freezing and thawing occur resulting in membrane lysis by the movement of ice crystals. Structural damage occurs causing exudation of fluid on drip when the meat is thawed (Grau and Macfarlane, 1980).

A variable amount of exudate is released when muscle is thawed following freezer storage and depends on muscle pH, rate of freezing, temperature, and time course of frozen storage and temperature fluctuations (Winger, 1984). Thaw exudate is increased in normal muscle when meat is frozen immediately postrigor (Ramsbottom and Koonz, 1940; Wierbicki *et al.*, 1957; Hamm, 1960; Lawrie, 1968), the pH is below 6.0 (Ramsbottom and Koonz, 1940; Lawrie, 1968), the meat is frozen slowly (Ramsbottom and Koonz, 1941; Brady *et al.*, 1942; Hiner and Hankins, 1947; Hamm, 1960; Crigler and Dawson, 1968; Jul, 1968), the frozen-storage temperature is high (Powrie, 1973) or the duration of frozen storage is long (Wierbicki *et al.*, 1957; Awad *et al.*, 1968; Winger and Fennema, 1976).

The influence of temperature fluctuations on quality during frozen storage has been studied by numerous workers. Some representative data for meat stored at -18°C compared to fluctuating temperatures are outlined in Table VII. All authors concluded that quality of meat stored at -18°C was superior to that of meat stored at fluctuating temperatures. This was even true for ground beef and pork stored in a fluctuating temperature environment between -18 and -29°C. Meat stored at constant -18°C was determined to be superior by McBee (1959) using several criteria of quality including weight loss, odor, flavor, and TBA values.

Besides exudation (drip), other important factors resulting in decrease in weight during frozen storage of meat are evaporation and freezer burn. The latter term describes the whitish patches on the surface of the meat caused by sublimation of ice crystals creating air pockets that scatter incident light (Lawrie, 1979). Appropriate packaging appears to be the best way to minimize moisture evaporation and freezer burn (Hiner *et al.*, 1951).

Berry (1983) reviewed several different factors that influence weight loss and palatability of meat products during storage. A summary of data cited by Berry is given in Tables VIII-X. Freezing increases evaporation and cooking loss. Protein denaturation increases with increased storage time. Tenderness and flavor decline during frozen storage. Freezing at lower temperature (-18 to -40°C) helps prevent structural changes, water loss, and changes in tenderness. Rapid freezing at low temperature (cryogenic tunnel) is the preferred method of freezing.

The major quality changes of meat during frozen storage besides changes in water binding and tenderness are changes in flavor or odor or those measured objectively as TBA values, free fatty acids, or peroxide values.

Microbiological changes are at a minimum below -10°C, and certainly there are very few at -18°C. The most important color changes result from oxidation of myoglobin to metmyoglobin, and obviously this depends on the actual storage temperature and the type of packaging material used.

McBee (1959) studied some oxidative changes during storage of beef and pork at temperatures from -7 to -18°C. Storage of meat samples for 12 months at -18°C was far superior to storage at -7°C with respect to odor, flavor desirability, and rancidity as measured by TBA. There were significant increases in TBA numbers and rancidity as measured by odor and flavor as storage time progressed at these temperatures.

2. Changes in Flavor during Frozen Storage

The major flavor change during frozen storage is rancidity development, and its rate differs from one species of meat to another. Palmer *et al.* (1953) found an inverse relationship between degree of unsaturation of pork fat and development of rancidity during frozen storage. They also reported decreased peroxide values as storage temperature declined for pork. Kopecky (1971) found that pork fat became rancid after storage for 1 month at -5°C, but similar changes required 1 year when the meat was stored at -18°C. Naumann *et al.* (1951) found wide animal variation in the time-temperature changes in rancidity of frozen pork, but most samples were "slightly rancid" after storage at -18°C for 9 months. The frozen-storage life of pork can be extended by lowering the temperature. Jeremiah (1981) was able to store pork for 28 weeks at -30°C in several different packaging materials without appreciable deterioration of the product.

Longwell (1949) found that beef fat had undesirable flavor after 9 months storage at -10 to -20°C. Chrystall (1972) likewise stated that beef stored for 9 months at -15°C had unacceptable flavor, and rancidity was discernible after storage for 6 months at this temperature.

Hiner *et al.* (1951) reported that vacuum packaging increased the time before rancid odors become objectionable. They found that vacuum-packaged lamb remained acceptable after 48 weeks at either -8 or -10°C. Unpackaged lamb was considered unpalatable within 12 weeks at these temperatures or within 30 weeks at -18°C.

Lundquist (1972) demonstrated that vacuum-packaged sirloin steaks, pork chops, lamb chops, and veal cutlets could be stored satisfactorily for over 20 months at -18°C.

3. Estimating Shelf-Life of Frozen Meat

As explained above, it should not be automatically concluded that lowering the temperature of meat products will extend the frozen-storage life because of the differences that might be anticipated from one product to another. Prestorage characteristics and quality may have a significant influence on frozen-storage quality. Methods used for predicting shelf-life may be only rough guidelines and sometimes are of very little value.

Several investigators (van Arsdel, 1969; Jul, 1969; Labuza, 1982; Bramsnaes, 1981; Varsanyi and Somogyi, 1983) have established time-temperature relationships that can be used to predict quality of refrigerated or frozen meat. Unfortunately, few experiments have been carried out to allow precise examination of these procedures under carefully controlled conditions.

Labuza (1982) published a series of time-temperature plots useful for predicting frozen shelf-life for meat stored at different temperatures including freezing. He also calculated activation energies and Q_{10} values from Arrhenius plots of the same data. These are useful methods for predicting shelf-life at individual temperatures when data at two or more temperatures are available.

Using these calculations and data from Pearce (1948), it can be predicted that the shelf-life of frozen pork at -18°C would be about 25 weeks and about the same from data analyzed by Labuza (1982) from Commonwealth Scientific Industrial Research Organization of Australia (1977). Storage time was extended by Jul (1969) by using packaging in polyethylene, and from the analysis of these data by Labuza (1982) it can be predicted that these pork samples could be stored for 71 weeks at -18°C. The latter data represent the extended shelf-life that might be obtained by using good product, processing, and packaging technologies.

Jul (1969) published a number of figures relating storage temperature and some measure of quality of many fresh and processed meat products. The shapes of the resulting curves were quite variable, some being very flat, others steep, and others curved or straight lines. Most of this lack of uniformity was due to the different conditions under which these products had been stored, packaged, and evaluated.

The shelf-life of cooked meat can also be predicted by the methods published by van Arsdel (1969), Jul (1969), and Labuza (1982). Cooked meat can have an appreciable shelf-life if precautions are taken to protect the product from oxidation. This is accomplished through use of antioxidants and cover sauces, particularly those containing Maillard browning reaction products used in conjunction with good packaging. Table XI is a summary of some data published by Jul (1969) for the shelf-life of cooked meats with and without gravy.

Overall, cooked meat has a shorter frozen shelf-life than uncooked meat, and meats cooked and stored with gravy or similar sauces have longer frozen shelf-lives than nonprotected meats. Data from Labuza (1982) summarizing work done by the Refrigeration Research Foundation (1973) indicated that the high-quality storage life (HQL) of meats stored with gravy from -10 to -20°C ranged from 17 to 54 weeks, whereas the practical storage life (PSL) of these products ranged from 49 to 140 weeks. Meats stored at these temperatures that were not protected by gravy had significantly lower shelf-lives. The maximal storage periods for HQL and PSL ranged from 10 to 37 weeks and from 30 to 94 weeks, respectively. These data confirm the results of Korschgen and Baldwin (1972), who found that beef, lamb, and pork precooked at low temperature (82-85°C) could be stored at -19°C at least 54 weeks if the meat were covered with gravy prepared from drippings during cookery.

Quality evaluation for most of the studies of cooked meat were by sensory analysis of general acceptability. Jul (1969) predicted that a number of cooked meat dishes could be stored at -18°C for over 175 weeks if the products were well formulated and adequately packaged. This prediction was based on data obtained by Dalhoff and Jul (1965) and by Bogh-Sorensen (1967), who was able to stabilize goulash for over 300 weeks at -18°C.

4. Summary of Shelf-Life of Frozen Meat

The keeping quality of frozen meat increases exponentially with decreasing temperature and shelf-life can be predicted from time-temperature data, but other very important features of quality during storage are previous history of the product and the procedures used for its processing, packaging, and freezing. Long shelf-life can be predicted for both raw and cooked meats that have been processed and packaged in carefully controlled environments.

Freezing increases evaporation, drip cooking loss, and protein denaturation, and decreases tenderness, color desirability, and flavor of fresh meat. These attributes can be improved by freezin and storage at low temperature. Freezing in cryogenic tunnels is a preferred method, and storage temperature should not be above -18°C. The temperature should be constant and not fluctuating even at lower levels.

The normal storage life of fresh beef at -18°C is ∿12 months, while pork may be somewhat shorter (6 months) and lamb somewhat longer (15 months). The shelf-lives of meat from all three species can be extended to 20 months by using good processing, packaging, and storage technologies.

The frozen shelf-life of cooked meat is short indeed unless protected by antioxidants, gravies, and/or vacuum packaging, and is generally much less than that of fresh meat. Some cooked meat dishes with gravy and other ingredients such as vegetables can be stored over 20 months by using exceptional processing and storage procedures.

ACKNOWLEDGMENT

The efforts of Patricia J. Scholes in typing and editing the manuscript are greatly appreciated.

TABLE I. Relationship between pH and flavor of steaks grilled by various methods

Cooking method	Number of panelists	Institute	Correlation coefficient
75°C center	14	Langford	-0.6**
70°C center	10	Dublin	-0.4
170°C (10 min)	9	Noskilde	-0.2
contact grill (6 min)	6	Kulmbach	0.1
50°C center	12	Theix	-0.6**

**P<0.01 Results compiled by Dransfield (1981)

TABLE II. Influence of storage temperature and initial counts on shelf-life of meat

Temperature °C	Shelf-life (Days for different counts)		
	$0\text{-}10^2$	$10^2\text{-}10^4$	$10^4\text{-}10^6$
0	16	11	6
5	9	6	4
10	5	3	2
15	3	2	1
20	2	1.5	<1

From International Commission on Microbiological Specifications for Food (1980).

TABLE III. Color and shelf-life of meat packaged in MA and vacuum packs

Packaging systems	Meat color	Shelf-life	Shelf-life limited by
MA packs with 80% oxygen + 20% carbon dioxide	Bright red	4-6 dys	Oxygen + bacteria
MA packs with nitrogen	Purple	1-3 wks	Bacteria
Vacuum packs	Purple	1-3 wks	Bacteria

Summarized from Hermansen (1983)

TABLE IV. Effect of CO_2 mixtures on shelf-life of packaged beef

Atmosphere	Days of storage	
	Odor	Color
Air	6	5
15% CO_2 - Air	16	8
15% CO_2 - 50% O_2- 35% N_2	17	12
15% CO_2 - 70% O_2- 15% N_2	19	14
15% CO_2 - 85% O_2	19	14

From Clark and Lentz (1973)

TABLE V. Shelf-life of beef and pork at 5°C

Product	Spoilage mode	Shelf-life: How determined	Shelf-life: Time (days)	Reference
Beef round	bacterial growth	odor	6	Clark and Lentz (1972)
Beef round	color	color	1	Snyder (1964)
Beef fillet	color	color	1	MacDougall and Taylor (1975)
Beef rump	color	color	1	MacDougall and Taylor (1975)
Beef sirloin	color	color	30	MacDougall and Taylor (1975)
Beef, ground	color	color	1	Rutgers University (1971)
Beef, ground	organoleptic	bacterial growth	3-4	Al-Delaimy and Stiles (1975)
Pork	bacterial counts	bacterial counts	4	Patterson and Edwards (1975)

Summarized from Labuza (1982)

TABLE VI. TBA values[a] and volatiles[b] of roast beef during storage at 4°C

Storage time (hr.)	0	3	72
TBA value	0.49	2.25	9.10
Volatile compounds	Area counts		
Aldehydes			
C_3	76	236	1,028
C_5	123	456	1,280
C_6	147	2599	15,693
C_7	74	148	661
C_8	64	92	415
C_9	15	35	186
2-Methyl-2-Butenal	0	5	22
2-Hexenal	0	13	84
2-Heptenal	15	25	272
2-Octenal	96	131	560
2-Nonenal	18	32	137
Benzaldehyde	22	66	412
Alcohols			
1-Penten-3-ol	0	8	35
1-Pentanol	10	88	383
1-Octen-4-ol	13	16	412
Miscellaneous			
2,3-Octanedione	0	44	880
2-Pentyl furan	6	37	232

[a]mg malonaldehyde/Kg meat

[b]Carbowax 20m Capillary GLC/MS

TABLE VII. Selected data where meat quality following storage at -18°C was compared with that during storage at fluctuating temperatures

Type of meat	Temp.	Fluctuating temp. °C	Reference
Pork	-18°C	-7°C to -18°C	Gortner *et al*. (1948)
Ground beef	-18°C	-12°C to -18°C	Ehrenkranz *et al*. (1952)
Pork	-18°C	-12°C to -18°C	Palmer *et al*. (1953)
Beef roast	-18°C	-12°C to -18°C	Gortner *et al*. (1948)
Ground beef	-18°C	-18°C to 12°C	Winter *et al*. (1952)
Ground beef & pork	-18°C	-18°C to 29°C	McBee (1959)

TABLE VIII. Influence of freezer conditions on water binding properties of beef

Product	Freezing conditions	Results	Reference
Beef loin steaks	Frozen at -8, -18, -23, -40 and -81°C	Lower freezing temp. resulted in more drip	Hiner *et al.* (1945)
Beef rib steaks	Frozen at -18 and -35°C	Steaks frozen at lower temp. have less drip	Ramsbottom and Koonz (1940)
Beef *sternomandibularis* m.	Frozen at -25°C vs. non-frozen	Freezing increased cooking loss	Locker and Darnes (1973)
Beef primals	Various frozen conditions	The slower the freezing rate the greater the evaporative loss	Malton and Cutting (1974)
Boxed beef	Frozen temp. -18, -21, -23°C	Lower temp. resulted in higher weight loss	Moleeratanond *et al.* (1981)
Beef patties	Cryogenic freezing at -74°C vs. blast air at -29°C	Blast air had higher freezing and cooking losses than cryogenic freezing	Sebranek *et al.* (1978)

(continued)

TABLE VIII. (continued)

Product	Freezing conditions	Results	Reference
Beef loin steaks	Air blast -30°C vs. still air -18°C	Air blast freezing less drip than still air	Kahn and Lentz (1977)
Beef loin steaks	Frozen at -10, -20, -30, -78°C	Total extractable proteins increased as temp. decreased	Petrovic and Rahelic (1981)
Ground beef	Frozen at -5 or -15°C	Greater WHC at -55°C than at -15°C	Deatherage and Hamm (1960)
Beef steaks	Storage up to 60 wk	Weight loss increased during storage and denaturation decreased at -10 and -20°C	Khan and Lentz (1977)
Boneless beef and pork	Storage up to 37 wk storage progressed	Greater thaw exudate as storage increased	Miller _et al_. (1980)

TABLE IX. Effects of type of freezing on palatability

Product	Freezing conditions	Results	Reference
Beef steak and roasts	Freezing at -29°C vs. non-freezing	Frozen product was more tender	Tressler _et al._ (1932)
Beef steaks	Freezing at -40°C vs. non-freezing	No difference in palatability	Tuma (1971)
Boxed beef	Freezing at -23, -21°C	Temperature had no effect on flavor	Moleeratanond _et al._ (1981)
Ground beef	Freezing at -50°C and -20°C	Freezing had no effect on palatability	Berry and Stiffler (1981)
Ground beef	Freezing in cryogenic tunnel vs. -29°C	Cryogenic freezing produced higher palatability ratings	Sebranek _et al._ (1978)
Lamb chops	Various air velocities -18°C, -29°C	No effect on palatability, shear force or cooking loss	Lind _et al._ (1971)

TABLE X. Effects of frozen storage on palatability of meat

Product	Storage time	Results	Reference
Beef loin steaks	1 & 9 mos.	9 mos product tougher than 1 mo	Dransfield (1974)
Beef loin steaks	1 & 9 mos	Flavor lower, tenderness lower. Shear values higher at 9 mos vs. 1 mo	Jakobsson and Bengtsson (1973)
Beef loin steaks and roasts	0, 5, and 10 mos	Tenderness and flavor decreased and shear increased during storage	McCoy *et al*. (1949)
Beef roasts and steaks	3, 6, 9 & 12 mos	Tenderness and flavor declined during storage	
Ground beef	2, 3 and 5 mos	Tenderness and flavor values declined during storage	Jakobsson and Bengtsson (1969)
Boxed beef	0, 3, 6 & 12 mos	Flavor declined after 6 mos storage	Hankins and Hiner (1941)

(continued)

TABLE X (continued)

Product	Storage time	Results	Reference
Pork loin	1, 21, 42 dys	Flavor decreased with storage	Berry *et al*. (1971)
Pork loin	0, 2, 4 mos	Tenderness increased and shear force decreased during storage	Kemp *et al*. (1976)
Pork patties	0, 2, 4 & 6 wks	No change in palatability or shear	Campbell and Mandigo (1978)
Pork chops	0, 1, 2, 3, 4, 6, 8, & 12 mos	Palatability decreased over storage period	Ramsbottom (1947)

TABLE XI. Shelf-life of cooked meats stored at -18°C with and without gravy

Meat product	Weeks of storage	Reference
Beef Goulash with milk sauce, carrots and onions	223	Jul (1969)
Cooked lean meat		Refrigeration Research Foundation (1973)
with gravy	71	
without gravy	14	
Cooked chopped beef with onions & gravy	120	Jul (1969)
Cooked pork		Dalhoff and Jul (1965)
with gravy	101	
without gravy	22	

Summarized from Jul (1969)

REFERENCES

Adams, J. R., and Huffman, D. L. (1972). *J. Food Sci.* *37*, 869.

Al-Delaimy, K. S., and Stiles, M. E. (1975). *Can. J. Public Health* *66*, 317.

Alford, J. A., and Pierce, D. A. (1961). *J. Food Sci.* *26*, 518.

Anderson, M. E., Marshall, R. T., Stringer, W. C., and Naumann, H. D. (1980). *J. Food Prot.* *43*, 568.

Asghar, A., and Pearson, A. M. (1980). *Adv. Food Res.* *26*, 53.

Awad, A., Powrie, W. D., and Fennema, O. (1968). *J. Food Sci.* *33*, 227.

Bem, Z., Helchelmann, H., and Leistner, L. (1976). *Fleischwirtschaft* *56*, 985.

Bendall, J. R., and Wismer-Pederson, J. (1962). *J. Food Sci.* *27*, 144.

Berry, B. W. (1983). "Effects of Freezing Rate on Meat Quality," Act. Rep. Vol. *35*, p. 113. Meat Sci. Res. Lab., USDA, ARS, Beltsville, Maryland.

Berry, B. W., and Stiffler, D. M. (1981). *J. Food Sci.* *46*, 1103.

Berry, B. W., Smith, G. C., Spencer, J. V., and Kroening, G. H. (1971). *J. Anim. Sci.* *32*, 636.

Bogh-Sorensen, L. (1967). Report 6/65. Dan. Meat Prod. Lab., Copenhagen.

Brady, D. E., Frei, P., and Hickman, C. W. (1942). *Food Res.* *7*, 383.

Bramsnaes, F. (1981). *Food Technol. (Chicago)* *35*, 38.

Briskey, E. J. (1963). *In* "Proceedings of the Meat Tenderness Symposium" (C. H. Krieger, ed.), p. 195. Campbell Soup Co., Camden, New Jersey.

Briskey, E. J. (1964). *Adv. Food Res.* *13*, 90.

Bryce-Jones, K. (1969). *Inst. Meat Bull.* *65*, 3.

Campbell, J. F., and Mandigo, R. W. (1978). *J. Food Sci.* *43*, 1648.

Campbell, R. J., Egan, A. F., Grau, F. H., and Shay, B. J. (1979). *J. Appl. Bacteriol.* *47*, 503.

Chin-Shing, C., and Parrish, F. C., Jr. (1978). *J. Food Sci.* *43*, 46.

Cho, M. J. (1976). Ph.D. Dissertation, University of Missouri, Columbia.

Chrystall, B. B. (1972). *Meat Ind. Res. Inst. N.Z.* [*Rep.*] *M.I.R.I.N.Z. 274*.

Chrystall, B. B., and Devine, C. E. (1978). *Meat Sci.* *2*, 49.

Clark, D. S., and Lentz, C. P. (1972). *J. Food Sci. Technol. 5*, 175.
Clark, D. S., and Lentz, C. P. (1973). *Can. Inst. Food Sci. Technol. J. 6*, 194.
Commonwealth Scientific Industrial Research Organization of Australia (1977). "Storage Temperature for Frozen Meats," Na. Provis. Vol. *12*. CSIRO, Australia.
Crigler, J. C., and Dawson, L. E. (1968). *J. Food Sci. 33*, 248.
Culp, G. R., Carpenter, Z. L., Smith, G. C., and Davis, G. W. (1973). *J. Anim. Sci. 37*, 258 (abstr.).
Dainty, R. H., Shaw, B. G., deBoer, K. A., and Scheps, E. S. J. (1975). *J. Appl. Bacteriol. 139*, 73.
Dalhoff, E., and Jul, M. (1965). *In* "Progress in Refrigeration Science and Technology" (W. T. Penzer, ed.), Vol. 1, p. 57. AVI, Westport, Connecticut.
Davey, C. L. (1983). *Proc. Annu. Reciprocal Meat Conf. 36*, 108.
Davey, C. L., and Gilbert, K. V. (1967). *J. Food Technol. 2*, 57.
Davey, C. L., Kuttel, H., and Gilbert, K. V. (1967). *J. Food Technol. 2*, 53.
Dayton, W. R., Lepley, R. A., and Schollmeyer, J. V. (1981). *Proc. Annu. Reciprocal Meat Conf. 34*, 17.
Deatherage, F., and Hamm, R. (1960). *Food Res. 25*, 623 (abstr.).
Diehl, H. C. (1969). *In* "Quality and Stability of Frozen Foods" (W. B. van Arsdel, M. J. Copley, and R. L. Olson, eds.), pp. IX and 285. Wiley (Interscience), New York.
Dransfield, E. (1974). *In* "Meat Freezing-Why and How?" Symp. No. 3, Meat Res. Inst., Langford, Bristol, England.
Dransfield, E. (1981). *In* "The Problem of Dark-Cutting in Beef" (D. E. Hood and P. V. Tarrant, eds.), p. 344. Martinus Nijhoff Publishers, The Hague, The Netherlands.
Ehrenkranz, F., Roberts, H., and Ross, S. (1952). *J. Home Econ. 44*, 441.
Empey, W. A., and Scott, W. J. (1939). *Bull.--C.S.I.R.O. (Aust.) 126*.
Eustace, I. J., Powell, V. H., and Bill, B. A. (1979). Meat Res. Rep. No. 3/79. Div. Food Res. Commonw. Sci. Ind. Res. Organ., Queensland, Australia.
Gill, C. O., and Newton, K. G. (1978). *Meat Sci. 2*, 207.
Gortner, W. A., Fenton, F., Volz, F. E., and Gleim, E. (1948). *Ind. Eng. Chem. 40*, 1423.
Grau, F. H., and Macfarlane, J. J. (1980). *CSIRO Food Res. Q. 40*, No. 3/4.

Hamm, R. (1960). *Adv. Food Res. 10*, 355.
Hankins, O. G., and Hiner, R. L. (1938). *Food Ind. 1984 12*, 49.
Hankins, O. G., and Hiner, R. L. (1941). *Refrig. Eng. 42*, 185.
Hedrick, H. B. (1981). *In* "The Problem of Dark-Cutting in Beef" (D. E. Hood and P. V. Tarrant, eds.), p. 213. Martinus Nijhoff Publishers, The Hague, The Netherlands.
Hermansen, P. (1983). *Proc. Annu. Reciprocal Meat Conf. 36*, 60.
Hess, E. (1980). *Fleischwirtschaft 60*, 1448, 1450, 1458, 1513.
Hiner, R. L., and Hankins, O. G. (1947). *Food Ind. 19*, 1078.
Hiner, R. L., Hadsen, L. L., and Hankins, O. G. (1945). *Food Res. 10*, 312.
Hiner, R. L., Gaddis, A. M., and Hankins, O. G. (1951). *Food Technol. 5*, 223.
International Commission on Microbiological Specifications for Food (1980). "Microorganisms in Food, Vol. III.
Jakobsson, B., and Bengtsson, N. E. (1969). *Proc. Eur. Meat Res. Workers 15*, 482.
Jakobsson, B., and Bengtsson, N. E. (1973). *J. Food Sci. 38*, 560.
Jensen, L. B. (1949). "Meat and Meat Foods," p. 47. Ronald Press, New York.
Jeremiah, L. E. (1981). *J. Food Qual. 5*, 73.
Jul, M. (1968). *In* "Low Temperature Biology of Foodstuffs" (J. Hawthorn and E. J. Rolfe, eds.), p. 413. Pergamon, Oxford.
Jul, M. (1969). *In* "Quality and Stability of Frozen Foods" (W. B. van Arsdel, M. J. Copley, and R. L. Olson, eds.), p. 191. Wiley (Interscience), New York.
Jungk, R. A., Snyder, H. E., Goll, D. E., and McConnell, K. G. (1967). *J. Food Sci. 32*, 158.
Kemp, J. D., Montgomery, R. E., and Fox, J. D. (1976). *J. Food Sci. 41*, 1.
Khan, A. W., and Lentz, C. P. (1977). *Meat Sci. 1*, 263.
Kim, M. K. (1974). Ph.D. Dissertation, University of Missouri, Columbia.
Kopecky, A. (1971). *Food Sci. Technol. 4*, 1543 (abstr.).
Korschgen, B. M., and Baldwin, R. E. (1972). *Food Prod. Dev. 6*, 39.
Kropf, D. H. (1980). *Proc. Annu. Reciprocal Meat Conf. 33*, 15.
Labuza, T. P. (1982). "Shelf-Life Dating of Foods." Food and Nutrition Press, Inc., Westport, Connecticut.

Lawrie, R. A. (1968). *In* "Low Temperature Biology of Foodstuffs" (J. Hawthorn and E. J. Rolfe, eds.), p. 359. Pergamon, Oxford.

Lawrie, R. A. (1979). "Meat Science," 3rd ed. Pergamon Press, Oxford.

Lentz, C. P. (1971). *Can. Inst. Food Sci. Technol. J. 23*, 30.

Lind, M. L., Harrison, D. C., and Kropf, D. H. (1971). *J. Food Sci. 36*, 629.

Locker, R. H. (1982). *Proc. Annu. Reciprocal Meat Conf. 35*, 92.

Locker, R. H., and Darnes, G. J. (1973). *J. Sci. Food Agric. 24*, 1273.

Longwell, J. H. (1949). *Res. Bull.--Mo., Agric. Exp. Stn. 440*.

Love, J. D. (1972). Ph.D. Thesis, Michigan State University, East Lansing.

Lundquist, B. R. (1972). "Package Engineering." Sept. MIR.

McBee, J. L. (1959). Ph.D. Dissertation, University of Missouri, Columbia.

McCoy, D. C., Hayner, G. A., Reiman, W., and Hockman, R. (1949). *Refrig. Eng. 57*, 971.

MacDougall, P. B., and Taylor, A. A. (1975). *J. Food Technol. 10*, 339.

Malton, R., and Cutting, C. C. (1974). *In* "Meat Freezing--Why and How?" Symp. No. 3, p. 36.1. Meat Res. Inst., Langford, Bristol, England.

Marriott, N. G., Naumann, H. D., Stringer, W. C., and Hedrick, H. B. (1967). *Food Technol. 21*, 104.

Martin, A. H., Freeden, H. T., and Weiss, G. M. (1970). *Can. J. Anim. Sci. 50*, 235.

Miller, A. J., Ackerman, S. A., and Palumbo, S. A. (1980). *J. Food Sci. 45*, 1466.

Moleeratanond, W., Ashby, B. H., Kramer, A., Berry, B. W., and Lee, W. (1981). *J. Food Sci. 46*, 829.

Naumann, H. D., and Balasundaram, K. K. (1975). *Proc. Eur. Meet. Meat Res. Workers 20*, 184.

Naumann, H. D., Brady, D. F., Palmer, A. Z., and Tucker, L. N. (1951). *Food Technol. 5*, 496.

Nichol, D. J., Shaw, M. K., and Ledward, D. A. (1970). *Appl. Microbiol. 19*, 937.

Niedzielski, Z., and Krala, L. (1981). *Acta Aliment. Pol. 7*, 35.

Palmer, A. Z., Brady, D., Naumann, H. D., and Tucker, L. N. (1953). *Food Technol. 7*, 90.

Park, R. J., and Murray, J. C. (1975). "Meat Research in CSIRO," p. 22.

Partmann, W., Bomar, M. T., and Bohling, H. (1975). *Fleischwirtschaft 55*, 1441.

Patterson, J. T., and Bolton, G. (1981). *In* "The Problem of Dark-Cutting in Beef" (D. E. Hood and P. V. Tarrant, eds.), p. 454. Martinus Nijhoff Publishers, The Hague, The Netherlands.
Patterson, J. T., and Gibbs, P. A. (1977). *Appl. Bacteriol. 43*, 25.
Patterson, R. L. S., and Edwards, R. A. (1975). *J. Food Agric. 26*, 1371.
Pearce, J. A. (1948). *Can. Food Ind. 19*, 18.
Pearson, A. M., and Gray, J. I. (1983). *In* "The Maillard Reaction in Foods and Nutrition" (G. R. Waller and M. S. Feather, eds.), p. 287. Am. Chem. Soc., Washington, D.C.
Pearson, A. M., Love, J. D., and Shorland, F. B. (1977). *Adv. Food Res. 34*, 1.
Petaja, E., Puolanne, E., and Vahervvo, I. (1981). *Suom. Elainlaakaril. 87*, 75, cited in *FSTA 15*, 3, 3S507 (1983).
Petrovic, L., and Rahelic, S. (1981). *Proc. Eur. Meet. Meat Res. Work 27*, A29.
Powrie, W. D. (1973). *In* "Low-Temperature Preservation of Foods and Living Matter" (O. R. Fennema, W. D. Powrie, and E. H. Marth, eds.), p. 282. Dekker, New York.
Ramsbottom, J. M. (1947). *Refrig. Eng. 53*, 19.
Ramsbottom, J. M., and Koonz, C. H. (1940). *Food Res. 5*, 423.
Ramsbottom, J. M., and Koonz, C. H. (1941). *Food Res. 6*, 571.
Refrigeration Research Foundation (1973). TTT and PPP Inf. Bull. No. 73-1, January.
Robson, R. M., and Huiatt, T. W. (1983). *Proc. Annu. Reciprocal Meat Conf. 36*, 116.
Roth, L. A., and Clark, D. S. (1972). *Can. J. Microbiol. 18*, 1761.
Rutgers University (1971). "Food Stability Survey." University Department of Food Science, for the State of New Jersey Department of Health in cooperation with the U.S. Department of Agriculture/Economic Research Service. Rutgers Univ., New Brunswick, New Jersey.
Sato, K., and Hegerty, G. R. (1971). *J. Food Sci. 36*, 1098.
Scott, W. J. (1937). *J. Counc. Sci. Ind. Res. (Aust.) 10*, 338.
Sebranek, J. G., Sang, P. N., Rust, R. E., Topel, D. G., and Kraft, A. A. (1978). *J. Food Sci. 43*, 842.
Seideman, S. C., Carpenter, Z. L., Smith, G. C., and Hoke, K. E. (1976). *J. Food Sci. 41*, 732.

Seideman, S. C., Cross, H. R., Smith, G. C., and Durland, P. R. (1983). *J. Food Qual. 6*, 211.
Sester, C. W., Harrison, D. L., Kropf, D. H., and Dayton, A. (1973). *J. Food Sci. 38*, 412.
Sleeth, R. B., and Naumann, H. D. (1960). *Food Technol. 14*, 98.
Smith, G. C., and Carpenter, Z. L. (1973). *Proc. Eur. Meet. Meat Res. Workers 19*, Part 1, 353.
Smith, G. C., Rape, S. W., Motycka, R. R., and Carpenter, Z. L. (1974). *J. Food Sci. 39*, 1140.
Snyder, H. F. (1964). *J. Food Sci. 29*, 535.
Sornay, J., Dumont, B. L., and Fournaud, J. (1981). *In* "The Problem of Dark-Cutting in Beef" (D. E. Hood and P. V. Tarrant, eds.), p. 363. Martinus Nijhoff Publishers, The Hague, The Netherlands.
Stringer, W. C. (1963). Ph.D. Dissertation, University of Missouri, Columbia.
Sutherland, J. P., Gibbs, P. A., Patterson, J. T., and Murray, J. C. (1976). *J. Food Technol. 11*, 171.
Tims, M. J., and Watts, B. M. (1958). *Food Technol. 12*, 240.
Topel, D. G., Miller, J. A., Berger, P. J., Rust, R. E., Parrish, F. C., Jr., and Ono, K. (1976). *J. Food Sci. 41*, 628.
Tressler, D. K., Birdseye, C., and Murray, W. T. (1932). *Ind. Eng. Chem. 24*, 242.
Tuma, H. J. (1971). *Proc. Meat Ind. Res. Conf.*, p. 53.
Van Arsdel, W. B. (1969). *In* "Quality and Stability of Frozen Foods" (W. B. van Arsdel, M. J. Copley, and R. L. Olson, eds.), p. 237. Wiley (Interscience), New York.
Varsanyi, I., and Somogyi, L. (1983). *Acta Aliment. Acad. Sci. Hung. 12*, 73.
Venugopal, B. (1970). Ph.D. Dissertation, University of Missouri, Columbia.
Wang, K. (1981). *J. Cell Biol. 91*, 3-5.
Watts, B. M. (1954). *Adv. Food Res. 5*, 1.
Watts, B. M. (1961). *In* "Proceedings of the Flavor Chemistry Symposium," p. 83. Campbell Soup Co., Camden, New Jersey.
Watts, B. M. (1962). Symposium on Foods: Lipids and Their Oxidation" (E. A. Day and R. O. Sinnhuber, eds.), p. 2. Avis Westport, Connecticut.
Wierbicki, E., Kunkle, L. E., and Deatherage, F. E. (1957). *Food Technol. 11*, 69.
Winger, R. J. (1984). *Food Technol. N.Z.*, October, p. 75.
Winger, R. J., and Fennema, O. R. (1976). *J. Food Sci. 41*, 1433.

Winter, J. D., Hustrulio, A., Nobel, I., and Ross, E. S. (1952). *Food Technol.* *6*, 311.
Wismer-Pedersen, J. (1976). *Proc. Eur. Meet. Meat Res. Workers* *22*, 1:BO:1.
Witte, V. C., Krause, G. F., and Bailey, M. E. (1970). *J. Food Sci.* *35*, 58.
Yamaguchi, M., Robson, R. M., Stromer, M. H., Cholvin, N. R., and Izumimoto, M. (1983). *Anal. Rec.* *206*, 345.
Younathan, M. T. M., and Watts, B. M. (1960). *Food Res.* *25*, 538.

CHAPTER 4

SHELF-LIFE OF FISH AND SHELLFISH

GEORGE J. FLICK, JR.
LEOPOLDO G. ENRIQUEZ
JANIS B. HUBBARD
Department of Food Science and Technology
Virginia Polytechnic Institute and State University
Blacksburg, Virginia

The spoilage of fish and shellfish under refrigerated, iced, and frozen conditions is a bacteriological, enzymatic, and oxidative phenomenon. A significant number of researchers have reported a series of diverse analyses in order to identify freshness or quality changes in fish and shellfish. An important fact that sometimes is overlooked in determining postharvest quality is that there usually are differences in the composition both within and between species of fish and shellfish. These variations depend on several factors such as genetics, environment, physiological state, and sex. Consequently, it is not uncommon to find in the literature contradictory results of analytical tests on the same species. The information contained in this chapter is a collection of chemical, biochemical, bacteriological, and sensory analyses performed on specific species of fish and shellfish. These analyses have been designed to determine the concentration of various components in fish and seafood and to obtain quantitative indices of freshness.

Each table contains the common name, scientific name, location of fish harvesting, and literature references (in parentheses). In addition, an index immediately following the tables contains references to all the objective and subjective observations in the tables, and also abbreviations. The numbers in this index refer to table numbers, not page numbers.

ABBREVIATIONS USED IN THE TABLES

A = optical density at 250 nm of a perchloric acid extract containing Hx, HxR, IMP, AMP, ADP, and AMP
AA-N = amino nitrogen
ADP = adenosine diphosphate
AMN = actomyosin nitrogen
AMP = adenosine monophosphate
AnPC = anaerobic plate count
APC = aerobic plate count
ATP = adenosine triphosphate
B = optical density at 250 nm of a 0.01 *N* NH_4Cl elution containing HxR and Hx
DMA-N = dimethylamine nitrogen
EDTA = ethylenediaminetetraacetic acid
EGTA = ethyleneglycol-bis(2-aminoethylether) tetraacetic acid
EPN = extractable protein nitrogen
ERV = extract-release volume
ESP = extractable sarcoplasmic protein
ETD = expressed thaw drip
FA = formaldehyde
FAA = free amino acids
FBN = formaline-bound nitrogen
FFA = free fatty acids
G-6-P = glucose-6-phosphate
Glyc = glycogen
g = gram
g % = grams per 100 g or ml of sample
Hx = hypoxanthine
HxR = inosine

IMP = inosine monophosphate
K (%) = $(B/A) \times 100$
kg = kilogram
LA = lactic acid
Leu = leucine
Lys = lysine
MBC = mesophilic bacterial count
meq = milliequivalent
mg = milligram
mg % = milligrams per 100 g or ml of sample
moles % = moles per 100 g or ml of sample
nm = nanometer
NAD = nicotinamide adenine dinucleotide
NPN = nonprotein nitrogen
OD = optical density
org. = organism
PAT = picric acid turbidity
PBC = psychrophilic bacterial count
P_i = inorganic phosphorus
ppm = parts per million
PV = peroxide value
RSW = refrigerated seawater
SAA = soluble active actin
SAN = soluble actin nitrogen
SEP = saline-extractable protein
SIA = soluble inactive actin
SPC = sulfide-producing count
SPN = sarcoplasmic protein nitrogen
SSP = salt-soluble protein
TAN = total actin nitrogen
TBA = thiobarbituric acid
TEP = total extractable protein
TMA-N = trimethylamine nitrogen
TMAO-N = trimethylamine oxide nitrogen
TPC = total plate count
TVA = total volatile acids
TVB-N = total volatile base nitrogen
TVN = total volatile nitrogen
Tyr = tyrosine
VRS = volatile reducing substances

Bass, *Lateolabrax japonicus*, Japan

TABLE 1. Changes in quality parameters during the iced and partially frozen storage (-3°C) of sea bass muscle (100)

	Storage time (days)				
	0	3	5	8	10
Iced					
Tyr (μg/g)	140	145	157	188	222
TMA-N (mg %)	0.30	0.35	0.45	4.00	5.70
K-value (%)	8.00	44.0	57.0	--	--
Frozen					
Tyr (μg/g)	140	140	139	143	140
TMA-N (mg %)	0.30	0.30	0.28	0.28	0.30

Bluefish, unspecified, United States

TABLE 2. Changes in average TBA numbers and TPC during the storage of bluefish fillets at 1°C (71)

	Storage time (days)						
	1	2	3	4	5	6	7
TBA no.[a]	2.50	1.70	1.50	1.80	5.00	5.20	8.50
TPC (log no.org./g)	3.95	4.32	6.11	5.99	--	6.53	7.96

[a] Measured as mg of malonaldehyde per kg of tissue

Bombay Duck, *Harpodon nehereus*, India

TABLE 3. Post-mortem changes of pH, Glyc, LA, Pi, ATP and TMA-N in Bombay duck during iced storage (73)

	mg %									
	Storage time (days)									
	0	2	3	5	6	7	8	10	11	12
State [a]	--	FR	FR	FR	S	S	S	S	SS	SS
pH	6.7	6.7	6.7	6.6	6.5	6.5	6.5	6.5	6.5	6.6
Glyc	20	7	7	6	5	5	4	4	3	3
LA	20	41	46	53	61	61	61	62	61	58
Pi	60	60	61	61	61	62	63	65	67	68
ATP [b]	38.2	33.4	30.1	27.2	24.8	20.8	17.2	11.2	7.7	4.9
TMA-N	0.14	0.19	0.19	0.19	0.19	0.22	0.24	0.24	0.32	0.34

a/ FR = full rigor; AR = almost full rigor; S = softening; SS = soft
b/ Measured as phosphate associated to ATP

Bream, _Rhabdosargus sarba_, England

TABLE 4. Changes in sensory evaluations, pH, TVB-N, TMA-N, PV, TBA number and Hx of gold-lined sea bream during storage at 0° and 10°C (28)

	Storage time (days)								
	3	7	10	14	17	21	24	28	31
0°C									
Taste panel score[a]	6.5	6.0	6.0	7.1	6.1	4.4	5.1	4.8	4.4
pH	6.2	6.1	6.4	6.5	6.5	6.7	6.7	6.9	6.7
TVB-N (mg %)	10.0	12.0	14.0	22.0	10.0	8.0	14.0	22.0	33.0
TMA-N (mg %)	7.0	0.0	0.0	8.0	0.0	0.0	0.5	5.0	17.0
PV (meq/kg)	2.4	2.4	1.9	3.1	2.0	3.7	4.2	3.6	3.0
Hx (μ moles/g)	0.3	0.3	0.6	1.7	1.4	1.8	2.6	3.6	3.3
TBA[b]	0.7	0.4	0.7	0.3	0.4	0.5	0.5	0.5	0.8
10°C									
Taste panel score	--	7.5	2.8	--	--	--	--	--	--
pH	--	6.5	7.5	7.5	--	--	--	--	--
TVB-N (mg %)	--	12.0	173.0	203.0	--	--	--	--	--
TMA-N (mg %)	--	4.0	34.0	39.0	--	--	--	--	--
PV (meq/kg)	--	3.2	2.4	3.9	--	--	--	--	--
Hx (μ moles/g)	--	1.0	10.5	8.2	--	--	--	--	--
TBA[b]	--	0.5	0.4	0.4	--	--	--	--	--

a/ 4 = Limit of acceptability
b/ In mg of malonaldehyde per kg sample

Bream, *Rhabdosargus sarba*, England

TABLE 5. Changes in bacterial count of gold-lined sea bream during storage at 0° and 10°C (28)

	log no. org./cm^2			
	Storage time (days)			
	8	16	24	32
Plate count agar				
Fish in ice				
4°C incubation				
Skin	3.3	4.8	8.6	8.5
Flesh	2.4	5.8	7.9	7.8
27°C incubation				
Skin	3.1	4.8	8.3	8.5
Flesh	2.2	5.1	7.8	7.7
Fish at 10°C				
10°C incubation				
Skin	6.6	7.9	--	--
Flesh	4.9	7.5	--	--
27°C incubation				
Skin	7.3	7.9	--	--
Flesh	6.9	7.5	--	--

TABLE 5. (continued)

	log no. org./cm^2			
	Storage time (days)			
	8	16	24	32
Seawater agar				
Fish in ice				
4°C incubation				
Skin	3.7	5.2	8.3	8.5
Flesh	2.8	4.2	7.0	7.5
27°C incubation				
Skin	3.5	4.5	6.5	8.2
Flesh	2.6	5.0	7.1	7.8
Fish at 10°C				
10°C incubation				
Skin	7.0	6.6	--	--
Flesh	6.2	6.5	--	--
27°C incubation				
Skin	6.6	6.6	--	--
Flesh	6.4	6.5	--	--

Bream, unspecified, Japan

TABLE 6. Changes in freshness-quality parameters of Japanese black bream during iced storage (39)

	Storage time (days)							
	0	4	6	8	10	12	14	16
TVB-N (mg %)	15.2	17.0	16.7	16.8	18.6	18.2	25.0	30.4
TMA-N (mg %)[a]	0.2	0.2	0.3	0.4	0.6	0.8	1.0	1.3
TMA-N (mg %)[b]	0.2	0.2	0.3	0.4	0.5	0.7	0.9	1.1
IMP (μmoles/g)	9.5	5.5	5.9	5.4	5.4	5.0	3.6	2.3
HxR + Hx (μmoles/g)	0.9	1.4	1.5	2.3	3.3	5.5	6.2	7.0
ATP + ADP + AMP (μmoles/g)	1.5	0.6	0.6	0.5	0.5	0.5	0.5	0.4
Bacterial count (log no. org./g)	1.1	1.1	1.1	1.1	1.2	1.6	2.0	4.0

a/ Conway's method
b/ Picrate method

Bream, unspecified, Japan

TABLE 7. Changes in freshness quality parameters of Japanese red bream during ice storage (39)

	Storage time (days)							
	0	2	6	8	10	12	14	16
TVB-N (mg %)	14.5	15.2	15.1	16.3	16.5	16.0	23.0	27.5
TMA-N (mg %)[a]	0.3	0.3	0.3	0.5	0.6	0.7	1.5	2.3
TMA-N (mg %)[b]	0.3	0.3	0.3	0.4	0.4	0.5	0.7	0.8
IMP (μ moles/g)	7.5	7.2	6.1	7.0	5.2	4.9	4.4	4.6
HxR + Hx (μ moles/g)	0.4	0.5	1.1	1.4	1.3	1.6	2.1	2.8
ATP + ADP + AMP (μ moles/g)	1.9	0.8	0.6	0.6	0.5	0.5	0.4	0.4
Bacterial count (log no. org./g)	1.3	1.3	1.2	1.2	1.2	3.2	3.8	4.1

a/ Conway's method
b/ Picrate method

Capelin, *Mallotus villosus*, Canada

TABLE 8. Changes in ungutted offshore nonspawning capelin during iced storage (15)

	Storage time (days)						
	0	2	4	7	10	13	16
Mean lipid content (g %)	12.00	14.13	9.65	9.55	10.37	9.36	10.41
Mean FFA (% total lipid)	1.78	1.43	1.63	1.27	1.87	2.43	2.06
Mean iodine no. (meq $Na_2S_2O_3$/kg fat)	130.0	124.9	129.1	124.3	127.2	128.0	121.7

Capelin, *Mallotus villosus*, Canada

TABLE 9. Changes in sensory evaluations, appearance, TBA number, Hx, DMA-N, TMA-N and moisture of nonspawning capelin stored at -23°C (17)

	Storage time (days)							
	0	90	180	270	360	450	540	630
Sensory Evaluation[a/]								
Odor	3.81	3.72	3.57	3.12	3.17	3.33	3.11	3.28
Flavor	3.78	3.67	3.64	3.26	3.28	3.31	3.06	3.26
Texture	3.81	3.68	3.58	3.25	3.17	3.29	3.08	3.32
Overall Acceptability	3.64	3.40	3.35	3.06	3.01	3.11	3.99	3.13
Frozen Immediately								
Appearance[a/]	3.93	4.20	3.95	3.80	3.40	3.55	3.10	3.60
TBA no.[b/]	1.55	1.90	2.15	1.40	1.50	1.70	2.40	1.70
Hx (mg/kg)	17.6	16.9	18.5	18.3	19.0	20.6	19.4	20.7
DMA-N (mg %)	0.24	0.29	0.33	0.25	0.30	0.45	0.89	0.88
TMA-N (mg %)	0.57	0.66	0.57	0.87	0.43	0.92	0.93	0.52
Moisture (%)	70.51	71.24	71.78	69.85	70.72	69.30	69.14	69.05

TABLE 9. (continued)

	Storage time (days)							
	0	90	180	270	360	450	540	630
Iced 5 days prior to freezing								
Appearance[a]	3.58	3.52	3.50	3.30	3.00	3.30	3.00	3.28
TBA no.[b]	2.05	1.70	2.80	1.60	1.40	2.40	3.00	2.00
Hx (mg/kg)	19.6	20.4	21.7	20.7	23.5	19.0	21.0	22.7
DMA-N (mg %)	0.17	0.17	0.47	0.33	0.33	0.63	0.83	0.48
TMA-N (mg %)	0.89	0.57	0.28	0.44	0.48	0.75	1.97	0.86
Moisture (%)	74.01	73.06	74.00	76.73	74.96	72.95	73.56	73.97
4 months at -23°C thawed and refrozen								
Appearance[a]	4.15	3.50	3.55	2.60	3.10	3.20	3.05	3.05
TBA no.[b]	1.45	1.70	1.45	1.80	1.75	2.60	1.85	1.80
Hx (mg/kg)	16.2	18.5	16.5	18.5	21.7	17.7	19.0	22.7
DMA-N (mg %)	0.20	0.23	0.19	0.34	0.50	0.51	0.73	0.61
TMA-N (mg %)	0.15	0.82	0.62	0.75	0.97	0.75	0.51	0.64
Moisture (%)	78.38	78.64	75.57	74.62	75.47	74.76	70.62	69.77

a/ 5 = highest quality; 1 = lowest quality
b/ In mg of malonaldehyde per kg of sample

Carp, Cyprinus carpio, Korea

TABLE 10. Changes in TVB-N and K value during the storage of carp muscle at 5°C (63)

	Storage time (days)				
	0	1	2	4	6
TVB-N (mg %)	5.33	12.1	--	8.77	11.8
K value (%)	16.7	24.2	--	65.8	83.0

Cod, *Gadus callarias*, England

TABLE 11. Changes in the concentration of nucleotides during the chilled storage of cod caught in February and July (56)

	μmoles/g								
	Storage time (days)								
	0	1	3	7	10	11	14	17	18
AMP									
Feb	--	0.18	0.12	0.14	0.1	--	0.13	0.14	--
July	0.58	--	0.15	0.14	--	0.1	--	--	0.4
ADP									
Feb	--	0.2	0.23	0.19	0.22	--	0.2	0.17	--
July	0.37	--	0.07	0.11	--	0.14	--	--	0.19
ATP									
Feb	--	0.18	0.16	0.17	0.14	--	0.15	0.17	--
July	0.24	--	0.11	0.08	--	0.09	--	--	0.03
NAD									
Feb	--	0.02	0.01	--	0.01	0.01	--	0.01	0.01
July	0.17	0.05	0	0	--	0	--	--	0
IMP									
Feb	--	5.7	3.1	--	0.9	0	--	0	0
July	3.9	--	--	1.3	--	0.3	--	--	0

TABLE 11. (continued)

	μmoles/g								
	Storage time (days)								
	0	1	3	7	10	11	14	17	18
HxR									
Feb	--	1.9	3.7	5.1	--	--	3.5	2.2	--
July	0	--	--	4.8	4.7	3.9	--	--	1.3
Hx									
Feb	--	0.01	0.03	0.06	0.1	--	2.	2.7	--
July	--	0.02	--	0.08	--	1.4	--	--	2.8

Cod, *Gadus morhua*, Canada

TABLE 12. Changes in EPN, DMA-N and TMA-N content during the iced and frozen storage (-5°C) of cod (24)

	Storage time (days)					
	0	3	6	9	13	15
Iced						
TMA-N (mg %)	0.40	0.85	12.0	25.5	--	37.5
DMA-N (mg %)	0.30	0.35	0.30	--	--	0.40
EPN (% muscle)	2.25	2.10	2.20	2.20	--	2.20
Frozen						
TMA-N (mg %)	0.30	0.30	0.60	--	--	0.50
DMA-N (mg %)	0.30	0.55	0.45	--	0.45	0.60
EPN (% muscle)	2.10	2.20	1.95	1.78	1.82	1.73

Cod, Gadus morhua, Canada

TABLE 13. Changes in EPN and formation of DMA-N of minced and filleted cod during storage at -5° and -10°C (36)

	Storage time (days)						
	10	20	30	40	50	60	70
Fillets -5°C							
DMA-N (mmoles %)	0.06	0.11	0.18	0.07	--	--	--
Minced -10°C							
DMA-N (mmoles %)	0.24	0.30	0.34	0.35	0.37	0.41	0.45
EPN (g %)	1.83	1.60	1.32	1.10	0.92	0.85	0.70

Cod, *Gadus morhua*, Canada

TABLE 14. Changes in the concentration of nucleotides during the iced storage of Atlantic cod at 0°C (37)

	% of total purines							
	Storage time (days)							
	0	1	3	8	9	10	11	14
ATP	70	18	9	5	--	--	--	--
ADP	20	13	9	--	9	9	9	9
IMP	10	65	34	18	17	15	12	4
HxR	0	13	50	68	70	70	56	13
Hx	0	--	7	10	10	10	10	10

Cod, *Gadus morhua*, Canada

TABLE 15. Changes in TMA-N and TPC of cod fillets during their storage at 0°C in two different types of containers (102)

		Storage time (days)			
		0	4	7	8
Wooden box					
	TMA-N (mg %)	1.27	4.24	28.8	42.4
	TPC (log no. org./g)	5.57	5.99	--	6.09
Plastic box	TMA-N (mg %)	1.27	4.03	30.0	42.4
	TPC (log no.org./g)	--	4.50	--	4.80

Cod, *Gadus morhua*, England

TABLE 16. Changes in pH and organoleptic parameters during the storage of cod fillets at -7°C (60)

	Storage time (days)					
	0	14	28	42	56	84
pH	6.63	6.67	6.67	6.70	6.51	6.60
Texture[a]	A	A	A	A	U	U
Flavor[a]	A	A	A	A	A	U

[a] A = acceptable; U = Unacceptable

Cod, unspecified, Belgium

TABLE 17. Changes in quality parameters during the storage of cod fillets at 0°C (32)

	Storage time (days)			
	0	2	4	6
TVN (mg %)	19.5	21.8	30.0	38.0
TMA-N (mg %)	0.10	0.10	7.60	15.0
FBN (mg %)	19.6	21.6	22.1	22.5

Cod, unspecified, Canada

TABLE 18. Changes in TBA number, TMA-N, EPN and FFA of cod frozen at -40°C for 1.5 hrs and stored at -12°C (21)

	Storage time (days)							
	0	25	50	75	100	125	150	175
TBA no.[a]	0.40	0.40	0.35	0.30	0.30	0.25	0.25	0.20
TMA-N (mg %)	0.4	0.5	2.0	2.6	3.1	3.3	3.3	3.3
EPN (%)	100	95	87	77	68	58	47	38
FFA (%)	5	21	28	35	38	42	45	47

[a] Measured as optical density at 532 nm

Cod, unspecified, Canada

TABLE 19. Changes in TMA-N value of cod fillets frozen at -40°C for 1.5 hrs and stored at different temperatures (21)

	mg %			
	Storage time (days)			
	100	200	300	400
- 3.3°C	3.2	--	--	--
-12°C	1.6	1.9	2.1	--
-26°C	0.6	0.6	0.6	0.6

Cod, unspecified, Canada

TABLE 20. Changes in FFA, EPN and TMA-N of cod fillets stored at -26°C (21)

	Storage time (days)					
	100	200	300	400	500	600
FFA (%)	10	12	13	16	17	18
EPN (%)	2.15	2.15	2.11	2.10	2.10	1.92
TMA-N (mg %)	0.51	0.55	0.55	0.50	0.49	0.48

Cod, unspecified, Canada

TABLE 21. Changes in DMA-N, K_2CO_3-amine and KOH-amine values of cod during storage at -5°C (22)

	Storage time (days)						
	0	50	102	160	196	206	378
DMA-N (mg %)	0.1	4.7	12.3	9.3	15.2	32.4	53.1
KOH-amine value[a]	0.78	0.92	1.19	1.28	1.36	2.64	2.07
K_2CO_3-amine value[a]	0.62	1.80	2.78	4.18	5.35	7.50	9.76

[a] Expressed as milliequivalents of TMA-N per 100g sample

Cod, unspecified, Canada

TABLE 22. Changes in DMA-N, KOH-amine value and K_2CO_3-amine value of cod frozen at -40°C for 90 min., then stored at -5°C (23)

	Storage time (days)					
	0	10	20	30	40	50
DMA-N (mg %)	0.2	0.3	1.2	2.2	0.6	2.9
KOH-amine value (mg %)	0.1	0.7	0.8	1.0	0.8	0.7
K_2CO_3-amine value (mg %)	0.3	1.0	2.4	3.9	2.1	2.2

Cod, unspecified, Denmark

TABLE 23. Changes in APC and TVN in iced cod (53)

	Storage time (days)					
	1	3	6	9	12	15
APC (log no. org./g)	4.00	4.35	5.05	5.90	6.95	7.75
TVN (mg %)	10.6	14.0	15.3	19.9	25.0	32.5

Cod, unspecified, Denmark

TABLE 24. Changes in TVB-N and TMA-N in cod stored at 2°C in air, vacuum and diverse levels of CO_2 (54)

	Storage time (days)					
	0	1	4	6	8	12
TVB-N (mg %)						
Air	10	15	21	32	53	82
Vacuum	10	15	23	32	46	77
40% CO_2	10	15	21	30	58	77
60% CO_2	10	15	20	24	32	65
100% CO_2	10	15	19	19	26	70
TMA-N (mg %)						
Air	1	7	10	20	37	60
Vacuum	1	7	10	20	43	50
40% CO_2	1	2	8	22	35	62
60% CO_2	1	5	6	11	21	49
100% CO_2	1	5	8	8	15	55

Cod, unspecified, Denmark

TABLE 25. Total viable microbial counts at 20°C (psychrophilic flora) and growth of H_2S-producing organisms, in cod stored at 2°C in air, vacuum and diverse levels of CO_2 (54)

	Storage time (days)				
	0	4	6	8	12
Psychrophilic flora (log no. org./g)					
Air	5.0	6.6	6.4	7.0	7.6
Vacuum	5.0	6.0	5.3	5.5	5.9
40% CO_2	5.0	6.0	5.3	5.7	5.5
60% CO_2	5.0	5.5	5.0	5.0	5.2
100% CO_2	5.0	5.0	4.0	4.4	5.7
H_2S-producing flora (log no. org./g)					
Air	2.3	5.5	5.3	6.5	7.0
Vacuum	2.3	5.0	4.8	5.2	6.0
40% CO_2	3.0	3.0	3.0	4.8	5.3
60% CO_2	3.0	3.0	3.3	4.2	4.8
100% CO_2	3.0	3.0	2.4	3.5	4.4

Cod, unspecified, Japan

TABLE 26. Changes in freshness-quality parameters of Pacific cod during ice storage (39)

	Storage time (days)						
	0	2	4	6	8	10	12
TVB-N (mg %)[a]	5.0	7.5	6.5	7.0	6.0	13.0	16.0
TMA-N (mg %)[a]	0.5	0.4	0.4	0.5	2.6	3.0	8.1
TMA-N (mg %)[b]	0.5	0.4	0.4	0.5	0.6	1.2	2.8
IMP (μmoles/g)	5.3	0.6	0.4	---	---	---	---
HxR + Hx (μmoles/g)	0.6	3.2	4.7	---	---	---	---
ATP + ADP + AMP (μmoles/g)	1.1	0.3	0.4	---	---	---	---
Bacterial count (log no. org./g)	1.3	1.2	1.3	1.8	2.2	3.0	3.1

a/ Conway's method
b/ Picrate method

Cod, unspecified, New Zealand

TABLE 27. Changes in total APC and TMA-N during iced storage of red cod fillets (96)

	Storage time (days)								
	1	4	6	8	11	14	16	18	20
APC									
(log no. org./g)									
5°C Incubation	--	4.68	5.30	6.08	7.23	7.74	8.04	8.46	8.48
25°C Incubation	--	5.77	6.25	6.50	7.30	7.90	8.32	8.60	8.48
TMA-N (mg %)	0.63	--	0.78	--	--	1.88	2.50	5.63	10.20

Cod, unspecified, Poland

TABLE 28. Influence of the initial concentration of TMAO-N on the accumulation of DMA-N and protein solubility in minced cod during frozen storage (83)

	Storage time (days)			
	10	20	30	40
30 mg TMAO-N/g				
DMA-N (mg %)	2	3	5	6
Soluble proteins[a]	98	94	84	80
230 mg TMAO-N/g				
DMA-N (mg %)	5	11	15	17
Soluble proteins[a]	84	58	36	22

[a] Percent of solubility before storage

Cod, unspecified, Scotland

TABLE 29. Development of FFA in cod flesh frozen at various rates, thawed and stored in ice (68)

	Storage time (days)			
	10	20	30	40
FFA (% total lipids)				
Very slow freezing rate	5	25	44	51
Intermediate freezing rate	5	23	42	50
Rapid freezing rate	8	25	43	51
Intermediate freezing rate				
7 day storage at -7°C	20	29	38	42
7 day storage at -7°C thawed and stored in ice	21	35	42	48

Cod, unspecified, Sweden

TABLE 30. Changes in sulfhydryl groups and soluble nitrogen of cod muscle stored frozen at -4°, -6° and -20°C (52)

	Storage time (days)							
	0	6	21	50	72	92	112	148
-4°C								
Sulfhydryl Groups								
Ferricyanide method[a/]	73.3	70.7	58.5	11.5	54.0	42.0	24.7	--
o-Iodosobenzoic acid method[a/]	59.1	67.0	54.2	8.9	43.9	38.1	26.8	50.0
Soluble N (% of total N)	96.6	103.8	39.6	40.1	36.9	36.7	31.2	35.9
-6°C								
Sulfhydryl Groups								
Ferricyanide method[a/]	69.1	66.2	71.5	5.9	45.2	69.9	57.8	--
o-Iodosobenzoic acid method[a/]	61.6	59.3	58.7	4.9	35.4	56.9	32.2	50.5
Soluble N (% of total N)	97.8	88.3	42.3	30.7	39.3	36.7	32.2	30.3

TABLE 30. (continued)

	Storage time (days)							
	0	6	21	50	72	92	112	148
-20°C (slow frozen)								
Sulfhydryl Groups								
Ferricyanide method[a]	68.4	69.3	70.1	52.8	67.0	76.9	48.9	--
o-Iodosobenzoic acid method[a]	50.8	35.9	29.8	17.4	33.4	49.8	42.5	37.8
Soluble N (% of total N)	96.2	106.1	70.3	77.5	79.0	66.6	57.0	50.0
-20°C (quick frozen)								
Sulfhydryl Groups								
Ferricyanide method[a]	62.0	39.5	63.1	41.1	45.7	57.2	44.5	--
o-Iodosobenzoic acid method[a]	50.9	46.2	50.0	41.7	50.0	53.5	38.3	30.7
Soluble N (% of total N)	98.5	81.8	72.1	84.4	87.9	78.3	52.0	42.3

[a] mg cysteine per 100 g sample

Cod, unspecified, United States

TABLE 31. Changes in concentration of nucleotides during the iced storage of Pacific cod muscle (92)

	Storage time (days)			
	0	3	7	14
IMP (μmoles/g)	6.20	1.00	0.10	0.0
Hx (μmoles/g)	0.0	0.25	0.85	3.00
HxR (μmoles/g)	0.0	3.00	4.30	2.30

Crab, *Callinectes sapidus*, United States

TABLE 32. Changes in quality parameters during the storage of pasteurized crab cake mix at 30°C (67)

	Storage time (days)					
	0	0.25	0.50	1.0	1.25	1.50
TMA-N (mg %)	1.20	1.10	1.20	1.35	1.50	1.95
VRS (meq/g)	26.0	25.2	28.5	35.0	44.5	49.5
TPC (log no. org./g)	5.40	5.45	5.90	7.30	8.20	8.35

Crab, *Cancer magister*, United States

TABLE 33. Changes in phenolic content and pH of live and retorted Dungeness crab stored at 1-3°C (5)

	Storage time (days)[a]				
	0	1	2	3	4
Live					
Phenolic content (μg/g)	140.8	169.5	188.3	194.9	322.5
pH	6.7	7.0	7.1	7.2	7.7
Retorted					
Phenolic content (μg/g)	157.5	163.0	161.0	--	--
pH	7.3	7.5	7.8	--	--

a/ Days denote storage period prior to processing

Croaker, *Micropogon undulatus*, United States

TABLE 34. Sensory scores and TMA-N content of tray-packed fresh croaker fillets during storage at 2°C (107)

	Storage time (days)		
	1	4	5
Color[a]	6.7	6.8	6.5
Texture[a]	6.5	6.7	6.2
Flavor[a]	5.4	6.1	4.5
Odor[a]	6.0	5.6	5.4
General acceptance[a]	5.3	5.6	4.1
TMA-N (mg %)	0.55	24.98	39.68

[a] 9 = like extremely; 1 = dislike extremely.

Croaker, *Micropogon undulatus*, United States

TABLE 35. Percent change of net weight, drip, gross weight and total moisture in tray-packed fresh croaker fillets during storage at 2°C (107)

	%			
	Storage time (days)			
	2	4	6	8
Decrease in net weight	2.2	2.8	3.5	4.0
Increase in drip	1.8	2.4	3.0	3.8
Decrease in gross weight	0.2	0.3	0.4	0.5
Increase in total moisture	0.5	1.3	2.2	3.1

Croaker, Pseudotolithus typus, Nigeria

TABLE 36. Changes in lipids, TMA-N and TVB-N content in West African long croaker during iced storage (12)

	Storage time (days)							
	1	4	7	11	14	18	21	25
Lipid (%)								
Gutted	0.8	0.5	0.4	0.6	0.3	0.4	0.6	0.2
Ungutted	1.6	1.5	0.5	1.4	1.0	0.4	0.5	0.3
FFA (%)								
Gutted	1.4	1.5	2.2	2.2	0.8	0.4	1.1	1.8
Ungutted	1.6	2.4	2.9	2.1	1.0	0.4	0.8	2.4
Protein (%)								
Gutted	19.7	19.6	10.7	16.0	16.6	14.1	15.4	14.9
Ungutted	15.5	19.4	19.5	17.4	17.9	17.0	17.5	17.5
TMA-N (mg %)								
Gutted	0.2	0.6	0.9	0.9	1.4	1.2	0.8	1.0
Ungutted	0.16	0.7	0.87	1.0	1.0	1.5	1.0	1.4
TVB-N (mg %)								
Gutted	20.1	13.4	16.8	20.2	23.5	26.9	26.9	33.6
Ungutted	26.8	20.1	18.8	26.2	23.5	20.3	33.6	60.5

Cusk, *Brosme brosme*, Canada

Table 37. Changes in DMA-N, KOH-amine value and K_2CO_3-amine value of cusk frozen at -40°C for 90 min., then stored at -5°C (23)

	Storage time (days)					
	0	10	20	30	40	50
DMA-N (mg %)	0.4	5.1	11.1	10.2	11.1	11.3
KOH-amine value (mg %)	0.2	1.0	1.2	1.0	1.2	1.4
K_2CO_3-amine value (mg %)	1.0	6.8	11.0	11.1	13.5	15.3

Cusk, _Brosme brosme_, Canada

TABLE 38. Changes in EPN and formation of DMA-N of minced and filleted cusk during storage at -5° and -10°C (36)

	Storage time (days)				
	10	20	30	40	50
Fillets -5°C					
DMA-N (mmoles/g)	0.41	0.77	0.73	0.82	0.83
Fillets -10°C					
DMA-N (mmoles %)	0.35	0.57	0.42	0.73	0.51
EPN (g %)	1.78	1.51	1.39	0.94	1.13
Minced -10°C					
DMA-N (mmoles %)	0.30	0.66	0.80	0.85	1.16
EPN (g %)	1.40	0.84	0.58	0.61	0.52

Dogfish, *Squalus acanthias*, Canada

TABLE 39. Changes in TMA-N, NH_3 content, texture, thaw drip, PV, TBA number, FFA and pH during the iced storage of gutted and ungutted spiny dogfish (11)

	Storage time (days)								
	1	4	6	8	11	13	15	18	20
Gutted									
TMA-N (mg %)	0.3	0.5	0.4	0.4	0.4	0.4	0.4	0.4	0.5
NH_3 (mg N %)	--	8.44	9.74	8.63	8.22	8.11	8.44	10.71	12.57
Texture[a/]	13.0	10.9	9.5	9.5	7.1	7.3	8.0	8.2	5.6
Thaw drip (%)	4.8	7.2	6.1	6.9	5.9	8.5	7.0	9.4	6.1
PV (meq/kg fat)	0.3	1.0	2.0	1.9	1.6	2.2	2.4	4.7	6.2
TBA no.[b/]	0.35	0.44	0.83	1.01	1.08	1.06	1.15	0.98	1.12
FFA (g %)	0.35	0.35	0.32	0.42	0.38	0.48	0.42	0.51	0.75
pH	6.1	6.0	6.2	6.2	6.1	6.1	6.1	6.2	7.3

TABLE 39. (continued)

	Storage time (days)								
	1	4	6	8	11	13	15	18	20
Not gutted									
TMA-N	0.3	0.5	0.4	0.4	0.4	0.5	0.4	1.0	2.0
NH_3 (mg N %)	9.48	8.07	10.17	8.97	8.92	12.05	25.10	47.97	66.99
Texture[a/]	13.0	8.8	8.0	8.0	6.5	7.6	8.2	6.1	5.0
Thaw drip (%)	4.8	8.5	8.7	8.7	8.0	7.0	6.9	6.6	5.1
PV (meq/kg fat)	0.3	1.0	1.5	1.5	1.8	2.3	1.7	2.8	3.0
TBA no.[b/]	0.35	0.60	0.92	1.05	0.81	1.00	1.20	1.72	1.18
FFA (g %)	0.35	0.31	0.29	0.32	0.35	0.45	0.39	0.31	0.37
pH	6.1	6.2	6.2	6.1	6.2	6.1	6.4	6.6	7.2

a/ Expressed as kg of shear press force
b/ In mg of malonaldehyde per kg of sample

Dogfish, _Squalus acanthias_, Canada

TABLE 40. Changes in TMA-N, NH_3 content, texture, thaw drip, PV number, TBA number, FFA and pH in ungutted spiny dogfish stored at 5°C (11)

	Storage time (days)				
	1	4	6	8	11
TMA-N (mg %)	0.3	0.8	3.5	7.0	14.0
NH_3 (mg N %)	--	8.7	25.97	130.22	181.32
Texture[a]	13.0	8.1	6.2	6.7	6.2
Thaw drip (%)	4.8	7.4	6.1	3.8	2.0
PV no. (meq/kg fat)	0.3	2.3	2.4	2.8	3.1
TBA no.[b]	0.35	0.50	0.88	0.52	0.48
FFA (g %)	0.35	0.37	0.34	0.33	0.45
pH	6.1	5.9	6.1	6.8	7.3

a/ Expressed as kg of shear press force
b/ In mg of malonaldehyde per kg of sample

TABLE 41. Changes in quality parameters during the storage of ungutted spiny dogfish at 10°C (11)

	Storage time (days)				
	1	4	6	8	11
TMA-N (mg %)	0.3	1.4	9.0	27.0	42.0
NH_3 (mg N %)	--	18.57	91.07	178.22	285.63
Texture[a]	13.0	7.0	5.0	6.9	7.0
Thaw drip (%)	4.8	5.2	4.9	2.0	1.9
PV no. (meq/kg fat)	0.3	2.0	2.3	1.9	2.7
TBA no.[b]	0.35	0.70	0.55	0.30	0.37
FFA (g %)	0.35	0.30	0.35	0.55	0.57
pH	6.1	6.2	6.6	8.1	8.3

a/ Expressed as kg of shear press force
b/ In mg of malonaldehyde per kg of sample

Flounder, *Pogonias cromis*, United States

TABLE 42. Sensory scores and TMA-N content of tray-packed fresh flounder fillets during storage at 2°C (107)

	Storage time (days)		
	1	4	5
Color[a]	7.0	7.6	7.0
Texture[a]	6.0	6.5	--
Flavor[a]	6.3	6.4	--
Odor[a]	7.1	6.9	1.0
General acceptance[a]	6.3	6.1	--
TMA-N (mg %)	1.53	25.13	29.20

[a] 9 = like extremely; 1 = dislike extremely

Flounder, *Pogonias cromis*, United States

TABLE 43. Percent change of net weight, drip, gross weight and total moisture in tray-packed fresh flounder fillets during storage at 2°C (107)

	%			
	Storage time (days)			
	2	4	6	8
Decrease in net weight	4.1	4.6	5.1	5.7
Increase in drip	4.0	4.3	4.7	5.1
Decrease in gross weight	0.1	0.2	0.3	0.5
Increase in total moisture	0.1	0.5	0.8	1.1

Flounder, unspecified, Canada

TABLE 44. Changes in KOH-amine value of flounder frozen at -40°C for 90 min., then stored at -5°C (23)

	Storage time (days)					
	0	10	20	30	40	50
KOH-amine value (mg %)	2.6	5.1	3.8	4.0	3.1	3.6

Flounder, unspecified, United States

TABLE 45. Direct comparison of psychrotroph growth, TMA-N production and taste profile of flounder stored at -2° and 1°C (103a)

	Storage time (days)						
	1	3	7	10	13	15	17
-2°C							
Psychrotrophs (log no. org./g)	2.83	2.97	3.50	4.22	6.16	7.60	7.78
TMA-N (mg %)	1.3	1.1	0.9	1.2	1.3	1.0	1.6
Taste Score[a]	6.0	5.0	6.2	6.1	5.7	6.1	5.6
1°C							
Psychrotrophs (log no. org./g)	3.24	2.71	5.63	7.27	8.70	--	--
TMA-N (mg %)	1.2	1.1	2.1	1.80	30.7	--	--
Taste Score[a]	6.3	6.3	5.0	2.9	2.1	--	--

a/ 7 = like extremely; 6 = like moderately; 5 = like slightly; 4 = neither like nor dislike; 3 = dislike slightly; 2 = dislike moderately; 1 = dislike extremely

Grenadier, *Coryphaenoldes rupestris*, Canada

TABLE 46. Changes in DMA-N and EPN concentration of minced roundnose grenadier stored at -10°C (36)

	Storage time (days)						
	10	20	30	40	50	60	70
DMA-N (mmoles %)	0.33	0.40	0.38	0.38	0.52	0.57	0.60
EPN (gm %)	0.48	0.36	0.30	0.23	0.20	0.19	0.19

Grenadier, *Macrourus rupestris*, Canada

TABLE 47. Changes in DMA-N, TMA-N, moisture, EPN and Hx content of roundnose grenadier during storage at -23°C (16)

	Storage time (days)						
	0	90	180	450	540	630	720
DMA-N (mg %)	0.78	1.27	1.39	1.47	1.35	1.24	2.69
TMA-N (mg %)	1.82	2.52	2.52	2.72	2.53	2.62	2.73
Moisture (%)	84.17	84.06	83.44	84.04	84.20	83.89	83.66
EPN (mg %)							
No treatment	2.66	2.35	2.11	1.86	1.87	1.91	1.53
Ungutted iced 8 days	3.11	2.42	2.25	2.48	1.82	1.87	1.31
Gutted iced 8 days	2.92	2.65	2.59	2.31	2.40	2.12	2.02
Ungutted iced 12 days	2.72	2.52	2.33	1.92	1.80	1.85	1.44
Gutted iced 12 days	2.95	2.79	2.80	2.31	2.41	2.22	1.81
Hx (mg %)							
No treatment	0.5	0.4	0.6	7.5	7.0	6.3	8.5
Ungutted iced 8 days	3.0	4.5	10.0	12.0	14.0	13.5	13.0
Gutted iced 8 days	2.0	3.0	4.5	10.0	9.5	9.0	11.0
Ungutted iced 12 days	9.5	9.5	9.5	20.5	21.0	21.5	24.5
Gutted iced 12 days	8.0	7.5	8.0	17.5	16.0	16.5	18.5

Grenadier, *Macrourus rupestris*, Canada

TABLE 48. Sensory evaluations of roundnose grenadier fillets during storage at -23°C (16)

	Storage time (days)							
	0	90	180	360	450	540	630	720
Appearance[a]	3.53	3.20	3.34	2.99	3.04	3.23	2.89	2.69
Odor[a]	3.47	3.20	3.13	2.78	2.78	2.99	2.82	2.68
Flavor[a]	3.56	3.18	3.24	2.84	3.08	3.04	2.87	2.72
Texture [a]	3.74	3.28	3.47	3.18	3.10	3.23	2.90	2.73
Overall acceptability[a]	3.60	3.24	3.38	2.93	3.10	3.09	2.88	2.83

a/ 5 = very good; 4 = good; 3 = fair; 2 = slightly spoiled; 1 = spoiled

Grenadier, Macruronus novaezelandiae, Australia

TABLE 49. Changes in TBA number, FFA, DMA-N, FA and SEP in both washed and unwashed minced blue grenadier blocks stored at -18°C (18)

	Storage time (days)					
	1	30	71	117	152	182
TBA no. (mg malonaldehyde/kg)						
Unwashed	0.57	0.47	0.67	0.57	0.83	0.70
Washed	0.70	0.40	0.57	0.50	0.50	0.70
FFA (mg %)						
Unwashed	82	75	--	172	240	275
Washed	78	75	--	164	243	206
DMA-N (ppm)						
Unwashed	100	440	635	--	800	900
Washed	50	175	220	300	350	415
FA (ppm)						
Unwashed	30	50	80	120	125	140
Washed	10	30	40	60	--	55
SEP (%)						
Unwashed	700	200	60	210	175	185
Washed	620	175	140	175	130	115

Grouper, *Epinephalus malabaricus*, India

TABLE 50. Post-mortem changes of pH, Glyc, LA, Pi, ATP and TMA-N in grouper during iced storage (73)

	mg %								
	Storage time (days)								
	0	2	6	10	12	14	16	18	20
State[a]	--	FR	FR	FR	FR	S	S	S	S
pH	6.9	6.8	6.7	6.5	6.4	6.4	6.4	6.4	6.5
Glyc	393	261	202	165	149	134	119	105	95
LA	168	253	315	408	438	466	489	504	510
Pi	125	130	134	137	137	137	140	141	143
ATP[b]	17.9	13.2	9.6	5.4	4.9	3.7	3.0	3.1	3.1
TMA-N	0.04	0.05	0.06	0.08	0.10	0.11	0.13	0.14	0.16

a/ FR = full rigor; AR = almost full rigor; S = softening; SS = soft
b/ Measured as phosphate associated to ATP

Gurnard, _Chelodonichthys kumu_, New Zealand

TABLE 51. Changes in TMA-N during ice storage of gurnard fillets (96)

	Storage time (days)						
	1	4	8	11	14	16	18
APC (log no.org./g)							
5°C incubation	3.63	4.40	6.30	6.92	7.30	8.48	8.96
25°C incubation	5.30	5.49	6.43	7.82	8.64	8.83	9.11
TMA-N (mg %)	0.31	--	0.31	0.47	--	--	--

Haddock, *Gadus aeglefinus*, Scotland

TABLE 52. Nucleotide degradation in haddock during iced storage (57)

	μmoles/g		
	Storage time (days)		
	0	10	20
ATP	0.2	0.1	0.0
ADP	0.2	0.2	0.0
AMP	0.1	0.0	0.0
NAD	0.1	0.0	0.0
IMP	6.5	3.3	0.0
HxR	0.1	3.1	0.7
Hx	0.1	0.8	2.4

Haddock, Melangogrammus aeglefinus, Canada

TABLE 53. Changes in the concentration of nucleotides during the iced storage of Atlantic haddock at 0°C (37)

	% of total purines									
	Storage time (days)									
	0	1	2	3	4	5	7	9	11	16
ATP	30	23	4	2	--	--	--	--	--	--
ADP	10	10	7	7	8	6	7	7	7	7
IMP	55	60	82	81	80	75	70	56	45	5
HxR	--	3	7	12	12	16	21	33	33	33
Hx	--	--	--	--	1	3	3	5	10	55

Haddock, unspecified, Canada

TABLE 54. Changes in DMA-N, KOH-amine value and K_2CO_3-amine value of haddock frozen at -40°C for 90 min., then stored at -5°C (23)

	Storage time (days)					
	0	10	20	30	40	50
DMA-N (mg %)	0.2	0.2	0.2	0.2	0.2	0.7
KOH-amine value (mg %)	0	0.8	0.7	0.8	1.0	0.8
K_2CO_3-amine value (mg %)	0.3	1.0	1.0	1.1	1.4	2.0

Hake, *Merluccius bilinearis*, Canada

TABLE 55. Changes in EPN, DMA-N and TMA-N content during the iced and frozen storage (-5°C) of hake (24)

	Storage time (days)					
	0	3	6	9	13	15
Iced						
TMA-N (mg %)	2.30	5.20	--	36.2	--	--
DMA-N (mg %)	7.50	19.5	12.5	47.2	43.2	--
EPN (% muscle)	1.60	0.87	0.65	0.65	--	0.75
Frozen						
TMA-N (mg %)	1.80	2.60	--	2.30	2.60	2.50
DMA-N (mg %)	7.50	20.0	20.0	18.5	43.0	40.0
EPN (% muscle)	1.60	0.80	0.90	0.65	0.56	0.60

Hake, *Merluccius bilinearis*, Canada

TABLE 56. Changes in DMA-N and EPN for minced and filleted silver hake during storage at -10°C (36)

	Storage time (days)				
	0	10	20	30	36
DMA-N (mmoles/g)					
Minced summer	0.00	0.20	0.66	1.31	1.40
Minced winter	0.10	0.20	0.32	0.42	0.63
Fillets winter	0.02	0.05	0.13	0.30	0.39
EPN (g %)					
Minced summer	1.98	1.02	0.50	0.35	0.30
Minced winter	1.75	1.11	0.78	0.62	0.55
Fillets winter	2.10	1.60	1.46	0.95	0.71

Hake, *Merluccius productus*, United States

TABLE 57. Changes in intact, vacuum-sealed hake stored at -20°C (4)

	Storage time (days)				
	0	30	60	85	120
TMAO-N (μg/g)	880	760	800	810	790
TMA-N (μg/g)	2.9	4.0	3.0	3.2	2.2
DMA-N (μg/g)	3.0	8.0	24.0	18.0	20.0
FA (μg/g)	7.8	14.0	15.5	10.0	12.0
TEP (%)	85.	72.	30.	27.	29.
ESP (%)	13.0	13.4	9.6	13.6	11.0

Hake, *Merluccius productus*, United States

TABLE 58. Changes in intact, air-sealed hake stored at -20°C (4)

	Storage time (days)				
	0	35	65	90	120
TMAO-N (μg/g)	880	800	790	762	790
TMA-N (μg/g)	3.0	4.3	3.2	1.8	2.1
DMA-N (μg/g)	2.5	7.5	12.0	17.5	19.7
FA (μg/g)	7.8	16.2	13.2	22.0	15.3
TEP (%)	85	40	43	27	22
ESP (%)	13.0	11.7	11.2	11.9	11.5

Hake, *Merluccius productus*, United States

TABLE 59. Changes in minced, vacuum-sealed hake stored at -20°C (4)

	Storage time (days)				
	0	20	60	90	120
TMAO-N (μg/g)	795	800	760	777	785
TMA-N (μg/g)	4.0	5.5	6.3	6.6	5.0
DMA-N (μg/g)	11.0	17.5	43.0	48.0	47.5
FA (μg/g)	12.0	15.6	22.0	24.0	24.4
TEP (%)	77.0	51.0	20.5	20.0	20.9
ESP (%)	12.5	10.3	10.8	12.2	11.8

Hake, *Merluccius productus*, United States

TABLE 60. Changes in minced, air-sealed hake stored at -20°C (4)

	Storage time (days)				
	0	35	65	90	120
TMAO-N (μg/g)	795	737	750	718	783
TMA-N (μg/g)	4.0	2.2	4.0	3.2	3.5
DMA-N (μg/g)	11.5	30.0	51.0	50.0	48.7
FA (μg/g)	12.5	22.0	21.3	34.0	25.7
TEP (%)	77.0	30.0	22.0	18.6	20.0
ESP (%)	12.7	11.0	10.8	10.3	10.6

Hake, Urophycis chuss, Canada

TABLE 61. Changes in EPN and formation of DMA-N of minced and filleted red hake during storage at -5° and -10°C (36)

	Storage time (days)					
	10	20	30	40	50	60
Fillets at -5°C						
DMA-N (mmoles/g)	0.57	0.78	1.17	1.01	0.90	--
Minced at -10°C						
DMA-N (mmoles %)	0.56	0.94	1.50	1.94	2.10	2.26
EPN (mg/g)	0.21	0.13	0.08	0.05	--	--

Hake, *Urophycis chuss*, Canada

TABLE 62. Changes in EPN and DMA-N formation of red hake during iced storage (36)

	Storage time (days)							
	2	4	6	8	10	12	14	16
EPN (g %)	1.95	1.91	1.87	1.82	1.80	1.74	1.70	1.66
DMA-N (mmoles %)	0.03	0.03	0.12	0.23	0.32	0.42	0.51	0.63

Hake, *Urophycis chuss*, United States

TABLE 63. Changes in EPN, TMAO-N, DMA-N, TMA-N and FFA of minced and fillet red hake during storage at -6°C (75)

	Storage time (days)							
	0	14	28	35	42	49	56	70
Minced/Unaged								
EPN (%)	--	25	18	15	16	15	15	11
TMAO-N (mmoles %)	5.27	1.06	0.87	0.30	0.38	0.20	0.03	0.04
DMA-N (mmoles %)	0.14	3.56	3.22	3.68	4.11	4.33	3.75	4.63
TMA-N (mmoles %)	0.01	0.20	0.18	0.20	0.26	0.20	0.31	0.26
FFA (mmoles %)	0.04	1.58	1.72	1.64	1.33	1.42	1.32	1.40
Minced/Iced 4 Days								
EPN (%)	--	27	20	13	16	15	17	14
TMAO-N (mmoles %)	2.51	1.04	0.47	0.19	0.09	0.08	0.0	0.0
DMA-N (mmoles %)	1.72	3.52	3.48	4.78	3.65	5.53	3.55	5.05
TMA-N (mmoles %)	0.07	0.24	0.21	0.27	0.24	0.21	0.29	0.27
FFA (mmoles %)	1.02	1.41	1.27	1.12	0.98	1.04	0.69	1.29
Filleted/Unaged								
EPN (%)	--	87	62	60	39	50	37	24
TMAO-N (mmoles %)	5.28	4.77	3.44	3.58	4.06	3.81	3.21	2.98
DMA-N (mmoles %)	0.08	1.24	1.34	1.34	2.04	2.01	1.78	3.15
TMA-N (mmoles %)	0.01	0.09	0.08	0.08	0.13	0.15	0.21	0.16
FFA (mmoles %)	0.03	0.58	0.54	0.56	0.61	0.76	0.76	0.88

TABLE 63. (continued)

	Storage time (days)							
	0	14	28	35	42	49	56	70
Filleted/Iced 4 Days								
EPN (%)	--	77	67	76	50	40	45	30
TMAO-N (mmoles %)	2.58	2.46	3.71	2.78	2.74	2.54	1.52	2.73
DMA-N (mmoles %)	0.26	0.84	1.31	1.36	1.30	1.93	1.80	2.22
TMA-N (mmoles %)	0.18	0.23	0.24	0.28	0.20	0.16	0.33	0.43
FFA (mmoles %)	0.05	0.16	0.41	0.37	0.34	0.39	0.27	0.84

Hake, <u>Urophycis chuss</u>, United States

Table 64. Changes in TMAO-N, TMA-N, DMA-N and FFA of minced red hake stored at -6°C (75)

	mmoles %							
	Storage time (days)							
	0	3	7	8	10	14	17	21
Minced/Unaged								
TMAO-N	10.5	9.01	10.4	--	8.38	8.59	7.66	7.30
DMA-N	0.05	1.06	3.03	--	4.00	4.63	5.26	6.10
TMA-N	0.09	0.08	0.21	--	0.19	0.15	0.23	0.24
FFA	0.04	0.29	0.81	--	1.09	1.37	0.92	1.11
Aged 6 Days Minced								
TMAO-N	8.11	0.0	--	0.0	0.0	1.50	0.0	0.0
DMA-N	1.14	3.26	--	3.83	1.83	1.94	1.34	2.23
TMA-N	5.20	11.5	--	9.94	12.4	7.54	9.83	12.2
FFA	0.04	0.11	--	0.26	0.02	0.12	0.16	0.35
Aged 6 Days Whole/minced								
TMAO-N	11.3	8.06	--	7.93	7.16	7.60	5.78	8.43
DMA-N	0.48	0.71	--	1.60	1.68	1.20	1.11	1.31
TMA-N	1.00	1.15	--	1.48	1.51	2.06	2.11	1.66
FFA	0.06	0.09	--	0.20	0.13	0.06	0.14	0.23

Hake, _Urophycis chuss_, United States

TABLE 65. Changes in DMA-N, TMA-N, EPN, bound FA, free FA and shear force of skinned red hake during storage at -18°C (59)

	Storage time (days)						
	0	28	56	84	112	140	168
-90°C Storage							
DMA-N (mmoles %)	0.1	0.1	0.1	0.1	0.1	0.1	0.1
TMA-N (mmoles %)	0.05	0.03	0.04	0.07	0.08	0.03	0.02
EPN (%)	100	100	100	100	100	100	100
Free FA (mmoles %)	0.05	0.04	0.04	0.02	0.11	0.04	0.05
Bound FA (mmoles %)	0.1	0.1	0.1	--	--	0.1	0.1
Shear force (kg)	0.45	0.69	0.56	0.55	0.52	0.60	0.49
1 Day on Ice							
DMA-N (mmoles %)	0.1	1.0	1.4	1.5	1.5	3.0	2.7
TMA-N (mmoles %)	0.08	0.07	0.13	0.27	0.2	0.20	0.14
EPN (%)	106	80	86	55	61	49	24
Free FA (mmoles %)	0.05	0.33	0.45	0.60	0.45	0.79	0.94
Bound FA (mmoles %)	0.01	0.6	0.8	0.8	0.9	1.9	1.7
Shear force (kg)	0.53	0.82	0.88	0.75	0.91	0.99	0.94

TABLE 65. (continued)

	Storage time (days)						
	0	28	56	84	112	140	168
5 Days on Ice							
DMA-N (mmoles %)	0.1	0.4	0.8	0.8	1.0	1.9	2.1
TMA-N (mmoles %)	0.07	0.04	0.08	0.14	0.15	0.20	0.10
EPN (%)	99	100	85	80	88	70	39
Free FA (mmoles %)	0.05	0.08	0.19	0.28	0.11	0.55	0.69
Bound FA (mmoles %)	0.1	0.2	0.6	0.5	0.8	1.2	1.5
Shear force (kg)	0.55	0.90	0.95	0.75	0.84	0.90	0.75
Saberized skinning[a/]							
DMA-N (mmoles %)	0.2	0.6	1.2	2.0	2.6	3.5	5.0
TMA-N (mmoles %)	0.04	0.06	0.12	0.23	0.27	0.11	0.10
EPN (%)	99	99	73	45	64	24	17
Free FA (mmoles %)	0.05	0.25	0.49	0.84	0.55	1.18	1.60
Bound FA (mmoles %)	0.1	0.3	0.7	1.2	2.1	2.2	3.2
Shear force (kg)	0.55	0.87	0.92	0.65	0.82	1.01	0.86

a/ Deep skinned to remove red muscle tissue

Hake, _Urophycis chuss_, United States

TABLE 66. Changes in TMAO-N, DMA-N, TMA-N and FFA of red hake fillets during storage at -20°C after aging 4 days in ice (75)

	mmoles %							
	Storage time (days)							
	0	28	56	84	140	168	196	224
Fillet (Skin-Off)								
TMAO-N	4.97	3.29	3.28	3.12	2.64	3.21	3.19	2.01
DMA-N	0.56	1.17	1.16	1.55	1.98	0.71	1.75	0.66
TMA-N	0.13	0.30	0.09	0.10	0.41	0.34	0.10	0.10
FFA	0.20	0.56	0.42	0.59	0.51	0.27	0.40	0.18
Fillet (Skin-On)								
TMAO-N	5.50	4.26	4.13	3.82	3.20	4.63	4.36	3.19
DMA-N	0.47	0.95	1.19	1.37	1.11	0.68	2.13	0.70
TMA-N	0.32	0.29	0.48	0.32	0.34	0.19	0.21	0.29
FFA	0.12	0.18	0.28	0.43	0.32	0.19	0.39	0.12

TABLE 66. (continued)

	mmoles %							
	Storage time (days)							
	0	28	56	84	140	168	196	224
Fillet (Gutted Fish)								
TMAO-N	3.03	3.74	4.13	3.90	3.17	3.33	3.66	3.98
DMA-N	0.48	0.28	0.47	0.86	0.73	0.39	0.75	0.25
TMA-N	0.09	0.21	0.06	0.12	0.34	0.27	0.18	0.08
FFA	0.10	0.09	0.12	0.16	0.17	0.09	0.16	0.07
Fillet (Whole Fish)								
TMAO-N	2.58	4.44	1.70	3.65	3.24	3.42	3.51	3.29
DMA-N	0.26	0.40	0.87	0.80	0.47	0.70	0.76	0.41
TMA-N	0.18	0.24	2.42	0.19	0.12	0.08	0.08	0.12
FFA	0.05	0.09	0.11	0.16	0.14	0.11	0.07	0.26

Hake, unspecified, Canada

TABLE 67. Changes in DMA-N, KOH-amine value and K_2CO_3-amine value of hake frozen at -40°C for 90 min., then stored at -5°C (23)

	Storage time (days)					
	0	10	20	30	40	50
DMA-N (mg %)	0.7	6.2	10.9	17.0	15.0	13.3
KOH-amine value (mg %)	0.4	0.9	1.3	1.7	1.9	2.1
K_2CO_3-amine value (mg %)	1.2	11.8	13.9	17.4	18.8	18.8

Hake, unspecified, South Africa

TABLE 68. Growth rates of bacteria and distribution of genera in chilled (5°C) wet and thawed frozen cape hake (86)

	Storage time (days)			
	0	1	3	6
Chilled wet				
TPC (log no. org./cm^2)				
Skin	4.555	5.040	6.286	7.817
Flesh	4.325	4.784	5.462	7.068
Achromobacter (%)	53.3	50.0	50.8	35.9
Pseudomonas (%)	35.0	42.0	49.2	64.1
Micrococcus (%)	7.8	4.0	--	--
Miscellaneous (%)	3.9	4.0	--	--
Frozen				
TPC (log no. org./cm^2)				
Skin	3.730	3.553	3.693	6.484
Flesh	3.554	3.260	3.155	5.733
Achromobacter (%)	57.8	62.0	66.7	73.0
Pseudomonas (%)	6.1	2.0	18.4	24.3
Micrococcus (%)	20.0	31.0	14.0	2.7
Miscellaneous (%)	16.1	5.0	0.9	2.7

Hake, unspecified, South Africa

TABLE 69. Growth rates of bacteria and distribution of genera in iced wet and thawed frozen cape hake at -29°C (86)

	Storage time (days)				
	0	4	7	11	14
Iced wet					
TPC (log no. org./cm^2)					
Skin	4.69	6.50	7.10	7.74	7.64
Flesh	4.34	5.31	6.71	6.97	7.30
Achromobacter (%)	70.9	79.5	67.1	65.0	85.0
Pseudomonas (%)	16.5	19.9	32.2	35.0	12.5
Micrococcus (%)	4.0	--	--	--	--
Miscellaneous (%)	8.6	0.6	0.7	--	--
Frozen					
TPC (log no. org./cm^2)					
Skin	4.36	4.23	6.27	7.30	7.87
Flesh	4.01	4.08	6.18	6.54	7.57
Achromobacter (%)	81.6	81.6	88.2	87.5	90.0
Pseudomonas (%)	1.4	8.1	11.2	12.5	10.0
Micrococcus (%)	10.2	4.4	0.6	--	--
Miscellaneous (%)	6.8	5.9	--	--	--

Hake, unspecified, South Africa

Table 70. Death rates of bacteria in cape hake during storage at different temperatures (86)

	log no. org./cm^2					
	Storage time (days)					
	0	1	28	56	84	140
0 days in ice						
Skin						
0 storage	5.095	--	--	--	--	--
- 7°C	--	5.088	3.726	3.799	--	--
-18°C	--	4.753	4.194	3.461	3.140	3.473
-29°C	--	5.027	4.298	3.283	3.260	3.727
Flesh						
0 storage	4.490	--	--	--	--	--
- 7°C	--	4.505	3.136	4.207	--	--
-18°C	--	4.140	3.647	4.538	2.438	3.810
-29°C	--	4.332	3.657	4.590	2.695	3.496
6 days in ice						
Skin						
0 storage	6.609	--	--	--	--	--
- 7°C	--	6.701	5.286	5.449	--	--
-18°C	--	6.473	5.975	6.390	5.824	5.337
-29°C	--	6.257	5.606	5.245	4.734	5.436

TABLE 70. (continued)

	log no. org./cm^2					
	Storage time (days)					
	0	1	28	56	84	140
6 days in ice						
Flesh						
0 storage	5.838	--	--	--	--	--
- 7°C	--	5.979	4.841	5.170	--	--
- 18°C	--	5.612	5.197	5.189	5.266	--
- 29°C	--	5.607	5.112	4.839	4.592	4.801
13 days in ice						
Skin						
0 storage	7.932	--	--	--	--	--
- 7°C	--	7.433	5.792	5.682	--	--
-18°C	--	7.214	6.217	7.056	6.736	5.719
-29°C	--	7.068	6.918	7.386	6.325	7.565
Flesh						
0 storage	6.999	--	--	--	--	--
- 7°C	--	6.505	5.301	5.313	--	--
-18°C	--	6.606	5.771	6.547	5.874	5.483
-29°C	--	6.252	6.066	6.672	5.625	6.791

Hake, unspecified, South Africa

TABLE 71. Death rates of bacteria and distribution of genera in cape hake during storage at -29°C (86)

	Storage time (days)				
	0	1	7	30	120
0 days in ice					
Bacterial count (log no. org./cm^2)					
Skin	4.68	4.60	4.31	3.84	3.50
Flesh	4.29	4.23	3.86	3.62	3.04
Achromobacter (%)	44.2	63.3	56.1	63.8	81.0
Pseudomonas (%)	27.8	15.5	6.0	3.8	5.0
Micrococcus (%)	17.1	14.1	19.2	17.5	12.7
Miscellaneous (%)	10.9	7.1	18.7	14.9	1.3
3-4 days in ice					
Bacteria Count (log no. org./cm^2)					
Skin	5.02	--	--	4.58	4.31
Flesh	4.56	--	--	3.74	4.53
6-7 days in ice					
Bacterial count (log no. org./cm^2)					
Skin	6.48	6.25	5.20	5.61	--
Flesh	5.69	5.63	4.79	5.18	--

TABLE 71. (continued)

	Storage time (days)				
	0	1	7	30	120
9-10 days in ice					
Bacterial count					
(log no. org./cm^2)					
Skin	7.13	6.31	6.66	6.46	6.21
Flesh	6.69	5.75	5.92	6.16	6.42
Achromobacter (%)	55.3	66.3	74.4	81.3	100.0
Pseudomonas (%)	43.8	30.0	20.7	16.2	--
Micrococcus (%)	2.5	2.5	3.8	1.7	--
Miscellaneous (%)	0.4	1.2	1.1	0.8	--
13-14 days in ice					
Bacterial count					
(log no. org./cm^2)					
Skin	7.92	7.65	7.08	7.23	--
Flesh	6.93	6.83	6.50	6.69	--

Hake, unspecified, United States

TABLE 72. Changes in texture scores, DMA-N, FA, TMAO-N, EPN, raw drip and pH of red hake during frozen storage at various temperatures (65)

	Storage time (days)								
	0	3.5	70	140	210	290	350	420	490
Texture[a/]									
+20°F	4.5	2.4	--	--	--	--	--	--	--
+10°F	4.7	3.5	2.4	1.2	--	--	--	--	--
+ 5°F	4.7	4.4	4.3	3.5	2.6	--	--	--	--
- 5°F	4.7	4.2	4.0	3.8	3.9	3.3	3.3	3.2	2.7
-20°F	4.7	4.4	4.4	4.4	4.2	3.9	3.9	3.7	3.4
-80°F	4.7	4.9	4.8	4.5	4.7	4.5	4.5	4.7	4.6
DMA-N (mg %)									
+20°F	2	60	90	--	--	--	--	--	--
+10°F	2	20	38	30	--	--	--	--	--
+ 5°F	2	10	10	11	21	--	--	--	--
- 5°F	2	8	7	9	14	11	18	24	17
-20°F	2	3	3	3	7	10	10	15	13
-80°F	2	2	2	2	2	2	2	2	3

TABLE 72. (continued)

	Storage time (days)								
	0	3.5	70	140	210	290	350	420	490
FA (%)									
+20°F	--	--	--	--	--	--	--	--	--
+10°F	1	10	10	22	--	--	--	--	--
+ 5°F	1	4	4	6	14	--	--	--	--
- 5°F	1	4	4	6	8	8	9	10	11.5
-20°F	1	1	2	2	3	4	4	7	8
-80°F	1	0.5	0.5	0.5	0.5	0.5	0.5	1	1
TMAO-N (mg %)									
+20°F	--	--	--	--	--	--	--	--	--
+10°F	85	57	55	18	--	--	--	--	--
+ 5°F	85	71	61	56	28	--	--	--	--
- 5°F	85	71	58	50	39	39	45	53	48
-29°F	85	60	72	52	50	53	48	53	52
-80°F	85	60	70	40	45	48	56	58	65
EPN									
+20°F	--	--	--	--	--	--	--	--	--
+10°F	100	31	29	25	--	--	--	--	--
+ 5°	100	70	72	49	25	--	--	--	--
- 5°F	100	77	67	38	37	37	29	22	16
-20°F	100	81	77	68	61	65	53	40	33
-80°F	100	95	94	75	90	83	76	73	86

TABLE 72. (continued)

	Storage time (days)								
	0	3.5	70	140	210	280	350	420	490
Raw Drip (%)									
+20°F	--	--	--	--	--	--	--	--	--
+10°F	43	49	46	52	--	--	--	--	--
+ 5°F	38	39	42	46	50	--	--	--	--
- 5°F	--	45	46	47	48	47	48	49	--
-20°F	--	39	44	45	45	47	50	48	--
-80°F	37	38	36	42	42	43	43	47	--
pH									
+20°F	--	--	--	--	--	--	--	--	--
+10°F	7.15	7.12	7.11	--	--	--	--	--	--
+ 5°F	7.15	7.00	6.97	7.07	7.04	--	--	--	--
- 5°F	7.15	7.04	6.65	6.82	7.17	7.03	7.16	7.25	7.24
-20°F	7.15	6.89	6.97	6.72	6.98	7.04	7.05	7.14	7.15
-80°F	7.15	6.89	6.87	6.83	6.92	7.08	6.99	6.94	7.07

a/ 1 = least quality; 5 = highest quality

Halibut, *Hippoglossus hippoglossus*, Canada

TABLE 73. Changes in the concentration of nucleotides during the iced storage of Atlantic halibut at 0°C (37)

	% of total purines								
	Storage time (days)								
	0	1	2	3	6	8	9	13	15
ATP	80	24	5	2	--	--	--	--	--
ADP	10	10	9	9	10	9	9	9	9
IMP	10	58	73	75	73	57	47	24	15
HxR	--	10	18	20	20	27	25	15	15
Hx	--	--	--	--	5	10	18	57	67

Halibut, Hippoglossus stenolepsis, Canada

TABLE 74. Mean flesh pH and TBA numbers in brine and plate frozen Pacific halibut stored at -30°C (14)

	Storage time (days)				
	98	217	315	434	567
pH					
Brine frozen	6.20	6.14	6.12	6.09	6.03
Plate frozen	6.21	6.17	6.16	6.15	6.03
TBA no.[a]					
Brine frozen (outside muscle)	0.09	0.16	0.23	0.26	0.10
Brine frozen (inside muscle)	0.06	0.06	0.06	0.04	0.07
Plate frozen (outside muscle)	0.10	0.09	0.04	0.18	0.06
Plate frozen (inside muscle)	0.06	0.08	0.05	0.03	0.03

a/ Measured as absorbance units at 256 nm

Halibut, unspecified, United States

TABLE 75. Changes in Hx, HxR, adenosine, NAD, ATP, ADP, AMP and IMP of halibut stored in melting ice (91)

	μ moles/g						
	Storage time (days)						
	0	1	3	7	14	17	21
Hx	0	0	0.20	0.58	1.5	1.9	3.1
HxR	0.10	0.25	0.85	1.0	1.6	1.8	0.98
Adenosine	0.11	0.04	trace	0	0	0	0
NAD	0.14	0.15	0.12	1.12	0.12	0.10	0.10
ATP	0.48	0.10	0.10	trace	trace	0	0
ADP	0.40	0.40	0.20	0.58	0.13	0.15	0.17
AMP	0.4	0.28	0.12	0.10	0.13	0.10	0.10
IMP	7.0	6.8	5.9	5.1	4.3	3.4	2.7

Halibut, unspecified, United States

TABLE 76. Changes in concentration of nucleotides during the iced storage of halibut muscle (92)

	Storage time (days)			
	0	3	7	14
IMP (μmoles/g)	7.00	6.25	5.60	3.90
Hx (μmoles/g)	0.0	0.45	1.1	2.00
HxR (μmoles/g)	0.0	0.70	1.7	2.00

Herring, Clupea harengus, Canada

TABLE 77. Changes in PV[a] during the frozen storage of Pacific herring at -28°C (10)

	Storage time (days)					
	28	112	140	196	280	380
Air						
Days in ice						
0	--	4	--	9	17	--
2	--	5	--	10	17	--
4	--	7	--	10	17	--
Days in RSW						
0	--	4	--	9	17	--
2	--	8	--	12	17	--
4	--	12	--	17	25	--
Cryovac						
Days in ice						
0	2	--	4	--	--	4
2	1	--	3	--	--	3
4	1	--	2	--	--	4
Days in RSW						
0	2	--	4	--	--	4
2	1	--	2	--	--	6
4	1	--	4	--	--	6

[a] In milliequivalents per kg of fat

Herring, Clupea harengus, Canada

TABLE 78. Changes in TBA number[a/] during the frozen storage of Pacific herring at -28°C (10)

	Storage time (days)					
	28	112	140	196	280	308
Air						
Days in ice						
0	--	1	--	2	12	--
2	--	2	--	2	7	--
4	--	1	--	2	9	--
Days in RSW						
0	--	1	--	2	12	--
2	--	3	--	4	7	--
4	--	2	--	2	8	--
Cryovac						
Days in ice						
0	0.5	--	1	--	--	4
2	0.5	--	0.5	--	--	1
4	0.5	--	0.5	--	--	2
Days in RSW						
0	0.5	--	1	--	--	4
2	0.5	--	0.5	--	--	6
4	1	--	1	--	--	3

a/ In milligrams of malonaldehyde per kg of muscle

Herring, *Clupea harengus*, England

TABLE 79. Amine production in herring stored in ice at 1°, 10° and 25°C (77)

	mg amine/100 g							
	Storage time (days)							
	0	1	2	3	4	7	14	17
1°C								
Putrescine	0.32	--	--	1.09	--	0.98	3.02	5.88
Histamine	0.08	--	--	0.12	--	5.82	25.37	48.83
Cadaverine	0.53	--	--	2.32	--	5.81	14.78	34.78
Spermidine	0.49	--	--	0.46	--	0.24	0.16	2.48
Spermine	1.18	--	--	1.26	--	--	0.57	1.41
10°C								
Putrescine	--	0.39	0.80	1.90	3.56	--	--	--
Histamine	--	0.61	2.19	15.32	47.18	--	--	--
Cadaverine	--	1.34	4.06	12.58	23.49	--	--	--
Spermidine	--	0.24	0.33	0.42	0.39	--	--	
Spermine	--	0.25	0.35	0.72	0.81	--	--	--
25°C								
Putrescine	--	0.76	5.28	17.29	--	--	--	--
Histamine	--	3.14	59.29	107.36	--	--	--	--
Cadaverine	--	4.9	22.72	49.33	--	--	--	--
Spermidine	--	0.16	0.47	0.75	--	--	--	--
Spermine	--	0.16	0.75	1.57	--	--	--	--

Herring, *Clupea harengus*, England

TABLE 80. Changes in quality parameters during the storage of herring muscle in ice and refrigerated sea water (89)

	Storage time (days)							
	0	1	3	5	7	8	10	13
Ice								
pH	6.57	6.48	6.36	6.38	6.50	6.65	6.60	6.80
TMAO-N (mg %)	54.3	38.7	36.9	40.2	34.7	49.4	12.0	3.50
TMA-N (mg %)	0.62	0.80	1.16	0.60	2.90	4.20	25.8	29.1
Hx (μmoles/g)	0.35	1.13	2.00	2.90	3.90	8.80	5.00	5.80
RSW								
pH	6.57	6.51	6.38	6.48	6.60	6.40	6.70	6.60
TMAO-N (mg %)	54.3	33.4	27.7	23.4	30.4	15.4	18.2	14.6
TMA-N (mg %)	0.62	0.83	1.93	4.60	7.60	12.5	27.2	35.1
Hx (μmoles/g)	0.35	1.47	3.03	5.60	6.00	1.10	7.20	8.20

Herring, Clupea harengus, Ireland

TABLE 81. Changes in odor score, TMA-N, Hx content, Torrymeter reading and number of organisms during storage of herring at 0°, 5° and 10°C (29)

	Storage time (days)								
	0	1	3	5	7	9	10	15	20
Odor scores[a]									
0°C	5.0	5.0	5.0	4.0	3.0	3.0	3.0	3.0	2.2
5°C	5.0	5.0	4.4	3.0	2.2	2.0	2.0	2.0	2.0
10°C	5.0	4.4	3.4	3.0	2.0	2.0	2.0	2.0	1.2
TMA-N (μmoles/g)									
0°C	0	--	--	20	--	--	25	50	90
5°C	0	--	--	25	--	--	60	145	270
10°C	0	--	--	30	--	--	100	280	1440
Hx (μmoles/g)									
0°C	0.25	--	--	0.35	--	--	0.45	0.45	0.55
5°C	0.25	--	--	--	--	--	--	--	--
10°C	0.25	--	--	--	--	--	--	--	--
log Torrymeter									
0°C	1.1	1.07	1.2	0.94	0.92	0.78	0.49	0.23	0.1
5°C	1.1	1.13	1.0	0.72	0.65	0.54	0.13	0.05	0.12
10°C	1.1	1.03	0.58	0.26	0.44	0.18	0.17	0.16	--

Table 81. (continued)

	Storage time (days)								
	0	1	3	5	7	9	10	15	20
APC (log no. org./g)									
0°C	4.4	--	--	5.3	5.9	--	--	6.6	6.9
5°C	4.4	--	--	5.8	6.4	--	--	6.9	8.2
10°C	4.4	--	--	6.4	6.7	--	--	7.7	5.3

a/ 1 = lowest; 5 = highest

Lingcod, *Ophiodon elongatus*, United States

TABLE 82. Changes in acetoin, TVB-N and sensory evaluations of lingcod stored at 34°F (45)

	Storage time (days)				
	0	2	4	6	8
Acetoin (mg %)	1	1	1	6	6
TVB-N (mg %)	8.4	9.1	10.2	23.1	27.6
Sensory Evaluation[a]	A	A	A	A	NA

[a] A = acceptable; NA = not acceptable

Lobster, *Nephrops norvegicus*, Scotland

TABLE 83. Changes in TMA-N, TVB-N, Hx and pH of Norway lobster during iced storage (95)

	Storage time (days)									
	0	1	3	5	6	7	8	10	11	12
TMA-N (mg %)	--	3.2	3.4	4.4	5.0	10.5	9.9	18.3	22.3	25.9
TVB-N (mg %)	27.9	35.6	32.6	39.4	43.5	47.3	58.4	65.9	76.6	--
Hx (μmol/g)	0.15	0.31	0.50	0.36	0.38	1.16	1.50	3.36	2.28	2.23
pH	6.32	6.51	6.90	6.85	7.06	7.15	7.13	7.30	7.65	7.90

Mackerel, *Rastrelliger kanagurta*, India

Table 84. Changes in quality parameters during the iced storage of Indian mackerel muscle (103)

	Storage time (days)				
	0	3	5	7	9
TPC (log no. org./g)	4.11	5.20	6.14	6.94	7.08
TVB-N (mg %)	6.11	5.84	6.81	9.32	13.4
TBA no.[a]	1.06	1.90	2.40	1.21	1.45

[a] Measured as mg of malonaldehyde per kg of tissue

Mackerel, Rastrelliger neglectus, Scotland

TABLE 85. Changes in quality parameters during the storage of unglazed and glazed chub mackerel muscle at -14°C (58)

	Storage time (days)			
	0	30	70	90
Unglazed				
TBA no.[a]	0.05	0.37	1.17	1.75
PV[b]	0.0	60.0	150.0	205.0
Iodine (g %)	153	138	121	113
Soluble protein (%)	72.0	63.0	54.0	42.0
Glazed				
TBA no.[a]	0.05	0.10	0.15	0.21
PV[b]	0.0	8.0	19.0	20.0
Iodine (g %)	153	142	140	129
Soluble protein (%)	72.0	70.0	58.5	42.5

a/ moles malonaldehyde per g sample
b/ moles hydroperoxide per g sample

Mackerel, *Scomber japonicus*, Japan

TABLE 86. Changes in the concentration of selected chemical and biochemical parameters during the storage of canned mackerel (97)

	Storage time (days)					
	0	365	730	1095	1460	3650
TMA-N (μg/g)	390	650	830	810	780	950
TMAO-N (μg/g)	0	0	0	0	0	0
Lys (mg/mgN)	0.55	0.58	0.45	0.43	0.42	0.34
Glucose (μg/g)	650	475	350	225	175	50
G-6-P (μg/g)	50	25	20	15	15	15
TBA no.[a]	0.90	1.65	1.75	2.15	2.60	2.75
Fe (μg/g)	35	45	55	80	100	145
Sn (μg/g)	45	60	65	80	95	110
Pb (μg/g)	10	20	23	25	30	45
Acid Value[b]						
Meat	1.9	2.0	3.3	4.1	7.8	9.6
Liquid	0.6	0.7	2.0	3.1	4.0	6.0

a/ Measured as mg of malonaldehyde per kg of tissue
b/ Absorbance units at 530 nm/g dry matter

Mackerel, Scomber japonicus, Japan

TABLE 87. Changes in ATPase activity of mackerel, measured at pH 6.0 and pH 7.0, during storage on ice (80)

	ΔPi μmol/min/mg					
	Storage time (days)					
	0	2	3	7	13	20
pH 7.0						
Ca^{2+}-ATPase	0.45	0.43	0.42	0.44	0.41	0.41
EDTA ATPase	0.65	0.65	0.65	0.67	0.65	0.60
Mg^{2+} (Ca^{2+}) ATPase	0.70	0.63	0.59	0.61	0.62	0.65
Mg^{2+} (EGTA) ATPase	0.19	0.20	0.19	0.26	0.34	0.39
pH 6.0						
Ca^{2+}-ATPase	0.45	0.35	0.35	0.37	0.34	0.25
EDTA ATPase	0.65	0.51	0.44	0.38	0.28	0.23
Mg^{2+} (Ca^{2+}) ATPase	0.75	0.63	0.56	0.57	0.58	0.45
Mg^{2+} (EGTA) ATPase	0.18	0.19	0.23	0.27	0.30	0.30

Mackerel, *Scomber japonicus*, Japan

TABLE 88. Changes in the concentration of TMA-N and DMA-N of ordinary mackerel muscle in iced and frozen (-6°C) storage (99)

	Storage time (days)						
	0	4	5	15	17	22	28
Iced							
TMA-N (mg %)	0.40	0.40	--	--	0.45	0.45	--
DMA-N (mg %)	0.30	0.30	--	--	0.35	0.35	--
Frozen							
TMA-N (mg %)	0.50	--	0.50	0.50	--	--	0.50
DMA-N (mg %)	0.20	--	0.20	0.20	--	--	0.20

Mackerel, Scomber japonicus, Japan

TABLE 89. Changes in the concentration of TMA-N and DMA-N during the iced and frozen storage (-6°C) of bloody mackerel muscle (99)

	Storage time (days)						
	0	4	5	15	17	22	28
Iced							
TMA-N (mg %)	0.40	2.60	--	--	8.95	9.10	--
DMA-N (mg %)	0.30	1.85	--	--	2.95	3.00	--
Frozen							
TMA-N (mg %)	0.50	--	3.50	6.10	--	--	6.90
DMA-N (mg %)	0.50	--	2.00	3.85	--	--	4.90

Mackerel, *Scomber scombrus*, England

TABLE 90. Amine production in mackerel stored in ice at 1°, 10°, and 25°C (77)

	mg amine %								
	Storage time (days)								
	0	1	2	3	7	14	17	21	28
1°C									
Putrescine	0.05	--	--	0.08	0.26	1.13	2.20	2.91	8.92
Histamine	0.01	--	--	0.41	2.16	5.25	12.04	32.44	57.94
Cadaverine	0.01	--	--	0.31	1.22	4.73	10.29	15.79	43.08
Spermidine	0.30	--	--	0.07	0.08	0.47	0.35	0.31	0.31
Spermine	0.37	--	--	0.12	0.09	0.57	0.43	0.41	0.55
10°C									
Putrescine	--	0.20	0.32	1.84	--	--	--	--	--
Histamine	--	0	0	0	--	--	--	--	--
Cadaverine	--	0.06	0.14	1.51	--	--	--	--	--
Spermidine	--	0.58	0.63	0.26	--	--	--	--	--
Spermine	--	0.92	0.56	0.43	--	--	--	--	--
25°C									
Putrescine	--	0.63	2.46	--	--	--	--	--	--
Histamine	--	0	5.90	--	--	--	--	--	--
Cadaverine	--	0.87	8.56	--	--	--	--	--	--
Spermidine	--	0.57	1.34	--	--	--	--	--	--
Spermine	--	0.68	0.0	--	--	--	--	--	--

Mackerel, Scomber scombrus, United States

TABLE 91. Histamine, TMA-N, APC and Hx content of dressed and steaked Atlantic mackerel during iced storage (55)

	Storage time (days)									
	0	2	6	10	13	14	15	16	22	25
Dressed										
Histamine (mg %)	--	1.8	1.9	2.0	--	--	10.0	--	--	--
TMA-N (mg %)	1.2	--	--	--	6.5	--	--	--	--	15
APC (log no. org./g)	3.4	4.0	5.6	--	--	7.0	--	--	--	--
Hx (mg %)	5	12	--	--	--	--	--	37.5	12	--
Steaked										
Histamine (mg %)	--	2.5	2.5	--	5.0	--	--	--	21.3	--
TMA-N (mg %)	1.2	--	1.2	--	11.	--	--	--	--	22
APC (log no. org./g)	3.6	3.8	--	5.9	--	7.0	--	--	--	--
Hx (mg %)	12	22	--	--	--	--	--	45	22	--

Mackerel, *Scomberomorus sierra*, Mexico

TABLE 92. Changes in quality parameters of salted Spanish mackerel cakes during storage (34)

	Storage time (days)			
	0	30	60	90
TPC (log no. org./g)	5.2	5.6	6.0	5.3
Malonaldehyde ($1x10^{-2}$ moles/g)	3.8	4.1	4.7	5.3
Moisture content (%)	45.7	16.6	8.3	8.0

Mackerel, Trachurus japonicus, Japan

TABLE 93. Changes in TVB-N, odor and bacterial counts of horse mackerel during storage at various temperatures (74)

	Storage time (days)						
	0	7	11	15	37	53	67
0°C							
TVB-N (mg %)	10.9	15.8	23.8	--	--	--	--
Odor[a/]	F	F	S	--	--	--	--
Bacterial count (log no. org./cm^2)	5.62	6.76	8.81	--	--	--	--
-3°C							
TVB-N (mg %)	10.9	11.9	11.6	14.3	--	--	--
Odor[a/]	F	--	--	F	F	F	--
Bacterial count (log no. org./cm^2)	5.62	--	--	5.20	6.14	7.68	--
-20°C							
TVB-N (mg %)	10.9	--	--	11.4	10.3	--	10.4
Odor[a/]	F	--	--	F	F	--	F
Bacterial count (log no. org./cm^2)	5.62	--	--	5.11	5.14	--	4.98

[a/] F = fresh; S = spoiled

Mackerel, *Trachurus japonicus*, Japan

TABLE 94. Changes in TVB-N and bacterial counts of thawed horse mackerel frozen at -3°C and -20°C, then stored at 0°C (74)

	Storage time (days)				
	0	3	6	12	13
Frozen 15 days at -3°C[a/]					
TVB-N (mg %)	11.1	--	19.4	--	--
Bacterial count (log no. org./cm^2)	5.20	--	8.65	--	--
Frozen 37 days at -3°C[b/]					
TVB-N (mg %)	11.6	18.6	--	--	--
Bacterial count (log no. org./cm^2)	6.15	8.43	--	--	--
Frozen 15 days at -20°C[c/]					
TVB-N (mg %)	11.4	--	--	20.2	--
Bacterial count (log no. org./cm^2)	5.11	--	--	9.04	--

TABLE 94. (continued)

	Storage time (days)				
	0	3	6	12	13
Frozen 67 days at -20°C[d]					
TVB-N (mg %)	10.4	--	--	--	22.5
Bacterial count (log no. org./cm^2)	3.98	--	--	--	8.69

a/ spoiled at day 6
b/ spoiled at day 3
c/ spoiled at day 12
d/ spoiled at day 13

Mackerel, Trachurus japonicus, Japan

TABLE 95. Changes in bacterial flora of horse mackerel during storage at 0°C (74)

	Storage time (days)		
	0	7	11
Pseudomonas I/II	0[a]	0	0
Pseudomonas III/IV-NH	0	2	2
Pseudomonas III/IV-H	2	10	10
Vibrio	4	2	3
Moraxella	0	0	0
Acinetobacter	3	0	0
Flavobacterium-Cytophaga	2	0	0
Micrococcus	2	0	0
Staphylococcus	2	0	0
PF-group[b]	0	0	0
Not determined	0	1	0

a/ Indicates number of strains
b/ A group of undetermined bacteria

Mackerel, _Trachurus japonicus_, Japan

TABLE 96. Changes in bacterial flora of horse mackerel during storage at -3°C (74)

	Storage time (days)		
	15	38	53
Pseudomonas I/II	2[a]	7	12
Pseudomonas III/IV-NH	0	0	0
Pseudomonas III/IV-H	0	0	0
Vibrio	0	0	0
Moraxella	5	0	1
Acinetobacter	2	0	0
Flavobacterium-Cytophaga	1	0	0
Micrococcus	2	1	0
Staphylococcus	2	1	0
PF-group[b]	0	5	2
Not determined	1	1	0

a/ Indicates number of strains
b/ A group of undetermined bacteria

Mackerel, *Trachurus japonicus*, Japan

TABLE 97. Changes in bacterial flora of horse mackerel during storage at -20°C (74)

	Storage time (days)		
	15	37	67
Pseudomonas I/II	0[a]	0	0
Pseudomonas III/IV-NH	0	0	0
Pseudomonas III/IV-H	0	0	0
Vibrio	0	0	0
Moraxella	4	3	4
Acinetobacter	0	0	0
Flavobacterium-Cytophaga	2	3	4
Micrococcus	3	4	4
Staphylococcus	5	3	3
PF-group[b]	0	1	0
Not determined	1	1	0

[a] Indicates number of strains
[b] A group of undetermined bacteria

Mackerel, *Trachurus japonicus*, Japan

TABLE 98. Changes in bacterial flora of thawed horse mackerel frozen 15 days at -3°C (74)

	Storage time (days)	
	0	6
Pseudomonas I/II	2[a]	2
Pseudomonas III/IV-NH	0	0
Pseudomonas III/IV-H	0	6
Vibrio	0	1
Moraxella	5	5
Acinetobacter	2	0
Flavobacterium-Cytophaga	1	0
Micrococcus	2	0
Staphylococcus	2	0
PF-group[b]	0	1
Not determined	1	0

[a] Indicates number of strains
[b] A group of undetermined bacteria

Mackerel, Trachurus japonicus, Japan

TABLE 99. Changes in bacterial flora of thawed horse mackerel frozen 37 days at -3°C (74)

	Storage time (days)	
	0	6
Pseudomonas I/II	7[a]	9
Pseudomonas III/IV-NH	0	1
Pseudomonas III/IV-H	0	0
Vibrio	0	0
Moraxella	0	0
Acinetobacter	0	0
Flavobacterium-Cytophaga	0	0
Micrococcus	1	2
Staphylococcus	1	0
PF-group[b]	5	2
Not determined	1	1

a/ Indicates number of strains
b/ A group of undetermined bacteria

Mackerel, *Trachurus japonicus*, Japan

TABLE 100. Changes in bacterial flora of thawed horse mackerel frozen 15 days at -20°C (74)

	Storage time (days)	
	0	12
Pseudomonas I/II	0[a]	9
Pseudomonas III/IV-NH	0	1
Pseudomonas III/IV-H	0	1
Vibrio	0	0
Moraxella	4	4
Acinetobacter	0	0
Flavobacterium-Cytophaga	2	0
Micrococcus	3	0
Staphylococcus	5	0
PF-group[b]	0	0
Not determined	1	0

a/ Indicates number of strains
b/ A group of undetermined bacteria

Mackerel, Trachurus japonicus, Japan

TABLE 101. Changes in bacterial flora of thawed horse mackerel frozen 67 days at -20°C (74)

	Storage time (days)	
	0	13
Pseudomonas I/II	0[a]	7
Pseudomonas III/IV-NH	0	0
Pseudomonas III/IV-H	0	0
Vibrio	0	0
Moraxella	4	7
Acinetobacter	0	0
Flavobacterium-Cytophaga	4	0
Micrococcus	4	0
Staphylococcus	3	0
PF-group[b]	0	1
Not determined	0	0

a/ Indicates number of strains
b/ A group of undetermined bacteria

Mackerel, *Trachurus novaezelandieae*, New Zealand

TABLE 102. Changes in IMP, HxR, Hx, K value, pH, TBA number and TMA-N in Jack mackerel during iced storage (78)

	Storage time (days)							
	0	1	2	7	12	14	16	23
IMP (μ moles/g)	10.4	8.3	7.7	4.5	3.2	3.0	2.5	1.2
HxR (μ moles/g)	0.3	0.5	1.0	1.8	2.0	2.2	3.0	2.7
Hx (μ moles/g)	0.2	0.1	0.2	0.4	0.7	1.3	1.0	1.7
K value (%)	3	4	8	20	36	41	46	64
pH	6.08	--	6.04	6.01	5.98	--	6.18	6.52
TBA no.[a/]	2.3	--	1.7	8.4	7.5	--	7.3	7.7
TMA-N (mg %)	0.7	--	0.5	1.5	1.8	2.5	5.3	8.3

a/ Measured as mg of malonaldehyde per kg of tissue

Mackerel, _Trachurus novaezelandieae_, New Zealand

TABLE 103. Changes in APC and sulphide-producing microbial counts of Jack mackerel during ice storage (78)

	log no. org./g								
	Storage time (days)								
	0	3	5	7	10	13	14	16	23
APC									
Surface	--	3.7	4.3	5.6	6.5	7.5	8.5	9.0	9.0
Flesh	0.3	2.5	3.3	3.7	4.9	6.2	6.4	7.0	8.8
SPC									
Surface	--	2.5	1.9	3.6	4.6	5.8	6.0	6.7	7.9
Flesh	--	1.0	0.8	1.4	3.0	4.2	4.5	5.1	7.5

Mackerel, unspecified, United States

TABLE 104. Changes in average TBA numbers and TPC during the storage of mackerel fillets at 1°C (71)

	Storage time (days)						
	1	2	3	4	5	6	7
TBA no.[a]	1.20	2.50	3.50	3.20	2.80	2.00	3.10
TPC (log no.org./g)	4.79	4.79	5.71	6.84	--	8.02	8.53

[a] Measured as mg of malonaldehyde per kg of tissue

Mullet, *Mugil cephalus*, Mexico

TABLE 105. Changes in quality parameters of salted mullet cakes during storage (34)

	Storage time (days)			
	0	30	60	90
TPC (log no. org./g)	3.9	5.5	6.2	5.4
Malonaldehyde (1×10^{-2} moles/g)	2.9	3.0	3.3	3.5
Moisture content (%)	44.8	16.2	8.1	8.0

Mullet, *Mugil cephalus*, United States

TABLE 106. Changes in average TBA numbers and TPC during the storage of mullet fillets at 1°C (71)

	Storage time (days)						
	1	2	3	4	5	6	7
TBA no.[a]	1.40	1.90	7.50	10.2	16.8	15.9	13.6
TPC (log no. org./g)	4.30	4.70	5.60	5.57	--	7.50	7.52

[a] Measured as mg of malonaldehyde per kg of tissue

Mullet, *Mugil dussumieri*, India

TABLE 107. Post-mortem changes of pH, Glyc, LA, Pi, ATP and TMA-N in mullet during iced storage (73)

	mg %								
	Storage time (days)								
	0	2	6	10	12	14	16	18	20
State[a]	--	FR	FR	FR	S	S	S	SS	SS
pH	6.8	6.7	6.6	6.5	6.5	6.5	6.6	6.6	6.6
Glyc	219	130	97	74	63	50	39	30	26
LA	273	319	387	414	425	435	436	442	448
Pi	106	110	108	114	116	119	118	120	123
ATP[b]	30.1	20.6	13.9	10.2	8.2	9.0	7.9	7.6	7.0
TMA-N	0.03	0.04	0.05	0.08	0.09	0.11	0.15	0.17	0.20

a/ FR = full rigor; AR = almost full rigor; S = softening; SS = soft
b/ Measured as phosphate associated to ATP

Mullet, Mugil spp., United States

TABLE 108. Changes in Hx, TMA-N, FAA and TBA number of mullet during storage at 0° and -2°C (62)

	Storage time (days)				
	0	6	11	13	16
Ice-unpackaged (0°C)					
Hx (μ moles/g)	2.4	3.0	3.4	4.3	5.0
TMA-N (mg %)	0.7	1.2	1.3	1.8	2.6
FAA (mmoles Leu eq %)	1.8	1.9	2.1	2.3	2.6
TBA no.[a/]	3.9	4.9	6.6	7.0	6.7
Ice-packaged (0°C)					
Hx (μ moles/g)	2.8	3.3	3.7	4.1	4.7
TMA-N (mg %)	0.7	1.3	1.6	1.8	2.4
FAA (mmoles Leu eq %)	1.9	2.1	2.1	2.2	2.3
TBA no.[a/]	3.8	4.8	6.8	6.1	--
Salt/ice-unpackaged (-2°C)					
Hx (μ moles/g)	1.1	1.4	2.3	2.5	3.8
TMA-N (mg %)	0.6	0.8	1.1	1.3	2.1
FAA (mmoles Leu eq %)	1.7	1.8	2.0	2.1	2.1
TBA no.[a/]	2.4	3.7	4.6	4.8	--

TABLE 108. (continued)

	Storage time (days)				
	3	6	11	13	16
Salt/ice-packaged (-2°C)					
Hx (μmoles/g)	1.6	1.6	1.9	2.8	3.4
TMA-N (mg %)	0.6	1.1	1.2	1.4	2.3
FAA (mmoles Leu eq %)	1.8	1.9	2.1	2.3	1.7
TBA no.[a/]	2.3	3.5	3.8	4.3	--
Propylene-glycol/ ice-unpackaged (-2°C)					
Hx (μmoles/g)	1.5	1.8	1.9	2.1	2.5
TMA-N (mg %)	0.7	0.8	1.0	1.4	2.4
FAA (mmoles Leu eq %)	2.0	2.1	2.1	2.4	1.9
TBA no.[a/]	--	2.5	5.2	4.9	--
Propylene-glycol/ ice-packaged (-2°C)					
Hx (μmoles/g)	0.9	1.2	1.4	1.9	2.4
TMA-N (mg %)	0.7	0.8	1.3	1.7	2.1
FAA (mmoles Leu eq %)	1.8	2.1	2.1	2.2	1.9
TBA no.[a/]	--	2.4	3.3	3.9	--

a/ In mg of malonaldehyde per kg of sample

Mullet, *Mugil* spp., United States

TABLE 109. Changes in physical properties and sensory evaluations of mullet stored at 0° and -2°C (62)

	Storage time (days)					
	1	3	5	7	10	14
Ice-unpackaged (0°C)						
Translucency[a/]	--	4.5	4.2	4.1	3.7	3.6
Color desirability[b/]	--	4.9	4.0	3.2	2.9	2.6
Indentation force (kg)	--	0.24	0.24	0.23	0.18	--
Taste[c/]	--	3.8	3.4	3.1	2.6	2.0
Texture[c/]	4.0	--	3.2	3.2	2.7	2.3
Odor[d/]	--	4.9	4.5	3.2	2.6	2.1
Eating quality[d/]	--	4.2	3.4	3.2	2.2	2.0
Ice-packaged (0°C)						
Translucency[a/]	--	5.0	4.0	3.9	4.0	3.4
Color desirability[b/]	--	5.0	4.5	3.5	2.5	2.0
Indentation force (kg)	--	0.24	0.22	0.21	0.29	--
Taste[c/]	--	3.7	3.4	3.4	3.0	2.6
Texture[c/]	3.8	--	3.3	3.2	3.1	2.6
Odor[d/]	--	4.8	4.4	3.5	3.0	2.4
Eating quality[d/]	--	3.8	3.2	3.0	2.6	2.0

TABLE 109. (continued)

	Storage time (days)					
	1	3	5	7	10	14
Salt/ice-unpackaged (-2°C)						
Translucency[a/]	--	4.0	3.1	3.7	3.6	2.1
Color desirability[b/]	--	4.0	3.5	3.1	2.6	2.0
Indentation force (kg)	--	0.23	0.18	0.16	0.15	--
Taste[c/]	--	3.8	3.7	3.6	3.4	3.0
Texture[c/]	3.8	--	3.7	3.7	3.6	3.1
Odor[d/]	--	4.8	4.8	4.5	4.1	3.0
Eating quality[d/]	--	4.4	4.2	4.0	3.7	3.2
Salt/ice-packaged (-2°C)						
Translucency[a/]	--	4.1	3.3	3.0	2.8	2.3
Color desirability[b/]	--	4.0	3.3	3.0	2.8	2.0
Indentation force (kg)	--	0.25	0.17	0.15	0.14	--
Taste[c/]	--	3.7	3.5	3.4	3.2	2.6
Texture[c/]	3.7	--	3.6	3.6	3.5	3.0
Odor[d/]	--	4.7	4.7	4.4	4.0	2.8
Eating quality[d/]	--	4.0	4.0	3.7	3.5	3.0

TABLE 109. (continued)

	Storage time (days)					
	1	3	5	7	10	14
Propylene-Glycol/ ice-unpackaged (-2°C)						
Translucency[a/]	--	4.0	2.9	2.4	2.0	1.5
Color desirability[b/]	--	4.0	3.1	2.5	1.6	1.3
Indentation force (kg)		0.20	0.18	0.16	0.14	--
Taste[c/]	--	3.2	2.7	2.6	2.5	1.6
Texture[c/]	3.8	--	3.5	3.5	3.3	3.0
Odor[d/]	--	4.6	4.5	4.0	3.5	2.8
Eating quality[d/]	--	4.2	3.8	3.0	2.8	2.0
Propylene-Glycol/ ice-packaged (-2°C)						
Translucency[a/]	--	4.0	3.0	2.5	2.1	2.0
Color desirability[b/]	--	3.9	3.2	2.9	2.3	1.3
Indentation force (kg)	--	0.19	0.16	0.14	--	--
Taste[c/]	--	3.8	3.7	3.5	3.0	2.6
Texture[c/]	3.8	--	3.7	3.7	3.5	3.0
Odor[d/]	--	4.8	4.7	4.5	3.6	3.0

a/ 1 = little; 3 = moderate; 5 = very
b/ 1 = undesirable; 3 = moderate; 5 = highly desirable
c/ 1 = poor; 3 = moderate; 5 = very good
d/ 1 = poor; 3 = moderate; 5 = highly acceptable

Mullet, *Mugil* spp., United States

TABLE 110. Changes in bacterial counts of mullet during storage under several treatments for two weeks (62)

	no. org. x 10^3/g						
	Storage time (days)						
	1	3	5	7	10	12	14
Ice-unpackaged							
Surface aerobes[a/]	1.1	4.5	22.0	140.0	--	--	--
Total aerobes	--	--	--	1.1	35.0	--	--
Sporeforming anaerobes	--	--	--	--	1.1	4.7	--
Total anaerobes	--	--	--	--	1.1	340.0	--
Ice-packaged							
Surface aerobes[a/]	--	--	1.1	8.5	220.0	--	--
Total aerobes	--	--	--	1.1	4.8	18.0	--
Sporeforming anaerobes	--	--	--	--	1.1	330.0	--
Total anaerobes	--	--	--	--	1.1	10.7	--
Salt/ice-unpackaged							
Surface aerobes[a/]	--	--	--	1.1	29.0	--	--
Total aerobes	--	--	--	--	1.1	2.5	--
Total anaerobes	--	--	--	--	1.1	10.3	9.0

TABLE 110. (continued)

	no. org. x 10^3/g						
	Storage time (days)						
	1	3	5	7	10	12	14
Salt ice-packaged							
Surface aerobes[a/]	--	--	--	--	5.0	17.0	--
Total aerobes	--	--	--	--	--	1.1	1.5
Total anaerobes	--	--	--	--	1.1	10.0	--
Propylene-glycol/ ice-unpackaged (-2°C)							
Surface aerobes[a/]	--	--	--	1.1	4.0	12.0	--
Total aerobes	--	--	--	--	--	1.1	6.0
Total anaerobes	--	--	--	--	1.1	7.0	--
Propylene-glycol/ ice-packaged (-2°C)							
Surface aerobes[a/]	--	--	--	1.1	3.7	9.5	--
Total aerobes	--	--	--	--	1.1	1.7	--
Total anaerobes	--	--	--	--	1.1	11.7	--

a/ 1 x 10^3 organisms per cm^2 of skin

Mullet, unspecified, United States

TABLE 111. Changes in FFA, PV and TBA number of mullet frozen in the round, skin on fillets and skin-off fillets during storage at -18°C, with 1 or 7 days pre-freezing iced storage (35)

	Storage time (days)				
	0	90	180	270	360
Frozen in the round					
1-day pre-freezing iced storage					
FFA[a/]	0.8	1.5	2.9	4.0	3.4
PV (meq/kg)	1.0	1.5	3.0	3.0	2.5
TBA no. (mg malonaldehyde/kg)	1.0	1.5	2.0	4.0	1.0
7 days pre-freezing iced storage					
FFA[a/]	1.0	3.4	5.7	7.0	3.5
PV (meq/kg)	1.0	4.0	9.0	11.0	7.0
TBA no. (mg malonaldehyde/kg)	1.0	1.0	3.0	8.0	3.0
Skin-on fillets					
1-day pre-freezing iced storage					
FFA[a/]	0.7	1.8	2.3	2.3	1.7
TBA no. (mg malonaldehyde/kg)	1.5	2.5	5.0	6.0	3.0

TABLE 111. (continued)

	Storage time (days)				
	0	90	180	270	360
7 days pre-freezing iced storage					
FFA[a]	1.0	2.1	5.2	7.0	3.2
TBA no. (mg malonaldehyde/kg)	1.5	3.0	5.0	5.0	3.5
Skin-off fillets					
1 day pre-freezing iced storage					
FFA[a]	0.8	1.3	2.4	3.3	4.1
TBA no. (mg malonaldehyde/kg)	1.5	10.5	17.5	20.5	11.0
7 days pre-freezing iced storage					
FFA[a]	1.0	2.3	3.1	4.0	2.0
TBA no. (mg malonaldehyde/kg)	1.5	6.0	12.0	20.0	27.0

a/ Percent of oleic acid in total fat

Mussel, *Perna canaliculus*, New Zealand

TABLE 112. Changes in bacterial count, flavor scores, ADP, AMP, Hx and HxR of live green-lipped mussel during storage at ambient temperature, ice and melting ice (19)

	Storage time (days)							
	2	4	6	7	8	10	12	14
Ambient temperature								
APC								
(log no. org./g)	4.3	6.3	--	--	--	--	--	--
Staphylococci								
(log no. org./g)	3.5	4.1	4.3	4.35	--	--	--	--
Flavor score[a]	3.1	2.0	--	--	--	--	--	--
ADP[b]	34.0	35.0	31.0	--	27.0	--	--	--
AMP[b]	--	--	--	--	18.0	--	--	--
Hx[b]	26.0	35.0	41.0	--	48.0	--	--	--
Ice								
Flavor Score[a]	3.5	2.8	2.1	1.3	0.5	--	--	--
ADP[b]	25.0	28.0	30.0	31.0	32.0	30.0	28.0	--
AMP[b]	2.0	2.0	2.0	3.0	3.0	4.0	5.0	--
Hx[b]	9.0	11.0	14.0	20.0	30.0	45.0	59.0	--

TABLE 112. (continued)

	Storage time (days)							
	2	4	6	7	8	10	12	14
Melting ice								
APC								
(log. no. org./g)	3.5	3.3	3.4	3.5	4.0	5.0	6.0	7.0
Flavor score[a]	4.5	4.1	3.6	3.5	3.3	2.8	2.4	--
ADP[b]	30.0	33.0	35.0	34.0	35.0	35.0	34.0	33.0
AMP[b]	1.0	--	2.0	--	--	--	--	--
Hx[b]	8.0	11.0	15.0	16.0	18.0	20.00	22.0	28.0

a/ Level acceptable for processing = 3
b/ Percentage of total nucleotides

Oyster, unspecified, United States

TABLE 113. Changes in quality parameters during the iced storage of oyster muscle (66)

	Storage time (days)							
	0	3	5	7	9	11	13	15
pH	6.69	6.49	6.32	6.24	6.21	6.11	6.05	5.96
TMA-N (mg %)	4.16	12.9	15.0	11.3	13.5	15.0	13.4	13.02
TVB-N (mg %)	19.7	28.3	32.1	27.2	27.8	28.1	28.8	30.1
Indole (meq %)	6.32	5.42	5.70	5.25	5.07	6.05	5.60	6.62
MBC (log no. org. %)	4.40	4.30	4.30	4.10	4.00	4.0	3.50	3.70
PBC (log no. org. %)	4.40	4.50	4.80	4.90	5.10	5.40	5.60	5.90

Perch, *Sebastodes alutus*, unspecified

TABLE 114. Changes in concentration of nucleotides during the iced storage of Pacific ocean perch muscle (92)

	Storage time (days)			
	0	3	7	14
IMP (μmoles/g)	6.05	4.50	1.80	0.0
Hx (μmoles/g)	0.0	0.75	2.30	5.65
HxR (μmoles/g)	0.0	1.90	3.00	0.05

Perch, *Sebastes marinus*, United States

TABLE 115. Accumulation of Hx and nucleotide degradation in ocean perch stored at various temperatures (90)

	Storage time (days)								
	1.5	2.5	3.5	5.5	7.5	9.5	30	60	120
-20°F									
Hx (mg/g)	0.12	--	--	--	--	--	0.11	0.12	0.11
32°F									
Hx (mg/g)	0.12	0.22	0.31	0.50	0.68	0.76	--	--	--
Nucleotide[a]	1.10	0.88	0.70	0.46	0.30	0.22	--	--	--
42°F									
Hx (mg/g)	0.12	0.50	0.58	0.85	--	--	--	--	--

a/ Optical density at 256 nm

Perch, *Sebastes marinus*, United States

TABLE 116. Sensory scores and TMA-N content of tray-packed fresh ocean perch fillets during storage at 2°C (107)

	Storage time (days)			
	1	2	3	5
Color[a]	6.8	7.5	7.0	6.0
Texture[a]	6.7	7.4	--	--
Flavor[a]	6.1	5.9	--	--
Odor[a]	6.2	6.3	5.0	4.0
General acceptance[a]	5.8	6.0	--	--
TMA-N (mg %)	4.43	7.93	--	38.80

[a] 9 = like extremely; 1 = dislike extremely.

Perch, *Sebastes marinus*, United States

TABLE 117. Percent change of net weight, drip gross weight and total moisture in tray-packed fresh ocean perch fillets during storage at 2°C (107)

	%			
	Storage time (days)			
	2	4	6	8
Decrease in net weight	2.9	3.2	3.5	3.9
Increase in drip	2.4	2.5	2.6	2.7
Decrease in gross weight	0.2	0.6	0.9	1.2
Decrease in total moisture	0.1	0.1	0.2	0.2

Plaice, Hippuglossoides platessoides fabricius, Canada

TABLE 118. Changes in the concentration of nucleotides during the iced storage of American plaice at 0°C (37)

	% of total purines						
	Storage time (days)						
	0	1	2	3	4	5	8
ATP	50	5	4	3	2	1	--
ADP	13	13	10	11	12	12	15
IMP	23	65	58	40	15	2	2
HxR	0	14	14	8	4	2	--
Hx	0	8	18	50	74	88	88

Plaice, *Pleuronectes platessa*, Scotland

TABLE 119. Nucleotide degradation in plaice during iced storage (57)

	μmoles/g			
	Storage time (days)			
	0	8	16	24
ATP	0.4	0.0	0.0	0.0
ADP	0.3	0.1	0.1	0.1
AMP	0.2	0.0	0.0	0.0
NAD	0.1	0.0	0.0	0.0
IMP	3.8	1.4	0.0	0.0
HxR	0.2	0.9	0.5	0.0
Hx	0.0	1.6	1.5	1.2
Guanine	0.3	0.3	0.0	0.0

Plaice, unspecified, Denmark

TABLE 120. Changes in APC and TVA in plaice during iced storage (53)

	Storage time (days)						
	1	3	6	9	12	15	18
APC (log no. org./g)	3.75	3.90	4.70	5.65	6.80	7.80	8.30
TVA (meq %)	0.05	0.04	0.06	0.10	0.17	0.25	0.35

Pollock, _Pollachius virens_, Canada

TABLE 121. Changes in TMA-N, DMA-N and EPN in pollock fillets during iced and frozen storage at -5°C (24)

	Storage time (days)					
	0	4	6	8	10	15
Iced						
TMA-N (mg %)	2.55	15.0	28.0	42.0	--	--
DMA-N (mg %)	2.00	1.40	1.40	2.10	4.90	--
EPN (%)	2.17	2.25	2.30	2.20	2.10	2.10
Frozen						
TMA-N (mg %)	0.80	0.60	0.60	1.30	4.90	--
DMA-N (mg %)	0.60	--	2.00	4.90	4.20	5.00
EPN (%)	2.20	1.82	--	1.76	--	1.63

Pollock, _Pollachius virens_, Canada

TABLE 122. Changes in EPN and formation of DMA-N of minced and filleted pollock during storage at -5° and 10°C (36)

	Storage time (days)					
	10	20	30	40	50	60
Fillets -5°C						
DMA-N (mmoles/g)	0.28	0.20	0.31	0.29	0.68	--
Minced at -10°C						
DMA-N (mmoles %)	0.20	0.25	0.35	0.35	0.36	0.42
EPN (g %)	1.01	0.88	0.65	0.66	0.58	0.50

Pollock, _Pollachius virens_, Canada

TABLE 123. Changes in the concentration of nucleotides during the iced storage of Atlantic pollock at 0°C (37)

	% of total purines								
	Storage time (days)								
	0	1	2	3	4	5	6	7	10
ATP	36	4	2	--	--	--	--	--	--
ADP	10	9	8	9	8	8	7	8	8
IMP	55	73	63	28	13	8	5	3	2
HxR	0	18	30	63	78	80	86	85	78
Hx	--	--	--	4	7	7	7	8	8

Pollock, *Theregra chalcogramma*, Japan

TABLE 124. Changes in various freshness-quality parameters of Alaska pollock

	Storage time (days)				
	0	2	4	6	8
TVB-N (mg %)	10.5	10.2	9.7	13.0	14.2
TMA-N (mg %)[a]	0.4	0.6	0.5	0.7	3.2
TMA-N (mg %)[b]	0.4	0.4	0.5	0.5	1.4
IMP (μ moles/g)	3.3	0.8	0.5	0.3	--
HxR + Hx (μ moles/g)	1.0	1.9	2.6	3.1	--
ATP + ADP + AMP (μ moles/g)	0.5	0.3	0.3	0.3	--
Bacterial count (log no. org./g)	1.1	1.0	1.2	1.9	3.0

a/ Conway's method
b/ Picrate method

Pollock, _Theregra chalcogramma,_ United States

TABLE 125. Changes in Torrymeter readings, pH, Hx, malonaldehyde and TMA-N composition in Alaska pollock during iced storage (3)

	Storage time (days)			
	1	3	5	8
Torrymeter readings	13.8	8.0	4.3	1.08
Hx (μmoles/g)	0.45	1.02	1.28	2.06
Malonaldehyde (μmoles %)	0.12	--	0.29	0.64
TMA-N (mg %)	0.26	0.61	0.67	2.20
pH	6.88	6.88	7.10	7.15

Pollock, *Theregra chalcogramma*, United States

TABLE 126. Changes in TMAO-N, DMA-N, EPN and ETD of Alaska pollock held at -18°C (6)

	Storage time (days)				
	0	60	120	180	240
Fillet					
TMAO-N (mg %)	57	58	59	57	56
DMA-N (mg %)	4	5	7	8	9
EPN (mg/g)	75	45	28	29	30
ETD (ml %)	--	31.5	32.0	34.5	36.5
Chopped fillet					
TMAO-N (mg %)	57	56	55	54	53
DMA-N (mg %)	4	5	7	8	9
EPN (mg/g)	75	40	31	30	27
ETD (ml %)	--	30.5	34.0	35.0	35.5
Chopped + 20% mince					
TMAO-N (mg %)	57	48	41	41	41
DMA-N (mg %)	4	18	27	24	22
EPN (mg/g)	75	26	22	21	20
ETD (ml %)	--	34.5	36.0	38.5	41.0

TABLE 126. (continued)

	Storage time (days)				
	0	60	120	180	240
Chopped + 50% mince					
TMAO-N (mg %)	57	46	37	31	26
DMA-N (mg %)	4	20	20	25	30
EPN (mg/g)	55	21	20	18	16
ETD (ml %)	--	39.5	42.5	44.0	45.5
Minced					
TMAO-N (mg %)	52	39	26	20	13
DMA-N (mg %)	4	28	30	33	38
EPN (mg/g)	55	19	17	16	15
ETD (ml %)	--	40.5	38.0	42.0	46.0

Pollock, unspecified, Canada

TABLE 127. Changes in DMA-N, KOH-amine value and K_2CO_3-amine value of pollock frozen at -40°C for 90 min., then stored at -5°C (23)

	Storage time (days)					
	0	10	20	30	40	50
DMA-N (mg %)	0.6	3.4	2.5	4.0	3.7	9.5
KOH-amine value (mg %)	0.9	1.8	2.0	2.0	2.1	1.4
K_2CO_3-amine value (mg %)	2.2	4.2	5.4	7.0	7.4	10.5

Prawn, *Macrobrachium rosenbergii*, Israel

TABLE 128. Changes in proximate composition of freshwater prawn stored at 0°C (1)

	%			
	Storage time (days)			
	2	4	7	14
Moisture	79.3	79.9	80.0	80.8
Fat	1.9	2.9	4.3	2.4
Ash	1.3	1.3	1.4	1.9
Crude protein[a]	17.5	15.9	14.3	14.9

[a] Determined by numerical difference

Redfish, unspecified, Canada

TABLES 129. Changes in TMA-N, IMP and Hx concentration of redfish fillets during iced storage (48)

	Storage time (days)							
	1	2	3	4	5	8	11	15
TMA-N (mg %)	--	--	--	--	0.2	4.9	25.0	64.0
IMP (μmoles/g)	3.5	1.2	0.2	0.1	--	--	--	--
Hx (μmoles/g)	0.6	2.3	3.9	4.5	4.9	4.8	4.8	4.7

Rockfish, _Roccus_ spp., United States

TABLE 130. Changes in TBA value and lipid composition during the iced storage of several species of rockfish (108)

	Storage time (days)			
	0	6	10	16
TBA no.[a]	0.12	--	--	0.27
Neutral lipid (mg/g)	636	626	628	588
Phospholipid (mg/g)	312	298	288	260
Cephalin (mg/g)	210	172	162	152
Lecithin (mg/g)	118	112	120	116
FFA (mg/g)	14	32	40	110

[a] Expressed as the optical density at 538 nm

Rockfish, *Sebastes melanops*, United States

TABLE 131. Changes in quality parameters of fillets from black rockfish during their storage in ice (27)

	Storage time (days)						
	0	4	6	8	10	12	14
Protein (%)	18.3	18.5	18.8	18.3	18.4	18.2	18.4
NPN (%)	0.34	0.33	0.31	0.31	0.31	0.30	0.30
TVA (meq H^+%)	0.06	0.06	0.06	0.07	0.10	0.11	0.10
TVB-N (mg %)	3.9	3.6	--	--	--	--	3.9
TMA-N (mg %)	0.20	0.41	0.41	0.49	0.59	0.62	0.82
DMA-N (mg %)	0.20	0.30	0.24	0.27	0.29	0.30	0.24

Rockfish, *Sebastes melanops*, United States

TABLE 132. Changes in quality parameters of fillets from black rockfish during their storage in modified refrigerated seawater (27)

	Storage time (days)						
	0	4	6	8	10	12	14
Protein (%)	18.3	18.8	19.0	19.0	19.2	19.1	19.0
NPN (%)	0.34	0.32	0.31	0.30	0.30	0.30	0.27
TVA (meq H^+ %)	0.06	0.06	0.07	0.06	0.06	0.07	0.08
TVB-N (mg %)	3.9	3.3	--	2.8	--	2.9	--
TMA-N (mg %)	0.20	0.37	0.43	0.49	0.56	0.67	0.69
DMA-N (mg %)	0.20	0.23	0.23	0.29	0.18	0.37	0.29

Rockfish, *Sebastes pinniger*, United States

TABLE 133. Changes in TMA-N and DMA-N concentration in ground flesh of Canary rockfish during storage at 0°C (72)

	ppm											
	Storage time (days)											
	0	3	5	6	7	8	9	10	11	13	15	20
TMA-N	35	38	42	49	57	140	260	335	540	675	750	--
DMA-N	15	19	20	20	20	21	21	23	24	25	25	35

Rockfish, _Sebastodes flavidus_, United States

TABLE 134. Changes in chemical and microbiological parameters during the storage of rockfish muscle in CO_2- treated and untreated refrigerated brine (8)

	Storage time (days)					
	0	3	8	10	14	17
Brine with CO_2						
pH (flesh)	6.70	5.80	6.40	--	6.50	6.40
Salt (%)	0.20	0.50	1.10	1.30	--	1.80
TPC (flesh) (log no.org./g)	4.08	--	4.20	--	4.14	4.30
TPC (brine) (log no.org./ml)	4.11	4.20	--	--	5.72	4.58
Brine without CO_2						
pH (flesh)	6.70	6.40	6.50	--	--	--
Salt (%)	0.20	0.60	1.00	--	--	--
TPC (flesh) (log no.org./g)	4.08	4.88	6.38	--	--	--
TPC (brine) (log no. org./ml)	4.11	4.08	6.44	--	--	--

Rockfish, *Sebastodes* sp., United States

TABLE 135. Changes in acetoin, TVB-N and sensory evaluations of rockfish stored at 34°F (45)

	Storage time (days)				
	0	2	4	6	8
Acetoin (mg %)	1	1	6	6	6
TVB-N (mg %)	8.5	11.5	31.8	57.7	66.5
Sensory evaluations[a/]	A	A	A	NA	NA

a/ A = acceptable; NA = not acceptable

Rosefish, Sebastes marinus, Canada

TABLES 136. Changes in FFA, extractable actomyosin and peroxide value in rosefish fillet stored at 10° and -10°F (38)

	Storage time (weeks)										
	7	63	91	147	182	231	294	350	420	490	560
FFA[a]											
+10°F	--	10.4	7.2	13.2	5.1	7.0	2.6	--	--	--	--
-10°F	6.2	--	--	2.7	1.7	6.3	5.9	--	--	--	--
Actomyosin[b]											
+10°F	1.7	1.5	1.5	1.5	1.5	1.5	1.5	1.5	1.5	1.5	1.5
-10°F	1.7	1.5	1.5	1.5	1.4	1.1	1.0	1.0	1.0	1.0	1.0
PV (meq %)											
+10°F	0.2	0.4	0.5	1.3	2.0	3.5	1.9	1.3	1.8	3.5	4.0
-10°F	0.2	0.3	0.3	0.3	0.4	0.4	0.4	2.3	1.1	1.0	1.4

a/ As percent of oleic acid in sample
b/ As percent of nitrogen in sample

Salmon, Oncorhynchus keta, United States

TABLE 137. Changes in chemical and microbiological parameters during the storage of chum salmon muscle in CO_2- treated and untreated refrigerated brine (8)

	Storage time (days)				
	0	3	9	11	18
Brine with CO_2					
pH (brine)	4.00	5.50	--	5.50	5.50
Salt (%)	0.30	0.60	1.30	--	1.30
TPC (skin) (log no. org./cm^2)	5.04	4.88	5.28	4.50	4.58
TPC (brine) (log no. org./ml)	4.11	4.97	5.14	5.32	4.30
Brine without CO_2					
pH (brine)	7.10	6.80	--	6.80	6.80
Salt (%)	0.30	0.60	1.20	--	1.40
TPC(skin) (log no. org. /cm^2)	5.04	--	6.38	6.00	6.52
TPC (brine) (log no. org./ml)	4.11	4.91	6.41	6.70	7.54

Salmon, *Oncorhynchus tshawytscha*, Canada

TABLE 138. Mean flesh pH and TBA values in brine and plate frozen Chinook salmon stored at -30°C (13,14)

	Storage time (days)				
	98	217	315	434	567
pH					
Brine frozen	6.05	6.06	6.02	5.88	6.02
Plate frozen	6.08	6.10	6.02	5.79	5.90
TBA no. [a/]					
Brine frozen (outside muscle)	0.16	0.54	0.34	0.18	0.12
Brine frozen (inside muscle)	0.06	0.10	0.08	0.04	0.07
Plate frozen (outside muscle)	0.08	0.14	0.18	0.08	0.13
Plate frozen (inside muscle)	0.06	0.06	0.08	0.05	0.06

[a/] Absorbance units at 530 nm

Sardine, Sardinops melanosticta, Japan

TABLE 139. Changes in ATPase activity of sardine, measured at several pH levels, during storage on ice (80)

	ΔPi μmol/min/mg					
	Storage time (days)					
	0	1	2	4	6	10
pH 7.5						
Ca^{2+}-ATPase	0.37	0.33	0.26	0.23	0.17	--
EDTA ATPase	0.67	0.11	0.17	0.15	0.10	--
Mg^{2+} (Ca^{2+}) ATPase	0.55	0.51	0.43	0.36	0.30	--
Mg^{2+} (EGTA) ATPase	0.34	0.45	0.37	0.34	0.29	--
pH 7.0						
Ca^{2+}-ATPase	0.35	0.30	--	0.31	0.33	0.32
EDTA ATPase	0.52	0.48	--	0.47	0.44	0.46
Mg^{2+} (Ca^{2+}) ATPase	0.54	0.50	--	0.50	0.49	0.51
Mg^{2+} (EGTA) ATPase	0.37	0.32	--	0.35	0.38	0.44
pH 6.0						
Ca^{2+}-ATPase	0.37	0.29	0.23	0.17	0.14	0.08
EDTA ATPase	0.52	0.39	0.23	0.15	0.12	0.05
Mg^{2+} (Ca^{2+}) ATPase	0.55	0.42	0.32	0.28	0.26	0.23
Mg^{2+} (EGTA) ATPase	0.38	0.26	0.25	0.22	0.19	0.18

Scad, *Trachurus trachurus*, England

TABLE 140. Changes in TMA-N, DMA-N, Hx and sensory evaluations of scad stored at chill and ambient temperatures (88)

	Storage time (days)								
	0	1	2	3.5	4.5	6.5	7.5	8.5	12.5
Iced									
Whole									
TMA-N (mg %)	1.4	1.4	1.5	1.6	1.6	1.5	1.8	2.5	3.4
DMA-N (mg %)	1.1	1.0	1.1	1.2	1.3	1.2	1.3	1.3	2.0
Hx (μ moles/g)	0.01	0.13	0.11	0.22	0.15	0.27	0.46	0.47	0.82
Overall acceptability[a/]	5.2	4.0	4.7	4.5	4.0	4.5	3.5	2.8	3.5
Gutted									
TMA-N (mg %)	1.4	1.4	1.4	1.5	1.5	1.9	2.1	2.0	2.8
DMA-N (mg %)	1.1	0.9	1.0	1.0	1.1	1.1	1.2	1.2	1.5
Hx (μ moles/g)	0.01	0.16	0.10	0.18	0.30	0.29	0.52	0.40	0.72
Overall acceptability[a/]	5.2	3.7	4.8	4.2	4.5	4.8	4.2	3.2	3.5

TABLE 140. (continued)

	Storage time (days)								
	0	1	2	3.5	4.5	6.5	7.5	8.5	12.5
Chilled Sea Water									
Whole									
TMA-N (mg %)	1.4	1.4	1.5	1.5	1.6	1.7	2.1	3.1	7.9
DMA-N (mg %)	1.1	1.0	1.0	1.1	1.2	1.2	1.3	1.4	1.8
Hx (μmoles/g)	0.01	0.07	0.19	0.15	0.19	0.21	0.39	0.32	0.61
Overall acceptability[a/]	5.2	4.0	4.0	4.7	4.8	4.8	4.8	5.3	2.8
Gutted									
TMA-N (mg %)	1.4	1.4	1.5	1.5	1.6	1.6	1.9	2.2	4.3
DMA-N (mg %)	1.1	1.1	1.0	1.0	1.0	1.1	1.1	1.2	1.6
Hx (μmoles/g)	0.01	0.09	0.14	0.20	0.18	0.23	0.25	0.37	0.74
Overall acceptability[a/]	5.2	4.0	4.5	4.2	4.8	5.0	4.4	5.2	1.8
Change of sea water									
Whole									
TMA-N (mg %)	1.4	1.4	1.5	1.4	1.5	1.7	1.8	1.9	3.9
DMA-N (mg %)	1.1	1.0	1.9	1.0	1.0	1.1	1.1	1.3	1.6
Hx (μmoles/g)	0.01	0.07	0.12	0.18	0.64	0.30	0.28	0.33	0.77
Overall acceptability[a/]	5.2	4.0	4.2	3.8	4.5	4.5	3.6	4.0	3.0

TABLE 140. (continued)

	Storage time (days)								
	0	1	2	3.5	4.5	6.5	7.5	8.5	12.5
Gutted									
TMA-N (mg %)	1.4	1.4	1.5	1.4	1.5	1.6	1.7	1.9	3.6
DMA-N (mg %)	1.1	1.1	0.9	0.9	1.1	1.0	1.0	1.2	1.6
Hx (μmoles/g)	0.01	0.09	0.12	0.21	0.24	0.18	0.26	0.33	0.42
Overall acceptability[a]	5.2	4.0	5.1	4.2	4.4	3.6	2.8	4.0	2.8
Ambient									
Whole									
TMA-N (mg %)	1.4	1.6	13.1	--	--	--	--	--	--
DMA-N (mg %)	1.1	1.1	1.2	--	--	--	--	--	--
Hx (μmoles/g)	0.01	0.25	7.09	--	--	--	--	--	--
Overall acceptability[a]	5.2	3.5	1.0	--	--	--	--	--	--
Gutted									
TMA-N (mg %)	1.4	1.8	12.5	--	--	--	--	--	--
DMA-N (mg %)	0.01	1.1	1.1	--	--	--	--	--	--
Hx (μmoles/g)	0.01	0.21	6.98	--	--	--	--	--	--
Overall acceptability[a]	5.2	3.7	1.0	--	--	--	--	--	--

a/ 7 = like much; 6 = like; 5 = like slightly; 4 = neither like or dislike; 3 = dislike slightly; 2 = dislike; 1 = dislike much

Scallop, *Chlamys opercularis*, England

TABLE 141. Changes in spoilage indicators of whole queen scallops during iced storage (98)

	Storage time (days)							
	0	2	4	6	8	10	12	14
TVB-N (mg %)	12.0	12.5	15.0	17.5	18.7	30.0	38.0	47.5
Hx (μmoles/g)	0.1	1.1	1.6	2.1	2.3	2.5	2.6	2.7
OD ratio (%)	20.0	42.0	61.0	65.0	69.0	71.0	72.0	72.0

Scallop, *Hinnites multirugosis*, United States

TABLE 142. Changes in bacterial counts, TBA value, NH_3 concentration, and pH during the storage of uncooked purple-hinged rock scallop at -18°C (70)

	Storage time (days)				
	1	15	90	120	150
Bacterial counts (log no. org./g)					
Homogenized	2.92	3.46	2.77	3.54	2.59
Rinsed	3.84	1.62	1.14	0.30	1.25
NH_3 conc. (mM)	$1.0x10^{-1}$	$1.0x10^{-1}$	$9.0x10^{-2}$	$9.0x10^{-1}$	$1.0x10^{-4}$
pH	6.35	6.31	6.35	6.39	6.30
TBA no. [a/]	0.69	0.47	0.25	0.27	0.37

[a/] Measured as optical density at 532 nm

Scallop, Hinnites multirugosis, United States

TABLE 143. Changes in bacterial counts, resazurin reduction time, NH_3 concentration, pH, TBA value and Hx content of uncooked purple-hinged rock scallop at 5°C (70)

	Storage time (days)				
	1	3	6	9	14
Bacterial counts (log no. org./g)					
Homogenized	3.04	2.20	1.84	2.70	4.71
Rinsed	1.90	1.30	2.83	5.48	7.35
Reduction time (hr)					
Homogenized	4-6.5	4-6.5	4-6.5	4-6.5	2.5
Rinsed	7	7	7	7	7
NH_3 conc. (mM)	$6.0x10^{-3}$	$3.0x10^{-2}$	$3.0x10^{-2}$	$3.0x10^{-2}$	$3.0x10^{-2}$
pH	6.46	6.42	6.68	6.48	6.45
TBA no.[a]	0.34	0.17	0.20	0.18	0.17
Hx (mg %)	14	18	10	12	16

[a] Measured as optical density at 532 nm

Scallop, *Placopecten magellanicus*, Canada

TABLE 144. Changes in nucleotide concentrations and pH in the adductor muscle of scallop during postmortem storage in ice (49)

	Storage Time (Days)								
	0	2	4	6	8	10	13	16	18
ATP (μmoles/g)	7.2	0.5	0.2	--	--	--	--	--	--
ADP (μmoles/g)	1.1	0.5	0.3	0.3	0.3	0.3	0.3	0.4	0.4
AMP (μmoles/g)	0.1	3.5	2.3	1.6	1.0	1.3	1.6	1.1	0.8
HxR (μmoles/g)	0.1	0.7	1.4	1.7	1.3	1.0	1.5	0.9	0.8
Hx (μmoles/g)	0.1	0.4	1.2	1.9	2.5	2.5	2.8	3.3	3.3
pH	6.9	6.4	6.7	6.3	6.4	6.4	6.3	6.2	6.4

Scallop, Placopecten magellanicus, Canada

TABLE 145. Changes in G-6-P, octopine and Hx of quick-frozen sea scallop meat during iced storage (50)

	Storage time (days)								
	0	2	4	6	8	10	12	14	16
G-6-P (μmoles/g)	0.2	1.4	2.1	5.0	6.5	7.5	8.0	8.5	9.0
Octopine (μmoles/g)	0	5	22	34	36	36	36	36	36
Hx (μmoles/g)	0.0	0.1	0.3	1.2	1.7	1.9	2.1	2.3	2.9

Scallop, *Placopecten magellanicus*, Canada

TABLE 146. Changes in the concentration of AMP, HxR and Hx during the storage of quick-frozen scallop muscle at -26°C (51)

	Storage time (days)					
	0	1	2	3	5	7
AMP (μmoles/g)	4.70	3.85	1.90	0.90	0.80	0.45
HxR (μmoles/g)	0.40	1.40	2.15	3.00	2.60	2.35
Hx (μmoles/g)	0.05	0.50	1.05	1.40	2.05	2.00

Scallop, Placopecten magellanicus, Canada

TABLE 147. Changes in the concentration of AMP, HxR and Hx during the storage of quick-frozen scallop muscle at -5°C (51)

	Storage time (days)						
	0	14	21	28	35	42	70
AMP (μ moles/g)	4.70	0.40	0.37	--	0.33	--	--
HxR (μ moles/g)	0.40	1.30	1.15	1.05	0.85	1.05	0.95
Hx (μ moles/g)	0.05	3.15	4.00	4.40	4.00	4.55	4.85

Scallop, _Platinopecten caurinus_, United States

TABLE 148. Changes in quality parameters during the storage of scallop muscle at 0°C (46)

	Storage time (days)							
	0	1	2	4	6	8	10	14
pH	7.00	6.50	6.30	5.90	--	--	5.70	5.90
Glyc (g %)	3.30	--	--	--	2.10	--	2.20	2.0
ATP (μmoles/g)	6.25	0.91	0.63	0.28	0.18	0.47	0.04	--
ADP (μmoles/g)	1.08	1.60	1.50	0.99	0.96	0.75	0.04	--
AMP (μmoles/g)	0.08	1.90	1.80	1.80	1.90	2.00	1.90	--

Scallop, unspecified, Canada

TABLE 149. Changes in DMA-N, K_2CO_3-amine, KOH-amine value, FFA and odor classification of scallop stored at -5°C (22)

	Storage time (days)							
	0	6	13	21	33	56	83	203
DMA-N (mg %)	0.18	0.34	0.27	0.53	0.38	0.63	0.56	1.38
K_2CO_3-amine value[a/]	0.17	0.43	0.90	1.54	2.20	3.20	3.67	9.90
KOH-amine value	8.00	7.80	8.10	8.80	10.20	10.10	11.17	17.50
FFA (%)	7.8	10.41	10.20	15.50	12.09	17.20	19.30	21.5
TBA no.[b/]	0.40	0.43	0.76	0.47	0.73	0.70	0.59	--
Odor[c/]	F	F	F	SO	SO	O	O	TO

a/ Expressed as mg of TMA-N per 100g sample
b/ In mmoles of malonaldehyde per g sample
c/ F = fresh; SO = slightly off; O = off; TO = strong off

Shark, _Cestracion zygoena_, Japan

TABLE 150. Changes in TVB-N, urea, NH_3, TMAO-N, amine-N and pH in shark muscle during storage at 12.0 - 16.2°C (85)

	Storage time (days)							
	0	2	4	6	10	12	14	16
TVB-N (mg %)	20	40	600	800	900	980	1000	1000
Urea (mg %)	700	680	440	200	30	0	0	0
NH_3 (mg % N)	0	20	550	730	780	780	780	0
TMAO-N (mg %)	200	190	150	120	70	40	0	0
Amine-N (mg %)	0	10	50	60	110	180	200	200
pH	5.3	5.8	8.1	8.2	8.2	8.2	8.2	8.2

Shark, *Mustelus antarcticus*, Australia

TABLE 151. Changes in quality parameters during the iced storage of gummy shark muscle (104)

	Storage time (days)						
	0	4	6	8	10	12	14
Hx (μmoles/g)	--	0.50	0.48	0.75	0.79	0.92	1.00
NH_3 (mg %)	8.00	17.0	20.0	21.0	30.0	--	--
pH	--	5.50	5.55	5.60	5.80	6.00	6.45
TPC (log no.org./cm^2)	--	2.45	3.40	3.70	4.60	4.65	4.95

Shark, *Mustelus mustelus*, Mexico

TABLE 152. Changes in quality parameters of salted shark cakes during storage (34)

	Storage time (days)			
	0	30	60	90
TPC (log no. org./g)	6.2	6.6	6.9	6.3
Malonaldehyde ($1x10^{-2}$ moles/g)	2.0	2.2	2.1	2.3
Moisture content (%)	53.9	23.8	14.1	12.0

TABLE 153. Changes in TVN, APC, pH and mechanical shear of frozen *Macrobrachium* shrimp stored at -20° and -40°C (47).

	Storage time (days)				
	0	30	90	180	270
TVN (mg %)					
-20°C tails, raw	8.16	12.87	13.04	10.03	26.22
-40°C tails, raw	8.16	14.40	12.79	10.80	25.72
-20°C whole, raw	8.16	14.29	12.60	12.82	20.09
-20°C whole, cooked	8.16	10.17	11.78	11.85	17.72
APC[a] (22°C)					
-20°C tails, raw	5.89	5.74	--	5.23	4.67
APC[a] (35°C)					
-20°C tails, raw	5.81	5.74	--	5.25	4.54
pH					
-20°C tails, raw	6.30	6.60	6.60	6.43	6.28
-40°C tails, raw	6.30	6.75	6.60	6.45	6.30
-20°C whole, raw	6.30	6.58	6.50	6.45	--
-20°C whole, cooked	6.30	6.60	6.90	6.65	6.55
Shear force (lbs)					
-20°C tails, raw	560	570	780	650	780
-40°C tails, raw	560	560	570	670	710
-20°C whole, raw	410	340	650	460	500
-20°C whole, cooked	410	420	440	450	470

a/ log no. organisms per g sample

Shrimp, Pandalopsis dispar, United States

TABLE 154. Changes in freshness-index nucleotides of sidestripe shrimp during iced storage (94)

	Storage time (days)									
	0	0.5	1	2	3	4	5	7	10	14
AMP (μmoles/g)	1.8	2.5	1.6	1.3	1.2	1.1	--	--	--	--
IMP (μmoles/g)	1.9	2.6	2.8	1.6	1.4	1.2	1.0	0.8	0.5	0.4
Hx (μmoles/g)	0.0	0.10	0.20	0.30	0.40	0.50	0.60	0.70	0.75	0.80

Shrimp, Pandalus borealis, United States

TABLE 155. Changes in the concentration of nucleotides of pink shrimp during iced storage (94)

	Storage time (days)									
	0	.5	1	2	3	4	5	7	10	14
AMP (μmoles/g)	1.8	2.6	1.7	1.4	1.3	1.2	--	--	--	---
IMP (μmoles/g)	1.8	2.6	2.0	1.4	1.2	1.0	0.8	0.6	0.5	0.4
Hx (μmoles/g)	0.05	0.10	0.20	0.40	0.60	0.80	1.00	1.10	1.25	1.30

Shrimp, *Pandalus jordani*, United States

TABLE 156. Postmortem quality changes in uncooked iced Pacific shrimp at 1°- 2°C (42)

	Storage time (days)							
	0	1	2	3	4	5	7	8
Total N (μg/g)	26.4	20.4	18.6	17.0	19.9	17.0	17.2	17.1
NPN (μg/16 mg N)	5.1	4.7	4.7	4.1	5.0	3.9	3.3	3.5
pH	7.7	7.8	8.1	8.3	8.2	8.4	8.6	8.8
Tyr (μg/16 mg N)	60	80	100	130	125	130	150	135
TMAO (μg/16 mg N)	290	490	435	400	418	300	260	215
TMA (μg/16 mg N)	2.0	2.8	4.8	4.3	8.7	8.2	10.7	15.0
DMA (μg/16 mg N)	3.0	5.0	20.0	15.8	25.3	30.1	37.0	40.2
FA (μg/16 mg N)	8.0	10.0	27.0	25.8	37.5	35.4	40.2	35.0

Shrimp, *Pandalus plytyceros*, United States

TABLE 157. Changes in freshness-index nucleotides of spot shrimp during iced storage (94)

	Storage time (days)									
	0	.5	1	2	3	4	5	7	10	14
AMP (μ moles/g)	2.0	3.0	2.7	2.3	2.0	1.6	--	--	--	--
IMP (μ moles/g)	1.0	1.4	2.0	1.6	1.4	1.1	0.9	0.9	0.8	0.7
Hx (μ moles/g)	0.0	0.10	0.20	0.30	0.35	0.50	0.60	0.70	0.75	0.90

Shrimp, *Penaeus aztecus*, United States

TABLE 158. Changes in deaminase activity, TVN and bacterial counts of brown shrimp during iced storage (25)

	Storage time (days)						
	1	3	6	10	13	16	20
Adenosine deaminase activity (μ moles/g/min)	0.18	0.14	0.13	0.12	0.06	0.05	0.04
AMP deaminase activity (μ moles/g/min)	0.09	--	--	0.05	0.02	0.01	--
TVN (mg %)	26.0	18.5	18.0	17.8	19.0	29.5	37.7
Bacterial count (log no. org./g)	3.0	3.5	--	7.1	--	8.5	8.7

Shrimp, Penaeus aztecus, United States

TABLE 159. Changes in biochemical metabolites and pH during the iced storage (0°C) of brown shrimp muscle (41)

	Storage time (days)						
	0	1	2	3	5	7	10
pH	7.35	7.50	7.72	7.78	8.10	8.17	8.23
Glyc (mg %)	150	130	110	100	100	103	65.0
LA (mg %)	155	250	275	300	335	350	370
IMP (μmoles/g)	4.40	4.30	4.85	5.00	4.90	4.75	3.05
HxR (μmoles/g)	--	0.85	2.45	3.10	4.50	5.00	5.60
Hx (μmoles/g)	--	--	0.50	1.05	3.10	3.90	4.70
ATP (μmoles/g)	0.60	0.90	0.93	0.85	0.80	0.70	0.73

Shrimp, *Penaeus aztecus*, United States

TABLE 160. Changes in quality parameters during the iced storage of brown shrimp at 5°C (106)

	Storage time (days)			
	0	3	9	16
pH	7.70	8.00	8.20	8.40
TMA-N (mg %)	0.00	0.80	4.50	14.5
Indole (μg/g)	0.0	0.0	30.4	194.0
PAT	0.44	0.46	0.50	0.50
APC (log no. org./g)	4.65	4.70	4.72	7.50
AnPC (log no. org./g)	3.85	4.30	4.00	3.80

Shrimp, *Penaeus duorarum*, United States

TABLE 161. Changes in deaminase activity, TVN and bacterial counts of pink shrimp during iced storage (25)

	Storage time (days)							
	1	2	4	6	8	11	17	20
Adenosine deaminase activity (μ moles/g/min)	0.29	--	--	--	--	0.11	0.07	0.04
AMP deaminase activity (μ moles/g/min)	--	0.09	0.04	--	0.03	--	--	--
TVN (mg %)	19.0	--	17.5	17.4	15.8	16.0	--	34.6
Bacterial count (log no. org./g)	--	2.6	--	3.3	4.9	7.0	8.8	9.0

Shrimp, *Penaeus merguensis*, Pakistan

TABLE 162. Changes in Hx, IMP and sensory evaluation of shrimp during iced storage (40)

	Storage time (days)					
	0	4	8	12	16	20
Hx (μ moles/g)	0.1	0.3	0.7	1.0	1.9	4.5
IMP (μ moles/g)	5.8	1.5	0.5	0.1	--	--
Sensory Evaluation [a]						
Flavor	8.8	8.2	6.4	4.6	3.8	2.8
Texture	8.6	8.1	7.6	6.3	5.2	4.0
Appearance	8.7	8.3	7.5	6.5	4.8	3.8
Total [b]	8.7	8.3	7.1	5.7	4.6	3.5

a/ 9 = extremely good; 8 = very good; 7 = moderately good; 6 = slightly good; 5 = neither good nor poor; 4 = slightly poor; 3 = moderately poor; 2 = very poor; 1 = extremely poor

b/ Combined average of flavor, texture and appearance

Shrimp, *Penaeus setiferus*, United States

TABLE 163. Changes in some quality parameters of white shrimp stored in commercial ice (26)

	Storage time (days)						
	1	3	5	7	12	14	15
AA-N (mM %)	57.0	65.0	56.5	59.8	55.0	38.5	55.0
TVN (mg %)	34.2	43.6	46.3	53.8	65.0	53.1	68.8
TVN/AA-N (mg/mM)	0.60	0.67	0.82	0.90	1.18	1.38	1.25
Bacterial count (log no. org./g)	3.15	3.00	2.75	3.20	5.45	6.40	6.70

Shrimp, *Penaeus setiferus*, United States

TABLE 164. Changes in some quality parameters of white shrimp during iced storage at 5°C (101)

	Storage time (days)							
	0	2	5	7	12	16	20	24
Bacterial count (log no. org./g)	4.7	--	4.9	--	6.2	6.1	5.9	7.1
ERV	9.9	13.3	13.7	15.7	13.5	13.0	5.0	0.1
pH	7.3	--	--	7.4	8.2	8.2	8.0	8.6

TABLE 165. Changes in physical and chemical properties of fresh shrimp during iced storage (9)

	Storage time (days)							
	1	3	5	8	10	12	15	19
Lot no. 1[a]								
Drip (ml)	--	8.5	9.8	10.0	9.5	--	--	--
Folin-Ciocalteu nitrogen (mg/ml)	--	1.5	2.9	2.0	1.5	--	--	--
pH	--	7.6	7.8	8.0	8.2	--	--	--
Lot no. 2[b]								
Drip (ml)	8.0	7.8	7.5	6.8	9.5	5.3	5.5	6.0
TMA-N (μg/ml)	2.0	2.0	2.0	2.4	4.1	9.9	17.6	35.3
pH	7.6	7.7	7.8	7.9	8.0	8.2	8.4	8.6

a/ The experiment was started approximately 3 days after the shrimp were caught
b/ The experiment was started approximately 1 day after the shrimp were caught

Shrimp, unspecified, United States

TABLE 166. Chemical and microbiological changes in Gulf shrimp during storage in crushed ice (20)

	Storage time (days)				
	0	4	8	12	16
Colony units (30°C) (log no. org./g)	4.40	3.93	4.04	5.40	7.00
Colony units (log no. org./g)	--	--	--	5.38	7.08
DMA-N (mg %)	--	0.55	0.93	2.11	3.19
TMA-N (mg %)	--	0.54	0.70	1.82	2.33
AA-N (mg %)	--	45.13	59.64	70.88	95.18
TVA[a]	--	5.23	6.36	7.18	9.24
FFA (% oleic acid)	--	0.27	0.67	0.85	1.35
Peroxide no.	--	0.00	0.00	0.12	0.36

a/ ml of 0.01 N NaOH per 100 g of tissue.

TABLE 167. Changes in bacterial composition during the storage of Gulf shrimp in crushed ice (20)

	% of total population				
	Storage time (days)				
	0	4	8	12	16
Achromobacter	27.2	31.3	46.0	67.0	82.0
Bacillus	2.0	0.6	2.0	0.0	0.0
Flavobacterium	17.8	13.1	18.0	2.0	1.5
Micrococcus	33.6	23.0	5.7	0.8	0.0
Pseudomonas	19.2	26.5	28.0	30.1	16.5
Miscellaneous	0.2	0.5	0.3	0.1	0.0

Skipjack, *Katsowonus pelamis*, Mexico

TABLE 168. Changes in quality parameters of salted skipjack cakes during storage (34)

	Storage time (days)			
	0	30	60	90
TPC (log no. org./g)	4.2	3.7	2.9	--
Malonaldehyde (1×10^{-2} moles/g)	4.0	4.9	5.0	5.8
Moisture content (%)	40.9	15.2	6.4	6.1

Skipjack, unspecified, Japan

TABLE 169. Changes in the concentration of TMA-N and DMA-N during the iced and frozen storage of ordinary skipjack muscle (99)

	Storage time (days)				
	0	6	12	20	28
Iced					
TMA-N (mg %)	0.50	0.60	0.55	0.75	0.70
DMA-N (mg %)	0.20	0.20	0.20	0.30	0.25
Frozen					
TMA-N (mg %)	0.50	0.50	0.55	0.60	0.60
DMA-N (mg %)	0.20	0.20	0.20	0.20	0.20

Skipjack, unspecified, Japan

TABLE 170. Changes in the concentration of TMA-N, DMA-N and FA during the iced and frozen storage of bloody skipjack muscle (99)

	Storage time (days)				
	0	6	12	20	28
Iced					
TMA-N (mg %)	5.50	5.55	5.70	5.30	5.25
DMA-N (mg %)	2.00	2.10	2.10	1.90	2.00
FA (mg %)	--	--	1.00	1.10	1.15
Frozen					
TMA-N (mg %)	5.50	5.70	6.00	6.90	7.00
DMA-N (mg %)	2.00	2.60	4.10	6.00	6.20
FA (mg %)	--	1.40	2.20	3.10	3.20

Snapper, *Litjanus blackfordii*, United States

TABLE 171. Sensory scores and TMA-N content of tray-packed fresh red snapper fillets during storage at 2°C (107)

	Storage time (days)			
	1	3	4	6
Color[a]	7.2	7.0	7.0	7.0
Texture[a]	7.1	6.9	--	--
Flavor[a]	6.9	6.7	--	--
Odor[a]	7.2	6.6	3.0	1.0
General acceptance[a]	7.2	6.6	--	--
TMA-N (mg %)	1.87	2.96	--	16.8

[a] 9 = like extremely; 1 = dislike extremely.

Snapper, *Litjanus blackfordii*, United States

TABLE 172. Percent change of net weight, drip, gross weight and total moisture in tray-packed fresh red snapper fillets during storage at 2°C (107)

	%			
	Storage time (days)			
	2	4	6	8
Decrease in net weight	3.4	4.3	5.1	6.0
Increase in drip	3.3	3.9	4.5	5.1
Decrease in gross weight	0.1	0.2	0.5	0.7
Decrease in total moisture	0.1	0.2	0.3	0.5

Sole, _Eopsetta jordani_, United States

TABLE 173. Changes in TMA-N, TPC and Pseudomonas counts during the storage of vacuum-packed petrale sole fillets at 0.5°C (76)

	Storage time (days)				
	0	2	4	7	9
TMA-N (mg %)	0.85	0.90	6.30	42.0	40.0
TPC (log no. org./g)	4.15	4.45	4.30	5.65	5.50
Pseudomonas (log no. org./g)	3.15	2.40	4.00	5.50	5.30

Sole, _Eopsetta jordani_, United States

TABLE 174. Accumulation of Hx and nucleotide degradation in petrale sole stored at 0°C (90)

	Storage time (days)							
	1.5	2.5	3.5	4.5	6.5	7.5	8.5	9.5
Hx (mg/g)	0.24	0.38	0.48	0.68	0.74	0.78	0.84	0.90
Nucleotides[a/]	0.96	0.78	0.58	0.42	0.26	0.24	0.22	0.22

[a/] Optical density at 256 nm

Sole, Parophrys vetulus, United States

TABLE 175. Changes in TMA-N content of English sole fillets during frozen storage following pre-freezing iced storage (44)

	mg N %				
	Storage time (days)				
	0	1	42	84	168
0 Days on ice	0.44	0.47	0.19	0.77	0.32
1.3 Days on ice	0.46	0.73	0.27	0.14	0.38
3.2 Days on ice	0.71	0.80	0.49	0.95	1.24
5.4 Days on ice	0.77	1.00	0.88	1.15	1.19
8.9 Days on ice	1.73	1.43	1.70	1.26	1.55
12.3 Days on ice	4.68	4.05	3.00	--	--
15.0 Days on ice	6.50	6.43	6.20	5.51	6.81

Sole, *Parophrys vetulus*, United States

TABLE 176. Changes in IMP during the storage at of pre- and post-rigor English sole muscle at 1.1°C (61)

	Storage time (days)						
	0	1	2	3	4	5	6
Pre-rigor IMP (μ moles/g)	5.40	3.10	2.50	1.40	0.30	0.20	0.10
Post-rigor IMP (μ moles/g)	4.70	2.97	2.35	1.10	0.80	0.13	--

Sole, *Parophrys vetulus*, United States

TABLE 177. Changes in Hx, nucleotides, TVB-N and TMA-N in English sole during storage in melting ice (90)

	Storage time (days)								
	0	2	4	6	8	10	12	14	16
Hx (mg/g)	0.01	0.19	0.36	0.51	0.62	0.72	0.78	0.80	0.82
Nucleotides[a/]	1.20	0.70	0.44	0.32	0.22	0.18	0.15	0.14	0.13
TVB-N (mg %)	1.5	1.9	2.2	2.4	2.4	2.6	3.6	19.1	--
TMA-N (mg %)	--	0.1	0.1	0.1	0.2	0.5	1.5	17.6	--

[a/] Optical density at 146 nm

Sole, *Parophrys vetulus*, United States

TABLE 178. Changes in concentration of nucleotides during the storage of English sole muscle in ice (92)

	Storage time (days)			
	0	3	7	14
IMP (μmoles/g)	5.50	0.80	0.0	0.0
Hx (μmoles/g)	0.0	2.60	4.90	--
HxR (μmoles/g)	0.0	3.20	0.0	0.0

Sole, *Pleuronectes microcephalus*, Scotland

TABLE 179. Nucleotide degradation in lemon sole during iced storage (57)

	μ moles/g			
	Storage time (days)			
	0	8	16	24
ATP	0.2	0.1	0.0	0.0
ADP	0.7	0.3	0.3	0.4
AMP	0.8	0.2	0.2	0.1
NAD	0.1	0.0	0.0	0.0
IMP	4.5	2.1	0.6	0.0
HxR	0.0	0.0	0.0	0.0
Hx	0.0	1.9	2.6	1.5

Sole, *Pseudopleuronectes dignabilis*, United States

TABLE 180. Changes in acetoin, TVB-N and sensory evaluations of English sole stored at 34°F (45)

	Storage time (days)				
	0	2	4	6	8
Acetoin (mg %)	1	1	1	8	6
TVB-N (mg %)	10.3	9.5	12.1	31.3	42.9
Sensory evaluation[a]	A	A	A	A	NA

[a] A = acceptable; NA = not acceptable

Sucker, _Catustomus_ spp., United States.

TABLE 181. Changes in TBA number[a/] during the frozen storage of the Great Lake sucker at -18°C (31)

	Storage time (days)				
	0	30	90	180	360
TBA no.					
Fish portion					
minced flesh	0.5	0.8	1.2	1.8	3.3
loin	0.5	0.4	0.3	0.4	0.6
belly flap	0.6	0.7	1.0	3.0	4.2
ground eviscerated	0.7	1.8	3.8	5.2	5.7
Fish location					
Lake Huron	0.4	0.5	0.7	1.0	2.0
Lake Michigan	0.4	0.7	1.6	2.8	9.0
Lake Superior	0.5	1.5	1.7	2.4	3.0
Fish harvest					
May	0.4	0.5	0.8	1.0	1.8
August	0.2	0.4	0.5	0.4	1.3
December	0.4	0.5	0.6	1.4	2.3
February	0.5	0.6	1.1	1.5	2.8

a/ In milligrams of malonaldehyde per kg of muscle

Tarakibi, unspecified, New Zealand

TABLE 182. Changes in APC and TMA-N during iced storage of tarakibi fillets (96)

	Storage time (days)						
	1	4	6	8	14	18	20
APC (log.org./g)							
5°C Incubation	4.54	4.95	5.11	6.00	7.48	9.30	--
25°C Incubation	6.08	5.94	5.78	6.38	8.50	9.30	--
TMA-N (mg %)	0.47	--	--	0.50	0.62	2.19	3.75

Trout, *Salmo gairdneri*, United States

TABLE 183. Changes in IMP during the storage of pre- and post-rigor rainbow trout muscle at 1.1°C (61)

	Storage time (days)						
	0	1	2	3	4	5	6
Pre-rigor IMP (μ moles/g)	5.95	5.25	4.55	3.40	2.95	2.50	2.05
Post-rigor IMP (μ moles/g)	5.45	4.50	4.05	3.50	3.00	2.55	1.75

Trout, *Salmo gairdneri irideus*, Korea

TABLE 184. Changes in TVB-N and K value during the storage of rainbow trout muscle at 5°C (63)

	Storage time (days)				
	0	1	2	4	6
TVB-N (mg %)	6.73	11.5	12.0	12.1	13.6
K value (%)	3.20	45.1	57.8	79.7	83.4

Trout, unspecified, United States

TABLE 185. Changes in average TBA numbers and TPC during the storage of trout fillets at 1°C (71)

	Storage time (days)						
	1	2	3	4	5	6	7
TBA no.[a]	2.80	1.50	1.90	1.30	2.40	1.50	1.40
TPC (log no.org./g)	4.32	4.89	5.70	5.65	--	6.92	6.72

[a] Measured as mg of malonaldehyde per kg of tissue

Tuna, unspecified, Japan

TABLE 186. Changes in the concentration of TMA-N, DMA-N and FA during the iced and frozen storage (-6°C) of ordinary bigeye tuna muscle (99)

	Storage time (days)				
	0	10	12	17	25
Iced					
TMA-N (mg %)	0.40	0.45	--	0.70	1.45
DMA-N (mg %)	0.25	0.30	--	0.30	0.30
FA (mg %)	--	--	--	--	--
Frozen					
TMA-N (mg %)	0.40	--	0.45	0.45	0.55
DMA-N (mg %)	0.20	--	0.25	0.30	0.30
FA (mg %)	--	--	--	--	--

Tuna, unspecified, Japan

TABLE 187. Changes in the concentration of TMA-N, DMA-N and FA during the iced and frozen storage (-6°C) of bloody bigeye tuna muscle (99)

	Storage time (days)				
	0	10	12	17	25
Iced					
TMA-N (mg %)	4.00	6.00	--	6.55	7.15
DMA-N (mg %)	1.80	2.20	--	2.50	2.75
FA (mg %)	--	1.10	--	1.10	0.95
Frozen					
TMA-N (mg %)	4.00	--	6.10	6.90	7.00
DMA-N (mg %)	1.80	--	3.05	3.85	4.65
FA (mg %)	--	--	1.20	1.75	2.25

Whitefish, *Clupea clupeiformis*, United States

TABLE 188. Changes in quality parameters during the frozen storage of whitefish muscle at -10°C (2)

	Storage time (days)				
	0	21	42	94	112
TEP (% total muscle)	93.3	85.9	79.6	56.8	42.6
pH thaw drip	6.20	6.20	6.35	6.60	6.70
Total lipid (g %)	3.84	3.55	3.77	3.82	3.80
PV (meq thiosulfate/kg lipid)	13.3	19.6	36.3	68.3	26.2

Whitefish, *Clupea clupeiformis*, United States

TABLE 189. Changes in Hx content during the refrigerated storage of whitefish muscle at different concentrations of carbon dioxide (105)

	μmoles/g			
	Storage time (days)			
	0	4	7	11
CO_2 (%)				
30	0.25	0.57	0.82	1.21
42	0.25	0.65	0.76	0.85
100	0.25	0.23	0.32	0.20

Whiting, *Merluccius bilinearis*, United States

TABLE 190. Changes in DMA-N content of minced and fillets during storage at various temperatures (64)

	mg %								
	Storage time (days)								
	0	61	122	183	243	304	365	426	487
Minced									
20°F	0	6	11.5	16	--	--	--	--	--
5°F	0	2	3	7	6	6.3	8	8.2	8.2
-5°F	0	0.5	1.0	2.0	2.2	3.0	3.0	4.2	5.
-22°F	0	0	0	0.4	0.3	0.3	0.8	0.7	0.8
Fillet									
20°F	0	4	9	11	--	--	--	--	--
5°F	0	1.2	1.0	3.0	3.0	2.5	--	5.0	4.6
-5°F	0	--	0.1	0.3	0.2	0.2	0.8	0.7	0.8

Whiting, *Merluccius productus*, United States

TABLE 191. Formation of DMA-N during the storage of Pacific whiting homogenates at 0° and -5°C (93)

	Storage time (days)								
	0	3	5	10	15	20	30	60	180
Stored at 0°C									
DMA-N (mg %)	2.60	4.00	4.30	4.60	7.45	7.65	--	--	--
Stored at -5°C									
DMA-N (mg %)	2.60	--	--	5.00	10.0	--	12.5	15.0	20.0

Whiting, Micromesistius poutassou, England

TABLE 192. Changes in TMA-N, DMA-N and Hx of blue whiting caught in February, during storage at chilled and ambient temperatures (87)

	Storage time (days)						
	0	2	4	6	8	10	11
Iced							
Whole							
TMA-N (mg %)	0.3	--	--	--	1.7	3.0	5.0
DMA-N (mg %)	0.58	0.38	0.73	1.11	1.43	1.78	3.00
Hx (μ moles/g)	0.48	0.55	0.60	0.82	1.09	1.65	1.40
Gutted							
TMA-N (mg %)	0.3	--	--	--	0.5	1.6	2.6
DMA-N (mg %)	0.58	0.55	0.45	0.95	0.97	0.98	1.12
Hx (μ moles/g)	0.48	0.53	0.67	0.92	1.07	1.21	1.85
Chilled Sea Water							
Whole							
TMA-N (mg %)	0.3	0.4	1.0	1.7	2.5	7.0	11.6
DMA-N (mg %)	0.58	0.37	0.60	1.79	2.28	3.26	3.32
Hx (μmoles/g)	0.48	0.54	0.46	0.81	0.94	1.92	2.46
Gutted							
TMA-N (mg %)	0.3	--	0.5	2.3	6.0	12.9	17.8
DMAN (mg %)	0.58	0.29	0.62	0.88	1.99	2.38	3.18
Hx (μ moles/g)	0.48	0.45	0.49	0.73	1.60	1.94	2.91

TABLE 192. (continued)

	Storage time (days)						
	0	2	5	6	8	10	11
Change of Sea Water							
Whole							
TMA-N (mg %)	0.3	--	--	--	1.6	5.0	7.5
DMA-N (mg %)	0.58	0.20	0.45	1.47	2.09	4.91	2.92
Hx (μ moles/g)	0.48	0.38	0.44	0.51	0.58	0.94	1.45
Ambient (9–13°C)							
Whole							
TMA-N (mg %)	0.35	2.91	--	--	--	--	--
DMA-N (mg %)	0.58	2.79	--	--	--	--	--
Hx (μ moles/g)	0.48	3.12	--	--	--	--	--

Whiting, *Micromesistius poutassou*, England

TABLE 193. Changes in TMA-N, DMA-N and Hx of blue whiting caught in March, during storage at chilled and ambient temperatures (87)

	Storage time (days)						
	0	2	4	6	8	10	11
Iced							
Whole							
TMA-N (mg %)	0.0	0.0	0.0	0.6	1.7	4.6	6.2
DMA-N (mg %)	0.92	1.50	1.91	2.31	2.75	4.63	3.68
Hx (μ moles/g)	0.83	1.29	0.63	2.11	1.70	2.08	1.68
Gutted							
TMA-N (mg %)	0.0	0.2	0.5	0.7	1.7	4.3	7.4
DMA-N (mg %)	0.92	1.14	1.95	4.01	3.65	4.95	5.65
Hx (μ moles/g)	0.83	0.40	0.60	0.61	1.98	1.82	1.95
Chilled Sea Water							
Whole							
TMA-N (mg %)	0.0	0.0	0.0	0.7	6.5	19.0	23.0
DMA-N (mg %)	0.92	1.71	3.40	2.61	5.88	7.30	6.02
Hx (μ moles/g)	0.83	0.35	0.43	1.03	1.26	1.07	1.71
Gutted							
TMA-N (mg %)	0.0	0.0	0.0	0.5	4.5	16.7	20.0
DMA-N (mg %)	0.92	1.31	1.36	2.28	2.99	3.65	3.78
Hx (μ moles/g)	0.83	0.47	0.47	0.93	1.22	1.35	1.52

TABLE 193. (continued)

	Storage time (days)						
	0	2	4	6	8	10	11
Change of Sea Water							
Whole							
TMA-N (mg %)	0.0	0.0	0.0	0.6	1.0	1.8	2.3
DMA-N (mg %)	0.92	1.02	1.67	2.14	3.91	3.59	3.55
Hx (μ moles/g)	0.83	0.22	0.35	0.36	1.04	0.95	0.84
Gutted							
TMA-N (mg %)	0.0	0.0	0.0	0.2	0.6	1.0	1.2
DMA-N (mg %)	0.92	1.03	1.40	2.10	2.64	2.98	2.08
Hx (μ moles/g)	0.83	0.20	0.39	0.30	0.62	0.66	0.46
Ambient							
Whole							
TMA-N (mg %)	0.06	1.98	--	--	--	--	--
DMA-N (mg %)	0.92	3.93	--	--	--	--	--
Hx (μ moles/g)	0.83	1.41	--	--	--	--	--

Whiting, *Micromesistius poutassou*, England

TABLE 194. Changes in TMA-N, DMA-N and Hx of blue whiting caught in April, during storage at chilled and ambient temperatures (87)

	Storage time (days)						
	0	2	4	6	8	10	11
Iced							
Whole							
TMA-N (mg %)	0.0	0.4	0.5	0.8	2.0	8.0	13.0
DMA-N (mg %)	1.86	2.12	2.32	3.36	4.13	9.05	11.58
Hx (μ moles/g)	0.58	0.69	0.93	1.23	1.40	1.62	2.10
Gutted							
TMA-N (mg %)	0.0	0.4	1.0	1.8	14.5	11.0	15.9
DMA-N (mg %)	2.08	2.33	3.42	3.95	5.64	6.33	8.42
Hx (μ moles/g)	0.58	0.79	0.96	1.26	1.56	2.32	2.59
Chilled Sea Water							
Whole							
TMA-N (mg %)	0.0	0.4	0.4	1.5	7.3	17.9	23.9
DMA-N (mg %)	1.86	2.59	2.47	5.29	4.76	7.49	7.01
Hx (μ moles/g)	0.58	0.54	0.66	0.69	1.02	1.23	1.56
Gutted							
TMA-N (mg %)	0.0	0.4	0.4	0.6	2.9	11.8	17.8
DMA-N (mg %)	2.08	2.05	3.05	3.59	4.43	5.29	6.63
Hx (μ moles/g)	0.58	0.68	0.68	0.85	0.99	1.49	1.84

TABLE 194. (continued)

	Storage time (days)						
	0	2	4	6	8	10	11
Change of Sea Water							
Whole							
TMA-N (mg %)	0.0	0.4	0.4	0.6	1.6	4.0	6.8
DMA-N (mg %)	1.86	1.57	2.21	2.25	3.84	6.66	6.41
Hx (μ moles/g)	0.58	0.52	0.58	0.58	0.62	0.67	1.23
Gutted							
TMA-N (mg %)	0.0	0.4	0.4	0.6	1.3	3.0	4.7
DMA-N (mg %)	2.08	2.39	2.17	2.89	3.28	4.49	5.16
Hx (μ moles/g)	0.58	0.59	0.47	0.47	0.59	0.60	0.82
Ambient (10–14°C)							
Whole							
TMA-N (mg %)	0.57	1.80	--	--	--	--	--
DMA-N (mg %)	1.86	4.76	--	--	--	--	--
Hx (μ moles/g)	0.58	1.24	--	--	--	--	--

Whiting, *Micromesistius poutassou*, England

TABLE 195. Changes in relative fluorescence[a/] intensity of blue whiting during frozen storage (30)

	Storage time (days)							
	0	20	40	60	80	100	120	140
-5°C	80	190	240	320	350	--	--	--
-10°C	80	140	160	230	290	320	380	410
-50°C	80	100	120	140	170	160	240	250
-20°C	80	80	90	110	120	110	120	140

a/ Arbitrary units

Yellowtail, *Seriola quinqueradiata*, Japan

TABLE 196. Changes in TVB-N, pH, NPN and creatine in yellowtail during iced storage (79)

	Storage time (days)				
	0	10	20	30	40
Dark muscle					
TVB-N (mg %)	8	18	20	23	44
pH	6.20	6.19	6.10	6.05	--
NPN (mg %)	250	280	290	290	--
Creatine (mg %)	170	170	170	170	--
White muscle					
TVB-N (mg %)	11	20	21	21	21
pH	5.86	5.80	5.80	5.78	5.75
NPN (mg %)	600	600	600	600	600
Creatine (mg %)	520	520	520	525	530

Unspecified, *Ophicephalus* spp., India

TABLE 197. Changes in muscle fractions during the iced storage of *Ophicephalus* species (7)

	Storage time (days)							
	1	4	5	6	9	11	13	15
SSP[a]	90.7	93.1	81.9	91.1	91.4	91.5	93.8	88.4
NPN[b]	10.2	9.6	9.8	10.1	9.7	10.1	9.5	9.5
AMN[a]	1.95	1.75	26.75	6.33	5.70	3.30	3.13	3.05
SPN[a]	26.5	28.6	29.5	30.0	29.2	28.9	32.0	31.1
TAN[a]	30.8	31.8	30.8	34.5	29.4	30.2	30.8	32.2
SAN[a]	19.2	21.7	15.0	18.8	17.2	19.9	28.8	20.9
SAA[a]	19.3	21.7	13.7	14.2	12.2	16.0	20.8	16.7
SIA[a]	--	--	1.25	4.60	5.00	3.90	8.00	4.20

a/ % total protein nitrogen
b/ % total nitrogen

Unspecified, *Prochilodus scrofa*, Brazil

TABLE 198. TVN and TBA values in ice stored fish with and without premortem struggling (69)

	Storage time (days)						
	5	4	7	11	14	18	21
TVN (mg %)							
Without struggling							
Whole	23.4	24.2	27.6	24.6	25.8	2.0	33.0
With struggling							
Whole	28.0	24.1	22.2	26.8	27.4	30.5	30.9
Eviscerated	27.2	25.7	24.2	27.1	29.0	32.9	36.0
TBA no. (mg malonaldehyde/Kg)							
Without struggling							
Whole	0.10	0.09	0.14	0.19	0.20	0.26	0.25
With struggling							
Whole	0.18	0.25	0.22	0.20	0.27	0.26	0.25
Eviscerated	0.19	0.28	0.24	0.24	0.37	0.38	0.35

Index to Tables*

*Numbers refer to table numbers, not page numbers.

REFERENCES

1. Angel, S., Basker, D., Kanner, J., and Juven, B. (1981). Assessment of shelf-life of freshwater prawns stored at 0°C. *J. Food Technol.* *16*, 357-366.
2. Awad, A., Powrie, W., and Fennema, O. (1969). Deterioration of freshwater muscle during frozen storage at -10°C. *J. Food Sci.* *34*, 1-9.
3. Babbitt, J. K. (1981). Measuring fresh fish quality. *Mar. Fish. Rev.* *43*, 28.
4. Babbitt, J. K., Crawford, D., and Law, D. (1972). Decomposition of trimethylamine oxide and changes in protein extractability during frozen storage of minced and intact hake (*Merluccius productus*) muscle. *J. Agric. Food Chem.* *20*, 1052-1054.
5. Babbitt, J. K., Law, D., and Crawford, D. (1973). Blueing discoloration in canned crab meat (*Cancer magister*). *J. Food Sci.* *38*, 1101-1103.
6. Babbitt, J. K., Koury, B., Groniger, H., and Spinelli, J. (1984). Observations on reprocessing frozen Alaska pollock (*Theragra chalcogramma*). *J. Food Sci.* *49*, 323-326.
7. Baliga, B. R., Moorjani, M. N., and Lahiry, N. L. (1969). Fractionation of muscle proteins of fresh water fish and changes during iced storage. *J. Food Sci.* *34*, 597-599.
8. Barnett, H. J., Nelson, R., Hunter, P., Baver, S., and Groniger, H. (1971). Studies on the use of carbon dioxide dissolved in refrigerated brine for the preservation of whole fish. *Fish. Bull.* *69*, 433-442.
9. Bethea, S., and Ambrose, M. E. (1961). Physical and chemical properties of shrimp drip as indices of quality. *Commer. Fish. Rev.* *23*, 9-14.
10. Bilinsky, E., Jonas, R. E., and Lau, Y. C. (1978). Chill storage and development of rancidity in frozen pacific herring (*Clupea harengus pallasi*). *J. Fish. Res. Board Can.* *35*, 473-477.
11. Bilinsky, E., Jonas, E. E., and Peters, M. D. (1983). Factors controlling the deterioration of the spiny dogfish (*Squalus acanthias*), during iced storage. *J. Food Sci.* *48*, 808-812.
12. Bossey, S. S. (1981). Some chemical and organoleptic assessment studies on the storage characteristics of the West African long croaker (*Pseudotolithus typus*). *Food Chem.* *7*, 169-174.
13. Botta, J. R., and Richards, J. F. (1973a). Thiobarbituric acid value, total long-cha'n free fatty acids,

and flavor of Pacific halibut (*Hippoglossus stenolepis* and Chinook salmon (*Oncorhynchus tshawytscha*) frozen at sea. *J. Fish. Res. Board Can. 30*, 63-69.

14. Botta, J. R., and Richards, J. F. (1973b). Flesh pH, color, thaw drip, and mineral concentration of Pacific halibut (*Hippoglossus stenolepis*) and Chinook salmon (*Onchorhynchus tshawytscha*) frozen at sea. *J. Fish. Res. Board Can. 30*, 71-77.
15. Botta, J. R., Noonan, P. B., and Laudes, J. T. (1978). Chemical and sensory analysis of ungutted offshore (Nonspawning) capelin (*Mallotus villosus*) stored in ice. *J. Fish. Res. Board Can. 35*, 976-980.
16. Botta, J. R., Downey, A. P., Lauder, J. T., and O'Neill, M. (1982). Chemical and sensory assessment of roundnose grenadier (*Macrourus rupestris*) subjected to long term frozen storage. *J. Food Sci. 47*, 1670-1674.
17. Botta, J. R., Lauder, J. T., Downey, A. P., and Saint, W. (1983). Chemical and sensory assessment of nonsprawning capelin (*Mallotus villosus*) subjected to long term frozen storage. *J. Food Sci. 48*, 1512-1515, 1536.
18. Bremer, H. A. (1980). Processing and Freezing of the flesh of the blue grenadier (*Macruronus novaezelandiae*). *Food Technol. Aust. 32*, 385-390.
19. Brooks, J. D., and Harvie, R. E. (1981). Quality changes during storage of the green-lipped mussel, *Perna canaliculus*. *Food Technol. Aust. 33*, 490-492.
20. Campbell, L. L., and Williams, O. B. (1952). The bacteriology of Gulf Coast shrimp. IV. Bacteriological, chemical, and organoleptic changes with iced storage. *Food Technol. 5*, 492-495.
21. Castell, C. H., Bishop, D. M., and Neal, W. (1968). Production of trimethylamine in frozen cod muscle. *J. Fish. Res. Board Can. 25*, 921-933.
22. Castell, C. H., Neal, W., and Smith, B. (1970). Formation of dimethylamine in stored frozen sea fish. *J. Fish. Res. Board Can. 27*(10), 1685-1690.
23. Castell, C. H., Smith, B., and Neal, W. (1971). Production of dimethylamine in muscle of several species of gadoid fish during frozen storage, especially in relation to presence of dark muscle. *J. Fish. Res. Board Can. 28*, 1-5.
24. Castell, C. H., Neal, W., and Sale, J. (1973). Comparsison of changes in trimethylamine, dimethylamine, and extractable protein in iced and frozen gadoid fillets. *J. Fish. Res. Board Can. 30*, 1246-1248.
25. Cheuk, W. L., Finne, G., and Nickelson, R. (1979). Stability of adenosine deaminase and adenosine monophosphate deaminase during iced storage of pink and brown

shrimp from the Gulf of Mexico. *J. Food Sci. 44*, 1625-1628.

26. Cobb, B. F., and Vanderzant, C. (1975). Development of a chemical test for shrimp quality. *J. Food Sci. 40*, 121-124.
27. Collins, J., Reppond, K., and Bullard, F. (1979). Black rockfish (*Sebastes melanops*): Changes in physical, chemical, and sensory properties when held in ice and carbon dioxide modified refrigerated seawater. *Fish. Bull. 77*, 865-870.
28. Curran, C. A., Nicolaides, L., and Poulter, R. (1980). Spoilage of fish from Hong Kong at different storage temperatures. 1. Quality changes in gold-lined sea bream (*Rhabdosargus sarba*) during storage at 0° (in ice) and 10°C. *Trop. Sci. 22*, 367-382.
29. Damoglou, A. P. (1980). Quality assessment. A comparison of different methods of freshness assessment of herring. *In* "Advances in Fish Science and Technology" (J. Connell, ed.), pp. 394-417. Fishing News Books Ltd., Surrey, England.
30. Davis, H. K. (1982). Fluorescence of fish muscle: Description and measurement of changes occurring during frozen storage. *J. Sci. Food Agric. 33*, 1135-1142.
31. Dawson, L. E., Uebersax, K. L., and Uebersax, M. A. (1978). Stability of freshwater sucker (*Catostomus* spp.) flesh during frozen storage. *J. Fish. Res. Board Can. 35*, 253-257.
32. Debevere, J. M., and Voets, J. P. (1972). Influence of some preservatives on the quality of prepacked cod fillets in relation to the oxygen permeability of the film. *J. Appl. Bacteriol. 35*, 351-356.
33. Debevere, J. M., and Voets, J. P. (1974). Microbiological changes in prepacked plaice in relation to the oxygen permeability of the film. *Lebensm.-Wiss. Technol. 7*, 73-75.
34. Del Valle, F. R., Hinojosa, J., Barrera, D., and De la Mora, R. (1973). Bacterial counts and rancidity estimates of stored quick-salted fish cakes. *J. Food Sci. 38*, 580-582.
35. Deng, J. C. (1978). Effect of iced storage on free fatty acid production and lipid oxidation in mullet muscle. *J. Food Sci. 43*, 337-340.
36. Dingle, J. R. (1978). Quality deterioration in the flesh of *Merluccius* species and other gadidae caused by the formation of formaldehyde during frozen storage. *FAO Fish. Rep. 203*, 70-83.
37. Dingle, J. R., and Hines, J. A. (1971). Degradation of inosine 5'-monophosphate in the skeletal muscle of

several North Atlantic Fishes. *J. Fish. Res. Board Can. 28*, 1125-1131.
38. Dyer, W. J., Morton, M. L., Fraser, D., and Bligh, E. G. (1956). Storage of frozen rosefish fillets. *J. Fish. Res. Board Can. 13*, 569-579.
39. Ehira, S., and Uchiyama, H. (1974). Freshness-lowering rates of cod and sea bream viewed from changes in bacterial count, total-volatile base- and trimethylamine-nitrogen, and ATP related compounds. *Bull. Jpn. Soc. Sci. Fish. 40*, 479-487.
40. Fatima, R., Farooqui, B., and Qadri, R. B. (1981). Inosine monophosphate and hypoxanthine as indices of quality of shrimp (*Penaeus merguensis*). *J. Food Sci. 46*, 1125-1127.
41. Flick, G. J., and Lovell, R. T. (1972). Post-mortem biochemical changes in the muscle of Gulf shrimp, (*Penaeus aztecus*). *J. Food Sci. 37*, 609-611.
42. Flores, S. C., and Crawford, D. L. (1973). Postmortem quality changes in iced pacific shrimp (*Pandalus jordani*). *J. Food Sci. 38*, 575-579.
43. Ghadi, S. V., Gore, M. S., and Kumta, U. S. (1974). Storage stability of "shell-on" semi-dried tropical shrimps (*Penaeus indicus, Metapenaeus affinis, Penaeus stylifera*). *Lebensm.-Wiss. Technol. 7*, 229-233.
44. Good, C. M., and Stern, J. A. (1955). Effects of iced and frozen storage upon the trimethylamine content of flounder (*Parophrys vetulus*) muscle. *Food Technol. 9*, 327-331.
45. Groniger, H. S. (1961). Formation of acetoin in cold and other bottom-fish fillets during refrigerated storage. *Food Technol. 15*, 10-12.
46. Groniger, H. S., and Brandt, K. R. (1970). Some observations on the quality of the weathervane scallop. *J. Milk Food Technol. 33*, 232-236.
47. Hale, M. B., and Waters, M. E. (1981). Frozen storage stability of whole and headless freshwater prawns, *Macrobrachium rosenbergii*. *Mar. Fish. Rev. 43*, 18-21.
48. Hiltz, D. F., Simpson, S. C., and Dyer, W. J. (1968). Very rapid accumulation of hypoxanthine in the muscle of redfish stored in ice. *J. Fish. Res. Board Can. 25*, 817-821.
49. Hiltz, D. F., and Dyer, W. J. (1970). Principal acid-soluble nucleotides in adductor muscle of the scallop, *Placopecten magellanicus*, and their degradation during post mortem storage in ice. *J. Fish. Res. Board Can. 27*, 83-92.
50. Hiltz, D. F., and Dyer, W. J. (1973). Hexose monophosphate accumulation and related metabolic changes in

unfrozen and thawed adductor muscle of the sea scallop (*Placopecten magellanicus*). *J. Fish. Res. Board Can.* *30*, 45-52.

51. Hiltz, D. F., Bishop, L. J., and Dyer, W. J. (1974). Accelerated nucleotide degradation and glycolysis during warming to a subsequent storage at -5°C of prerigor, quick-frozen adductor muscle of the sea scallop (*Placopecten magellanicus*). *J. Fish. Res. Board Can.* *31*, 1181-1187.
52. Husaini, S. A., and Alm, F. (1955). Denaturation of Proteins of egg white and fish and its relation to the liberation of sulfhydryl groups on frozen storage. *Food Res* *20*, 264-272.
53. Huss, H. H., Dalsgaard, D., Hansen, L., Ladefoged, L., Pedersen, A., and Zittan, L. (1974). The influence of hygiene in catch handling on the storage life of iced cod and plaice. *J. Food Technol.* *9*, 213-221.
54. Jensen, M. H., Petersen, A., Røge, E. H., and Jepsen, A. (1980). *In*. "Advances in Fish Science and Technology" (J. Connell, ed.), pp. 294-303. Fishing News Books Ltd., Surrey, England.
55. Jhaveri, S. N., Leu, S., and Constantinides, S. M. (1982). Atlantic mackerel (*Scomber scombrus, L.*): Shelf life in ice. *J. Food Sci.* *47*, 1808-1810.
56. Jones, N. R., and Murray, J. (1962). Degradation of adenine- and hypoxanthine-nucleotide in the muscle of chill-stored trawled cod (*Gadus callarias*). *J. Sci. Food Agric.* *13*, 475-480.
57. Kassemsarn, B., Perez, B. S., Murray, J., and Jones, N. R. (1963). Nucleotide degradation in the muscle of iced haddock (*Gadus aeglefinus*), lemon sole (*Pleuronectes microcepholus*), and plaice (*Pleuronectes platessa*). *J. Food Sci.* *28*, 28-37.
58. Keay, J. N., Rattagool, P., and Hardy, R. (1972). Chub mackerel of thailand (*Rastrelliger neglectus, van Kampen*): A short study of its chemical composition, cold storage and canning properties. *J. Sci. Food Agric.* *23*, 1359-1368.
59. Kelleher, S. D., Buck, E., Hultin, H., Parkin, K. L., Licciardello, J., and Damon, R. (1981). Chemical and physical changes in red hake blocks during freezing storage. *J. Food Sci.* *47*, 65-70.
60. Kelly, T. R. (1969). Quality of frozen cod and limiting factors on its shelf life. *J. Food Technol.* *4*, 95-103.
61. Kemp, B., and Spinelli, J. (1969). Comparative rates of IMP degradation in unfrozen and frozen-and-thawed fish. *J. Food Sci.* *34*, 132-135.

62. Lee, C. M., and Toledo, R. T. (1984). Comparison of shelf-life and quality of mulle stored at zero and subzero temperature. *J. Food Sci. 49*, 317-321.
63. Lee, E., Ohshima, T., and Koizumi, C. (1982). High performance liquid chromatographic determination of K value as an index of freshness of fish. *Bull. Jpn. Soc. Sci. Fish. 48*, 255.
64. Licciardello, J. J., Ravesi, E. M., and Allsup, M. G. (1980). Frozen storage characteristics of whiting blocks. *Mar. Fish. Rev. 42*, 55-60.
65. Licciardello, J. J., Ravesi, E. M., Lundstrom, R. C., Whilhelm, K. A., Correia, F. F., and Allsup, M. G. (1982). Time-temperature tolerance and physicochemical quality tests for frozen red hake. *J. Food Qual. 5*(3), 215-234.
66. Liuzzo, J. A., Lagarde, S., Grodner, R., and Novak, A. (1975). A total reducing substance test for ascertaining oyster quality. *J. Food Sci. 40*, 125-128.
67. Loaharanu, P., and Lopez, A. (1970). Bacteriological and shelf-life characteristics of canned, pasteurized crab cake mix. *Appl. Microbiol. 19*, 734-741.
68. Lovern, J. A., and Olley, J. (1962). Inhibition and promotion of post-mortem lipid hydrolysis in the flesh of fish. *J. Food Sci. 27*, 551-559.
69. Maia, E. L., Rodriguez-Amaya, D., and Moraes, M. A. (1983). Sensory and chemical evaluation of the keeping quality for the Brazilian freshwater fish, *Prochilodus scrofa*, in ice storage. *J. Food Sci. 48*, 1075-1077.
70. Maxwell-Miller, G., Josephson, R. V., Spindler, A. A., Holloway-Thomas, D., Avery, M., and Phleger, C. F. (1982). Chilled (5°C) and frozen (-18°C) storage stability of the purple-hinge rock scallop, *Hinnites multirugosus Gale*. *J. Food Sci. 47*, 1654-1661.
71. Mendenhall, V. T. (1972). Oxidative rancidity in raw-fish fillets harvested from the Gulf of Mexico. *J. Food Sci. 37*, 547-550.
72. Miller, A., Scanlan, R. A., Lee, J. S., and Libbey, L. M. (1972). Voltaile compounds produced in ground muscle tissue of canary rockfish (*Sebastes pinniger*) stored on ice. *J. Fish. Res. Board Can. 29*, 1125-1129.
73. Nazir, D. J., and Magas, N. G. (1963). Biochemical changes in fish muscle during rigor mortis. *J. Food Sci. 28*, 1-7.
74. Okuzumi, M., Shimizu, M., and Matsumoto, A. (1981). Spoilage and bacterial flora of thawed fishes, partially frozen at -3°C, during chill storage. *Bull. Jpn. Soc. Sci. Fish. 47*, 239-242.

75. Parkin, K. L., and Hultin, H. (1982). Some factors influencing the production of dimethylamine and formaldehyde in minced and intact red hake muscle. *J. Food Process. Preserv. 6*, 73-97.
76. Pelroy, G. A., and Seman, J. P. (1969). Effect of EDTA treatment on spoilage characteristics of petrale sole and ocean perch. *J. Fish. Res. Board Can. 26*, 2651-2657.
77. Ritchie, A. H., and Mackie, I. M. (1980). The formation of diamines and polyamines during storage of mackerel (*Scomber scombrus*). *In* "Advances in Fish Science and Technology" (J. Connell, ed.), pp. 489-494. Fishing News Books Ltd., Surrey, England.
78. Ryder, J. M., Buisson, D. H., Scott, D. N., and Fletcher, G. C. (1984). Storage of New Zealand Jack mackerel (*Trachurus novaezelandieae*) in ice: Chemical, microbiological, and sensory assessment. *J. Food Sci. 49*, 1453-1456.
79. Sakaguchi, M., Murata, M., and Kawai, A. (1982). Changes in free amino acids and creative contents in yellowtail (*Seriola quinqueradiata*) muscle during ice storage. *J. Food Sci. 47*, 1661-1666.
80. Seki, N., Oogane, Y., and Watanabe, T. (1980). Changes in ATPase activities and other properties of sardine Myofibrillar proteins during ice-storage. *Bull. Jpn. Soc. Sci. Fish. 46*, 607-615.
81. Sidhu, G. S., Montgomery, W. A., and Brown, M. A. (1974a). Postmortem changes in spoilage in rock lobster muscle. I. Biochemical changes and rigor mortis in *Jasus novae-hollandiae*. *J. Food Technol. 9*, 357-370.
82. Sidhu, G. S., Montgomery, W. A., and Brown, M. A. (1974b). Postmortem changes and spoilage in rock lobster muscle. II. Role of amino acids in bacterial spoilage and production of volatile bases in the muscle of *Jasus novae-hollandiae*. *J. Food Technol. 9*, 371-380.
83. Sikorski, T., and Kostuch, S. (1982). Trimethylamine N-oxide demethylase: Its occurrence, properties, and role in technological changes in frozen fish. *Food Chem. 9*, 213-222.
84. Simidu, W., and Hibiki, S. (1953a). Studies on putrefaction of aquatic products. VIII. On putrefaction of squid muscle. *Bull. Jpn. Soc. Sci. Fish. 19*, 877-881.
85. Simidu, W., Hibiki, S., and Hujita, M. (1953b). Studies on putrefaction of squatic products. Supplement on putrefaction of shark muscle. IX. On volatile base developing in shark muscle. *Bull. Jpn. Soc. Sci. Fish. 19*, 882-885.

86. Simmonds, C. K., and Lamprecht, E. C. (1980). *In* "Advances in Fish Science and Technology" (J. Connell, ed.), pp. 417-425. Fishing News Books Ltd., Surrey, England.
87. Smith, J. G., Thomson, A. B., Young, K. W., and Parsons, E. (1980). *In* "Advances of Fish Science and Technology" (J. Connell, ed.), pp. 299-303. Fishing News Books Ltd., Surrey, England.
88. Smith, J. G., McGill, A. S., Thomson, A. B., and Hardy, R. (1980). *In* "Advances in Fish Science and Technology" (J. Connell, ed.), pp. 303-307. Fishing News Books Ltd., Surrey, England.
89. Smith, J. G. M., Hardy, R., McDonald, I., and Templeton, J. (1980). The storage of herring (*Clupea harengus*) in ice, refrigerated sea water and at ambient temperature. Chemical and sensory assessment. *J. Sci. Food Agric. 31*, 375-385.
90. Spinelli, J., Eklund, M., and Miyauchi, D. (1964). Measurement of hypoxanthine in fish as a method of assessing freshness. *J. Food Sci. 29*, 710-714.
91. Spinelli, J. (1967). Degradation of nucleotides in ice-stored halibut. *J. Food Sci. 32*, 38-41.
92. Spinelli, J. (1971). Biochemical basis of fish freshness. *J. Process Biochem. 6*, 36-37, 54.
93. Spinelli, J., and Koury, B. J. (1981). Some new observations on pathways of formation of dimethylamine in fish muscle and liver. *J. Agric. Food Chem. 29*, 327-331.
94. Stone, F. (1971). Inosine monophosphate (IMP) and hypoxanthine formation in three species of shrimp held on ice. *J. Milk Food Technol. 34*, 354-356.
95. Stroud, G. D., Early, J. C., and Smith, G. L. (1982). Chemical and sensory changes in iced *Nephtops norvegocis* as indices of spoilage. *J. Food Technol. 17*, 541-551.
96. Sumner, L., and Perry, I. R. (1978). Spoilage of some New Zealand fish species. 1. Held in ice. *Food Technol. N.Z. 12*, 19, 23, 25.
97. Taguchi, T., Tanaka, M., Okubo, S., and Suzuki, K. (1982). Changes in quality of canned mackerel during long-term storage. *Bull. Jpn. Soc. Sci. Fish. 48*, 1765-1769.
98. Thompson, F. B., Davis, H. K., Early, J. C., and Burt, J. R. (1974). Spoilage and spoilage indicators in queen scallops (*Chlamys opercularis*). I. Held on ice. *J. Food Technol. 9*, 381-390.
99. Tokunaga, T. (1970). Trimethylamine oxide and its decomposition in the bloody muscle of fish. II. Formation of DMA and TMA during storage. *Bull. Jpn. Soc. Sci. Fish. 36*, 510-515.

100. Uchiyama, H., and Kato, N. (1974). Partial freezing as a means of preserving fish freshness. 1. Changes in free amino acids, TMA-N, ATP and its related compounds, and nucleic acids during storage. *Bull. Jpn. Soc. Sci. Fish. 40*, 1145-1154.
101. Vanderzant, C., and Nickelson, R. (1971). Comparison of extract-release volume, pH, and agar plate count of shrimp. *J. Milk Food Technol. 34*, 115-118.
102. Varga, S., Hirtle, W. A., and Anderson, W. E. (1974). A comparison of the storage life and bacteria counts of chilled cod fillets packaged in wooden boxes and heat-sealed plastic containers. *J. Fish. Res. Board Can. 31*, 234-237.
103. Venugopal, V., Savagaon, K. A., Kumta, U. S., and Sreenivasan, A. (1973). Extension of shelf-life of Indian mackerel (*Rastrelliger kanagurta*) by irradiation. *J. Fish. Res. Board Can. 30*, 305-309.
103a. Virginia Tech (1984). Demonstration of a quality maintenance program for fresh fish products. Final report. Submitted to Mid-Atlantic Fisheries Development Foundation, Inc., Suite 600, 2200 Somerville Road, Annapolis, Maryland 21401.
104. Waller, F. (1980). Spoilage and spoilage indicators on shark held in ice. *Food Technol. Aust. 32*, 184-187.
105. Warthesen, J. J., Waletzko, P., and Busta, F. (1980). High-pressure liquid chromatographic determination of hypoxanthine in refrigerated fish. *J. Agric. Food Chem. 28*, 1308-1309.
106. Waters, M. E., and Hamdy, M. K. (1970). Effect on nitrofuran and chlorotetracycline on the microbial population of shrimp. *J. Milk Food Technol. 33*, 232-236.
107. Williams, S. K., Martin, R., Brown, W. L., and Bacus, J. N. (1983). Moisture loss in tray-packed fresh fish during eight days storage at 2°C. *J. Food Sci. 48*, 168-171.
108. Wood, G., and Hintz, L. (1971). Decomposition in foods. Lipid changes with the degradation of fish tissue. *J. Assoc. Off. Anal. Chem. 54*, 1019-1023.

CHAPTER 5

SHELF-LIFE OF FRUITS

DAVID C. LEWIS
TAKAYUKI SHIBAMOTO
Department of Environmental Toxicology
University of California
Davis, California

I. INTRODUCTION

The shelf-life of a fruit or fruit product refers to the period after harvest or manufacture during which the product remains of salable quality. The maximum possible shelf-life attainable with reasonable expenditure is desired. Relative to many other foodstuffs, the shelf-lives of fresh fruits are short, frequently only a few days.

In 1977 more than $3 billion worth of fresh fruits were sold in the United States, and the market for processed fruits is even larger (Sacharow and Griffin, 1980). Losses of fruit between harvest and sale represent a major economical loss for suppliers and consumers alike (Table 1). In less developed countries, which can ill afford to lose essential food supplies, losses are estimated to be even more staggering (Table 2).

That such major losses still occur despite years of scientific investigation of the problem indicates the particular difficulties involved in transporting and storing living, respiring fruit prior to sale. Losses can be curtailed greatly, however, with proper handling methods.

Fruit products can be broadly classed as either perishable or shelf-stable. Fresh fruits are perishable to varying degrees; that is, to avoid spoilage they can be stored only a limited time. Techniques such as refrigeration, freezing, storage in a modified atmosphere, and chemical treatment are commonly used to extend shelf-lives of fruits from a matter of days or weeks to, in some cases, many months. Semiperishable or shelf-stable fruit products, such as some dried fruit products, are relatively immune to microbial spoilage and may be stored at room temperature for months or even years (Table 3).

II. CRITERIA DETERMINING SHELF-LIFE

A. Changes in Food Safety

Spoiled fruits will generally be recognized as clearly inedible before reaching a stage in which the fruit has become a health hazard. Spoiled fruit will usually be discarded or trimmed by the producer, the marketer, or the consumer. However, the threat of food poisoning from spoiled fruit cannot be completely discounted.

In some cases fungal toxins have been detected in fruit products. Much attention has been directed at the fungal toxin patulin, 4-hydroxy-4-H-furo(3,2C)pyran-2-6H-one. The occurrence and toxicity of patulin has been well reviewed (Ciegler *et al.*, 1971; Davis and Diener, 1978; Salunkhe *et al.*, 1980; Scott, 1974). A wide variety of fungi are known to produce patulin, notably *Penicillium expansum*, which causes blue rot mold in stored apples and has been linked to patulin contamination of apples, cherries, grapes, and pears (Buchanan *et al.*, 1974; Davis and Diener, 1978). Patulin in moderate amounts causes acute and chronic toxicity in experimental animals, but of particular concern are possible teratogenic, mutagenic, and carcinogenic properties of this contaminant (Ciegler *et al.*, 1971, 1976). Patulin contamination may be avoided by discarding or trimming moldy and rotted fruit. Buchanan and associates (1974) found mold-infected areas of fruits contained as much as 125 ppm patulin, while mold-free areas of the same fruit had undetectable levels of the toxin. Patulin appears to be particularly stable to storage in apple juice because of the low levels of sulfhydryl groups relative to other juices such as orange juice.

Ascorbate and ascorbic acid increase patulin degradation in stored apple juice (Brackett and Marth, 1979). Sulfur dioxide has been reported to degrade patulin by opening the lactone ring (Pohland and Allen, 1970), but the effectiveness of this treatment at SO_2 levels below 200 ppm has been questioned (Burroughs, 1977). Passing apple cider through activated-charcoal filters is effective in removing patulin, but has the added effect of removing color from the cider (Sands *et al.*, 1976; Walton *et al.*, 1976). Procedures for decreasing the occurrence of patulin in apple juices and other products, including preharvest fungicide treatment, low-temperature storage, and trimming of rotted fruit have been proposed by the Processed Apple Institute (1977).

Aflatoxin is an occasional cause of fruit spoilage. Salunkhe *et al.* (1980) noted that dried figs are sometimes rejected due to aflatoxin production. Ochratoxin has been reported to occur on citrus fruits, but contamination of fruits by others of the many known fungal toxins appears to be rare (Davis and Diener, 1978).

B. Nutritional Changes

The health hazard due to microorganisms in fresh and processed fruits is minimal, and thus much research interest has been directed at prolonging the nutritional and aesthetic qualities of stored fruits (Woodroof, 1975b).

Sugar-amino acid browning reactions may result in a loss of nutritionally essential amino acids. Freeze-dried citrus juices stored at 25°C lost 18-25% of their amino acid content, whereas those stored at 5°C lost only 5-6% (Nicolosi Asmundo *et al.*, 1980). Vitamin losses in stored and processed fruits may be significant (Table 4).

Ascorbic acid stability varies inversely with pH. Therefore, ascorbic acid levels may be maintained better in more acid fruits such as citrus fruits. Citric acid in citrus fruits can chelate metallic ions such as copper and inhibit ascorbic acid degradation (Kealey and Kinsella, 1979). Nutritive losses can occur during processing. Ramasastri (1974) reported Indian gooseberries (*Emblica officinalis*) lost 30% of their ascorbic acid content during sun-drying, and dehydrated gooseberry powder lost an additional 58% after 48 weeks of storage. Pickling of the same fruit caused a 50% loss of ascorbate during preparation and an additional 40% loss after 4 weeks' storage.

C. Aesthetic Changes

While loss of nutritional quality during storage may only become apparent upon analyses of fruit, many other changes are immediately recognized by the producer or consumer and can lead to rejection of the fruit. Changes in flavor, color, texture, and overall appearance are important forms of deterioration in stored fruit, and attempts to extend shelf-life must consider these factors.

Loss of attractive appearance may lead to rejection of the fruit even though it may still be safe and nutritious. Citrus fruits with green rinds result in low pack-outs. Degreening treatments may in some cases actually increase decay losses while increasing salability of the citrus (Brown, 1980).

Culling of unacceptable fruit prior to processing or sale includes an evaluation of aesthetic qualities. For example, citrus fruits are examined not only for the absence of molds and rots, but for such characteristics as texture, firmness, soundness of skin, and absence of injury. Sensory evaluation of fruit may involve a "taste panel," such as consumer panels conducted under the auspices of the USDA Processed Products Branch and the Florida Citrus Processors Association for flavor analysis of citrus products (Fellers, 1980).

1. Color Changes

The most important color changes in stored fruits and fruit products are caused by chemical reactions known as browning reactions. Browning of fruits may be enzymatic or nonenzymatic. Changes in color, flavor, and nutritional quality have all been associated with browning reactions. In some cases these changes are desirable, such as aroma compounds formed during cooking or browning of raisins, prunes, sultana grapes, and dates. In most cases, however, browning reactions cause undesirable effects, and considerable effort is expended to reduce these changes.

Polyphenoloxidases (PPO) and peroxidases found in fruit tissues can catalyze oxidation of certain endogenous phenolic compounds to quinones that polymerize to form intense brown pigments (Vamos-Vigyazo, 1981; Walker, 1975). Ripening agents, ionizing radiation, and some trace elements in fertilizers may increase PPO activity. Inhibitors of enzymatic browning fall into two groups: (1) compounds that reduce *o*-quinones back to *o*-hydroxyphenols and (2) compounds that react with quinones formed by fruit polyphenoloxidases to form stable, colorless compounds. Reducing agents are lost as they perform their function, and thus except in high concentrations, are capable of providing only temporary inhibition of browning. Compounds functioning as reducing agents in fruits include ascorbic acid, sulfur dioxide, potassium metabisulfite, and other sulfites. Cysteine and glutathione are examples of the second class of compounds that form addition compounds with quinones (Vamos-Vigyazo, 1981).

Some fruits such as apples or bananas discolor readily following injury, while others such as oranges and lemons do not. Fruits resistant to discoloration have higher ascorbic acid levels and/or lower PPO activity. In the presence of sufficient ascorbic acid, *o*-quinones are reduced back to catechol forms and brown polymers are not formed (Bauerfeind and Pinkert, 1970). Darkening in orange juice parallels loss of ascorbic acid (Kealey and Kinsella, 1979).

Three separate processes may be involved in nonenzymatic browning of fruits: (1) Maillard condensation of sugars with amino acids, (2) ascorbic acid polymerization, and (3) furfuraldehyde condensation with nitrogen compounds (Shewfelt, 1975). Pink discoloration of apple sauce may occur as a result of amino acid-sugar reactions. Nutritionally essential amino acids are lost in such sugar-amino acid browning reactions. Sulfur dioxide reduces amino acid-sugar browning reactions as does low-temperature storage of fruit products (Woodroof, 1975b).

Other types of reactions lead to color changes in stored fruits. Anthocyanins are important natural constituents responsible for normal fresh fruit coloration. Anthocyanins in stored fruit products such as strawberry preserves can degrade to brown-colored anthocyanidins (Woodroof, 1975b). Anthocyanins are degraded by PPO-produced quinones, but this process is inhibited by ascorbic acid, which has a lower oxidation-reduction potential and is preferentially oxidized (Pifferi and Cultrera, 1974). Red anthocyanins can be formed from colorless proanthocyanins leading to pink discoloration of canned apples, peaches, or pears. Discoloration of canned fruits can also result from colored chelates of anthocyanins with Sn^{2+}, Al^{3+}, and Fe^{3+} ions. Similar flavone and flavonol chelates with Al^{3+} can cause color changes (Vamos-Vigyazo, 1981).

2. Flavor Changes

Limonin, a complex di-lactone that can develop in stored or heat-processed navel and early-season Shamouti orange juice, imparts a bitter taste leading to rejection of the juice. The maximum acceptable level of limonin in navel orange juice is 5-6 ppm (Ponting *et al.*, 1973). Maier *et al.* (1971) reported that exposing navel oranges to an atmosphere containing 20 ppm ethylene for 3 hr followed by 5 days' storage in normal atmosphere resulted in a limonin reduction from 12 to 8 ppm.

Dehydrated fruit products are quite susceptible to flavor and color changes during storage. In dehydrated orange juice crystals, Maillard-type amine-sugar nonenzymatic browning reactions are thought to be responsible for flavor degradation. Ascorbic acid and *d*-fructose have been hypothesized to be the precursors of off-flavors in this product (Shaw *et al.*, 1967; Tatum *et al.*, 1967, 1969).

III. FACTORS AFFECTING SHELF-LIFE

A. Temperature

Decay processes in fruits are affected by temperature, and each fruit has an ideal storage temperature (Table 5). Bananas develop a black discoloration if chilled below 10°C (Sacharow and Griffin, 1980). Berries represent a particular problem in storage and distribution due to their high perishability. Herregods (1979) reported that berries, including blackberries and raspberries, kept only 1 day (mean) at 15°C. However, storage life was increased to 1 to 3 days at 2 to 5°C and to more than 3 days (up to 21 days for gooseberries) at 0 to 1°C. Blueberries stored at 4.4°C had much longer shelf-life than those stored at 10°C (Singh and Heldman, 1975).

B. Moisture

Fruits removed from the tree have lost their source of moisture. Since fruits are 75-95% water and have equilibrium humidities of as much as 98%, they are susceptible to rapid drying in a normal atmosphere (Sacharow and Griffin, 1980). Water loss can result in decay, especially near the stem end. However, the water activity a_w of fruits renders them susceptible to microbial spoilage. Bacterial growth is inhibited in fruits at a_w values <0.90. Most yeasts are inhibited at water activities <0.88. Molds are the most resistant to drying, retaining viability at an a_w of 0.80, and some molds, such as *Aspergillus glaucus*, will grow at a_w values <0.70 (Hsu, 1975).

C. Light

Most fruits exhibit discoloration due to light exposure, and vitamin C levels may also be reduced. The length and intensity of illumination are important, and turning off case lighting at night will reduce discoloration. Opaque or colored packaging may also be used. Light wavelength, oxygen, and humidity may affect light deterioration of fruits (Woodroof, 1975b).

D. pH

A notable difference between fruits and other foods is the natural high acidity of fruits. Most fruits have pH values <4, and some (cranberries, grapefruits, lemons, limes, and raspberries) have a pH <3 (Splittstoesser, 1978). Less acidic fruits include figs (pH 4.8-5), mangos (3.8-4.7), and pears (3.7-4.6). Most fruits, therefore, are sufficiently acidic to inhibit growth of bacteria and yeasts. As with most food products, a major concern is the possibility of contamination with *Clostridium botulinum*. FDA regulations require canned figs to have a pH of <4.9 to prevent growth of this dangerous species (Ito *et al.*, 1978). Lemon juice or citric acid are allowable additives to achieve this. Citric acid is also added to canned prunes (Gardner, 1966). Musk-melons have a high pH and may be canned with low-pH fruits. Banana puree requires organic acid additives to reduce pH (Hauschild, 1980).

E. Market Disorders

Market disorders are distinct from spoilage in that deterioration is due to internal alterations rather than infection (Salunkhe and Wu, 1974; Table 6). Market disorders may be linked to physical handling injury, freezing, chilling, excessive heat, or chemicals such as ethylene or ammonia. Some disorders may represent inevitable changes associated with senescence and the loss to the fruit of its sustaining tree. Fruit yellowing results from aging and may be hastened by exposure to ethylene. Thus some strategies for delaying denescence involve limiting exposure to ethylene.

Apple scald is one of the most important market diseases of this fruit. The fruit develops a burned appearance followed by softening under the scald. Its cause is not well understood but has been hypothesized to be associated with gases produced by the apple itself. Chilling injury is the major market disorder of bananas. Many other fruits are susceptible to chilling damage (Table 7). Similar peel blackening also occurs in bananas stored at <80% relative humidity. Pitting, characterized by pitted blotches on the rind, is the major physiological disorder in citrus fruits. Chill-damaged pineapple fails to develop normal ripe shell coloration. Bruised cherries are prone to scald formation associated with demethylation of pectin (Buch *et al.*, 1961).

F. Spoilage

Deterioration and spoilage are major factors in the relatively short shelf-life of fruits. Postharvest decay of fruits is responsible for major economic loss, with some fruits more perishable than others. Bacteria, molds, and yeasts are normal features of the surface of fruits prior to harvest. Molds are few in number at this time, but become the predominant pathogen after harvest. Molds and yeasts are the primary causes of spoilage. Molds are particularly common on citrus fruits but are responsible for significant decay of all fruits (Table 8). Yeasts cause fermentation and the development of disagreeable flavors (Hsu, 1975), but they have no means of invading intact fruit and thus are of secondary importance (Splittstoesser, 1978). Yeasts are most important in spoilage of fruit juices and crushed fruits (Walker and Ayres, 1970). Bacteria are an insignificant cause of fresh fruit spoilage because of the high acidity of most fruits.

IV. METHODS TO INCREASE SHELF-LIFE

A. Strain Selection

Improved storage life is one of many characteristics used in the selection of superior fruit cultivars. MacLachlan and Gormley (1974) examined storage life of 20 different strawberry cultivars. The Cambridge Favorite had high yield and shelf-life. However inferior flesh color and freezing performance were observed. Thus selection for optimum shelf-life may need to be balanced against other desirable strain qualities. Moore *et al.* (1975) reported that the Cardinal strawberry cultivar had good shelf-life as well as flavor, ascorbic acid content, color appearance, and processing characteristics.

B. Handling and Packaging

Proper handling and packaging of fruit are important for preventing losses and extending shelf-life. Postharvest washing, frequently combined with fungicide, is standard practice for many fruits. Washing to remove pesticides may extend shelf-life of fruits (Brown, 1976).

Wax coatings are applied to some fruits to improve water retention and inhibit fruit deterioration. Coating Amoun oranges with Britex wax reduced decay and water loss (Abd El-Latief *et al.*, 1975). Shelf-lives of lemons and oranges were increased by 8 to 10 weeks by coating with natural waxes (Morales and Lomelin, 1974). Sealing citrus fruit with resins high in low molecular weight hydrocarbons has been extensively employed (Lueck, 1980). Chandhry *et al.* (1979), in a study comparing chemical treatments and wax coatings as a means of increasing shelf-life of mangoes, found wax emulsion to be the optimum treatment. However, wax coating of blueberries did not increase shelf-life (Singh and Heldman, 1975).

Packagers classify fruits as being hard or soft. Soft fruits (such as berries, cherries, grapes, and plums) generally have a shorter shelf-life due to damage and decay. They are susceptible to anaerobic spoilage and bruising. Desirable packaging methods include semirigid containers with a film cover and good ventilation (Sacharow and Griffin, 1980). Ayres and Denisen (1958) observed that spoilage of raspberries and strawberries could be reduced with disposable plastic rather than reusable wooden containers. Hard fruits have shelf-lives measured in weeks instead of days, and have less critical packaging requirements (Sacharow and Griffin, 1980).

Seal-packaging of individual citrus fruits with polyethylene film doubled the storage life. Weight loss was decreased by 80%, chilling injury to susceptible fruits was decreased, and flavor and appearance were retained for longer periods. Mold growth was inhibited, although stem-end rots increased. Use of fungicides, ripening inhibitors, or ripening accelerators in the film was suggested (Ben-Yohoshua, 1978). Polyethylene bags were also effective in extending shelf-life of purple passion fruit stored under ambient conditions of 27 to 30°C and 30 to 50% relative humidity (Ganapathy and Singh, 1976). Lazar and Hudson (1976) reported that sodium sulfite-dipped, steam-blanched, sliced apples packaged in flexible pouches were stable to 34°F storage exceeding 3000 days. Shelf-life for room temperature storage was only 2-4 weeks.

C. Chemical Additives and Treatments

Many chemical treatments have been applied on a commercial or experimental basis with the goal of extending the shelf-life of fresh and processed fruits. Chemical treatments may require much less capital investment compared with cooling and may be applied before or after harvest (Kavanagh and Kenny, 1978).

1. Senescence-Retarding Chemicals

A number of chemicals are known that either accelerate or delay ripening in fruits (Salunkhe and Wu, 1974). A major strategy involves preventing accumulation of the endogenous ripening agent ethylene in the storage atmosphere of fruits. N^6-Benzyladenine-treated sweet cherries showed increased weight and chlorophyll retention during 7 days' refrigerated storage (Tuli *et al.*, 1962), but in another study this compound was ineffective in lengthening shelf-life of the same fruit (Salunkhe *et al.*, 1962). Abdel-Gawad and Romani (1968) reported that gibberellic acid did not retard senescence in pears. However, this compound has been successfully applied to delay ripening of navel oranges (Eilati *et al.*, 1969) and kaki fruits (Kitagawa *et al.*, 1966). Storage deterioration and losses of Coorg mandarins were decreased with preharvest application of 25 ppm chlorophenoxyacetic acid or β-naphthoxyacetic acid (Rodrigues and Subramanyam, 1966). Spraying with isopropyl *n*-phenyl carbamate delayed ripening and decreased spoilage of stored sapota fruits (Lakshminarayana *et al.*, 1967). 2-Methyl-3-phytyl-1,4-naphthoquinone (vitamin K_1) and 2-methyl-1,4-naphthoquinone (vitamin K_3) delay ripening of bananas (Beccari, 1969). Cycloheximide inhibits senescence processes of unripe pears (Frenkel *et al.*, 1969). Treatment of McIntosh apples with postharvest calcium chloride dips inhibited low-temperature and senescent breakdown (Blanpied, 1981). Guava fruit dipped for 30 min in a 1% calcium nitrate solution showed delayed senescence and remained organoleptically acceptable for up to 9 days (Singh *et al.*, 1981).

2. Antioxidants and Browning Inhibitors

Several types of nonmicrobial degradative changes are known to occur in fruits during storage. Antioxidants protect foods from oxidative degradation. Other additives can inhibit nonoxidative browning reactions. Both of these types of deterioration in fruits can lead to discoloration, loss of acceptable organoleptic qualities, and loss of nutritional content. Ascorbic acid and ascorbate salts, sulfites and bisulfites, and stannous chloride have been used as fruit antioxidants and browning inhibitors. Additionally, citric acid, phosphoric acid, and EDTA salts have received use as scavengers of metal ions that catalyze oxidative reactions.

Ascorbic acid and sulfur dioxide are frequently used to inhibit oxidative reactions that lead to browning. Ascorbic acid is one of the most effective and widely used antioxidants. It can have a protective effect on flavor retention during storage as well as color. Ascorbic acid is used in frozen fruits, especially sliced peaches, at levels of 300 to 500 ppm. It is also added to apple juice (100-500 ppm), artificially sweetened preserves and jellies (1000 ppm), frozen concentrated grape juice, and other fruit drinks (National Academy of Sciences, National Research Council, 1965).

However, ascorbic acid is not as inexpensive as some other additives, and is in some cases ineffective. Browning has been reported to be enhanced in ascorbic acid-fortified concentrated lemon juice (Pintauro, 1974). Sulfur dioxide and sulfites are commonly used to inhibit browning. EDTA has also been used as an antibrowning adjunct. Also, sliced fruit in heavy syrup exhibits much slower browning (Shewfelt, 1975). Addition of stannous chloride to citrus juices has been the subject of a patent by Higby and Prichett (1965). Shelf-life was studied using colorimetrically determined juice discoloration to set an end point. Significant improvement was noted for citrus juices (Table 9).

A number of PPO inhibitors other than the widely used ascorbic acid have been tried as treatments to prevent enzymatic browning. Treatment of d'Anjou pears with 1 to 2% solutions of 2-mercaptobenzthiazole was highly effective in preventing friction discoloration (Wang and Mellenthin, 1974). This compound also inhibits banana PPO (Palmer and Roberts, 1967). 2,3-Naphthalenediol was an effective PPO inhibitor in apple slices and peaches (Harel *et al.*, 1966; Mayer *et al.*, 1964). Diethyldithiocarbamate, at 0.01 to 1 m*M* concentrations, causes up to 87% inhibition of apple PPO (Walker, 1964). Grape PPO is also inhibited by this compound (Hazel *et al.*, 1965).

3. Preservatives

Preservatives act by inhibiting growth of microorganisms in fruits and fruit products (Table 10). Some act selectively on a range of bacterial or fungal species, whereas others, such as thioisomaltol, have a broad-spectrum antimicrobial activity. Decay in stored fruits is frequently caused by fungal growth, and less commonly, by bacterial or yeast contamination. Salunkhe *et al.* (1962) suggested that antifungal agents be added to fruits as preservatives. The use of chemical treatments to control fruit spoilage has been reviewed by many authors (Lueck, 1980; Salunkhe and Wu, 1974; Smith, 1962).

D. Refrigeration

Cooling of fruit after harvest has a beneficial effect on shelf-life. Cooling may be in the form of precooling of the produce to remove field heat and decrease ripening, mold growth, and moisture loss. Hydrocooling with refrigerated water is frequently used. This method allows the addition of fungicides and disinfectants to the water. Chemicals employed include chlorine, benomyl, botran, and captan (Woodroof, 1975a).

Refrigeration as a means of extending shelf-life has been detailed in numerous reviews (Ginsburg *et al.*, 1975;

Ryall and Pentzer, 1974; Woodroof, 1975a). Maximum refrigerated storage times of fruits may be on the order of days, weeks, or months, depending on their perishability (Table 11). Perishable fruits require cooling to <40°F. Most fruits stored for long periods (>1 month) are kept at 31 to 32°F. Many fruits with a short storage life, especially citrus fruits, are sensitive to chilling injury and are stored at higher temperatures. Tropical and subtropical fruits should be cooled to 50 to 55°F (Ryall and Pentzer, 1974). Furthermore, the rate of cooling can affect shelf-life. Hudson and Tietjen (1981) detailed the effect of cooling rate on decay and shelf-life of highbush blueberries. Brown (1980) concluded that cold storage is currently the most effective method for preserving citrus fruits, except lemons and limes.

E. Freezing

Freezing fruits greatly extends their shelf-life (Table 12). Frozen fruits typically have a shelf-life of greater than 1 year at 0°F, although this decreased with storage at 10°F (Woodroof, 1975b). The frozen product market in 1977 exceeded $100 million (Sacharow and Griffin, 1980). Freezing has received less application for fruit storage than for many other foods because of problems with loss of texture, discoloration, and drip loss (Boyle and Wolford, 1968). In 1979, frozen strawberries, red tart cherries, apples, peaches, blackberries, red respberries, and blueberries accounted for 80% of frozen fruit production (Luh *et al.*, 1975). The predominance of berries in the frozen fruit market is accounted for by their extreme perishability.

Even after freezing, fruits such as apples, bananas, peaches, and pears are susceptible to enzymatic browning. Freezing in nitrogen prevents this discoloration. Freezing, if not sufficiently rapid, also causes the formation of large ice crystals. Resulting tissue damage leads to loss of texture or mushiness in thawed fruits. Charlampowicz *et al.* (1971) reported 10% loss of frozen strawberry color after 3 months' storage. Spoilage can occur even in frozen fruits. Ricci (1972) isolated numerous yeasts from soured figs stored at 0°C.

F. Controlled-Atmosphere Storage

Modified-atmosphere (MA) or controlled-atmosphere (CA) storage techniques have been used for more than 50 years. Storage life of apples is increased in a high-CO_2, low-O_2

atmosphere. This principle is applicable to other respiring fruits such as citrus, peaches, pears, grapes, and cherries (Luh *et al.*, 1975). Recommended CA storage conditions are presented in Table 13. CA storage capacity for apples now exceeds 2 billion lb, and this technology is in large part responsible for the availability of fresh apples throughout the year in the United States (Doores, 1983). The effects of atmospheric conditions on fruit deterioration and shelf-life are shown in Table 14.

Hypobaric (subatmospheric pressure) storage has been employed to increase storage life considerably. Preventing accumulation of the endogenous ripening agent ethylene in the storage atmosphere of fruits is one effect of this method. Fungal growth is greatly inhibited in fruit stored at or below 278 mm Hg, with degree of suppression inversely related to the atmospheric pressure. A 2.7% oxygen and 97.3% nitrogen CA was not as effective as hypobaric storage in preventing fungal growth (Salunkhe and Wu, 1974). These authors have detailed the effects of hypobaric storage on fruit shelf-life (Salunkhe and Wu, 1973, 1974; Table 15). Storage life of cherries was increased from 63 to 90 days with storage at 102 mm Hg. Mold growth, brown discoloration, and sugar losses were inhibited. Shelf-lives of Golden Delicious apples were extended by 2.5 months, and Red King apples by 3.5 months at 278 mm Hg. Apricot shelf-life was similarly increased from 3 to 90 days at 102 mm Hg. Papaya transported over a 21-day period in hypobaric containers (20 mm Hg, 10°C, and relative humidity 90-98%) exhibited slowed ripening and spoilage. One shipper reported abnormal softening in 5 to 45% of fruits. Fungicide waxing decreased spoilage further (Alvarez, 1980).

G. Canning

Canned fruits are second only to fresh fruits in overall proportion of the total crop. In 1971, canned Clingstone peaches accounted for 36% of the canned fruit pack. Canned apple sauce, mixed fruits, and pears collectively represented an additional 53% of production (Luh and Kean, 1975).

Canned fruits suffer deterioration relative to their storage temperature. The typical sequence of quality impairment is (1) loss of flavor and/or development of off-flavors, (2) discoloration, (3) texture changes, and (4) nutrient losses. Flavor changes may involve browning reactions, loss of volatile flavor components, or staling. Burned or bitter flavors are sometimes associated with browning, particularly in canned fruits in syrup (U.S. Congress, Office of Technology

Assessment, 1979). Red discoloration of canned peaches is lessened by alkali immersion peeling (van Blaricom and Hair, 1973).

Vitamin losses take longer to develop. Losses of ascorbic acid have frequently been noted in canned fruits. Elkins (1979) studied vitamin retention in canned peaches stored in ambient air for 18 months. Carotene, niacin, and riboflavin losses were less than 10%, but ascorbic acid and thiamine concentrations were decreased by 28 and 30% respectively.

Temperature extremes have adverse effects on canned fruits. In particular, freezing is usually detrimental to canned fruit quality (Cecil and Woodroof, 1963). Browning in canned peaches can occur at temperatures >37°C. Container deterioration can occur, and is more common with acidic canned fruit than with other food products. Grose (1978), of the United Kingdom Consumers' Association, cites high metal contamination in canned fruit as an important problem.

H. Drying and Dehydration

The commercial production of dried fruits exceeds 400,000 tons/year (Salunkhe *et al.*, 1971). Sun-dried apricots, dates, figs, grapes (raisins), peaches, pears, and plums (prunes) are produced in California and in Mediterranean countries. Fruits commonly sun-dried may alternatively be dehydrated. Dehydration in hot-air ovens offers advantages over sun-drying, notably fewer flavor and color changes as well as decreased spoilage. Other dehydrated fruits include apples, bananas, and citrus fruits. Sawatditat *et al.* (1979) reported a dry steam dehydration method for bananas that gave good flavor retention as well as a 6-month shelf-life.

Dried fruits have low a_w levels that retard microbial growth and spoilage (King *et al.*, 1968). Therefore, with correct storage conditions the shelf-life of dried fruits can extend to months and years rather than days. Proper packaging is important in obtaining maximum shelf-life of dried fruits. Ideal packaging will protect the fruit product from moisture, air, light, microorganisms, and so forth, without contamination of the product, and will be of reasonable cost. Fruit products with low moisture content are hygroscopic, and care must be taken to preclude water uptake during and following packaging (Heiss and Eichner, 1971a,b). Some products are packed in an inert atmosphere to reduce oxygen content to less than 2%. Dessicant envelopes, such as silica gel, are sometimes added to dried fruits. Microbial spoilage will not occur in dried fruits with less than 25%

moisture content, due to high sugar osmotic pressure (Somogyi and Luh, 1975).

The primary forms of deterioration of dried fruits are discoloration and flavor changes (Table 16). Browning reactions are a major problem with dried fruits, especially sliced apples. Soaking cut, peeled fruit in ascorbic acid-organic acid or salt-acidulant solutions prior to drying will retard browning (Gardner, 1966).

Dried-fruit powders have been developed, and may find particular application in areas where refrigerated storage is not widely available. Freeze-dried orange and guava juices have long shelf-lives (Foda *et al.*, 1970), Sharma *et al.* (1974) studied the keeping qualities of canned spray-dried mango milk powder stored at 30 ± 1°C or at room temperature for 1 year. No changes in flavor or color of the reconstituted product were noted at either temperature. Off-flavors can occur in freeze-dried avocado exposed to air. This has been particularly controlled with addition of butylated hydroxyanisole (BHA) and refrigeration (Gomez and Bates, 1970).

I. Irradiation

Irradiation of fruits and fruit products has been used for preservation and insect deinfestation. Extensive testing has been conducted of this technology as an economical means of preventing food spoilage without possible risks associated with the use of chemical treatments. However, much concern has been expressed about possible adverse effects of irradiation on food safety. Questions about effects on food quality have also been raised. Irradiation of fruits has attracted a good deal of research interest. This technology offers a means of inhibiting fungal and bacterial growth without significantly raising internal fruit temperature. However field trials in California led Maxie and associates (1971) to conclude that irradiation was impractical for commercial preservation of most fruits, and that other more effective and inexpensive treatments were available.

α- and β-radiation doses of 1 to 5×10^5 rads can inhibit fungal and bacterial spoilage. Radiation of dates with 1 to 27×10^5 rads increased shelf-life by 5 days without detectable loss of sensory qualities or protein, amino acid, or sugar content (Auda, 1980). However, fruit softening and degradation of vitamins, proteins, and carbohydrates have been associated with irradiation (Salunkhe and Wu, 1974). Ripe Bartlett pears show adverse effects from irradiation, although green pears were not harmed by doses of 2.5×10^5 rads.

No adverse effects were observed in 90-day trials of rats provided with irradiated dates as 60% of the diet (Auda, 1980). Studies with rhesus monkeys fed irradiated apricots and raisins similarly did not result in observable toxicity (Zaytsev *et al.*, 1975).

One alternative deinfestation method involves the use of ultrasound. Horner (1979) has explored this technology for destruction of fruit flies in Hawaiian papayas. The method of choice has been ethylene dibromide (EDB); however, EDB has been classed as a suspect carcinogen and its future is in doubt.

TABLE 1. Post-harvest Losses of Fresh Fruits[a]

Fruit	Post-harvest Losses (%)
Pineapple	40-50
Mango	20-33
Orange	15-30
Peach	15-24
Strawberry	14-18

[a] Adapted from Salunkhe *et al.* (1980).

TABLE 2. Fruit Production and Post-harvest Losses in Less Developed Countries[a]

Fruit	Production (1,000 tons)	Estimated Loss (%)
Bananas	36,898	20-80
Citrus fruits	22,040	20-95
Grapes	12,720	27
Apples	3,677	14
Peaches, apricots, and nectarines	1,831	28
Avocados	1,020	43
Papayas	931	40-100
Raisins	475	20-90

[a] From National Academy of Sciences (1978) and Food and Agriculture Organization (1981).

TABLE 3. Mode of Deterioration, Environmental Factors and Shelf-Life of Fruit Products[a]

Type of Fruit Product	Mode of Deterioration	Critical Environmental Factors	Average Shelf-Life (weeks)
Fresh	Microbial decay, nutrient loss	Temperature, light, oxygen, handling, relative humidity	b
Dried	Browning, nutrient loss	Moisture, light, temperature, oxygen	1-24
Frozen	Flavor loss, nutrient loss	Temperature	6-24
Frozen juice concentrate	Nutrient loss, flavor loss, yeast growth	Temperature	18-30
Canned	Flavor loss, nutrient loss	Temperature	12-36

[a] From United States Congress, Office of Technology Assessment (1979).

[b] Highly variable depending on fruit type and storage conditions.

TABLE 4. Vitamin Losses from Fruits Preserved by Freezing, Canning, or Drying[a]

Method of Preservation		Vitamin Loss (%)				
		Ascorbic Acid	Niacin	Riboflavin	Thiamin	Vitamin A
Frozen, not thawed[b]	Mean	18	16	17	29	37
	Range	0-50	0-33	0-67	0-66	0-78
Canned, solids and liquid[b]	Mean	56	42	57	47	39
	Range	11-86	25-60	33-83	22-67	0-68
Dried, uncooked[c]	Mean	39	0	0	55	6
	Range	0-65	--	--	11-90	0-18

[a] Adapted from Fennema (1975).

[b] Fruits examined were apples, apricots, blueberries, cherries, orange juice, peaches, raspberries, and strawberries.

[c] Fruits examined were apples, apricots, orange juice, and peaches.

TABLE 5. Shelf-life and Recommended Storage Conditions for Deciduous Fruits, Berries, and Grapes[a]

Fruit	Storage Temperature °F	Relative Humidity %	Approximate Shelf-Life	Highest Freezing Point °F
Deciduous Fruits				
Apples	30-40	90-95	3-8 months	29.3
Apricots	31-32	90-95	1-2 weeks	30.1
Cherries				
(sour)	32	90-95	3-7 days	29.0
(sweet)	30-31	90-95	2-3 weeks	28.8
Figs	31-32	85-90	7-10 days	27.6
Nectarines	31-32	90-95	2-4 weeks	30.4
Peaches	31-32	90-95	2-4 weeks	30.3
Pears	29-31	90-95	2-7 months	29.2
Persimmons				
(Japanese)	30	90-95	3-4 months	28.1
Plums and				
Prunes	31-32	90-95	2-4 weeks	30.1
Quinces	31-32	90-95	2-3 months	28-4
Berries				
Blackberries	31-32	90-95	2-3 days	30.5
Blueberries	31-32	90-95	2 weeks	29.7
Cranberries	36-40	90-95	2-4 months	30.4
Currants	31-32	90-95	1-2 weeks	30.2
Gooseberries	31-32	90-95	2-4 weeks	30.0
Raspberries				
(red)	31-32	90-95	2-3 days	30.9
(black)	31-32	90-95	2-3 days	30.0
Strawberries	31-32	90-95	5-7 days	30.6
Grapes				
Labrusca	31-32	85-90	2-8 weeks	29.7
Muscadine	31-32	85-90	2-3 weeks	--
Vinifera	30-31	90-95	3-6 months	28.1

[a] From Lutz and Hardenburg (1968) and Ryall and Pentzer (1974).

TABLE 6. Major Market Disorders of Pome and Citrus Fruits[a]

Disorder	Affected Fruit	Cause	Control Measures
Pome Fruits			
Ammonia injury	Apples, pears	Ammonia exposure	Prevent exposure
Bitter pit	Apples	Calcium deficiency	Calcium salts
Brown core	Apples	Senesence	38-40°F controlled atmosphere storage (CA)
Bruising	Apples, pears	Handling injury	Proper handling
Copper injury	Pears	Moisture and copper exposure	Pack dry
Core breakdown	Pears	Senescence	Various
Diphenylamine (DPA) injury	Apples	Excess DPA exposure	Substitute ethoxyquin
Freezing injury	Apples, pears	Freezing	Avoid freezing
Internal breakdown	Apples	Senescence, moisture	Various
Internal browning	Apples	Low temperature storage	38-40°F CA storage
Jonathan spot	Apples	Unknown	CA storage
Modified atmosphere (MA)	Apples, pears	MA storage	Increase O_2 level
Ripening disorders	Pears	Long storage	30-31°F storage
Scalds	Apples, pears	Senescence	Scald inhibitors
Sodium o-phenyl-phenolate (SOPP)	Apples, pears	Excess SOPP exposure	Correct application
Water core	Apples	Unclear	Early harvest

TABLE 6. (continued) Major Market Diseases of Pome and Citrus Fruits[a]

Disorder	Affected Fruit	Cause	Control Measures
Citrus Fruits			
Aging	Citrus	Water loss	High humidity
Albedo browning	Lemons	Chilling	Proper storage
Brown staining	Citrus	Chilling	Proper storage
Creasing	Citrus	Unknown	None
Endoxerosis	Lemons	Unknown	None
Exanthema	Citrus	Copper deficiency	Copper application
Freezing injury	Citrus	Freezing	Avoid freezing
Granulation	Citrus	Rapid fruit growth	Reduce fertilization
Membranous stain	Lemons	Chilling	Proper storage
Modified atmosphere (MA)	Citrus	MA storage	Increase O_2 level
Oil spotting	Citrus	Bruise injury	Proper handling
Pitting	Citrus	Chilling	Proper storage
Puffiness	Oranges, tangerines	Water loss	High humidity
Rumple	Lemons	Unknown	40°F storage
Stylar-end breakdown	Limes	Unknown	Various
Watery breakdown	Grapefruit	Chilling	Proper storage

[a] Data from Ryall and Pentzer (1974).

TABLE 7. Fruits Susceptible to Chilling Injury[a]

Fruit	Minimum Safe Storage Temperature (°F)	Characteristic Chilling Injury
Apples	36-38	Internal browning, brown core, soft scald, soggy breakdown.
Avocados	40-45	Grayish brown discoloration of flesh.
Bananas, green or ripe	53-56	Dull skin color, browning of flesh, failure to ripen.
Cranberries	36	Rubbery texture, red flesh.
Grapefruit	50	Pitting, scald, watery breakdown.
Lemons	52-55	Pitting, membranous stain, red blotch.
Limes	45-48	Pitting.
Mangos	50-55	Grayish scald-like discoloration of skin, uneven ripening.
Olives	45	Internal browning.
Oranges[b]	38	Pitting, brown stain.
Papayas	45	Pitting, failure to ripen, off flavor, decay.
Pineapples	45-50	Dull green when ripened.

[a] From Lutz and Hardenburg (1968) and Ryall and Pentzer (1974).

[b] California and Arizona oranges.

TABLE 8. Some Mold Genera Responsible for Significant Spoilage of Fruits[a]

Fungal Genus	Apples	Bananas	Berries	Citrus	Grapes	Pears	Stone Fruits[b]
Alternaria	+		+	+			
Aspergillus				+			+
Botryodiplodia		+					
Botrytis	+		+	+	+	+	+
Cladiosporium			+		+		+
Colletotrichum				+			
Diaporthe				+			
Diplodia				+			+
Fusarium				+			
Geotrichum				+			+
Gloeodes	+			+		+	+
Gloeosporium	+	+				+	
Guignardia				+			
Monilinia			+				+
Nigrospora		+					
Penicillium	+			+		+	+
Phomopsis				+			
Phytophthora	+		+	+			
Rhizopus	+		+		+	+	+
Sclerotinia			+	+			
Septoria				+			
Trichoderma				+			
Venturia	+					+	

[a] From Splittstoesser (1978).
[b] Stone fruits include apricots, cherries, nectarines, peaches, and plums.

TABLE 9. Increased Shelf-life Following Addition of Stannous Chloride to Citrus Juices[a]

Fruit	Sn^{2+} (ppm)	Temp. (°C)	Untreated Shelf-life (Days)	Treated Shelf-life (Days)	Increased Shelf-life (%)
Grapefruit concentrate	80	82	69	150	117
Lemon concentrate	80	82	62	183	195
Lemon juice	150	82	97	299	208
Orange juice	100	100	30	>90	>200

[a] From Pintauro (1974).

TABLE 10. Chemical Preservatives Used to Extend Fruit Shelf-Life[a]

Chemical Preservative	Treated Fruits	Type of Application
5-Acetyl-8-hydroxyquinoline salts	Fresh and processed fruits	Surface or additive
Amino acids	Fruit juices, processed fruits	Additive
Benomyl	Fruits before harvest	Pre-harvest
Biphenyl	Fresh citrus	Surface
Captan	Fruits before harvest	Pre-harvest
Cobalt salts	Fresh fruit	Surface
Dehydroacetic acid	Fresh fruits	Surface
Dibromotetrachloroethane	Fresh fruits	Surface (wax emulsion)
Diethylpyrocarbonate	Fruit juices	Additive
Diphenylamine	Fresh apples	Surface
Eugenol	Fresh fruits	Surface
Folpet	Fruits before harvest	Pre-harvest
Iodine	Fresh fruits	Surface (coated wrappers)
Methyl bromide	Fresh fruits	Fumigation
Parabens, ethyl and propyl	Dried fruits, fruit juices	Additive
O-phenylphenol & sodium salt	Fresh citrus	Surface
Pimaricin	Fresh fruits, orange juice	Additive
Propylene oxide	Dried prunes	Additive
Sodium benzoate	Fruit juices	Additive
Sorbic acid, sorbates	All processed fruits	Additive
Sulfur dioxide, sulfites	All processed fruits	Additive
Thiabenazole	Fresh citrus, bananas	Surface
Thioisomaltol	Fresh and processed fruits	Additive
Vitamin K_5	Fruit juices	Additive

[a] Data from Furia (1972); Lueck (1980); Pintauro (1974); and Salunkhe and Wu (1974).

TABLE 11. Refrigerated Storage of Fruits[a]

Fruit	Maximum Storage Period (Weeks)
Apples, Delicious	35
Apples, summer	4
Apricots	2
Blueberries	2
Cherries	1-3
Grapefruit	6
Grapes, American	8
Grapes, Vinifera	26
Lemons	26
Limes	8
Nectarines	4
Oranges	8-12
Peaches	4
Pears	26-31
Plums and prunes	4
Raspberries	3 (days)
Tangerines	4

[a] From Ryall and Pentzer (1974) and Lutz and Hardenburg (1968).

TABLE 12. Approximate Average Shelf-Life of Frozen Fruits Stored at Different Temperatures[a]

Temperature (°C)	Approximate Shelf-Life (Weeks)
0	1
-5	6
-10	21
-15	85
-20	>200

[a] Adapted from Kramer (1979).

TABLE 13. Recommended Controlled Atmosphere Storage Conditions and Potential for Benefit[a]

Fruit	Controlled Atmosphere O_2 (%)	CO (%)	Potential for benefit
Deciduous tree fruits			
Apple	2-3	1-2	Excellent
Apricot	2-3	2-3	Fair
Cherry, sweet	3-10	10-12	Good
Fig	5	15	Good
Grape	--	--	Slight or none
Kiwifruit	2	5	Excellent
Nectarine	1-2	5	Good
Peach	1-2	5	Good
Pear	2-3	0-1	Excellent
Persimmon	3-5	5-8	Fair
Plum and prune	1-2	0-5	Good
Strawberry	10	15-20	Excellent
Subtropical and tropical fruits			
Avocado	2-5	3-10	Good
Banana	2-5	2-5	Excellent
Grapefruit	3-10	5-10	Fair
Lemon	5	0-5	Good
Lime	5	0-10	Good
Olive	2-5	5-10	Fair
Orange	10	5	Fair
Mango	5	5	Fair
Papaya	5	10	Fair
Pineapple	5	10	Fair

[a] Adapted from Kader (1980).

TABLE 14. Effects of Modified Atmospheres on Fruits

Fruit	Atmospheric Conditions	Effect	Reference
Pears	2.5% O_2, 5–10% CO_2	Doubled shelf-life	a
Cherries	2.5% O_2, 10.5% CO_2	Extended shelf-life	b
Apples	2.5% O_2, 1.5% CO_2	Extended shelf-life	b
Pears	2.5% O_2, 1.5% CO_2	Extended shelf-life	b
Shamouti oranges	2.5–5.0% O_2, at 0°C	Inhibited fungal growth and rotting	c
Cranberries	Nitrogen gas	Inhibited microbial spoilage	d
Nubian plums	7.8% CO_2, 11.0% O_2 in polyethylene bags	Extended shelf-life	e
Red Delicious apples	30% CO_2 at 20°C	Inhibited PPO activity by 70%	f
Apricots	3% O_2, 5% CO_2	Extended storage life by 10 days	g

a: Allen and Claypool (1948), b: Littlefield (1968), c: Aharoni and Latter (1972), Brown and McCormack (1972), d: Lockhart et al. (1971), e: Couey (1965), f: Chaves and Tomas (1976), g: Ghena et al. (1978).

TABLE 15. Effect of Subatmospheric Storage on the Shelf-Lives of Fruits[a]

Fruit	Control Shelf-Life (days)	Subatmospheric Atmosphere Shelf-Life (days)	Increase (%)
Apricots	53	90	70
Cherries	60	93	55
Peaches	66	93	41
Pears	3.5 (months)	8 (months)	134

[a] Adapted from Salunkhe and Wu (1973).

TABLE 16. Color and Flavor Deterioration in Stored Dried Fruits[a]

Fruit	Storage Temperature (°F)	Time Required for Detectable Color Change (Weeks)	Time Required for Detectable Flavor Change (Weeks)
Figs	50	<6	56-60
	70	2-4	40-42
	90	1	9-10
Prunes	50	--	60-65
	70	8-10	48-50
	90	4-5	9-10
Raisins	50	8-10	16-18
	70	2-4	14-16
	90	1-2	9-10

[a] From Nury *et al.* (1973).

REFERENCES

Abdel-Gawad, H. S., and Romani, R. J. (1968). *Am. Soc. Hortic. Sci., 65th Annu. Market. Prog.*, p. 122.
Abd El-Latief, F., El-Azzouni, M. M., and Khattab, M. M. (1975). *Beitr. Trop. Landwirtsch. Veterinaer med. 13*, 323-328.
Aharoni, Y., and Latter, F. S. (1972). *Phytopathology 73*, 371-374.
Allen, F. W., and Claypool, L. L. (1948). *Proc. Am. Soc. Hortic. Sci. 52*, 192.
Alvarez, A. M. (1980). *HortScience 15*, 517-518.
Auda, H. (1980). *Food Irradiat. Inf. 10*, 34-40.
Ayres, J. C., and Denisen, E. L. (1958). *Food Technol. 12*, 562.
Bauernfeind, J. C., and Pinkert, D. M. (1970). *Adv. Food Res. 4*, 219-315.
Beccari, F. (1969). *Proc. Conf. Trop. Subtrop. Fruits* p. 93.
Ben-Yehoshua, S. (1978). *Proc. Int. Soc. Citric. 1977, Vol. 2*, pp. 110-115.
Blanpied, G. D. (1981). *HortScience 16*, 525-526.
Boyle, F. P., and Wolford, E. R. (1968). *In* "Freezing Preservation of Foods" (D. K. Tresler, W. B. van Arsdel, and M. J. Copley, eds.), pp. 70-112. AVI, Westport, Connecticut.
Brackett, R. E., and Marth, E. H. (1979). *J. Food Prot. 42*, 864.
Brown, G. E. (1980). *In* "Citrus Nutrition and Quality" (S. Nagy and J. A. Attaway, eds.), pp. 193-224. Am. Chem. Soc., Washington, D.C.
Brown, G. E., and McCormack, A. A. (1972). *Plant Dis. Rep. 56*, 909-912.
Brown, R. F. (1976). *U.S. Patent* 3,945,170.
Buch, M. L., Satori, K. G., and Hills, C. H. (1961). *Food Technol. 15*, 526.
Buchanan, J. R., Sommer, N. F., Fortlage, R. J., Maxie, E. C., Mitchell, F. G., and Hsieh, D. P. H. (1974). *J. Am. Soc. Hortic. Sci. 99*, 262-265.
Burroughs, L. F. (1977). *J. Assoc. Off. Anal. Chem. 60*, 100.
Cecil, S. R., and Woodroof, J. G. (1963). *Food Technol. 17*, 131-138.
Chandhry, A. R., Chandri, A. B., and Akhtar, M. P. (1979). *Pak. J. Sci. Res. 31*, 279-288.
Charlampowicz, Z., Skret, B., and Sobiech, W. (1971). *Pr. Inst. Lab. Badaw. Przem. Spozyw. 21*, 251-263.

Chaves, A. R., and Tomas, J. O. (1976). *Rev. Agriquim. Tecnol. Aliment. 16*, 114.
Ciegler, A., Kadis, S., an Ajl, S. J., eds. (1971). "Microbial Toxins: A Comprehensive Treatise," Vol. 6. Academic Press, New York.
Ciegler, A., Beckwith, A. C., and Jackson, L. K. (1976). *Appl. Environ. Microbiol. 31*, 664.
Couey, H. M. (1965). *Proc. Am. Soc. Hortic. Sci. 86*, 166.
Davis, N. D., and Diener, U. L. (1978). *In* "Food and Beverage Mycology" (L. R. Beuchat, ed.), pp. 397-444. AVI, Westport, Connecticut.
Doores, S. (1983). *CRC Crit. Rev. Food Sci. Nutr. 19*, 133-150.
Eilati, S. K., Goldschmidt, E. E., and Monselise, S. P. (1969). *Experientia 25*, 209.
Elkins, E. R. (1979). *Food Technol. 33*, 66-70.
Fellers, P. J. (1980). *In* "Citrus Nutrition and Quality" (S. Nagy and J. A. Attaway, eds.), pp. 319-340. Am. Chem. Soc., Washington, D.C.
Fennema, O. (1975). *In* "Nutritional Evaluation of Food Processing" (R. S. Harris and E. Karmas, eds.), 2nd ed. AVI, Westport, Connecticut.
Foda, Y. H., Hamed, M. G. E., and Abdalla, A. (1970). *Food Technol. 24*, 74-80.
Food and Agriculture Organization (1981). "Food Loss Prevention in Perishable Crops." United Nations Food and Agriculture Organization, Rome.
Frenkel, C., Klein, I., and Dilley, D. R. (1969). *Phytochemistry 8*, 945.
Furia, T. E. (1972). "Handbook of Food Additives." CRC Press, Cleveland, Ohio.
Ganapathy, K. M., and Singh, H. P. (1976). *Indian J. Hortic. 33*, 220-223.
Gardner, W. H. (1966). "Food Acidulants." Allied Chemical, New York.
Ghena, N., Ceausescu, M. E., Bogdan, M., Ionescu, L., Burloi, N., and Panait, E. (1978). *Acta Hortic. 85*, 343-350.
Ginsberg, L., Truter, A. B., Visagie, T. R., and Combrink, J. (1975). *S. Afr. Food Rev. 2*, 117-129.
Gomez, R. F., and Bates, R. P. (1970). *J. Food Sci. 35*, 472-475.
Grose, D. H. (1978). *IFST Proc. 11*, 141-143.
Harel, F., Mayer, M., and Shain, Y. (1966). *J. Sci. Food Agric. 17*, 389.
Hauschild, A. H. W. (1980). *In* "Safety of Foods" (H. D. Graham, ed.), 2nd ed., pp. 68-107. AVI, Westport, Connecticut.

Hazel, E., Mayer, A. M., and Shain, Y. (1965). *Phytochemistry* *4*, 783.
Heiss, R., and Eichner, K. (1971a). *Food Manuf.* *46*(5), Part 1, 53-65.
Heiss, R., and Eichner, K. (1971b). *Food Manuf.* *46*(6), Part 2, 37-42.
Herregods, M. (1979). *Fruit Belge* *47*, 79-82.
Higby, W. K., and Prichett, D. E. (1965). U.S. Patent 3,219,458, assigned to Sunkist Growers, Inc.
Horner, M. E. (1979). M.S. Thesis, University of Hawaii, Honolulu.
Hsu, E. J. (1975). *In* "Commercial Fruit Processing" (J. G. Woodroof and B. S. Luh, eds.), pp. 113-140. AVI, Westport, Connecticut.
Hudson, D. E., and Tietjen, W. E. (1981). *HortScience* *16*, 656-657.
Ito, K. A., Chen, J. K., Seeger, M. L., Unverferth, J. A., and Kimball, R. N. (1978). *J. Food Sci.* *43*, 1634-1635.
Kader, A. (1980). *Food Technol.* *34*, 51-53.
Kavanagh, T., and Kenny, A. (1978). *Ir. J. Food Sci. Technol.* *2*, 117-122.
Kealey, K. S., and Kinsella, J. E. (1979). *CRC Crit. Rev. Food Sci. Nutr.* *11*, 1-40.
King, A. D., Fields, R. K., and Boyle, F. P. (1968). *Food Eng.* *40*, 82-83.
Kitagawa, H., Suguiura, A., and Sugiyama, M. (1966). *HortScience* *7*, 59.
Kramer, A. (1979). *Food Technol.* *33*, 58-61, 65.
Lakshminarayana, S., Subramanyam, H., and Surendranath, V. (1967). *J. Food Sci. Technol.* *4*, 66.
Lazar, M. E., and Hudson, J. S. (1976). *Food Prod. Dev.* *10*, 87-89.
Littlefield, N. A. (1968). Ph.D. Dissertation, Utah State University, Logan.
Lockhart, C. L., Forsyth, F. R., Stark, R., and Hall, I. V. (1971). *Phytopathology* *63*, 335-336.
Lueck, E. (1980). "Antimicrobial Food Additives." Springer-Verlag, Berlin and New York.
Luh, B. S., and Kean, C. E. (1975). *In* "Commercial Fruit Processing" (J. G. Woodroof and B. S. Luh, eds.), pp. 141-265. AVI, Westport, Connecticut.
Luh, B. S., Feinberg, B., and Chung, J. I. (1975). *In* "Commercial Fruit Processing" (J. G. Woodroof and B. S. Luh, eds.), pp. 266-373. AVI, Westport, Connecticut.
Lutz, J. M., and Hardenburg, R. E. (1968). "Commercial Storage of Fruits, Vegetables, and Florist and Nursery Stock," Agric. Handb. No. 66. U.S. Dept. of Agriculture, Washington, D.C.

MacLachlan, J., and Gormley, R. (1974). *J. Sci. Food Agric.* *25*, 165-177.
Maier, V. P., Brewster, L. C., and Hsu, A. C. (1971). *Citrograph 56*, 373-375.
Maxie, E. C., Sommer, N. F., and Mitchell, F. G. (1971). *HortScience 6*, 202-204.
Mayer, A. M., Hazel, E., and Shain, Y. (1964). *Phytochemistry 3*, 447.
Moore, J. N., Bowden, H. L., and Sistrunk, W. A. (1975). *HortScience 10*, 86.
Morales, G. J. C., and Lomelin, G. J. M. (1974). *Tecnol. Aliment. (Mexico City) 9*, 115-124.
National Academy of Sciences (1978). "Post-Harvest Losses in Developing Countries." Nat. Acad. Sci., Washington, D.C.
National Academy of Sciences, National Research Council (1965). "Chemicals Used in Food Processing." Nat. Acad. Sci., Washington, D.C.
Nicolosi Asmundo, C., Cataldi Lupo, M. C., and Zamorani, A. (1980). *Tecnol. Aliment. (Buenos Aires) 3*, 33-38.
Nury, F. S., Brekke, J. E., and Bolin, H. R. (1973). *In* "Food Dehydration" (W. B. van Arsdel, M. J. Copley, and A. I. Morgan, eds.), 2nd ed., Vol. 2, pp. 158-198. AVI, Westport, Connecticut.
Palmer, J. K., and Roberts, J. B. (1967). *Science 157*, 200.
Pifferi, P. G., and Cultrera, R. (1974). *J. Food Sci. 39*, 786.
Pintauro, N. D. (1974). *Food Technol. Rev.* No. 17.
Pohland, A. E., and Allen, R. (1970). *J. Assoc. Off. Anal. Chem. 53*, 688.
Ponting, J. D., Stanley, W. L., and Copley, M. J. (1973). *In* "Food Dehydration" (W. B. van Arsdel, M. J. Copley, and A. I. Morgan, eds.), 2nd ed., Vol. 2, pp. 199-245. AVI, Westport, Connecticut.
Processed Apple Institute, Patulin Technical Committee (1977). "Patulin: A Guide to Action." Processed Apple Institute, Atlanta, Georgia.
Ramasastri, B. V. (1974). *Indian J. Nutr. Diet. 11*, 134-136.
Ricci, P. (1972). *Ann. Phytopathol. Soc. Jpn. 4*, 109-117.
Rodrigues, J., and Subramanyam, H. (1966). *J. Sci. Food Agric. 17*, 425.
Ryall, A. L., and Pentzer, W. T. (1974). "Handling, Transportation, and Storage of Fruits and Vegetables," Vol. 2, AVI, Westport, Connecticut.
Sacharow, S., and Griffin, R. C. (1980). "Principles of Food Packaging." AVI, Westport, Connecticut.
Salunkhe, D. K., and Wu, M. T. (1973). *J. Am. Soc. Hortic. Sci. 98*, 113-116.
Salunkhe, D. K., and Wu, M. T. (1974). *CRC Crit. Rev. Food Technol. 5*, 15-24.

Salunkhe, D. K., Cooper, G. M., Dahliwal, A. S., Boe, A. A., and Rivers, R. L. (1962). *Food Technol. 16*, 119.
Salunkhe, D. K., Brennand, C. P., and Bolin, H. R. (1971). *Utah Sci. 32*, 123.
Salunkhe, D. K., Wu, M. T., Do, J. Y., and Maas, M. R. (1980). *In* "Safety of Foods" (H. D. Graham, ed.) 2nd ed., pp. 198-264. AVI, Westport, Connecticut.
Sands, D. C., McIntyre, J. L., and Walton, G. S. (1976). *Appl. Environ. Microbiol. 32*, 288.
Sawatditat, A., Sundhagul, M., and Varangoon, P. (1979). *In* "Monographs on Appropriate Industrial Technology," No. 7, pp. 102-106. United Nations Industrial Development Organization, New York.
Scott, P. M. (1974). *In* "Mycotoxins" (I. F. H. Purchase, ed.), pp. 383-403. Am. Elsevier, New York.
Sharma, S. P., Bhanumurthi, J. L., and Srinivasan, M. R. (1974). *J. Food Sci. Technol. 11*, 171-174.
Shaw, P. E., Tatum, J. H., and Berry, R. E. (1967). *Carbohydr. Res. 5*, 266.
Shewfelt, A. L. (1975). *In* "Commercial Fruit Processing" (J. G. Woodroof and B. S. Luh, eds.), pp. 497-538. AVI, Westport, Connecticut.
Singh, B. P., Singh, H. K., and Chauhan, K. S. (1981). *Indian J. Agric. Sci. 51*, 44-47.
Singh, R. P., and Heldman, K. R. (1975). *Trans. ASAE 18*, 156-162.
Smith, W. L. (1962). *Bot. Rev. 28*, 411.
Somogyi, L. P., and Luh, B. S. (1975). *In* "Commercial Fruit Processing" (J. G. Woodroof and B. S. Luh, eds.), pp. 374-429. AVI, Westport, Connecticut.
Splittstoesser, D. F. (1978). *In* "Food and Beverage Mycology" (L. R. Beuchat, ed.), pp. 83-110. AVI, Westport, Connecticut.
Tatum, J. H., Shaw, P. E., and Berry, R. E. (1967). *J. Agric. Food Chem. 15*, 773.
Tatum, J. H., Shaw, P. E., and Berry, R. E. (1969). *J. Agric. Food Chem. 17*, 38.
Tuli, V., Dedolph, R. R., and Wittwer, S. H. (1962). *Mich., Agric. Exp. Stn., Q. Bull. 45*, 223.
U.S. Congress, Office of Technology Assessment (1979). "Open Shelf-Life Dating of Foods." U.S. Govt. Printing Office, Washington, D.C.
Vamos-Vigyazo, L. (1981). *CRC Crit. Rev. Food Sci. Nutr. 15*, 49-127.
Van Blaricom, L. O., and Hair, B. L. (1973). *Tech. Bull.--S. C. Agric. Exp. Stn. 1043*.
Walker, H. W., and Ayres J. C. (1970). *In* "The Yeasts" (A. M. Rose and J. S. Harrison, eds.), Vol. 3, pp. 463-527. Academic Press, New York.

Walker, J. R. L. (1964). *Aust. J. Biol. Sci. 17*, 360.
Walker, J. R. L. (1975). *Enzyme Technol. Dig. 4*, 89.
Walton, G. S., Sands, D. C., and McIntyre, J. L. (1976). *Proc. Am. Phytopathol. Soc. 3*, 255.
Wang, C. Y., and Mellenthin, W. M. (1974). *HortScience 9*, 196.
Woodroof, J. G. (1975a). *In* "Commercial Fruit Processing" (J. G. Woodroof and B. S. Luh, eds.), pp. 32-77. AVI, Westport, Connecticut.
Woodroof, J. G. (1975b). *In* "Commercial Fruit Processing" (J. G. Woodroof and B. S. Luh, eds.), pp. 595-626. AVI, Westport, Connecticut.
Zaytsev, A. N., Shillinger, Y. I., and Kamaldinova, Z. M. (1975). *Food Irradiat. Inf. 5*, 43-54.

CHAPTER 6

CHOCOLATE

IAN HORMAN
UMBERTO BRACCO
Nestlé Research Laboratories
Vevey, Switzerland

I. COCOA AND CHOCOLATE COMPOSITION

In addition to the constituents shown in Table I, cocoa contains tocopherols: these are liposoluble vitamin E compounds and are also antioxidants. They are retained in cocoa butter and therefore in chocolate (see Table II), where they confer a natural protection against oxidation during storage. The low water content of processed chocolate ensures that fungal growth is not a direct problem, but cocoa beans themselves absorb much water if stored in humid conditions (Fig. 1), so the raw products are subject to microbiological spoilage. The major constituents of chocolate are sugars and fats (Table II), and it is in particular physical changes in the fat fraction that dictate the behavior of chocolate during storage. Table III shows the distribution of triglycerides and individual fatty acids in cocoa butter. The triglycerides are mainly monounsaturated with oleic acid in the 2-position of the glycerol bridge.

II. TRIGLYCERIDE CRYSTAL STRUCTURE

Solid tristearin was first observed to have two (Heintz, 1849), then three (Duffy, 1853) distinct melting points. This behavior is common among triglycerides, the different melting points relating to different crystal structures as illustrated in Figs. 2-4. The degree of order in the structures increases through the series $\alpha < \beta' < \beta$, resulting in increased crystal densities along with corresponding increases in melting points and heats of fusion.

Figure 5 shows the molecular structure and proposed packing structure of 2-oleodistearin (SOS), one of the principal cocoa butter triglycerides: the cis double bond of the oleyl residue gives a bend in the chain. Figure 6 gives a more complete picture of the polymorphic behavior of triglycerides.

III. COCOA BUTTER POLYMORPHS

Cocoa butter is the lipid fraction of the cocoa mass. Table IV summarizes the melting points, heats of fusion, and densities of the different cocoa butter polymorphs reported in the literature, and shows the corresponding nomenclatures used by different authors to describe the various polymorphic forms. The melting points of the individual triglycerides of cocoa butter both isolated and in the cocoa butter melt are shown in Fig. 7. Figure 8 illustrates the melting and crystallization temperature ranges for the different polymorphs of whole cocoa butter; the dynamic stability of these polymorphs is shown in Table V, and the corresponding solid fat index (SFI) values, which express the proportions of fat in the solid and liquid phases, are summarized in Table VI.

The form in which cocoa butter crystallizes is of utmost importance in chocolate production. Tempering involves holding the chocolate mass at about 30°C (see Fig. 7) for a short time, and promotes crystallization in the β and β' forms. This prolongs the storage lifetime of chocolate products by delaying the onset of fat bloom (see Section IV), and also results in a maximum contraction of the chocolate mass on cooling, thus ensuring that the solid chocolate leaves the molding trays cleanly.

IV. STORAGE DEFECTS

Fat bloom is the major problem encountered during chocolate storage. Figures 9 and 10 illustrate examples of bloom as it is seen on tempered and on untempered chocolate, respectively. Table VII summarizes the characteristics, the causes, and the prevention of bloom: β crystals of cocoa butter grow on the chocolate surface, apparently as a result of migration of the monounsaturated triglycerides through the liquid phase of the chocolate mass. The amount of liquid triglyceride present (Table VII) is therefore important, and this in turn is governed by the degree of tempering and the storage temperature. The effects of storage temperature on bloom formation are described in Table VIII.

Bloom can also arise as a result of using fats in chocolate that are incompatible with cocoa butter. Figure 11 shows the evolution of the melting point of tempered and untempered cocoa butter as a function of time at different storage temperatures, and Fig. 12 the effect of admixing nut oil or milk fat. Milk fat is physically and chemically incompatible with cocoa butter, and changes both the crystalline form and the texture of the solid chocolate (see Table IX). However, its incorporation almost eliminates fat bloom (Kleinert, 1961). Contrary to this, the incorporation of most other oils and fats enhances bloom, and this limits severely the number of fats that can be used to replace cocoa butter in chocolate. Table IX shows three classes of fats and oils used in chocolate. The compatible fats can be used to replace part of the cocoa butter (cocoa butter equivalents); incompatible fat like palm kernel stearines can be used as a total replacement for the whole cocoa butter fraction, as it presents the same physical characteristics (cocoa butter alternative); other incompatible oils and fats, like milk fat and nut oil, are also used when it is desired to change the physical characteristics of the final product, for example in soft chocolate centers.

Other storage problems are summarized in Table X. Sugar bloom is another defect of appearance: on casual inspection it looks similar to fat bloom but is composed of a layer of sugar crystals on the surface. Oxidative and lipolytic rancidity are flavor defects, the former coming from oxidation of unsaturated fats and the latter from enzymatic hydrolysis of short- and medium-chain triglycerides.

V. PACKAGING

Efficient packaging obviously plays an important role in shelf-life. It must be as impervious as possible to light, oxygen, humidity, foreign odors, pests, and so on. The principal packaging materials are laminates of aluminum foil, waxed paper, and plastic films as described in Table XI.

VI. SUMMARY OF STORAGE DEFECTS AND CONDITIONS

Chocolate is sold in many forms (e.g., dark, milk, white, as coated confectioneries) where the product often contains not only chocolate but other ingredients. These ingredients, such as fruit, nuts, and caramel, influence the likelihood of occurrence of the different storage defects, as illustrated semiquantitatively in Table XII. The optimum temperature and humidity conditions during the different phases of manufacture and storage are to be found in Table XIII, and defects in handling and storage are summarized in Table XIV.

TABLE I

The composition of Cocoa mass, Cocoa shells and Germs

	Mass		Shells		Germs
	a	b	a	b	a
Water	2.0	2.1	10	3.8	7 - 10
Fat	55.7	54.7	2-4	3.4	2.5-5.0
Ash	2.7	2.7	8	8.1	5.7-6.9
Nitrogen :					
Total	2.2	2.2	2.7	2.8	5
Soluble	0.8				
Protein N		1.3	2.0	2.1	4
Purine N	0.4		0.7		
NH_3 Nitrogen		0.04		0.04	
Protein	11.9				
Theobromine	1.2	1.4		1.3	2.5
Caffeine	0.2	0.07		0.1	
Carbohydrates :					
Glucose,fructose (reducing)	1	0.1	0.1	0.1	2
Sucrose(non red.)	0.5	0		0	
Starch	6.2	6.1	3	2.8	0.5
Pectines		4.1	7	8.0	
Fibre		2.1	16	18.6	2.5
Cellulose	9.3	1.9	27	13.7	4
Pentosans	1.5	1.2	7	7.1	
Mucilage & Gums		1.8		9.0	
Polyhydroxyphenols	6.2	6.2		3.3	
Acids :	1.4				
Volatiles(as acetic)					
free	0.2	0.14		0.1	
bound	0.2				
Non volatiles (as citric)	0.7			0.7	
oxalic	0.4	0.3		0.3	

[a] Fincke, H. (1965).

[b] Powell, B.D. and Harris, T.L. (1964).

TABLE II

Composition of Chocolate[a]

			White Chocolate			Milk Chocolate			Plain Chocolate		
			a	b	c	a	b	c	a	b	c
Total calories			545		21.8	498		19.2	470		18.8
from lipids			302		46.5	297		39.6	249		38.3
from carbohydrates			208		13.1	217		13.7	214		13.5
Proteins				7.9	12.2		7.2	11		4.11	6.3
Carbohydrates				54.4	13.6		57.2	14.3		58.1	14.5
Lipids				35.9	47.8		30.0	39.9		29.7	39.6
Polyunsaturated				1.6	26.5		1.4	23.2		1.6	26.2
Calcium		mg		289	28.9		221.8	22.18		23.8	2.4
Magnesium		mg		30.6	7.6		48.7	12.12		81.0	20.3
Phosphorus		mg		225	22.5		209.5	20.95		128.3	12.8
Vitamins	A	IU		337	6.7		306.6	6.12		39.2	0.78
	B1	mg		0.08	5.3		0.08	5.3		0.06	4.0
	B2	mg		0.40	23.5		0.31	19.23		0.06	3.5
	C	mg		1.54	2.5		1.14	1.9			
	D	IU		79.9	20.0		68.3	17.07		54.0	13.5
	E	IU		2.51	8.3		2.0	6.67		2.7	9.0
Iron		mg		0.17	0.94		0.82	4.56		2.05	11.4
Copper		mg		0.30	15.0		0.42	21.0		0.7	35

[a] Per 100 g dry matter. [b] % w/w [c] As % of daily requirements

TABLE III
Triglycerides and Fatty Acids of Cocoa Butter

Triglycerides		97 - 99 %
trisaturated		3%
monounsaturated [a]		83%
POP	21 - 23 %	
POS	55 - 58 %	
SOS	19 - 22 %	
di- and triunsaturated		14%
Fatty Acids		
free (oleic acid)		0.5 - 2 %
combined in triglycerides		
C12:0		0.1 %
C13:0		traces
C14:0		0.2 %
C15:0		traces
C16:0		25 - 30 %
C17:0		traces
C18:0		32 - 37 %
C20:0		1 %
C22:0		0.2 %
C16:1		1 %
C18:1, n-9		30 - 37 %
C18:2, n-6		2 - 4 %
C18:3, n-3		0.3 %
C20:1, n-9		traces
Unsaponifiables		0.6 - 1 %
Phospholipids		0.02 - 0.2 %

[a]POP = 2-oleodipalmitin ;
POS = 1-palmito-2-oleo-3-stearin ;
SOS = 2-oleodistearin

TABLE IV

Melting points, Heats of fusion and Densities of Cocoa Butter Polymorphs

(Polymorphs[a])					Melting Points °C[h]					Heat of Fusion °C		Density
b	c	d	e	f	b	c	d	e	f	b	g	j
I	1	γ	-	1	17.3	(14.9 (16.1	(16.0 (18.0	-	(16.0 (18.0	-		
II	2	α	α	2	23.3	(17.0 (23.2	(21.0 (24.0	-	(21.0 (24.0	(20.6 (20.8	19.0	0.9720 (7)
III	3	-	β_2^1	3	25.5	(22.8 (27.1	-	-	(25.5 (27.1	(26.5 (26.9		
IV	4	β''	β_1^1	4	27.5	(25.1 (27.4	(27.0 (29.0	-	(27.0 (29.0	(28.1 (29.0	28.0	0.9903 (8.3)
V	5	-	pre-β	5	33.8	(31.3 (33.2	-	34.5	(30.0 (33.8	(32.7 (33.8		
VI	6	β	β	6	36.3	34.5	34.0 35.0	36.2	34.0 36.3	35.4 35.4	36.0	1.0026 (9.6)

a Corresponding nomenclatures used by different authors. b Wille, R.S. & Lutton E.S. (1966).
c Huyghebaert, A. & Hendrick H (1971) d Vaeck, S.V.(1961) e Witzel, H & Becker K (1969)
f Johnston, G.M. (1972) g Vaeck, S.V.(1951) h Clear point j At 0°C. Value in parentheses is contraction between 30°-10°C. Vaeck, S.V. (1961)

TABLE V
Stability of Cocoa Butter States[a]

Temp. °C		STATE					
		I	II	III	IV	V	VI
-30	b	> 4 hrs					
	c						
0	b	15 secs	5 hrs				
	c	15 mins					
5	b		< 2 hrs	5 days	>1 week		
	c		16 hrs				
10	b			1 day			
	c			3 days[d]			
16	b	2 secs	< 1 hr	< 4 hours	2 days	>14 wks	
	c				2 wks		
21	b				3 hrs	7 wks	
	c				1 day	>18 wks	
26	b		melted[e]	melted[e]	<1 hour	4 days	
	c				1 hour	3 wks	stable

[a] Wille, R.L. & Lutton, E.S. (1966)

[b] Approx. time at which transformation to next higher state begins

[c] Approx. time to complete transformation.

[d] Transformed in solid state directly to V.

[e] Resolidified within 30 min as V.

TABLE VI

Solid Fat Index of the Different States of Cocoa Butter [a]
(by dilatometry)

Form	II	III	IV	V	VI
10°C	63.5	70	73	81.5	(71)
15°C	(60)	68	70	82.5	(72)
20°C	(35)	55	60	81	(73)
25°C	0	0	30	73	75
28°C	0	0	0	68	70
30°C	0	0	0	53	43
35°C	0	0	0	0	8
40°C	0	0	0	0	0

[a] Wille, R.L. and Lutton, E.S. (1966)

Sample size : 2 g

Form II : melt, chill 1 h at 0°C (32°F)

Form III : melt, chill 1 h at 0°C, store 2 h at 10°C(50°F)

Form IV : melt, chill 1 h at 0°C, store 2 h at 15.6°C(60°F)

Form V : melt, chill 1 h at 0°C, store 1 day at 21.1°C (70°F), 1 day at 26,7°C(80°F), overnight at 10°C (50°F)

Form VI : melt, chill 1 h at 0°C, store 1 day at 21.1°C, 3 weeks at 26.70°C, overnight at 10°C.

TABLE VII
Fat Bloom in Chocolate

Appearance :	- generally as a greyish or whitish film on the chocolate surface
	- sometimes as bigger crystal clusters
	- looks mouldy
Explanation :	- bloom crystals scraped from chocolate surface melt at 34.3 - 34.6°C, very close to mp. of 34.5°C observed for pure β-form of cocoa butter.[a]
	- evidence for migration of triglycerides, particularly SOS, POP, POS, to surface of chocolate
Causes :	- incorrect or inadequate tempering
	- incorrect cooling procedures, such as covering unheated centers or differential cooling
	- storage at warm temperatures (see table)
	- admixture of fats incompatible with cocoa butter
	- covering soft centers having a high fat content
	- prolonged storage
Prevention :	- tempering chocolate ; this involves temperature jump of 40°C liquid chocolate to 29°C, at the same time applying high shear forces, followed by a 1° temperature jump increase, and gives large numbers of β and β' crystal nuclei
	- use of cocoa butter - compatible fats (see table)
	- use of anti-bloom agents, such as butter fat (dehydrated cow butter) and hydrogenated arachis oils [b]
	- avoidance during production of conditions known to promote bloom

[a] Minifie, B.W. (1980)

[b] Kleinert, J. (1961)

TABLE VIII

Effect of Storage Temperature and Tempering on Fat Bloom Formation[a]

	18°C	23°C	27°C	29.5°C
Tempered	Glossy surface. No bloom.	A few white specks. Surface dull.	Surface covered with white specks.	Whole surface covered with white bloom.
Untempered	Mottled surface but glossy.	Mottled surface dull.	Very mottled.	Very discolored large white blotches

[a]Minifie, B.W. (1980). Cocoa butter mixed with carbon black to give dark surface against which bloom can be seen.

TABLE IX

Cocoa Butter Compatible and non-compatible Fats and Oils used in Confectionary Products

	Common Name	MP(°C)	Fatty Acids						Trigl. Composit.		
		MP °C	C_{14}	C_{16}	C_{18}	C_{18}=	C_{18}==	C_{20}	Sat.	Mono. unsat	Poly. unsat
Compatible[a]:	Sal Oil	30-35	-	5	44	42	3	6	2	64	34
	Illipe Butter	34-38	-	20	41	36	1	-	4	79	17
	Kokum Butter	38-42	1	4	53	40	2	-	2	74	24
	Shea Butter	32-34	-	6	40	50	4	-	-	52	48
Incompatible[b]:											
	Palm Kernel[d] stearines	32-34	19	8	10	6	2	-	90	10	0
Incompatible[c]:											
	Milk fat	28-34	12	23	7	42	19	-	40	45	15
	Almond oil	-	1	5	77	17	-	-	-	10	90
	Hazelnut oil[e]	-	-	5	2	84	9	-	-	10	90
Cocoa butter		33-35	-	26	33	36	3	1	3	78	19

[a] Physically and chemically compatible with cocoa butter

[b] Same physical characteristics but cannot be mixed with cocoa butter

[c] Chemically and physically different from cocoa butter [d] C_{12}: ∿52% [e] C_{18}===: 0.5%

TABLE X
Other Storage Problems

Sugar Bloom	- caused by humidity - use of unrefined sugar having high moisture content and hygroscopic activity
Oxidative rancidity	- not a problem in pure cocoa butter because of natural antioxidants in cocoa solids - can arise from lipids in nuts (almond, brazil, cashew, hazelnut, peanut, plean, walnut); fruits (raisin, currant); soya bean lecithin used to decrease chocolate viscosity ; eggs.
Lipolytic rancidity	- milk fat or milk powder can give butyric acid which imparts a cheesy note. - palm kernel oil or coconut oil can give lauric acid which imparts a soapy taste.
Defects in other ingredients	- starch, through humidity, or liquid sugar, through crystallization, can give rise to graininess or sandiness. - mould growth in nuts.
Absorption of odours	- if packaging inadequate, liposoluble odours can be absorbed. Can be a problem in kiosks in filling stations.

TABLE XI
Principal Packaging Materials[a]

	Laminate	Uses	Comments[b]
1.	Aluminium foil / Heat seal plastic	Moisture proof, insect proof wrap for chocolate blocks where closed wrapping is required.	Low water vapour permeability but fragile. Good odour barrier.
2.	Aluminium foil / waxed tissue	Close wrapping without heat seal	Suitable for warm climates:prevents fat-staining of outer package and also protects against odour
3.	Aluminium foil / paper/polythene or Paper/aluminium foil/polythene	Strong wrap for very hygroscopic products where heat-sealing is required	Good odour barrier
4.	Paper/PVDC[c]/ polythene or Paper/polythene/ PVDC	Similar to 3, but cheaper	Good odour barrier

[a] Adapted from Minifie, B.W. (1970)

[b] Water vapour permeability, at 75% relative humidity and 25°C = 0.1 - 0.2 g/m^2/day; at 90% relative humidity and 38°C = 0.2 - 0.5 g/m^2/day.

[c] PVDC = polyvinyldichloride

TABLE XII
Different Types of Chocolate, and Likelihood of a Given Defect

Chocolate Type	Fat Bloom	Sugar Bloom	Oxidative rancidity	Lipolytic rancidity	Sandiness
Dark	XXX	XX	X	X	a
Milk	X	XX	XX	XX	a
White	X	X	XX	XX	a
Coating					
-on fat filling	XXXX	XX	X	XXX	-
-on sugar filling	XXX	XXX	X	X	XX
Chocolate					
-with fruit	XXX	XX	XXX	X	-
-with nuts	XXX	X	XXXX	X	-
-with cereals	XXX	X	XXXX	X	-

[a] Sometimes occurs through development of large β-form triglyceride crystals.

TABLE XIII
Optimum Air Conditions for Storage of Confectionary Products[a]

For raw materials :	Dry Bulb Temperature °C	Relative Humidity
Nuts (insect inhibition)	7	65 - 70
Nuts (rancidity)	1-3	85 - 80
Eggs	-1	85 - 90
Chocolate	16	50 - 55
Butter	-7	
Dates, Figs, etc.	4-7	75 - 65
Corn Syrup	32-35	
Liquid Sugar	24-27	40 - 30
During processing :		
Chocolate pan supply	13-17	55 - 45
Enrober room	29-33	25
Chocolate cooling tunnel supply air	4-7	95 - 80
Moulded goods cooling	4-7	85 - 70
Chocolate packing room	18	50
Chocolate finished stock storage	10	50
Centers tempering room	24-27	35 - 30
Sanded gum drying	38	25 - 40
Gum finished stock storage	10-18	65
Sugar pan supply air	29-40	30 - 20
Polishing pan supply air	21-27	50 - 40
Hard candy cooling tunnel supply air	16-21	55 - 40
Hard Candy packing	21-24	40 - 35
Hard Candy storage	18-22	35 - 40
Caramel rooms	21-27	40

[a]Anon. (1957)

TABLE XIV

Summary of Problems in Handling and Storage[a]

1. Gray Bottoms:

   (a) Boxed goods :
   Coating is untempered in the Bottomer.
   Chill belt is too cold
   Tunnel belt is too cold at point of transfer from enrober belt.

   (b) Bar goods :
   Any of the above
   Heat sealer applied to label is too hot.

2. Dull Surface:

   Enrober operating temperature is too low
   Tunnel is too cold near the entrance
   Centers are too cool
   Blower air is too cool
   Relative humidity is too high in the tunnel, enrober room or in storage. The latter causes dulling gradually; the former two immediately.
   Coating is permeable and center moisture has penetrated to the surface. Permeability may be caused by improper temper or hardening.

3. Greasy Surface:

   Insufficient or unstable temper in the coating.
   Blower air is too warm.
   Tunnel is too cold near the entrance, causing case-hardening.
   There may be too small a temperature gradient between tunnel entrance and discharge end.
   Tunnel belt may be insufficiently loaded with dipped goods.
   Chocolate coating may be "contaminated" with Confectioners Coating
   An improper substitute fat was added to the chocolate to thin it down.
   If a Confectioners Coating, an incompatible hard butter or one of a different melting point was used.

4. Streaky Surface :

   Too high a ratio of untempered to tempered coating.

Insufficient premixing of the above.
Non-uniform center temperatures
Adding cocoa butter at the enrober or cooking kettle.
Blower air is too warm or too cool.
Using a substitute fat or, in Confectioners Coating, the wrong hard butter.

5. Finger Prints :

Sweaty hands at the packing table. They may cause either fat bloom or sugar bloom.
Condensation that hasn't completely dried off before pieces are handled. The relative humidity in the tunnel may be too high.
Greasy surfaces as described in 3 above will fingerprint easily.
Packers may be holding dipped pieces too long.

6. Fat Bloom :

Failure to completely melt coating before tempering.
Unstable or insufficient temper.
Coating was seeded with crystals too large, even though stable
Coating was cooled too rapidly.
Centers were wet on the surface. This is particularly dangerous where the centers contain fat that is not well emulsified into the center. Use an emulsifier such as lecithin.
Blower air is too cold
Tunnel is too cold at or near its entrance.
Packaging materials are too warm.
Heat sealer is too hot or product is not sufficently protected.
Storage temperatures are too high or are fluctuating
Product was shipped too soon after manufacture in warm weather. Beta prime crystals melt and migrate to the surface before transforming to the Beta form, instead of going through this stage in solid form and in their original position.

7. Sugar Bloom :
High relative humidity in packing room or storage.
Condensation occurs as over-chilled product enters packing room.

Sweaty hands at the packing table.

8. Off-Flavors in Finished Goods :

Coating was over-heated.
Blower air is carrying odors.
Tunnel is not clean : moldy or fermenting material on its floor.
In Confectioners Coating, moisture from centers has permitted hydrolysis of coating fats. Contact supplier.
In Confectioners Coating, marshmallow goods have been excessively wetted by the tip flattening roller. May result in rancid, soapy and musty flavor.
In hand dipping, nail polish odors and perfumes are absorbed.
Condensation on centers has permitted mold growth under the coating.
Adhesives have leaked thru label overlap before drying: volatile solvent is absorbed.
Label glues have fermented or become moldy.
Goods have been stored, either as coating or as a finished product, near highly aromatic items.
Boxes, cartons or paper were manufactured with odoriferous fungicides.
Odoriferous inks were used on labels or they were too freshly printed.

---

[a] Adapted from Cook, L.R. (1963)

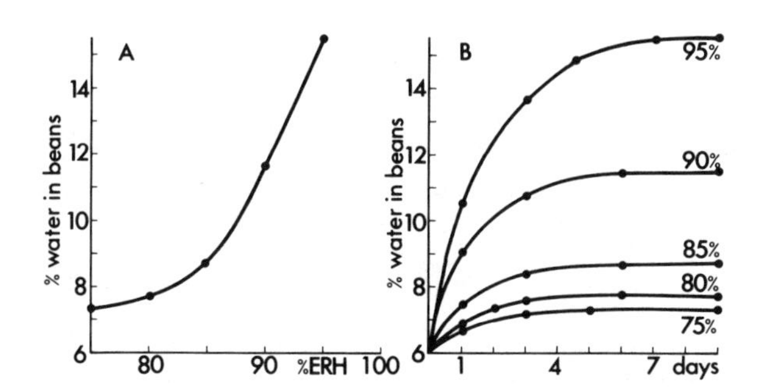

Fig. 1. Absorption isotherms of cocoa beans. Percentage water content as a function (A) of equilibrium relative humidity (ERH) and (B) of time and ERH (redrawn from Burle, 1962).

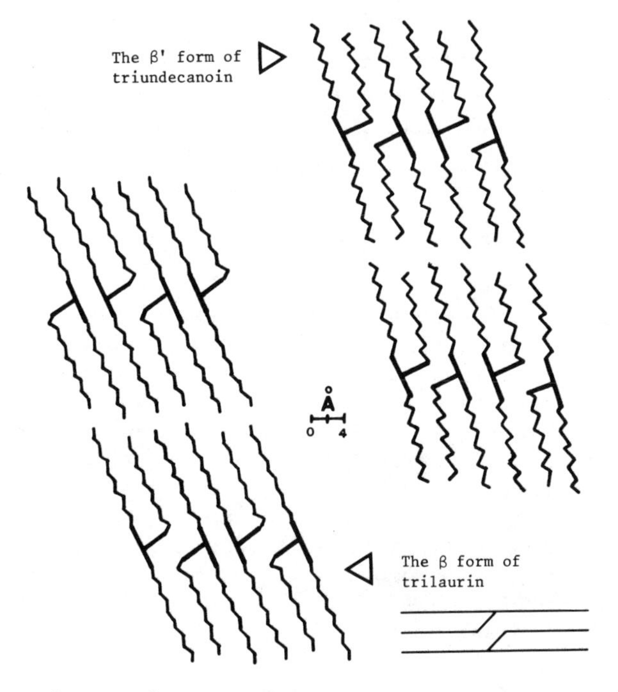

Fig. 2. Triglyceride crystal structures. Comparative structural arrangements of fatty acid chains in β' and β forms of triglyceride crystals. Note in each case the "tuning fork" conformation of individual triglycerides, and the basic packing structure (see insert) as proposed by Clarkson and Malkin (1934), from X-ray powder diffraction (from data presented by Hernqvist, 1984).

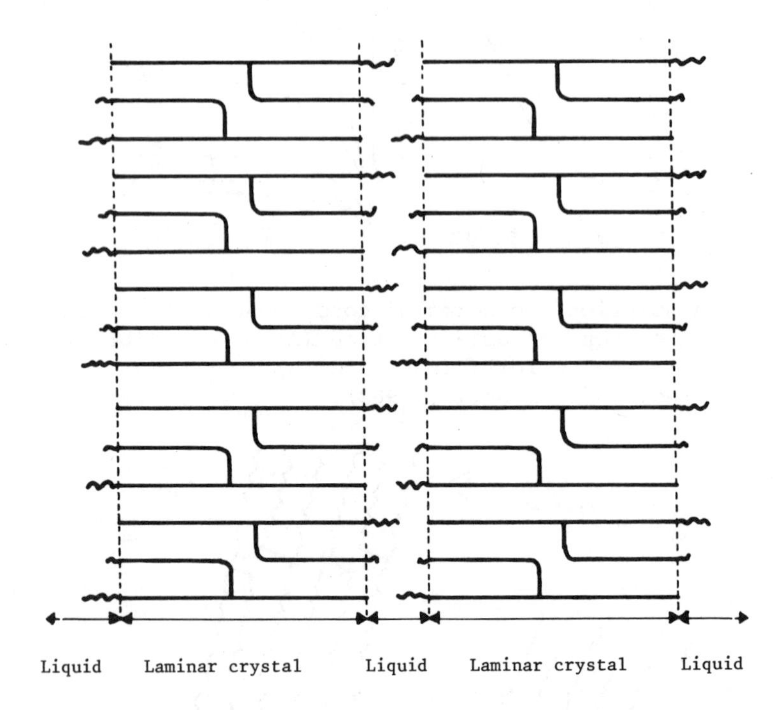

Fig. 3. Triglyceride crystal structures. Schematic representation of the α form of triglyceride crystals. The proposed structure consists of laminar crystals separated by a liquidlike region in which the methyl-C-terminal CH_2 and CH_3 groups are highly mobile. The structure shows a lower degree of order than in the β' or β forms (from data presented by Hernqvist, 1984). An increase in the mobility of CH_2 groups with increasing distance from the glycerol center is reported in bilayer membranes formed in oil-water emulsion, as shown using deuterium NMR by Stockton and Smith (1976).

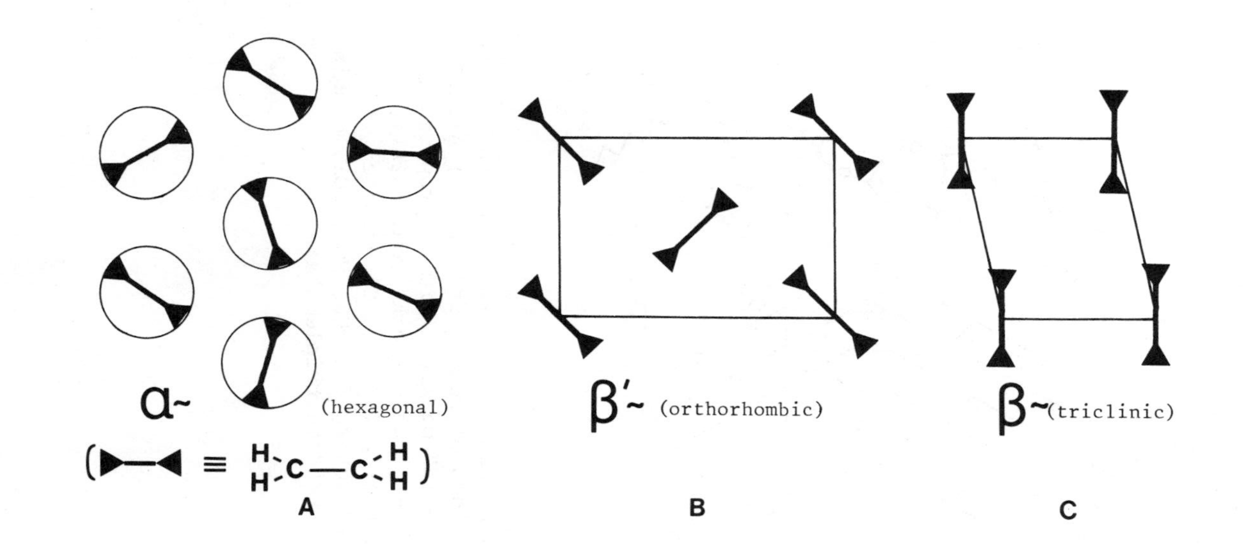

Fig. 4. Triglyceride crystal structures. Cross section through adjacent chains in the α, β′, and β polymorphic forms. (A) Dimethylene groups in adjacent chains retain high degree of torsional mobility and bear no fixed spatial geometry relative to each other. (B) Increased order in dimethylene groups gives increased packing density between chains and less chain mobility. (C) Dimethylene groups have achieved maximum packing density (from data presented by Hernqvist, 1984).

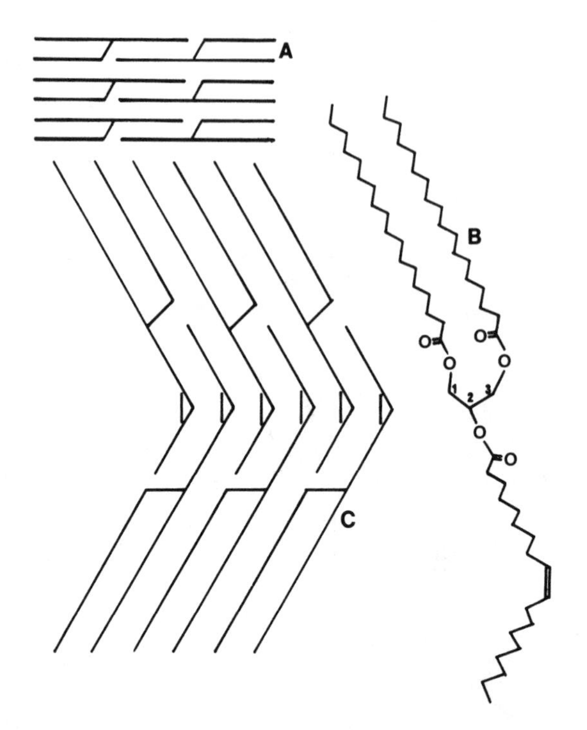

Fig. 5. Proposed packing structures in monounsaturated triglycerides of cocoa butter. (A) Alternative "triple-chain" packing form as proposed by Lutton (1948). (B) "Tuning-fork" configuration of 2-oleodistearin (SOS). (C) Proposed "bent" crystal structure for POS, POP, SOS, combining "tuning forks" in triple chain.

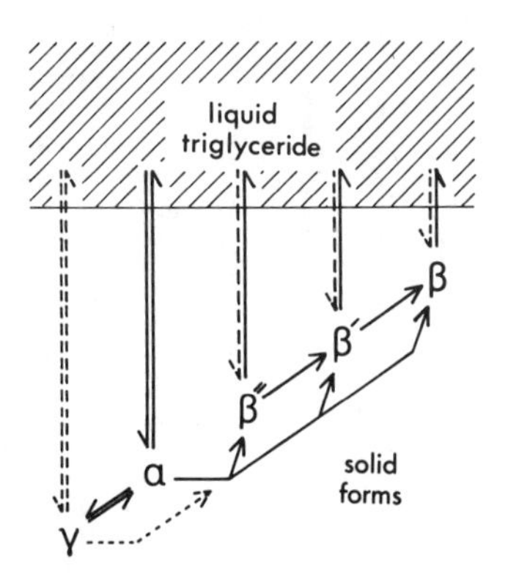

Fig. 6. Polymorphic transitions of triglycerides. Rapid cooling of the melted triglycerides generally yields the α form, which then undergoes a series of polymorphic transitions to finish in the stable β form. All transitions are irreversible, except $\alpha \rightleftharpoons \gamma$. Dotted lines show the possibility of direct crystallization in β forms depending on the holding temperature and cooling speed of the cooled liquid. Transition rates are temperature dependent. Not all triglycerides give clearly defined crystals for all of the polymorphic forms.

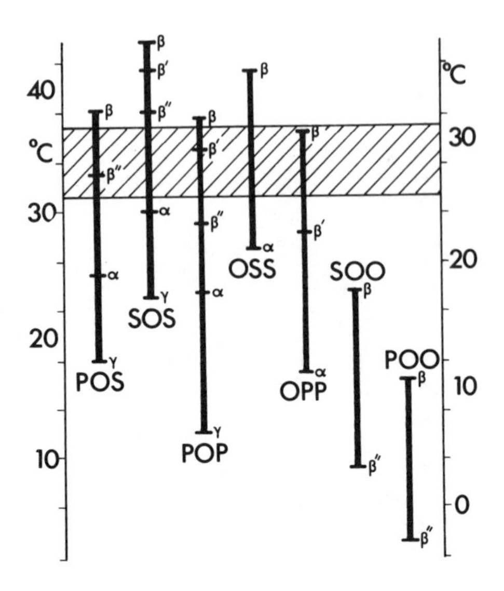

Fig. 7. Polymorphism of cocoa butter and tempering zone. (left scale) Melting points of different polymorphic forms of individual cocoa butter triglycerides. (right scale) Melting points of triglycerides in whole cocoa butter. Shaded area corresponds on right scale to tempering zone, where liquid chocolate mass is "seeded" with predominantly β' or β crystals, which continue to grow as chocolate is further cooled in molds to about 18°C. POS = 1-palmito-2-oleo-3-stearin; SOS = 2-oleodistearin; POP = 2-oleodipalmitin; OSS = 1-oleodistearin; OPP = 1-oleodipalmitin; SOO = 1-stearodiolein; POO = 1-palmitodiolein. γ Form corresponds to sub-α (data from Fincke, 1965).

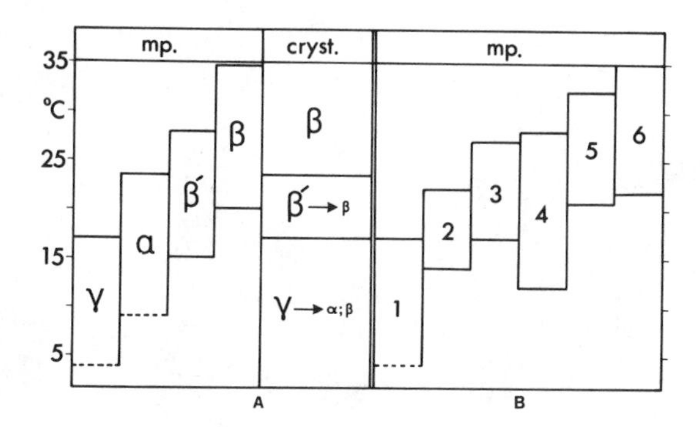

Fig. 8. Melting and crystallization ranges of pure cocoa butter polymorphs. (A) Crystallization from the melt (data from Vaeck, 1960). (B) Data from Johnston (1972).

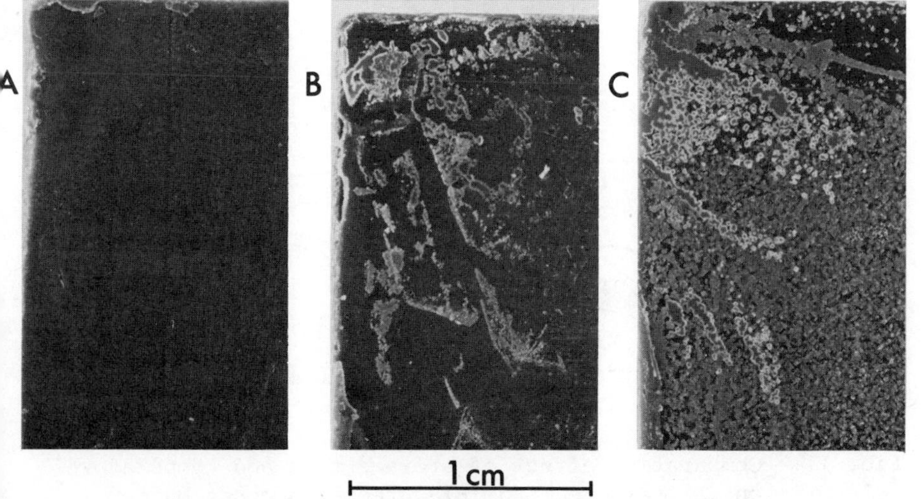

Fig. 9. Development of fat bloom on tempered plain chocolate stored at room temperature (21 ± 2°C). (A) Fresh chocolate surface; (B) after 6 months; (C) after 10 months.

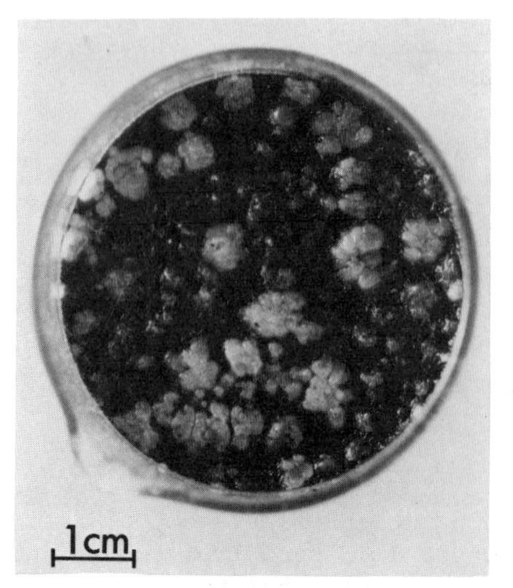

Fig. 10. Cocoa butter fraction crystallizing on the surface of a soft chocolate confectionery center, stored untempered at room temperature for 4 months. Crystal growth starts under the chocolate surface. The latter recedes as the crystals become larger. The different stages of crystal growth are clearly seen.

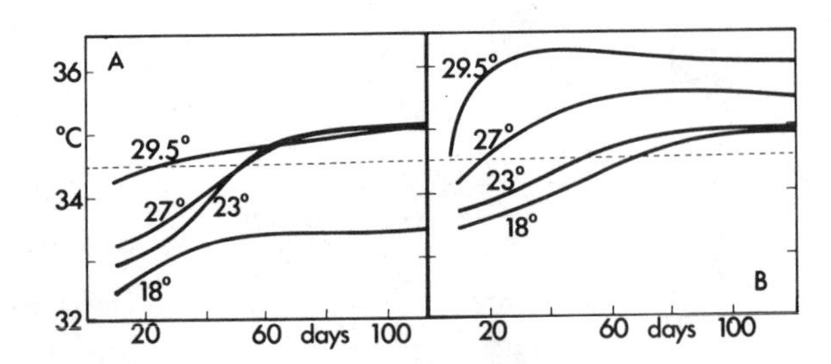

Fig. 11. Comparative effect of storage time and temperature on (A) tempered and (B) untempered cocoa butter. Samples stored for 90 days at different temperatures, then all of 18°C from 90 days onward. Horizontal broken line indicates melting point of β form of cocoa butter (from Minifie, 1980).

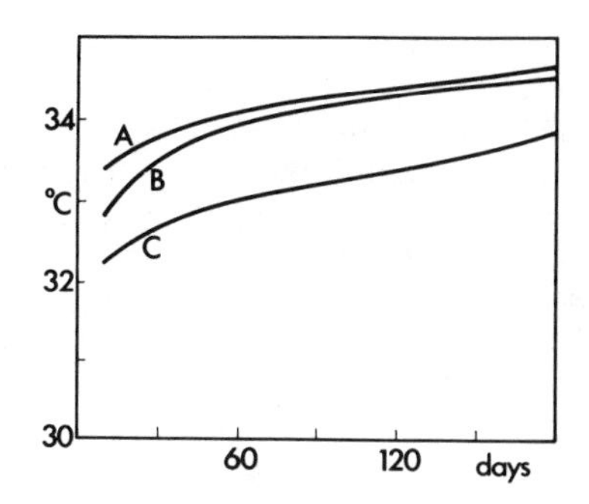

Fig. 12. Effect of added nut oil or milk fat on the melting point of cocoa butter. (A) Cocoa butter; (B) plus 15% nut oil; (C) plus 15% milk fat (dehydrated cow butter) (from Minifie, 1980).

REFERENCES

Anonymous (1957). *MC, Manuf. Confect.* *57*(7), 28.
Burle, L. (1962). "Le Cacoyer," p. 332. Maisonneuve, Paris.
Clarkson, C. E., and Malkin, T. (1934). *J. Chem. Soc.*, p. 666.
Cook, L. R. (1963). "Chocolate Production and Use." Magazines for Industry Inc., New York.
Duffy, P. (1853). *J. Chem. Soc.* *5*, 197.
Fincke, H. (1965). "Handbuch der Kakaoerzeugnisse." Springer-Verlag, Berlin and New York.
Heintz, W. (1849). *Jahresbericht* *2*, 342.
Hernqvist, L. (1984). Doctoral Dissertation, Lund University, Sweden.
Huyghebaert, A., and Hendrick, H. (1971). *Lebesm.=Wiss. Technol.* *4*, 59.
Johnston, G. M. (1972). *J. Am. Oil Chem. Soc.* *49*, 462.
Kleinert, J. (1961). *Rev. Int. Choc.* *16*, 201.
Lutton, E. S. (1948). *J. Am. Chem. Soc.* *70*, 248.
Minifie, B. W. (1970). "Chocolate, Cocoa and Confectionery. Science and Technology." Churchill, London.
Minifie, B. W. (1980). "Chocolate, Cocoa and Confectionery. Science and Technology," 2nd ed., pp. 494-518. AVI, Westport, Connecticut.
Powell, B. D., and Harris, T. L. (1964). *In* "Kirk-Othmer Encyclopedia of Chemical Technology," 2nd ed., Vol. 5, pp. 363-402. Wiley (Interscience), New York.
Stockton, G. W., and Smith, I. C. P. (1976). *Chem. Phys. Lipids* *17*, 351.
Vaeck, S. V. (1951). *Rev. Int. Choc.* *6*, 350.
Vaeck, S. V. (1960). *Fette, Seifen, Anstrichm.* *62*, 709.
Vaeck, S. V. (1961a). *Rev. Tech. Ind. Aliment.* *8*, 53.
Vaeck, S. V. (1961b). *Rev. Int. Choc.* *16*, 490.
Wille, R. S., and Lutton, E. S. (1966). *J. Am. Oil Chem. Soc.* *43*, 491.
Witzel, H., and Becker, K. (1969). *Fette, Seifen, Anstrichm.* *71*, 507.

Further Reading

Shelf-life of Chocolate
Bungard, G. (1973). *Gordian* *73*, 431.
Graham, A. S. (1969). *Candy Ind. Confect. J.* *132*, 7.
Preservation and Storage of Chocolate
Kleinert, J. (1970). *Rev. Int. Choc.* *25*, 354.
Kleinert, J. (1972). *Ind. Aliment.* *11* (3), 117.

Thermodynamic Stability of Cocoa Butter and Storage Stability

Witzel, H., and Heiss, R. (1969). *Fette, Seifen, Anstrichm. 71*, 663.

Tempering and Fat Bloom

Becker, K. (1974). *Proc. Int. Congr. Cocoa Choc. Res. 1st*, 1974, pp. 257-262;

Bueb, M., Becker, K., and Heiss, R. (1972). *Fette, Seifen, Anstrichm. 74*, 101;

Campbell, L. B., and Keeney, P. G. (1968). *Food Technol. 22*, 1150.

Cerbulis, J. (1969). *J. Food Technol. 4*, 133.

Effect of Processing Season on Free Fatty Acids of Chocolate and Butter Fat

El-Demersah, O., Didenko, R. A., and Kopylova, V. V. (1974). *Molochn. Prom-st. 6*, 18.

Changes in Levels of 2-Phenylethylamine During Processing and Storage

Saxby, M. J., Chaytor, J. P., and Reid, R. G. (1981). *Food Chem. 6*, 281.

CHAPTER 7

CANDY AND SUGAR CONFECTIONERY

THOMSEN J. HANSEN
Department of Nutrition and Food Sciences
Drexel University
Philadelphia, Pennsylvania

Handbook of Food and Beverage Stability: Chemical, Biochemical, Microbiological, and Nutritional Aspects

I. INTRODUCTION

The words candy and confectionery are often used interchangeably to mean a sweet sugar- or chocolate-based snack or dessert food. Occasionally, confection is used to denote an especially fancy or expensive piece of candy. However, these definitions are not universally accepted. Strictly defined, the word candy refers to products containing sugar as the dominant component, and does not include chocolate products. Confection is a more general term referring to any sweet product manufactured from sugar and other ingredients. In this sense, confectionery includes chocolate-based products and additionally flour-based products such as cakes, cookies, and pastries. Sugar is a common ingredient in all these products, and any discussion of candy and confections must of necessity include the role of sugar. However, flour confections are usually treated as bakery products, and chocolate confections are treated as a separate group, since these are based primarily on ingredients other than sugar. This chapter deals with candy and other sugar confectionery, with sugar playing its usual dominant role in the discussion, as it does in production, quality, and stability of these products.

The term sugar can itself be subject to different interpretations. The main sugar used in confectionery, including chocolate and flour confectionery, is sucrose. This is the common cane or beet sugar, and the term sugar usually means sucrose, both in the home and in manufacturing. However, many other ingredients of confections are or contain chemical components classified as sugars. Principal among these sugars are glucose, fructose, lactose, and maltose. Commercially, these are available as pure sugars or in the form of corn or other syrups (which contain glucose, maltose, and polymers of glucose, and is sometimes called glucose syrup), honey, invert sugar, and high-fructose corn syrup (all three primarily an equal mixture of glucose and fructose), milk and milk products (which contain lactose), and malt and malt syrup (which contain maltose and polymers of glucose). Sucrose is a disaccharide composed of a glucose unit and a fructose unit linked together. Sucrose can be hydrolyzed to a mixture of the two monosaccharides, and this mixture is invert sugar. Glucose and fructose are often called dextrose and levulose, respectively. The different meanings for different ingredients, especially the use of the name glucose for both a sugar and a syrup, obviously are a source of confusion. In this chapter, the individual chemical components will be called sucrose, glucose, fructose and so on. The term sugar will refer to all of them collectively, although often sucrose will be the major sugar involved. The term sweetener includes not only the sugars and their commercial sources, but also sugar alcohols such as sorbitol, mannitol, and xylitol, and the artificial sweeteners.

Except where noted, metric units of measurement are used. Particularly note that temperatures are given in degrees Celcius. Conversion to the American system can be made by the appropriate conversion factors, or in many cases by referring to the works cited. Much of the information presented here was originally obtained using American units, and later converted to metric in the references or for this chapter. Water activities are given in decimal form, and can be converted to percentage equilibrium relative humidity by multiplying by 100. When concentrations are expressed as percentages, these are percentages by weight.

I wish to point out two items that are not discussed herein, so that readers do not think they have been overlooked. One is humectance, a term I have avoided because it means different things to different people and in different applications. Humectance is associated with water absorbance or retention, so it can be replaced in any situation by discussion of water activity and relative humidity. The second item is refractometry, which is widely used in the sugar refining and confectionery industries to determine dissolved solids content. During production, refractometry is a simple and useful method. On the other hand, the major application of refractometry in determining storage stability of confections is to ascertain resistance to microbial growth. In this regard, measurement of water activity is both more accurate and easier than measurement of refractive index. A fuller discussion of this area is given by Minifie (1980), which includes refractometric techniques for those who wish to use them.

The general composition of various confectionery product types is given in Table I (Martin, 1955). The major differences are in amounts and types of sugars and in amount of water. In addition to differences in composition, the physical nature of the sucrose defines two broad and somewhat overlapping groups. In one group, sucrose is present in crystalline form, usually as microscopic crystals dispersed in syrup. This group includes fondant cremes and other crystallized, or grained, products. Many of the examples in this chapter will refer to fondant. The other group contains uncrystallized sugars in a very viscous solution, which is handled like a solid at room temperature. This group includes the hard candies, which are extremely viscous owing to their very high sugar content, and soft, chewy products in which fat, protein, or gelling agents provide high viscosity. The differences in ingredients and crystallization necessitate different storage conditions for the various products. One important difference, as will be discussed, is that crystallized products tend to dry out under normal storage conditions, while uncrystallized products tend to pick up moisture from the air.

An additional storage concern is that confectionery products are largely seasonal. While confectionery is consumed throughout the year, peaks of demand occur at holidays such as Easter, Halloween, and Christmas. The types of products in demand are different for the different holidays with, for example, egg-shaped cremes at Easter and hard candies at Christmas. In order to spread out production, a manufacturer must have the capability to store products for several months without loss of quality. During this storage period, some control over temperature and humidity can be achieved. Fortunately, the amount of time during which the product is exposed to uncontrolled conditions is short, as the time between distribution and consumption is usually no more than a few weeks.

This chapter deals little with the technology of confectionery manufacture. Interested readers should refer to some of the many books on this subject. Among the best of these are those by Lees and Jackson (1975), Minifie (1980), and Pratt *et al.* (1970). While there is still a great deal of art in candy making, confectionery manufacture, like all food production, is much more of a science now than it was a few decades ago. Sucrose is one of the purest chemical substances used in foods, and confectionery products are composed primarily of sucrose and water. Because of this relatively simple chemical composition, study of the chemical and physical properties of confectionery should be amenable to rigorous scientific treatment. A great deal of information on the fundamental scientific properties of sucrose was developed by and for the sugar refining industry, and can be used by the confectionery industry. Basic data on sucrose and other sugars can be found in publications from the U.S. National Bureau of

Standards (Bates, 1942; Swindells *et al.* (1958), and in Honig (1953), Hoynak and Bollenback (1966), Pancoast and Junk (1980), and van Hook (1961). Many more data are needed on the properties of other confectionery ingredients, in order to allow manufacturers to provide more and better products to meet consumer demands.

II. PROPERTIES OF SUCROSE AND OTHER INGREDIENTS

A. Solubility

Manufacture of most confectionery begins with a mixture of sucrose and other ingredients in water, which is then heated to produce a solution at elevated temperature. The composition of this solution determines many of the properties of the resulting product. If the sucrose is allowed to crystallize during cooling, the resulting product is a dispersion of sucrose crystals in the remaining syrup, with the syrup being saturated with sucrose. The extent of crystallization is determined by the solubility of sucrose in water, as influenced by the other ingredients. For pure sucrose, solubility in water as a function of temperature is described by the following equation for the temperature range from 0 to 90°C (Charles, 1960):

$$S = 64.397 + 0.07251T + 0.0020569T^2 - 0.000009035T^3 \quad (1)$$

where S = solubility(g/100g solution) and T = temperature (°C).

The values in Table II are calculated from Eq. (1) for the range 5-35°C (41-95°F), which confectionery products can be reasonably expected to encounter during storage. Solubility values in Eq. (1) and Table II are expressed in grams sucrose per 100g solution, or weight percent. This is the common unit for expression of sugar concentrations, both in the confectionery and sugar refining industries. Weight percent is a dimensionless unit, so that the same value applies even if a weight unit other than grams is used, as long as the same unit is used for both the sucrose and solution weights. Also in Table II are solubility values expressed in grams sucrose per 100g water. These units, given the symbol S', will simplify the formula for calculation of crystallization. These two quantities are related by the following equation.

$$S'/100 = S/(100 - S) \quad (2)$$

In order to convert weight percent to volumetric concentrations (grams per milliliter), it is necessary to know the solution densities. Density varies with both concentrations and temperature, and several compilations of sucrose solution densities are available, most based on extensive studies by the U.S. National Bureau of Standards (Bates, 1942). At a fixed concentration, density decreases with increasing temperature. On the other hand, at a given temperature, density increases with increased concentration. Since the concentration of a saturated solution increases as temperature increases, the two factors somewhat offset each other, with the concentration effect being greater. For example, if a saturated solution at 5°C were warmed to 20°C with no change in concentration, the density would decrease from 1.315 to 1.314. Addition of enough sucrose to bring the solution at 20°C to saturation increases the density to 1.329. As seen in Table II, the density of a saturated sucrose solution in the range 5-35°C increases slightly. For conversion to molar concentration (moles per liter), it is also necessary to know the molecular weight of sucrose, which is 342.30g/mol.

In a crystallized product, the extent of crystallization (grams crystal per 100g product) at a given temperature is defined as follows:

$$P_c = P_t - P_w S'/100 \quad (3)$$

where P_c = percentage crystal, P_t percentage total sucrose, and P_w = percentage water.

Again, since these percentages are dimensionless, weight units other than grams can be used. If the system is composed of only sucrose and water, $P_t + P_w = 100$. Thus a solution of 88g sucrose and 12g water, when cooled to 20°C (S' = 199.4) under conditions that allow crystallization, will contain 64.1g sucrose crystals and 35.9g saturated syrup. The syrup consists of the remaining 23.9g sucrose dissolved in the 12g water, giving a syrup concentration of 66.6% sucrose by weight. Crystallization is also accompanied by a volume change. The density of pure crystalline sucrose is 1.588g/ml (Hirschmuller, 1953). The total volume of the crystals and syrup in this example is 67.5ml. Using similar calculations, the change in crystal content and volume with change in temperature or water content can be determined. This is an important factor in storage of confectionery.

The addition of other ingredients changes the solubility of sucrose, though not in an obvious manner. Fructose and glucose, for example, compete with sucrose for solvent water but also interact directly with sucrose. The compositions of sucrose-invert sugar-water mixtures saturated with sucrose have been studied at various temperatures and invert sugar contents (Bates, 1942). Values are given in Table III. Solubility in this case is expressed as grams sucrose per 100g water (S') at various levels of invert sugar, also given as grams per 100g water. As the amount of invert sugar increases, the solubility of sucrose decreases, but because all of the invert sugar stays in solution, the total amount of dissolved sugars increases. The total dissolved solids per 100g water is obtained by adding the respective values for sucrose and invert sugar. The amount of dissolved solids expressed as weight percent can be obtained from the latter value using Eq. (2). Solubility after replacement of sucrose by glucose or fructose alone closely follow that of invert sugar. Corn syrup effects sucrose solubility in the same way, if only the solids content of the syrup is considered (Lees, 1965). Equation (3) can still be used to calculate percentage crystal from the solubility values in Table III, but now P_t and P_w total less than 100. A solution consisting of 78.4g sucrose and 9.6g invert sugar in 12g water has 88% total solids, as in the first example that used only sucrose and water. In the new case, P_t = 78.4 and P_w = 12, and with 80g invert sugar per 100g water, S' = 172.9 at 20°C. After crystallization, this product would consist of 57.7g crystal and 42.3g solution. This sucrose-invert sugar fondant is used as a basis for discussion of storage changes in Section IV. The crystal phase is assumed to consist of only sucrose, while the solution contains all the water and invert sugar, in addition to the remaining 20.7g sucrose. Total dissolved solids content of this solution is then 71.7%. Even though sucrose is less soluble in the invert sugar solution than in pure water, less crystallization occurs because there is less total sucrose in the second example.

The density of a solution of sucrose and invert sugar is always less than that of a sucrose solution of the same solids content. Densities can be determined using the equations developed by Flavell (Pancoast and Junk, 1980). At a concentration of 71.7% at 20°C, the density of a solution of sucrose is 1.355g/ml (Bates, 1942). When the ratio of invert sugar to total dissolved solids is 9.6:30.3, the calculated density is 1.354g/ml. Applying this value to the above product, the total volume is 67.6ml. By comparing the two examples, it is seen that if sucrose is partially replaced by invert sugar, the amount of crystal decreases while the total volume of the product remains about the same.

Calculations such as these can be performed for other mixtures of confectionery ingredients. Although effects on sucrose solubility are seldom studied, other ingredients probably act similarly to invert sugar, and density differences can probably be ignored. However, better data, for example on solubility of sucrose in corn syrups, are needed to predict accurately the properties and behavior of fondants and other confectionery products.

Glucose is less soluble than sucrose, and will form crystals under certain conditions. Many confectionery solutions are supersaturated with respect to glucose. In a sucrose-based product, glucose crystallization is undesirable. Fortunately, glucose does not crystallize readily unless seed crystals are added. Some fondant-type products are made using glucose as the sugar of crystallization (Goodall, 1968). At 20°C, a saturated solution contains 47.8% glucose (91.6g/100g water), and has a density of 1.211g/ml (McDonald, 1953). Fructose is more soluble than sucrose, with a solubility of 78.94% (374.8g/100g water) at 20°C (McDonald, 1953). Even should this solubility limit be reached, it is quite difficult to crystallize. Glucose and fructose each have a molecular weight of 180.16g/mol.

B. Water Activity

The second most important ingredient in confectionery products, after sugar, is water. Some would even consider water more important than sugar (Hinton, 1970), since the most important changes that occur during storage involve water. If the water content of the product changes, this will affect, among other things, the percentage crystal in the product as shown by Eq. (3). The water activity a_w of the product, compared to the relative humidity of the surrounding air, determines whether the product will tend to lose or gain moisture. In addition, the a_w determines whether or not the product is susceptible to microbial spoilage. Water activity depends only on the solution portion of a confection, and the presence of crystals or other insoluble material has no effect. A fondant composed of only sucrose and water will have the same a_w as a saturated sucrose solution. Thus despite the low water content of a crystallized fondant, a_w may be high enough to support microbial growth.

An initial approximation of a_w, based on thermodynamics of ideal solutions, can be determined using Raoult's law, which predicts that the activity of water is equal to its mole fraction in the solution. At 20°C, 100g of a saturated sucrose solution contains 66.6g (0.195 mol) sucrose and 33.4g (1.856 mol) water. The mole fraction of water is thus 1.856/(0.195 + 1.856) = 0.91. An underlying assumption of Raoult's law, however, is that there is no interaction between the water and the solute, which is not true for sucrose solutions. Because of sucrose-water interaction, a_w for a saturated sucrose solution is somewhat lower than the mole fraction of water. A more exact equation, applicable to crystallized confectionery products or sugar syrups, was developed by Norrish (1966) using thermodynamic principles and empirical observation.

$$\log a_w = \log x_w - x_s^2 K_s \tag{4}$$

where x_w = mole fraction of water, x_s = mole fraction of solute, and K_s = a constant for the solute.

If $K_s = 0$, Eq. (4) becomes equivalent to Raoult's law. All K values determined by Norrish are positive (see Table IV), so all resulting a_w values are lower than those estimated by Raoult's law. For the sucrose example given above, $x_w = 0.91$ and $x_s = 0.09$ for sucrose, which has $K_s = 2.60$. Equation (4) yields $a_w = 0.86$, the actual a_w of a saturated sucrose solution at 20°C. This will inhibit bacterial growth, but many yeasts and molds can tolerate a_w values this low (Nickerson and Sinskey, 1972). Also, such a product would lose water if exposed to an atmosphere with relative humidity less than 86%. This will be the case in all but the most humid environments.

Since a saturated sucrose solution is unstable with respect to both microbial browth and drying, additional material must be dissolved to stabilize the product. One method of achieving this is to allow a concentrated sucrose solution to cool without crystallization. This is the method of preparation of hard or chewy candies. Equation (4) is not applicable to such high concentrations, and Raoult's law, while more accurate, still predicts values higher than those observed. Hard candies have a_w values in the range 0.1-0.3 (Sacharow and Griffin, 1970). This is quite resistant to microbial spoilage, but such a product will now pick up moisture from the air under all but the driest conditions of relative humidity.

Another approach to lowering a_w, which must be done for crystallized material, is to add other soluble ingredients in addition to sucrose. If sucrose is partially replaced by invert sugar, the a_w of a saturated solution is decreased, both because the total amount of soluble solids increases and because the monosaccharide components of invert sugar have a lower molecular weight than sucrose. Equation (4) can be generalized to apply to multicomponent solutions by summing the contribution of each solute, weighted by its solute constant (Norrish, 1966)

$$\log a_w = \log x_w - [\Sigma(x_s\sqrt{K_s})]^2 \tag{5}$$

Table IV gives solute constants for various confectionery ingredients, including for corn syrups of different dextrose equivalents (DE). For the sucrose-invert sugar example given above, the solution contains 20.7g (0.060 mol) sucrose and 9.6g (0.053 mol) invert sugar in 12g (0.667 mol) water. This gives a mole fraction of water of 0.88 and an a_w from Eq. (5) of 0.79. This is low enough to protect against spoilage by most yeast and molds, which become inactive below $a_w \cong 0.80$. Such a product would still dry out in an atmosphere of less than 79% relative humidity, but at a slower rate than a solution of sucrose alone. Table V gives a_w values for several saturated sucrose solutions. Solution compositions are taken from Table III, and values are calculated from Eq. (5).

C. Diffusion

To make a crystallized confection, sucrose and other ingredients are mixed with water and heated to dissolve the sugar and evaporate some of the water. When this hot solution is cooled, the solution becomes supersaturated and crystal growth, if desired, is initiated by stirring or seeding. As the crystals grow, sucrose molecules in the supersaturated solution move to the surface of existing crystals where they are incorporated into the crystal lattice. The rate of mass transfer from the bulk solution to the growing crystal surfaces is the limiting factor in the rate of crystal growth (Smythe, 1967; van Hook, 1953, 1961). Rapid growth results in large crystals, which are generally undesirable because they produce a gritty feeling in the mouth. For this reason, production conditions are set up so that sucrose diffusion, and thus crystal growth, is slow. Likewise, during product storage, sucrose molecules continually dissolve from crystals, diffuse through the syrup, and are reincorporated into other crystals. This recrystallization process, which can also lead to undesirably large crystals, is also delayed by conditions that slow sucrose diffusion. Recrystallization will be discussed in Section IV,E. The rate of mass transfer is affected by supersaturation level, presence of existing crystals, surface tension, stirring rate, viscosity, and diffusion coefficient. Temperature affects solubility, viscosity, and diffusion coefficient, so has an indirect but substantial effect on mass transfer. One way to control the rate of mass transfer of sucrose, and thus the rate of crystal growth during both manufacture and storage, is to control the temperature. Low temperatures lead to slower crystal growth. In hard candies, unagitated cooling keeps mass transfer so slow that crystals cannot grow at all. Ingredient selection also affects crystal growth. Glucose and fructose, when used to replace sucrose partially, decrease supersaturation and so decrease the rate of crystal growth. Corn syrup is an excellent crystallization controller, as it both decreases supersaturation and lowers the diffusion coefficient of sucrose. It is the polysaccharide components of corn syrup that inhibit sucrose diffusion. In chewy candies, crystal growth is prevented by temperature control and by the use of polymers such as gelatin, pectin, or albumin (from egg white). In fondant production, diffusion coefficient is a very important determining factor for rate of mass transfer, and so of crystal growth. Diffusion coefficient of sucrose in solution depends on both concentration and temperature (English and Dole, 1950; Schneider *et al.*, 1976), and data are presented in Table VI. Unfortunately, quantitative data for diffusion coefficients are not extensively available. Especially lacking are values for diffusion coefficient of sucrose in solutions containing ingredients that are used to control crystal growth in confectionery.

D. Viscosity

The viscosity of the syrup portion of a confection affects its quality and storage properties. High viscosity retards sucrose crystal growth, although the impact of viscosity change is not as important as changes in diffusion coefficient (Lees, 1965; van Hook, 1953). In fondant, the solution in which the crystals are dispersed must be sufficiently viscous to avoid separation of the crystal and syrup phases. The viscosity needed to achieve this depends on the density of the syrup, the density of the sucrose crystals (which is constant), and the crystal size. If the syrup is too viscous, on the other hand, it will produce a gummy feeling in the mouth. A viscous syrup is also needed for foam formation and stability, important in aerated products such as marshmallow. Uncrystallized products are actually very viscous solutions that do not flow under the force of their own weight. Hard candies have extremely high viscosities, and are considered glassy solids.

The viscosities of solutions of pure sugars have been investigated extensively (Swindells *et al.*, 1958). Table VII gives viscosities of sucrose solutions in the temperature and concentration range of Table II. Temperature has a greater effect than concentration, so that the viscosity of a saturated sucrose solution increases as temperature decreases, even though the sucrose concentration decreases. Replacement of sucrose with other sugars has a definite, though slight, effect on viscosity. The viscosity of sucrose-invert sugar solutions is always less than that of a sucrose solution of the same total solids content (Pancoast and Junk, 1980).

In most confectionery, especially those where high viscosity is desired, viscosity is determined not so much by the sugar content as by the inclusion of polymeric components such as corn syrup polysaccharides, egg albumin, and gelatin. As with diffusion coefficient, viscosity studies applicable to confectionery are not extensive. A study of the effects of dextran on viscosity of sucrose solutions produced the following equations (Greenfield and Geronimos, 1978):

$$v/v_0 = 1 + 0.006MW^{0.35}c^{1.21} \tag{6}$$

$$\ln(v/v_0) = 0.0157MW^{0.25}c^{0.92} \tag{7}$$

where v = viscosity, v_0 = viscosity of similar solution with no dextran, MW = dextran molecular weight, and c = dextran concentration (g/100ml).

The two equations were based on two different theoretical treatments. They give slightly different results, especially at high dextran concentrations and molecular weights. Viscosities were measured over the temperature range 30-65°C, sucrose concentrations 67-78%, dextran molecular weights 6×10^4-2×10^7, and dextran concentrations from 0.1 to 10g/100ml. Table VIII gives viscosities calculated from the average of Eqs. (6) and (7), using v_0 values from Table VII. A comparison of Tables VII and VIII illustrates the effect of the polymer on viscosity. Dextran is a linear polymer of glucose, so viscosities of sucrose solutions containing a high-amylose starch would probably be close to those in Table VIII.

It is impossible to measure directly viscosities on the order of those present in hard candies at typical storage temperatures. An acceptable limit for the viscosity at which a solution make the transition from a liquid to a glass is 10^{15}cP (White and Cakebread, 1966).

E. Sweetness

An obvious sensory feature of confectionery products is their sweetness. Sweetness is one of the four basic taste stimuli, and there would be little demand for confectionery products if people did not derive satisfaction from sweet-tasting foods. Sugar confectionery is unavoidably sweet, and in fact oversweetness is sometimes a problem. Sweeteners differ in their intensity, and while there is no absolute scale for sweetness, they are usually quantified in relation to sucrose. Such a relationship is given in Table IX (Inglett, 1981). Even relative sweetness varies with concentration, temperature, pH, and other properties of the solution. Physical state, such as crystal size, also affects sweet taste perception. One notable comparison is between sucrose and invert sugar. Sucrose is slightly sweeter than an equivalent amount of invert sugar at room temperature, and this difference is greater at elevated temperatures, but reversed at low temperatures. Degree of sweet taste can be varied by formulation and can change during storage, especially if sucrose is hydrolyzed to invert sugar or crystal size changes.

F. Heat of Solution

Crystalline sugars have a negative heat of solution, meaning that when a sugar is dissolved in water, heat is absorbed and the temperature of the resulting solution is lower than that of the components before dissolution. Values for heat of solution have been determined to be -2437 cal/mol for sucrose (Tikhomiroff *et al.*, 1965), and -5325 cal/mol for glucose (Taylor and Rowlinson, 1955). These are values going from the crystalline sugar to a saturated solution. The glucose value is for α-glucose hydrate, the form present in glucose fondants. Heat absorption occurs in the mouth as well as in the laboratory, and is responsible for the cool sensation produced when a crystalline sugar dissolves in saliva. On the other hand, sugar solutions have a positive heat of dilution. This means that the saturated solution initially formed as the sugar dissolves becomes warmer as the solution is further diluted. Experimental values are +982 cal/mol for sucrose (Higbie and Stegeman, 1950; Tikhomiroff *et al.*, 1965) and +345 cal/mol for glucose (Taylor and Rowlinson, 1955).

Values for other materials, such as the sugar alcohols (Emodi, 1982), are of similar sign and magnitude. The temperature effect produced in the mouth will thus depend on the extent of crystallization and on the composition of both the crystal and syrup phases. Crystal size also affects the temperature sensation. Small crystals are more likely to dissolve completely, and so seem cooler than large crystals.

III. PROCESSING EFFECTS ON QUALITY AND STABILITY

A. Crystal Size

Crystallization of sucrose from solution is often not spontaneous but must be induced by adding seed crystals, air bubbles, or other centers on which crystals can grow. The method of crystal initiation, along with the product composition and the temperature and stirring rate, determine the size of the crystals in the final product. If crystallization is desired, it must go to completion. Otherwise the product received by the consumer may be quite different from that originally put in the package. If crystallization is not desired, care must be taken that crystals do not form during production or during storage and distribution.

The extent of crystal formation is a chemical property determined by the composition and temperature of the product. The size of these crystals, however, is determined by processing conditions. Even at a fixed weight of crystals, the number and size can vary, and this variation results in product differences, desirable or not. The human tongue can discriminate particles if they are larger than $\sim 30\mu m$. Crystals in a confection should be no larger than this to avoid a gritty texture. Since the crystals actually are spread over a range of sizes, it is difficult to determine the maximum crystal size in a sample. Fondant is considered acceptable with regard to crystal size if the average size is less than $\sim 20\mu m$ (Minifie, 1980; Woodruff and van Gilder, 1931). The total weight of crystals is related to the size of an individual crystal by the density of crystalline sucrose and by the number of crystals. A 57.7-g mass of crystalline sucrose, as in the second example in Section II,A, can be achieved by a single cubic crystal 3.3 cm on a side, 1×10^9 crystals of $30\mu m$, or 3×10^{10} crystals of $10.7\mu m$. The very large number of crystals in a fondant allow considerable variation in the extent of crystallization with only minor changes in crystal size. If the 3×10^{10} crystals were uniformly distributed throughout 42.3g of syrup, there would be about 2.5μm between crystals.

B. Bulk Viscosity

The viscosity of a confectionery product as a whole, as distinguished from the viscosity of the liquid portion, is affected not only by the composition but also by the physical state of the item. This includes the number, size, and size distribution of crystals. A quantitative relationship between sucrose crystal size and bulk viscosity was developed by Kelly (1958) but has not found widespread application. Nevertheless, the underlying principles are correct, namely, that an increase in bulk viscosity can be produced by an increase in percentage crystal or syrup viscosity, or by a decrease in crystal size. Since crystal size tends to increase on storage, bulk viscosity tends to decrease. Changes in percentage crystal, such as due to water loss or gain, also have a significant effect. Any change that decreases percentage crystal (e.g., sucrose hydrolysis) or decreases syrup viscosity (e.g., higher temperature) would lower bulk viscosity.

One example of a change in bulk viscosity is the so-called maturing or ripening of fondant. Fondant, when stored up to a day or so after manufacture, becomes softer and more plastic, and the surface becomes glossy. No completely satisfactory explanation of this change has been developed (Minifie, 1980). It probably involves completion of crystallization, but it is not clear how this change produces the observed effects. During production, as crystallization nears completion and supersaturation decreases, the rate of crystal growth slows. When the manufacturing process is completed, the crystals probably continue to grow slowly, over several hours, as the slightly supersaturated syrup finally becomes only saturated. The resulting increase in crystal size results in a lower bulk viscosity, although this effect is probably small. In addition, the smallest of the crystals, which contribute most to the bulk viscosity, may redissolve due to recrystallization. The viscosity of the syrup would also decrease as the amount of dissolved sucrose decreases. All of these factors would result in a softer product.

The bulk viscosity and density of a product are also affected by the extent and manner of air incorporation. Air bubbles affect the bulk viscosity in a manner similar to the effects of sucrose crystals. Bulk density, of course, is lowered by air incorporation.

C. Coatings

One of the most important changes that take place on storage of confections is uptake or loss of water. Packaging materials help prevent this and are discussed in Section IV. The same effect can be provided by an edible barrier that becomes part of the product. Pieces of soft or moist confectionery are usually coated with crystalline sugar or an edible wax (Minifie, 1980) or with chocolate or other fat (Feuge, 1970). This not only prevents drying, but also provides structural support. The coating of chocolate around a creme center or the coating of sugar around a jellybean are excellent packages if well made and kept intact. If a sugar coating comes in contact with a bit of water, the saturated sucrose solution formed will be of high a_w and will lose water again quite readily.

D. Browning

Brown pigments are formed when sugars are heated, either alone or in the presence of proteins (Fennema, 1985). This process, when applied to confectionery products in which it is desirable, is called carmelization. In other products, such as marshmallow, browning is usually undesirable. Sucrose, which is a nonreducing sugar, does not participate directly in these reactions. The other common sugars in confectionery are reducing sugars, meaning they have a free or easily accessible carbonyl group that can participate in the chemical reactions that lead to brown colors. If the sucrose molecule is hydrolyzed, the glucose and fructose formed are able to form brown products. Sugar alcohols are produced by reducing the carbonyl group of a sugar, so the sugar alcohols do not allow browning. Browning occurs more rapidly in the presence of proteins, such as egg albumin. Besides the nature of the sugar and other reactants, the extent of browning depends on temperature and time. While browning can readily occur at the high temperatures encountered during confectionery production, it is unlikely to occur at storage temperatures unless storage time is considerably longer than usual, such as over a year.

IV. CHANGES DURING STORAGE

A. Water Transfer

When the a_w of a confectionery product is different from the relative humidity of the surrounding air, water transfer will occur. This changes the physical properties of the product, and in the case of water uptake by the product increases the risk of microbial growth. The a_w can be varied to some extent by varying of the composition, and humidity in a warehouse can be controlled. Even so, different confectionery products have different water activities, and humidity cannot be controlled after product distribution. Since it is impossible to keep each product at its equilibrium relative humidity, the only way to prevent significant water transfer is to surround the product with a barrier. Edible coatings accomplish this to some extent. Packaging of confections is also extremely important. Table X lists several packaging materials used for confectionery and rates their resistance to water vapor permeation (Sacharow and Griffin, 1970). Fondants generally have an a_w higher than that of air, so they tend to dry out. This results in increased crystallization. The sucrose-invert sugar fondant described in Section II,A, if it lost 2g of its water in a uniform manner, would change properties as shown in Table XI. Table XI gives percentage crystal calculated from Eq. (3), volume calculated from densities, crystal size assuming 3×10^{10} uniform crystals/100g, percentage solids and viscosity of the syrup, and a_w calculated from Eq. (5). One of the most significant changes from loss of water is the decrease in volume, due simply to the loss of mass. If this product were coated with chocolate, this volume decrease would cause either collapse of the coating or an air space inside the coating. Both these consequences are undesirable. A collapsed coating is more apparent, but an air space can allow moisture condensation, which would lead to a localized area of high a_w where microbial growth could occur. Note also that the total amount of crystal increases, while the size of an individual crystal increases only slightly. The syrup viscosity also increases. This combination increases bulk viscosity, yielding a harder product.

The values in Table XI assume that the water loss is uniform throughout the fondant. In reality, water loss occurs mostly at the surface. Besides the other changes, surface drying can cause white spots to appear. As the distance between crystals at the surface decreases to the wavelength of visible light, ~0.8μm, a group of crystals looks like a single white mass. This is a physical change in the system, not a chemical change. References to the white spots as "sucrose hydrate" are erroneous. While surface drying of fondant is not desirable, it does inhibit further drying. Rapid surface moisture loss can form a barrier to further water loss from the interior. Over the course of a few months, a surface-hardened fondant in a very dry atmosphere can actually have a higher overall water content than a similar fondant stored in a less dry atmosphere where drying is initially slower but more continuous throughout the storage period (Mullineaux and Hansen, 1984).

In the event a fondant is stored in an unusually humid environment, it could pick up water. Changes in properties that would occur are given in Table XI. These changes are, as expected, opposite those of drying. There is a volume increase and a decrease in percentage crystal and syrup viscosity, compared to the original fondant. The bulk viscosity would decrease, giving a more liquid product, despite the small decrease in calculated crystal size. The a_w increases, but if the water is distributed uniformly the change is not significant. Localized wet spots, however, could have a high enough a_w to allow microbial growth.

Hard candies, which usually are lower in a_w than the air, tend to pick up moisture on the surface. The immediate effect is only at the surface, since the high solids concentration of the product inhibits movement of water to the interior. Unfortunately, this is enough to have undesirable results. The water dilutes the solution at the surface, lowering the viscosity to the point where the solution flows under finger pressure, giving a sticky candy. If the candies are not individually wrapped, they may stick together. Also, the diluted solution allows more rapid mass transfer of sucrose, and it becomes possible for crystals to grow (Makower and Dye, 1956). As sucrose begins to crystallize, the sugar concentration in the remaining solution becomes still lower, making crystal growth still more rapid. Within hours, the entire surface may be covered with crystalline material, a defect known as graining.

A confectionery product can also pick up moisture by condensation if it is transferred to an environment where the dew point is above the product temperature. This is most likely to occur if the product is moved from a cool production or storage area into a warm, humid atmosphere. Another factor to consider is that in a composite confection, portions that have different water activities will tend to exchange water with each other, even if the overall water content remains constant.

B. Fermentation

Fermentation of confectionery can occur when the a_w is above ~0.8 (Nickerson and Sinskey, 1972). The organisms responsible are osmotolerant yeasts and molds. Fermentation gives rise to off-flavors and off-odors, along with changes in appearance and texture. If the product is tightly enclosed, as by a coating of chocolate, gas production can cause a buildup of pressure to the point where the enclosure bursts, spreading the fermenting contents throughout the package. If mold growth occurs, there is additionally a safety concern, as many molds produce mycotoxins that are carcinogenic at low levels (Wogan, 1965). Fortunately, most confectionery production involves a cooking step in which the microbial load is reduced. Good sanitation practices can prevent introduction of organisms subsequent to cooking. Even if the a_w of the product as a whole is low enough to inhibit fermentation, microbial growth can occur in small areas where the a_w is higher. Condensation due to volume or temperature changes, surface moisture in a humid environment, and insufficient mixing can all lead to isolated spots where fermentation can occur. As an additional concern, there are some osmophilic yeasts that are capable of growing at a_w values as low as 0.62 (Tilbury, 1981). It is not practical to achieve such a low value in many confectionery products. The best way to prevent fermentation by these organisms is to prevent their introduction into the product. This is achieved by good sanitation practices and quality control of incoming ingredients.

C. Sucrose Hydrolysis

Sucrose is a disaccharide that can be hydrolyzed to its component monosaccharides according to the equation

$$\text{sucrose} + \text{water} = \text{glucose} + \text{fructose} \tag{8}$$

This rather straightforward chemical reaction can have enormous effects on the physical properties of a confectionery product. The desirability of this reaction depends on the product. The equilibrium for Eq. (8) lies far to the right, with a free-energy change of −5.5 kcal/mol under standard conditions, equivalent to an equilibrium constant of above 10,000 (Lehninger, 1975). Sucrose hydrolysis is often called inversion, and the product mixture called invert sugar, because the hydrolysis products rotate a beam of polarized light in the opposite direction from the rotation of a sucrose solution. Inversion is accelerated by heat, and is catalyzed by acid or by the enzyme invertase, which is commercially available from yeast. If sucrose is used as the only sugar in a confection, some inversion must be allowed during production in order to produce a good-quality product. The invert sugar so produced provides slower crystal growth and lower a_w than sucrose alone. This practice was common in the past, and some recipes still used require inversion during production. This is accomplished by slow cooking and/or by addition of acid. Inversion in this manner is very difficult to control. Production and use of commercial invert sugar and corn syrups make this practice no longer necessary.

After production, inversion is generally not desirable. Effects of inversion are illustrated in Table XI. The water content after inversion is lower than before, not by drying but because water is consumed in the reaction. The decreased water content and higher dissolved solids level produce a significantly lowered a_w. This helps stabilize the product against drying and microbial growth. The lower percentage crystal phase makes the product after inversion much more liquid, even though the syrup viscosity increases and the calculated crystal size slightly decreases.

If the desired product is a chocolate-coated liquid center, inversion during storage is a convenient method. In this process, the fondant center is produced and then coated with chocolate as with an ordinary creme center. The high percentage crystal and bulk viscosity of the original fondant allow it to be made and coated like a solid. If invertase is incorporated in the fondant, sucrose is hydrolyzed and amount of crystal decreases. Since the enzyme is heat sensitive, it must be added after cooking. By the time the product reaches the consumer, a week or more after manufacture, the bulk viscosity of the center is low enough to be considered liquid. Even though the equilibrium for Eq. (8) lies far to the right, inversion only goes to ∿80% dissolved solids (Minifie, 1980). The enzyme becomes less active as the reaction proceeds and the a_w becomes lower.

D. Temperature Effects

The most important effect of change in temperature on confectionery is the effect on bulk viscosity. Elevated temperature decreases the amount of crystal, and lowers the viscosity of syrup, as shown in Table XI. Both of these changes decrease the bulk viscosity, so increased temperature leads to a more liquid product. In addition, increased temperature increases the rate of mass transfer, which leads to more rapid formation of large crystals by recrystallization. Spoilage microorganisms and enzymes are also more active at higher temperatures, at least within the range of temperatures commonly encountered. Temperatures high enough to inhibit microbial growth would cause other serious product defects. Elevated temperatures also have undesirable effects on coatings, such as melting of a chocolate coating. Low temperatures are not detrimental to confectionery products, as long as freezing temperatures are avoided and care is taken to prevent condensation when storage temperature is changed.

E. Recrystallization

In crystallized products, the sucrose cannot spontaneously revert to solution, but the size of the crystals can increase. At the molecular level, there is a dynamic equilibrium at the crystal-solution interface. Sucrose molecules continuously dissolve from crystal surfaces, diffuse through the solution, and are reincorporated into other crystals. If the rates of dissolving and reincorporation are equal, the total mass of crystal remains constant. However, sucrose tends to move from the smaller crystals to larger ones. Whereas supersaturation is the driving force for crystallization, surface tension is the thermodynamic force that drives recrystallization (Fennema, 1985; van Hook, 1961). The dynamic equilibrium at the molecular level leads to an observable increase in crystal size. Once a small crystal becomes completely dissolved, it cannot re-form. This process of recrystallization eventually can lead to a product with unacceptably large crystal size, even if the crystal size was initially small. Recrystallization, like crystal growth, is slowed by factors that retard mass transfer, namely high dissolved solids level, polymer components, and low temperature.

The tendency toward large crystal size is true independent of changes in water content (Mullineaux and Hansen, 1984). If there is no water transfer, crystals grow by recrystallization. If the product picks up moisture, the smallest crystals are the first to dissolve. If the product loses moisture, additional sucrose is incorporated from the solution into existing crystals. The crystal size decreases calculated in Table XI are therefore not observed in actual samples. Some data on crystal size increase during storage are given in Table XII (Woodruff and van Gilder, 1931). While only the observed minimum and maximum crystal sizes were reported, there was a clear increase in crystal size during storage in all four samples. The investigators correctly pointed out that production conditions affect crystal size to a greater extent than does storage. The main reason is that factors that effectively control crystal size during production are also effective at slowing the rate of recrystallization.

One of the more puzzling observations of crystal size in stored fondant was that, in a surface-dried fondant, the surface crystals were smaller than those in the interior (Lee, 1965). Crystals at the surface were about 4μm, while those in the interior were about 7μm. This suggested that drying was accompanied by a decrease in crystal size. This violates the fundamental chemical properties of the system. Rather, I believe that the differences in crystal size resulted from recrystallization rather than simply from drying. The initial crystal size throughout the product was probably 4μm or lower. As the surface dried, the size of the crystals at the surface did not greatly increase, as drying itself causes only a slight increase (see Table XI). However, drying slowed mass transfer at the surface, so that the crystals there could not grow further, essentially locking the surface crystals at their initial size. Meanwhile, the crystals on the interior, at the original water content and mass transfer rate, were free to grow by recrystallization. The overall effect, with interior crystals larger than surface crystals, would be that observed.

F. Foam Collapse

If the syrup viscosity of an aerated product is too low, because of either improper formulation or changes during storage (water pickup, temperature increase), air bubbles dispersed through the product can escape or coalesce. Bubbles can also coalesce or escape by diffusion through the intervening film of liquid (Mansvelt, 1970). Foams are thermodynamically unstable and cannot re-form once broken. This causes changes in the physical properties of the product, the most notable being increased density. If the product is coated, the resulting decrease in volume of the product can create an air pocket between the center and the coating, similar to the result from water loss. This can lead to coating collapse from moisture condensation.

G. Flavor Changes

Flavor changes in confectionery can be divided into two classes, both of which are undesirable. The first is loss of desired flavor components, and the second is development of off-flavors. Aside from the sweet taste provided by sugar, other desirable flavors are generally due to added flavoring ingredients to give, for example, fruit or mint flavors. These flavors are due to volatile organic compounds in natural or synthetic flavoring ingredients. Since the flavor chemicals are volatile, as they must be to have their desired effect, there is the possibility of evaporation during storage. Evaporation not only leads to a general decrease in flavor intensity, but can lead to flavor changes, as flavors are due to mixtures of chemical components. If the more volatile components are lost, the flavor of the residual, less volatile components can be quite different from that of the original mixture. During production, such ingredients are added as late as possible, preferably after all heating steps, to minimize evaporation. During storage, flavor evaporation can be slowed by encapsulation of the flavor chemicals or by proper packaging. Table X rates packaging materials as barriers to oxygen permeation, and the same ratings are generally applicable to flavor permeation. Oxygen uptake is itself a contributor to loss of desirable flavors, as many flavor chemicals are susceptible to oxidation with resultant change in flavor. A good barrier to oxygen and/or the use of chemical antioxidants will help prevent oxidative flavor loss.

Oxygen is also involved in one of the major sources of off-flavors, namely oxidation of unsaturated fatty acids. Polyunsaturated fatty acids, and to a much smaller degree monounsaturated fatty acids, can react with molecular oxygen to begin a series of reactions that lead to rancid odor (Fennema, 1985). The odor is caused primarily by volatile aldehydes. The most important confectionery fats, namely cocoa butter, milk fat, and coconut oil, are low in polyunsaturated fatty acids, so oxidative rancidity is not a great problem. However, peanut oil contains a relatively large amount of polyunsaturated fatty acids, so confections containing peanuts or peanut butter are very susceptible to rancidity. As with oxidative flavor loss, oxidative rancidity can be minimized by good packaging and antioxidants. In addition, good-quality ingredients are extremely important. An oil in which the series of reactions leading to rancidity has begun will not itself have an off-flavor, but subsequent development of off-flavor is almost impossible to prevent. The use of good-quality peanuts, combined with proper packaging and storage, should still provide a high-quality product after many months, even without the use of antioxidants.

Fats can also be converted to off-flavor derivatives by hydrolysis. This process, also called lipolysis or hydrolytic rancidity, produces free fatty acids. The ingredients most commonly associated with this problem are milk products. This is so both because hydrolysis of milk fat produces a significant amount of volatile, foul-smelling fatty acids, such as butyric acid, and because milk products sometimes contain residual activity of lipase, an enzyme that catalyzes the hydrolysis. Milk derivatives used in confectionery should be free of lipase activity, or else any activity should be destroyed during the cooking steps of the confectionery process. Lipases can also be introduced into confectionery products by mold growth. Mold growth is also undesirable for other reasons, and is best controlled by control of a_w.

H. Color Fading

Many confections, notably the hard candies, are brightly colored. Color stability is therefore an important aspect of the storage quality of these products. The main cause of color fading during storage is the bleaching effect of light. The best way to avoid this would be to enclose the product in a light-proof container. This is fine for storage prior to distribution, but at the retail level a light-proof wrapper negates the attractiveness of the colored candy on display. Therefore, the colors must be light stable for at least a few weeks, although not necessarily for several months. Natural colors are notoriously unstable to both light and oxygen, and artificial colors are virtually always used in hard candies (Kearsley and Rodriguez, 1981).

V. NUTRITIONAL AND SAFETY CONCERNS

A. Energy and Calories

The most popular nutritional and safety accusations against confectionery are rooted in two bases: that candy provides calories but no other nutrients, and that refined sugars contribute to various health problems. Unemotional discussion of these issues is rare. Only a brief summary is presented here. More complete discussion is contained in the references cited. Since these issues are based on the overall composition of the products, particularly on the amount of sugar, storage changes have little influence on the discussion.

Sucrose and other sugars, like all available carbohydrates, supply about 4 kcal/g. A confection that contains 80% sugar would provide about 400 kcal/100g product, or about 100 kcal/oz. While most confections provide some other nutrients, mainly from milk, nuts, or other nonsugar ingredients, it is unjustified to claim that the amounts are nutritionally significant. Therefore, the "empty calorie" claim is basically true. However, this should not be used as proof that confectionery consumption causes obesity or nutrient deficiencies. Energy and nutrient imbalances are the result of an individual's total dietary pattern, and cannot be blamed on any one item. It is true that many individuals have difficulty meeting other nutrient requirements without exceeding their energy needs. Such individuals would be well advised to limit intake of confectionery and other "empty calorie" foods. For most people, confectionery products can be a pleasurable part of a balanced diet.

An interesting point has been made that cultivation of sugar cane or sugar beets provides the highest level of food energy per unit area of land cultivated (Hockett, 1955). Accordingly, a society deficient both in food energy and in arable land (but nothing else) should consider sucrose as an important agricultural product. Interestingly, sugar cane is becoming an important source of fuel energy in several countries.

B. Nutrient Stability

Since confectionery is not a particularly valuable source of nutrients, it should not be surprising that the issue of nutrient stability in confectionery is rather unimportant. Thiessen (1975) came to the same conclusion with regard to nutrient stability in foods, such as jams, that have sugar added as a means of preservation. If anything, the high sugar content and resulting low a_w probably slows reactions that lead to nutrient destruction.

An interesting related and much more important topic is the stability of nutrient supplements, in which sugar is often used as a coating or tableting agent. A sugar coating on a nutrient supplement has the same effect as on a soft candy center. Namely, it adds structural stability and inhibits water and oxygen transfer. It should be expected, then, that a sugar coating stabilizes oxygen-sensitive nutrients, such as vitamin C. The sugar, then, plays a more important role than merely to make the supplement easier to swallow.

C. Dental Caries

One disorder toward which sugars have a rather clear contribution is dental caries, or tooth decay (DePaola, 1982). The decay process begins when oral bacteria ferment sugar that adheres to the tooth surface. The acid produced by fermentation leads to demineralization and, eventually, to caries. While any substance fermentable by the bacteria can assist the cariogenic process, the chemical and physical forms have a great influence. Sucrose appears to have a special role in promoting dental caries, even more than other sugars. More important than the chemical nature of the material, however, is the length of time that it remains in the mouth. The longer the sugar is on the tooth surface, the greater the acid production. Time in the mouth is decreased if the caries-promoting substance is in liquid form or if it is followed by other components of a meal. Thus the greatest cariogenic risk seems to come from consuming a sticky sucrose product between meals or at the end of a meal. Unfortunately, this set of circumstances often occurs. The traditional admonitions, to cut down on between-meal sweets and to brush your teeth after every meal, are excellent ways of reducing risk of dental caries. It should be mentioned, however, that other methods, such as fluoride treatment, are also quite effective. Interestingly, some unidentified component or components of chocolate have been found to protect against the cariogenicity of sucrose (Paolino *et al.*, 1979). The practical significance of this observation is not clear, but the area certainly deserves further study.

D. Metabolic Effects

Refined sugars have been implicated as causative agents or contributing factors to a whole array of disorders. Most of the evidence is anecdotal or circumstantial. However, there is some clinical and experimental evidence that sugar plays a role in diabetes and cardiovascular diseases (Reiser, 1982). In these, the contribution, if any, appears due to the rapid increase in blood glucose that follows a high intake of carbohydrates. This, in turn, triggers an increase in serum levels of both insulin and triglycerides. As with dental caries, sucrose appears to play an especially significant role, but the overall diet is probably more important than any one component (Crapo *et al.*, 1976).

Some individuals have a metabolic error that makes them intolerant to fructose (Hue, 1974). They must avoid even moderate amounts of fructose and of sucrose, which yields fructose on hydrolysis in the intestines. These people should be under medical and dietetic supervision.

E. Intestinal Absorption

In an attempt to appeal to consumers who are concerned about the negative images of sugar, confectionery manufacturers have developed several sugar-free products. These are usually sweetened with the sugar alcohols sorbitol, mannitol, or xylitol. Much less common are artificially sweetened confections containing saccharin or aspartame. The sugar alcohols are not well utilized by oral bacteria, so do not promote dental caries. They are considered caloric sweeteners, because they are metabolized as carbohydrates, but provide fewer calories than an equivalent amount of sugar. One reason is that they are not completely absorbed from the intestine. In addition, they are metabolized only slowly, so that even the amount absorbed is not completely converted to calories. Since the sugar alcohols are poorly absorbed, a large amount of a given dose passes unchanged through the digestive system. Excessive intake can cause diarrhea due to the presence of a large amount of the sugar alcohol, and associated water, in the large intestine. Recommended maximum daily intakes of these sugar alcohols are about 20g for sorbitol, 15g for mannitol, and 100g for xylitol (Ellis and Krantz, 1941; Emodi, 1982). However, as little as 5g of sorbitol has been reported to cause abdominal distress (Hyams, 1982).

F. Safety of Nonsugar Components

Unlike the high sugar levels in confectionery, which generate much controversy, the safety of other minor components is rather unquestioned. There has been some recent concern about the safety of compounds formed during the development of caramel color. Model systems for the browning process have been found to contain material that is mutagenic to bacteria (Perkins *et al.*, 1982; Shibamoto, 1982). Since most bacterial mutagens are also animal carcinogens, such findings cannot be ignored. Candies are not the only dietary source of caramel, and storage time probably has little effect on intake level. The actual health risk from caramel or from the many other food-borne mutagens (Ames, 1983) is unknown.

The greatest potential health risk from a nonsugar component of confectionery products is probably from mycotoxins. These substances, many of which are carcinogenic, are produced by molds that may infest foods (Wogan, 1965). Aflatoxin, which is a very potent carcinogen, is produced by the mold *Aspergillus flavus*. Peanuts provide a suitable medium on which this mold can grow and produce aflatoxin.

A confectionery manufacturer should use the appropriate quality control measures to be sure that incoming peanut products are free from excessive amounts of aflatoxin. Because of their composition, it is unlikely that aflatoxin would be produced in confectionery products after production.

TABLE I
Range of Cooking Temperatures, Moisture Contents, and Proportions of Sugars of Principal Types of Candy

Candy	Final Cooking Temp. Range °F	Final Moisture Content Range, %	Sugar Ingredients Range, %			Principal Other Ingredients	
			Sucrose	Invert Sugar	Corn Syrup Solids	Ingredient	Range, %
Hard							
Plain	275-338	1.0-1.5	40-100	0-10	0-60	—	—
Butterscotch	240-265	1.5-2.0	40- 65	—	35-60	Butter	1-7
Brittle	290-295	1.0-1.5	25- 55	—	20-50	—	—
Cremes							
Fondant	235-244	10.0-11.5	85-100	5-10	0-10	Starch	0-1
Cast	235-245	9.5-10.5	65- 75	—	24-40	Egg albumin	0-0.05
Butter	235-247	9.5-11.0	50- 65	—	25-40	Butter	1-15
Fudge	240-250	8.0-10.5	30- 70	0-17	12-40	Milk solids	5-15
						Fat	1-5
Caramel	240-265	8.0-11.5	0- 50	0-15	0-50	Milk solids	15-25
						Fat	0-10
Nougat	255-270	8.0-8.5	20- 50	0-15[a]	30-60	Fat	0-5
Marshmallow							
Grained	240-245	12.0-14.0	50- 78	0- 5	15-40	Gelatin	1.5-3
Soft	225-230	15.0-18.0	25- 54	0-10	40-60	Gelatin	2-5
Jellies							
Starch	230-235	14.5-18.0	25- 60	0-10	28-65	Starch	7-12
Pectin	220-230	18.0-22.0	40- 65	—	30-48	Pectin	1.5-4

[a]Honey is often used
Source: Martin (1955) with permission American Chemical Society.

TABLE II
Concentration and Density
of Saturated Sucrose Solutions

Temperature (°C)	Solubility g/100 g Solution[a]	Solubility g/100 g Water[b]	Density (g/mL)[c]
5	64.81	184.2	1.315
6	64.90	184.9	
7	65.00	185.7	
8	65.10	186.6	
9	65.21	187.4	
10	65.32	188.3	1.320
11	65.43	189.3	
12	65.55	190.3	
13	65.67	191.3	
14	65.79	192.3	
15	65.92	193.4	1.324
16	66.05	194.5	
17	66.18	195.7	
18	66.32	196.9	
19	66.46	198.1	
20	66.60	199.4	1.329
21	66.74	200.7	
22	66.89	202.0	
23	67.04	203.4	
24	67.20	204.8	
25	67.35	206.3	1.334
26	67.51	207.8	
27	67.68	209.4	
28	67.84	211.0	
29	68.01	212.6	
30	68.18	214.3	1.339
31	68.35	216.0	
32	68.53	217.7	
33	68.71	219.5	
34	68.89	221.4	
35	69.07	223.3	1.344

[a]From equation (1)
[b]From equation (2)
[c]From Bates (1942)

TABLE III
Solubility of Sucrose in Invert Sugar Solutions

g Invert Sugar / 100 g Water	Solubility (g/100 g water) at:			
	20°	23.15°	30°	50°
0	199.4[a]	208.6	213.6[b]	260.4
20	191.7	198.8	206.7	252.1
40	185.6	190.8	200.1	245.0
80	172.9	177.2	188.1	233.2
160	--	--	167.3	213.9
200	--	--	159.3	207.9
300	--	--	147.3	196.0
400	--	--	140.4	188.5

[a]Values at 20° are extrapolated.
[b]The value from Table II (214.3) is probably more accurate.

Source: Bates (1942)

TABLE IV
Solute Constants (K) for Calculation of Water Activity

Ingredient	K	Mol. Wt.
Sucrose	2.60	342
Glucose, Fructose, Invert sugar	0.70	180
Corn syrups:		
32.8 DE	2.48	505
42 DE	2.31	460
55 DE	2.25	396
64 DE	1.96	353
83.4 DE	1.64	260
90.7 DE	1.30	225
Sorbitol	0.85	182
Glycerol	0.38	92

Source: Norrish (1966) by permission Institute of Food Science and Technology

TABLE V
Water Activity of Sucrose/Invert Sugar/Water Solutions Saturated with Sucrose

g invert sugar / 100 g water	water activity at: 20°	23.15°	30°	50°
0	0.86	0.85	0.85	0.81
20	0.84	0.84	0.83	0.79
40	0.82	0.82	0.81	0.76
80	0.79	0.79	0.78	0.74
160	—	—	0.72	0.68
200	—	—	0.69	0.65
300	—	—	0.61	0.58
400	—	—	0.55	0.52

Source: Calculated from equation 5 using composition from table III.

TABLE VI
Diffusion Coefficient (D) of Sucrose in Aqueous Solution

g Sucrose / 100 g Solution	$D \times 10^7$ cm^2/s at 25°
61.05	10.71
66.22	7.80
70.44	5.42
72.21	4.51
74.50	3.14
	at 35°
61.05	15.88
65.95	12.00
70.40	8.61
72.20	7.33
72.77	6.70

Source: English and Dole (1950) by permission American Chemical Society

TABLE VII
Viscosity of Aqueous Sucrose Solutions

g Sucrose / 100g Solution	Viscosity (cP) at			
	10°	20°	30°	40°
60	110.9	58.49	33.82	21.04
61	133.8	69.16	39.32	24.11
62	163.0	82.42	46.02	27.80
63	200.4	99.08	54.27	32.26
64	249.0	120.1	64.48	37.69
65	313.1	147.2	77.29	44.36
66	398.5[a]	182.0	93.45	52.61
67	513.7	227.8	114.1	62.94
68	672.1	288.5	140.7	75.97
69	892.5	370.1	175.6	92.58
70	1206	481.6	221.6	114.0
71	1658	636.3	283.4	142.0
72	2329	854.9	367.6	178.9
73	3340	1170	484.3	228.5
74	4906	1631	648.5	296.0
75	7402	2328	884.8	389.5

[a]Values at 10° above 65% are extrapolated.

Source: Swindells et al. (1958).

TABLE VIII
Viscosity of Aqueous Sucrose Solutions Containing Dextran (5 g /100 mL)

g sucrose / 100 g solution	dextran molecular weight 6.9×10^4 viscosity (cP) at 20°	30°	5×10^5 viscosity (cP) at 20°	30°
60	179	104	334	193
61	212	121	395	225
62	253	141	471	263
63	304	167	566	310
64	369	198	686	368
65	451	237	844	441
66	559	287	1040	534
67	699	350	1300	652
68	885	432	1650	803
69	1140	539	2110	1000
70	1480	680	2750	1270
71	1950	870	3630	1620
72	2620	1130	4880	2100
73	3590	1490	6680	2770
74	5010	1990	9310	3700
75	7140	2720	13300	5050

Calculated from equations (6) and (7).

TABLE IX
Relative Sweetness of Various Sweeteners

Sweetener	Sweetness (Sucrose=1)
Sucrose	1
Lactose	0.4
Maltose	0.5
D-Glucose	0.7
D-Fructose	1.1
Invert sugar	0.7-0.9
Sorbitol	0.5
Mannitol	0.7
Cyclamate	30-80
Aspartame	100-200
Sodium saccharin	200-700

Source: Inglett (1981) by permission Institute of Food Technologists.

TABLE X
Packaging Materials for Candies

Material	Rating as a Barrier to		
	Moisture	Odor and Flavor	Oils and Fats
1.5 mil aluminum foil	Excellent	Excellent	Excellent
0.5 mil aluminum foil-1.0 mil polyethylene	Excellent	Excellent	Excellent
1.0 mil aluminum foil-1.0 mil polyethylene	Excellent	Excellent	Excellent
Coated 1.0 mil aluminum foil	Excellent	Excellent	Excellent
0.75 mil PVDC (Saran) film	Excellent	Excellent	Excellent
1.0 mil rubber hydrochloride film	Very Good	Very Good	Excellent
1.5 mil coated cellophane	Very Good	Good	Excellent
2.0 mil polyethelene film	Good	Good	Very Good
0.35 mil aluminum foil-tissue paper	Good	Good	Very Good
42 lb/ream bleached Kraft paper-8 lb/ream polyethylene	Good	Fair	Good
Glassine paper	Poor	Poor	Good

source: Sacharow and Griffin (1970) by permission AVI Publishing Co.

TABLE XI
Storage Changes in Fondant

	change				
	None	Minus 2g water	Plus 2g water	Invert 8g sucrose	Raise Temp to 30°
wt % sucrose	78.4	80.0	76.9	70.4	78.4
wt % invert sugar	9.6	9.8	9.4	18.0	19.6
wt % water	12.0	10.2	13.7	11.6	12.0
wt % crystal	57.7	63.3	53.4	53.9	55.8
wt % syrup	42.3	36.7	46.6	46.1	44.2
volume (mL/100g)	67.6	65.5	69.5	67.5	67.7
crystal size (μm)[a]	10.7	10.9	10.5	10.4	10.6
wt % solids in syrup	71.7	72.0	70.6	74.9	72.8
syrup viscosity (cP)[b]	580	640	440	1100	250[c]
water activity	0.79	0.78	0.80	0.74	0.78

[a]assuming 3 x 10 crystals/g; [b] from Pancoast and Junk (1980) ; [c] estimated
source: calculated, for details see text

TABLE XII
Change in Size of Sucrose Crystals on Fondant Storage

	crystal size (μm) range after storage		
Sample[a]	1 day	7 days	21 days
16	5.0-25.5	5.8-28.3	6.1-26.1
4a	5.2-15.5	5.2-18.3	6.1-20.0
4b	3.6-12.5	4.0-13.3	3.3-14.4
4c	2.8-13.2	3.3-15.8	2.8-17.5

[a]Sample codes assigned by the original authors
Source: Woodruff and van Gilder (1931) by permission American Chemical Society.

REFERENCES

Ames, B. N. (1983). *Science 221*, 1256-1265.

Bates, F. J. (1942). "Polarimetry, Saccharimetry and the Sugars," Cir. No. 440. Nat. Bur. Stand., Washington, D.C.

Charles, D. F. (1960). *Int. Sugar J. 62*, 126-131.

Crapo, P., Reaven, G., and Olefsky, J. (1976). *Diabetes 25*, 741-747.

DePaola, D. P. (1982). *In* "Food Carbohydrates" (D. R. Lineback and G. E. Inglett, eds.), pp. 134-152. AVI, Westport, Connecticut.

Ellis, F. W., and Krantz, J. C. (1941). *J. Biol. Chem. 141*, 147-154.

Emodi, A. (1982). *In* "Food Carbohydrates" (D. R. Lineback and G. E. Inglett, eds.), pp. 49-61. AVI, Westport, Connecticut.

English, A. C., and Dole, M. (1950). *J. Am. Chem. Soc. 72*, 3261-3267.

Fennema, O. R., ed. (1985). "Food Chemistry," 2nd ed. Dekker, New York.

Feuge, R. O. (1970). *In* "Twenty Years of Confectionery and Chocolate Progress" (C. D. Pratt *et al.*, eds.), pp. 449-455. AVI, Westport, Connecticut.

Goodall, H. (1968). "The Physical Properties of Dextrose Fondants," Res. Rep. No. 141. Br. Food Manuf. Ind. Res. Assoc., Leatherhead, Surrey, England.

Greenfield, P. F., and Geronimos, G. L. (1978). *Int. Sugar J. 80*, 67-72.

Higbie, H., and Stegeman, G. (1950). *J. Am. Chem. Soc. 72*, 3799.

Hinton, C. L. (1970). *In* "Twenty Years of Confectionery and Chocolate Progress" (C. D. Pratt *et al.*, eds.), pp. 635-646. AVI, Westport, Connecticut.

Hirschmuller, H. (1953). *In* "Principles of Sugar Technology" (P. Honig, ed.), Vol. 1, pp. 18-74. Am. Elsevier, New York.

Hockett, R. C. (1955). *Adv. Chem. Ser. 12*, 114-124.

Honig, P. (1953). "Principles of Sugar Technology," 2 vols. Am. Elsevier, New York.

Hoynak, P. X., and Bollenback, G. N. (1966). "This is Liquid Sugar," 2nd ed. Refined Syrups and Sugars, Inc., Yonkers, New York.

Hue, L. (1974). *In* "Sugars in Nutrition" (H. L. Sipple and K. W. McNutt, eds.), pp. 357-371. Academic Press, New York.

Hyams, J. S. (1982). *J. Pediatr. 100*, 772-773.

Inglett, G. E. (1981). *Food Technol.* *35*(3), 37-41.
Kearsley, M. W., and Rodriguez, N. (1981). *J. Food Technol.* *16*, 421-431.
Kelly, F. H. C. (1958). *Sharkara 1*, 37-45.
Lees, R. (1965). "Factors Affecting Crystallization in Boiled Sweets, Fondants and Other Confectionery," Sci. Tech. Surv. No. 42. Br. Food Manuf. Ind. Res. Assoc., Leatherhead, Surrey, England.
Lees, R., and Jackson, E. B. (1975). "Sugar Confectionery and Chocolate Manufacture." Chem. Publ. Co., New York.
Lehninger, A. L. (1975). "Biochemistry," 2nd ed. Worth, New York.
McDonald, E. J. (1953). *In* "Principles of Sugar Technology" (P. Honig, ed.), Vol. 1, pp. 75-127. Am. Elsevier, New York.
Makower, B., and Dye, W. B. (1956). *J. Agric. Food Chem.* *4*, 72-77.
Mansvelt, J. W. (1970). *In* "Twenty Years of Confectionery and Chocolate Progress" (C. D. Pratt *et al.*, eds.), pp. 557-577. AVI, Westport, Connecticut.
Martin, L. F. (1955). *Adv. Chem. Ser.* *12*, 64-69.
Minifie, B. W. (1980). "Chocolate, Cocoa and Confectionery: Science and Technology," 2nd ed. AVI, Westport, Connecticut.
Mullineaux, E., and Hansen, T. J. (1984). *Proc. PMCA Prod. Conf. 38th, 1984*, pp. 34-38.
Nickerson, J. T., and Sinskey, A. J. (1972). "Microbiology of Foods and Food Processing." Am. Elsevier, New York.
Norrish, R. S. (1966). *J. Food Technol.* *1*, 25-39.
Pancoast, H. M., and Junk, W. R. (1980). "Handbook of Sugars," 2nd ed. AVI, Westport, Connecticut.
Paolino, V. J., Kashket, S., Cooney, C. L., and Sparagna, C. A. (1979). *J. Dent. Res.* *48*, 372.
Perkins, E. G., Becher, M. G., Genthner, F. J., and Martin, S. E. (1982). *In* "Food Carbohydrates" (D. R. Lineback and G. E. Inglett, eds.), pp. 458-482. AVI, Westport, Connecticut.
Pratt, C. D., de Vadetzsky, E., Langwill, K. E., McCloskey, K. E., and Schuemann, H. W., eds. (1970). "Twenty Years of Confectionery and Chocolate Progress." AVI, Westport, Connecticut.
Reiser, S. (1982). *In* "Food Carbohydrates" (D. R. Lineback and G. E. Inglett, eds.), pp. 170-205. AVI, Westport, Connecticut.
Sacharow, S., and Griffin, R. C. (1970). "Food Packaging." AVI, Westport, Connecticut.

Schneider, F., Emmerich, A., Finke, D., and Panitz, N. (1976). *Zucker 29*, 222-229.
Shibamoto, T. (1982). *Food Technol. 36*(3), 59-62.
Smythe, B. M. (1967). *Aust. J. Chem. 20*, 1087-1095.
Swindells, J. F., Snyder, C. F., Hardy, R. C., and Golden, P. E. (1958). "Viscosities of Sucrose Solutions at Various Temperatures," Circ. No. 440, Suppl. Nat. Bur. Stand., Washington, D.C.
Taylor, J. B., and Rowlinson, J. S. (1955). *Trans. Faraday Soc. 51*, 1183-1192.
Thiessen, R. (1975). *In* "Nutritional Evaluation of Food Processing" (R. S. Harris and E. Karmas, eds.), 2nd ed., p. 382. AVI, Westport, Connecticut.
Tikhomiroff, N., Pultrini, F., Heitz, F., and Gilbert, M. (1965). *C. R. Acad. Sci. 261*, 334-337.
Tilbury, R. H. (1981). *In* "Biology and Activities of Yeasts" (F. A. Skinner, S. M. Passmore, and R. R. Davenport, eds.), pp. 153-179. Academic Press, New York.
van Hook, A. (1953). *In* "Principles of Sugar Technology" (P. Honig, ed.), Vol. 1, pp. 149-187. Am. Elsevier, New York.
van Hook, A. (1961). "Crystallization Theory and Practice." Van Nostrand-Reinhold, Princeton, New Jersey.
White, G. W., and Cakebread, S. H. (1966). *J. Food Technol. 1*, 73-82.
Wogan, G. N., ed. (1965). "Mycotoxins in Foodstuffs." MIT Press, Cambridge, Massachusetts.
Woodruff, S., and van Gilder, H. (1931). *J. Phys. Chem. 35*, 1355-1367.

CHAPTER 8

GRAIN LEGUMES

LOUIS B. ROCKLAND
THOMAS M. RADKE
Food Science Research Center
Chapman College
Orange, California

Dry beans and other grain legumes are important sources of food protein, carbohydrates, minerals, fiber, and several B-complex vitamins. Legume protein content and nutritional qualities represent a compromise between animal protein (i.e., milk, meat, fish, and other muscle proteins) and proteins derived from cereal grains (i.e., rice, wheat, corn). Dry beans generally contain a relatively high percentage of protein, a deficiency of the amino acid methionine, and an excess of lysine. In contrast, cereal grains contain a slightly larger proportion of methionine and a lower proportion of lysine. The amino acid deficiencies in legumes and grains are obviated during normal food preparation by supplementation with small amounts of animal protein or by combining legume and cereal grain proteins in proportions that optimize their complementary nature (Rockland and Nishi, 1979).

Handbook of Food and Beverage
Stability: Chemical, Biochemical,
Microbiological, and Nutritional Aspects

Dry beans must be cooked, preferably in boiling water, to maximize their nutritional qualities (i.e., protein efficiency ratio or PER, and digestibility, Table I) and provide products suitable for human consumption (Rockland and Radke, 1981). The cooking process, which induces several physical and chemical changes, includes the following:

1. Partial release of calcium and magnesium into the cooking water
2. Rapid intracellular starch gelatinization
3. Gradual plasticization or partial solubilization of components of the middle lamella and separation of bean cells along the planes of the cell wall without cell rupture
4. Progressive slow denaturation of protein (Hahn *et al.*, 1977; Rockland *et al.*, 1977; Rockland, 1978).

These processes facilitate softening of the cotyledon and seedcoat, and elution and/or inactivation of antinutritional factors (Table II). Since the cooking process is required in order to provide nontoxic, digestible products, nutrient data available for raw legumes are academic. However, contemporary legal restrictions mandate the inclusion of nutrient data on raw legumes as marketed, without concern or regard for the nutritional qualities of the final cooked product as consumed. Compositional and nutrient data for both raw and cooked beans, as reported in the literature, represent random samples and a variety of soaking and cooking procedures. Therefore, in the present text, no attempt has been made to assemble and evaluate previously published data. The tables presented herein represent the evaluation of authentic samples of each bean variety prepared under standard conditions (Table III). The compositional and nutrient data can be compared without concern for cultivar variations or the relative reliability of published data. In addition to data for raw beans, the tables provide new standardized data on the composition and nutrient qualities of both standard (STD) and quick-cooking (QC) beans (Tables IV to X).

The standard beans were prepared by soaking dry beans in water overnight and cooking in fresh boiling water for 1 to 2 hr, or until tender. Quick-cooking beans were prepared by soaking dry beans in a mixed-salt formulation for 24 hr, draining and cooking from 8 to 15 min depending upon the variety (Rockland and Metzler, 1967; Rockland, 1972; Iyer *et al.*, 1980). Compared with the time and energy required for cooking standard water-soaked beans, quick-cooking beans represent a saving of from 85 to 90% of the time and energy normally required to cook beans. In addition, the cooked, quick-cooking beans have an improved smooth texture and an enhanced natural flavor.

Tables XI to XIV contain United States Recommended Dietary Allowance (USRDA) values for raw beans as well as nutrient fractions retained after cooking. For those constituents that have no assigned USRDA levels, values have been assigned on the basis of Maximum Estimated Safe and Adequate Intake (MESAI).

Tables XV to XVII contain conversion factors that permit convenient translation of equivalences for dry, soaked, and cooked beans in units of cups, pounds, and kilograms.

TABLE I

Protein Efficiency Ratios and Digestibilities of Dry Beans

Variety	Protein Efficiency Ratio		Digestibility[a]	
	Raw	Cooked	Diet	Nitrogen
Small white	0	1.3	91	78
Pink	0	1.2	90	74
Pinto	0	1.2	91	75
Lt red kidney	0	1.2	92	74
Dk red kidney	0	1.4	91	75
Baby lima	0	1.4	93	79
Large lima (65)	0	1.6	91	78
Blackeye	1.2	1.6	92	80
Garbanzo	2.2	2.4	87	76

[a]Percent digestibilities for cooked beans.

TABLE II

Naturally Occurring Undesirable and Toxic Factors in Raw Dry Beans

Factor	Active Principal	Major Effects	Heat Effect[a]
Anti-trypsin	Protein	Decreased digestibility, pancreas hypertrophy	+
Anti-amylase	Protein	Uncertain	+
Haemagglutinins	Protein	Growth inhibition	+
Essential-metal binding	Protein, phytate	Inhibits metal absorption	+
Cyanogenetic glycoside	Linimarin, phaseolunatin	HCN toxicity	+
Protein quality	Methionine deficiency	Growth inhibition	(+)
Nitrogen utilization	Unknown	Growth inhibition	(+)
Food digestibility	Unknown	Inefficient food utilization	(+)
Flatulence	Unknown	Gastrointestinal discomfort	(-)

[a]Cooked in boiling water or autoclaved.
+ heat improves nutritional quality.
(-) unaffected by heat.
(+) heat improves nutritional quality, except if overheated, in which case quality is reduced.

TABLE III

Common Commercial Varieties of Dry Beans

Scientific Name	Common Name	Cultivar
Phaseolus vulgaris	Small white	California
	Pink	Sutter
	Pinto	Commercial
	Light red kidney	California
	Dark red kidney	California
Phaseolus lunatus	Baby lima	Mescela
	Large lima	White Ventura 65
Vigna unguiculata	Blackeye pea or bean, cowpea	No. 5
Cicer arietinum	Garbanzo, chickpea, Bengal gram	California

TABLE IV

Proximate Composition of Raw and Cooked Dry Beans[a]

Variety	Protein			Fat			Fiber			Ash		
	Raw	Cooked		Raw	Cooked		Raw	Cooked		Raw	Cooked	
		STD[b]	QC[c]		STD[b]	QC[c]		STD[b]	QC[c]		STD[b]	QC[c]
	%	%	%	%	%	%	%	%	%	%	%	%
Small white	26.5	27.3	26.9	1.4	1.7	1.6	5.7	6.8	5.9	4.5	3.4	4.5
Pink	24.3	25.2	25.5	1.4	1.4	1.6	4.2	4.7	4.8	4.4	3.5	4.0
Pinto	24.4	24.2	24.1	1.6	1.8	2.0	4.3	4.4	5.0	4.3	3.5	4.0
Lt red kidney	26.3	26.5	26.6	1.6	1.9	2.2	4.1	5.0	4.7	3.9	2.7	3.7
Dk red kidney	25.4	25.7	25.6	1.2	1.8	2.0	4.6	5.4	5.6	4.5	3.3	4.1
Baby lima	24.4	25.9	25.1	1.1	1.5	1.5	5.3	6.0	5.8	4.7	2.9	4.0
Large lima (65)	25.8	25.7	25.6	1.1	1.9	1.9	5.8	7.7	6.3	5.0	4.1	5.0
Blackeye	24.6	25.0	24.4	1.4	1.8	2.2	3.0	2.9	3.2	3.9	3.0	3.8
Garbanzo	19.2	18.8	19.0	7.7	9.0	8.9	2.9	2.9	3.0	3.2	1.7	2.5

[a] Moisture-free basis.
[b] Standard rehydrated dry beans.
[c] Quick-cooking processed dry beans (Rockland, 1978).

TABLE V

Sugars, Total Carbohydrate and Total Calories in Raw and Cooked Dry Beans[a]

Variety	Sugars			Carbohydrate[b]			Total Calories[c]		
	Raw	Cooked		Raw	Cooked		Raw	Cooked	
		STD[d]	QC[e]		STD[d]	QC[e]		STD[d]	QC[e]
	%	%	%	%	%	%	Kcal	Kcal	Kcal
Small white	5.6	2.7	2.1	61.9	60.8	61.1	370	370	370
Pink	5.7	2.4	2.1	65.7	65.2	64.1	370	370	370
Pinto	6.9	---	3.3	65.4	66.1	64.9	370	370	370
Lt red kidney	6.1	2.5	2.1	64.1	63.9	62.8	370	370	370
Dk red kidney	4.7	2.0	1.7	64.3	63.8	62.7	370	370	370
Baby lima	5.7	1.9	1.2	64.5	63.7	63.6	370	370	370
Large lima (65)	6.1	3.1	3.1	62.3	60.6	61.2	370	370	370
Blackeye	6.2	3.0	2.7	67.1	67.3	66.4	370	370	370
Garbanzo	4.8	2.1	2.8	67.0	67.6	66.6	415	425	425

[a]Moisture-free basis.
[b]By difference.
[c]Per 100 gm.
[d]Standard rehydrated dry beans.
[e]Quick-cooking processed dry beans (Rockland, 1978).

TABLE VI

Thiamine, Riboflavin and Niacin in Raw and Cooked Dry Beans[a]

Variety	Thiamine			Riboflavin			Niacin		
	Raw	Cooked		Raw	Cooked		Raw	Cooked	
		STD[b]	QC[c]		STD[b]	QC[c]		STD[b]	QC[c]
	mg %	mg %	mg %	mg %	mg %	mg %	mg %	mg %	mg %
Small white	1.10	0.55	0.35	0.15	0.08	0.09	1.50	0.55	0.65
Pink	0.95	0.60	0.50	0.19	0.12	0.12	2.10	0.95	0.95
Pinto	0.75	0.50	0.30	0.15	0.10	0.12	1.90	0.85	0.75
Lt red kidney	0.90	0.50	0.35	0.19	0.11	0.13	2.85	1.20	1.25
Dk red kidney	0.90	0.55	0.30	0.17	0.11	0.12	3.25	1.00	1.40
Baby lima	0.70	0.35	0.15	0.10	0.08	0.07	1.45	0.80	0.65
Large lima (65)	0.75	0.40	0.40	0.16	0.09	0.10	2.45	1.15	1.20
Blackeye	1.45	0.65	0.85	0.20	0.12	0.10	3.15	1.45	1.65
Garbanzo	0.70	0.30	0.55	0.13	0.08	0.10	1.90	0.55	0.85

[a]Moisture-free basis.
[b]Standard rehydrated dry beans.
[c]Quick-cooking processed dry beans (Rockland, 1978).

TABLE VII

Calcium, Magnesium and Phosphorus Content of Raw and Cooked Dry Beans[a]

Variety	Calcium			Magnesium			Phosphorous		
	Raw	Cooked		Raw	Cooked		Raw	Cooked	
		STD[b]	QC[c]		STD[b]	QC[c]		STD[b]	QC[c]
	mg %	mg %	mg %	mg %	mg %	mg %	mg %	mg %	mg %
Small white	160	160	160	230	180	140	760	640	780
Pink	120	100	100	220	150	140	700	600	550
Pinto	160	120	120	---	---	---	480	400	460
Lt red kidney	90	80	70	180	130	110	490	430	430
Dk red kidney	70	70	70	180	160	130	760	700	710
Baby lima	100	70	50	240	150	110	470	430	560
Large lima (65)	70	60	40	180	140	140	530	470	570
Blackeye	70	60	40	220	180	170	580	600	620
Garbanzo	100	100	70	120	100	100	380	380	400

[a]Moisture-free basis.
[b]Standard rehydrated dry beans.
[c]Quick-cooking processed dry beans (Rockland, 1978).

TABLE VIII

Sulphur, Manganese and Nickel in Raw and Cooked Dry Beans[a]

Variety	Sulphur			Manganese			Nickel		
	Raw	Cooked		Raw	Cooked		Raw	Cooked	
		STD[b]	QC[c]		STD[b]	QC[c]		STD[b]	QC[c]
	mg %	mg %	mg %	mg %	mg %	mg %	mg %	mg %	mg %
Small white	300	200	210	1.3	1.4	1.1	1.1	0.4	0.5
Pink	270	190	190	1.4	1.2	1.2	1.3	0.5	0.3
Pinto	260	200	200	1.0	1.2	---	---	---	---
Lt red kidney	210	170	160	1.3	1.3	1.3	0.6	0.3	0.2
Dk red kidney	290	220	220	0.9	1.0	1.0	0.5	0.2	0.1
Baby lima	200	190	200	2.0	2.0	1.4	0.4	0.2	0.2
Large lima (65)	220	200	220	2.0	1.6	1.5	1.2	0.6	0.7
Blackeye	220	210	200	1.1	1.2	1.1	0.7	0.3	0.3
Garbanzo	240	200	220	3.2	3.7	2.9	0.9	0.5	0.8

[a]Moisture-free basis.
[b]Standard rehydrated dry beans.
[c]Quick-cooking processed dry beans (Rockland, 1978).

TABLE IX

Iron, Copper and Zinc Content of Raw and Cooked Dry Beans[a]

Variety	Iron			Copper			Zinc		
	Raw	Cooked		Raw	Cooked		Raw	Cooked	
		STD[b]	QC[c]		STD[b]	QC[c]		STD[b]	QC[c]
	mg %	mg %	mg %	mg %	mg %	mg %	mg %	mg %	mg %
Small white	14	11	10	1.0	0.5	0.5	3.6	3.5	1.6
Pink	11	10	10	0.8	0.6	0.6	3.1	2.8	1.7
Pinto	9	8	8	1.2	1.2	1.2	2.6	2.5	1.4
Lt red kidney	11	11	9	0.8	0.8	0.9	4.1	3.5	2.1
Dk red kidney	9	8	9	0.8	0.8	0.6	2.9	2.5	1.8
Baby lima	10	7	6	0.5	0.5	0.4	3.0	2.6	1.5
Large lima (65)	9	8	8	1.1	0.9	0.9	3.8	3.2	2.5
Blackeye	10	8	8	1.0	0.7	0.7	4.3	4.3	4.0
Garbanzo	7	7	6	1.2	1.1	1.2	3.7	4.0	3.6

[a]Moisture-free basis.
[b]Standard rehydrated dry beans.
[c]Quick-cooking processed dry beans (Rockland, 1978).

TABLE X

Potassium, Sodium and Chloride Content of Raw and Cooked Dry Beans[a]

Variety	Potassium			Sodium			Chloride		
	Raw	Cooked		Raw	Cooked		Raw	Cooked	
		STD[b]	QC[c]		STD[b]	QC[c]		STD[b]	QC[c]
	mg %	mg %	mg %	mg %	mg %	mg %	mg %	mg %	mg %
Small white	1900	950	550	5	5	1000	5	5	350
Pink	1900	1200	550	5	5	1000	5	5	520
Pinto	1700	850	450	5	5	800	5	5	440
Lt red kidney	1500	700	400	10	10	900	5	5	320
Dk red kidney	1400	1100	550	20	20	1000	5	5	540
Baby lima	2000	1200	550	20	20	1100	5	5	540
Large lima (65)	2400	1600	1200	5	5	900	5	5	430
Blackeye	1500	800	700	5	5	1000	5	5	420
Garbanzo	1000	550	400	10	10	600	5	5	300

[a]Moisture-free basis.
[b]Standard rehydrated dry beans.
[c]Quick-cooking processed dry beans (Rockland, 1978).

TABLE XI

Percentages of USRDA[a] for Thiamine, Riboflavin and Niacin and Fractions Retained After Cooking

Variety	Thiamine			Riboflavin			Niacin		
	USRDA Raw	Fract. ret. Cooked		USRDA Raw	Fract. ret. Cooked		USRDA Raw	Fract. ret. Cooked	
		STD[b]	QC[c]		STD[b]	QC[c]		STD[b]	QC[c]
	%			%			%		
Small white	73	0.50	0.32	9	0.53	0.60	8	0.37	0.43
Pink	63	0.63	0.53	11	0.63	0.63	11	0.45	0.45
Pinto	50	0.67	0.40	9	0.67	0.80	10	0.45	0.39
Lt red kidney	60	0.56	0.39	11	0.58	0.68	15	0.42	0.44
Dk red kidney	60	0.61	0.33	10	0.65	0.71	17	0.31	0.43
Baby lima	47	0.50	0.21	6	0.80	0.70	8	0.55	0.45
Large lima (65)	50	0.53	0.53	9	0.56	0.63	13	0.47	0.49
Blackeye	97	0.45	0.59	12	0.60	0.50	17	0.46	0.52
Garbanzo	47	0.43	0.79	8	0.62	0.77	10	0.29	0.45

[a]United States Recommended Dietary Allowance per 100 gm raw, dry beans, moisture-free basis.
[b]Standard rehydrated dry beans.
[c]Quick-cooking processed dry beans (Rockland, 1978).

TABLE XII

Percentage of USRDA[a] for Calcium, Magnesium and Phosphorus and Fraction Retained After Cooking

Variety	Calcium			Magnesium			Phosphorus		
	USRDA Raw	Fract. ret. Cooked		USRDA Raw	Fract. ret. Cooked		USRDA Raw	Fract. ret. Cooked	
		STD[b]	QC[c]		STD[b]	QC[c]		STD[b]	QC[c]
	%			%			%		
Small white	16	1.00	1.00	58	0.78	0.61	76	0.84	1.03
Pink	12	0.83	0.83	55	0.68	0.64	70	0.86	0.79
Pinto	16	0.75	0.75	--	--	--	48	0.83	0.96
Lt red kidney	9	0.89	0.78	45	0.72	0.61	49	0.88	0.88
Dk red kidney	7	1.00	1.00	45	0.89	0.72	76	0.92	0.93
Baby lima	10	0.70	0.50	60	0.63	0.46	47	0.91	1.19
Large lima (65)	7	0.86	0.57	45	0.78	0.78	53	0.89	1.08
Blackeye	7	0.86	0.57	55	0.82	0.77	58	1.03	1.07
Garbanzo	10	1.00	0.70	30	0.83	0.83	38	1.00	1.05

[a]United States Recommended Dietary Allowance per 100 gm raw, dry beans, moisture-free basis.
[b]Standard rehydrated dry beans.
[c]Quick-cooking processed dry beans (Rockland, 1978).

TABLE XIII

Percentages of USRDA[a] for Iron, Zinc and Copper and Fractions Retained After Cooking

Variety	Iron			Zinc			Copper		
	USRDA Raw	Fract. ret. Cooked		USRDA Raw	Fract. ret. Cooked		MESAI[b] Raw	Fract. ret. Cooked	
		STD[c]	QC[d]		STD[c]	QC[d]		STD[c]	QC[d]
	%			%			%		
Small white	78	0.79	0.71	24	0.97	0.44	17	0.50	0.50
Pink	61	0.91	0.91	21	0.90	0.55	20	0.75	0.75
Pinto	50	0.89	0.89	17	0.96	0.54	37	1.00	1.00
Lt red kidney	61	1.00	0.82	27	0.85	0.51	27	1.00	1.13
Dk red kidney	50	0.89	1.00	19	0.86	0.62	27	1.00	0.75
Baby lima	56	0.70	0.60	20	0.87	0.50	17	1.00	0.80
Large lima (65)	50	0.89	0.89	25	0.84	0.66	30	0.82	0.82
Blackeye	56	0.80	0.80	29	1.00	0.93	23	0.70	0.70
Garbanzo	39	1.00	0.86	25	1.08	0.97	37	0.92	1.00

[a] United States Recommended Dietary Allowance per 100 gm raw, dry beans, moisture-free basis.
[b] Maximum Estimated Safe and Adequate Intake per 100 gm raw, dry beans, moisture-free basis.
[c] Standard rehydrated dry beans.
[d] Quick-cooking processed dry beans (Rockland, 1978).

TABLE XIV

Percentages of MESAI[a] for Potassium, Sodium and Chloride and Fractions Retained After Cooking

Variety	Potassium			Sodium			Chloride		
	MESAI Raw	Fract. ret. Cooked		Percentages of MESAI Raw	Percentages of MESAI Cooked		Percentages of MESAI Raw	Percentages of MESAI Cooked	
		STD[b]	QC[c]		STD[b]	QC[c]		STD[b]	QC[c]
	%			%	%	%	%	%	%
Small white	34	0.50	0.29	0	0	30	0	0	7
Pink	34	0.63	0.29	0	0	30	0	0	10
Pinto	30	0.50	0.26	0	0	24	0	0	9
Lt red kidney	27	0.47	0.27	0	0	27	0	0	6
Dk red kidney	25	0.79	0.39	1	1	30	0	0	11
Baby lima	36	0.60	0.28	1	1	33	0	0	11
Large lima (65)	43	0.67	0.50	0	0	27	0	0	8
Blackeye	27	0.53	0.47	0	0	30	0	0	8
Garbanzo	18	0.55	0.40	0	0	18	0	0	6

[a] Maximum Estimated Safe and Adequate Intake per 100 gm beans, moisture-free basis.
[b] Standard rehydrated dry beans.
[c] Quick-cooking processed dry beans (Rockland, 1978).

TABLE XV

Converting One Cup of Dry Beans to Cups, Pounds and Kilograms of Soaked and Cooked Beans

Variety	Soaked Beans			Cooked Beans		
	Cups	Pounds	Kilos	Cups	Pounds	Kilos
Small White	2 1/3	0.90	0.41	2 3/4	1.05	0.48
Pink	2 1/2	0.85	0.39	2 2/3	0.95	0.43
Pinto	2 1/2	0.85	0.39	2 3/4	1.00	0.45
Lt red kidney	2 1/4	0.80	0.36	2 1/2	0.90	0.41
Dk red kidney	2 1/2	0.85	0.39	2 2/3	0.95	0.43
Baby lima	2 1/2	0.85	0.39	2 2/3	0.95	0.43
Large lima (65)	2 2/3	0.95	0.43	2 2/3	1.00	0.45
Blackeye	2 1/3	0.75	0.34	2 2/3	0.90	0.41
Garbanzo	2 1/3	0.80	0.36	2 1/2	0.90	0.41

TABLE XVI

Converting One Pound of Dry Beans to Cups, Pounds and Kilograms of Soaked and Cooked Beans

Variety	Soaked Beans			Cooked Beans		
	Cups	Pounds	Kilos	Cups	Pounds	Kilos
Small White	5 1/3	2.0	0.91	6 1/3	2.4	1.09
Pink	6	2.1	0.95	6 2/3	2.4	1.09
Pinto	6 1/2	2.2	1.00	6 3/4	2.5	1.13
Lt red kidney	5 2/3	2.0	0.91	6 1/3	2.3	1.04
Dk red kidney	6 1/4	2.2	1.00	6 3/4	2.5	1.13
Baby lima	6	2.1	0.95	6 1/3	2.4	1.09
Large lima (65)	6 3/4	2.4	1.09	7	2.6	1.18
Blackeye	6 3/4	2.1	0.95	6 2/3	2.5	1.13
Garbanzo	5 2/3	2.0	0.91	6	2.2	1.00

TABLE XVII

Converting One Kilogram of Dry Beans to Cups, Pounds and Kilograms of Soaked and Cooked Beans

Variety	Soaked Beans			Cooked Beans		
	Cups	Pounds	Kilos	Cups	Pounds	Kilos
Small White	11 1/2	4.41	2.0	14	5.29	2.4
Pink	13	4.63	2.1	14 1/2	5.29	2.4
Pinto	14	4.85	2.2	15	5.51	2.5
Lt red kidney	12 1/2	4.41	2.0	14	5.07	2.3
Dk red kidney	13 1/2	4.85	2.2	15	5.51	2.5
Baby lima	13	4.63	2.1	14	5.29	2.4
Large lima (65)	15	5.29	2.4	15	5.73	2.6
Blackeye	15	4.63	2.1	14 1/2	5.51	2.5
Garbanzo	12 1/2	4.41	2.0	13	4.85	2.2

REFERENCES

Hahn, D. M., Jones, F. T., Akhavan, I., and Rockland, L. B. (1977). Light and scanning electron microscope studies on dry beans: Intracellular gelatinization of starch in cotyledons of large lima beans *(Phaseolus lunatus)*. *J. Food Sci. 42*(5), 1208.

Iyer, V., Salunke, D. K., Sathe, S. K., and Rockland, L. B. (1980). Quick-cooking beans (*Phaseolus vulgaris* L.). *Qual. Plant.--Plant Foods Hum Nutr. 30*, 27.

Rockland, L. B. (1972). "Quick-cooking Legumes," PAG Bull. 2, No. 3, p. 52. Protein Advisory Group of the United Nations System, New York.

Rockland, L. B. (1978). Relationship between fine structure and composition and development of new food products from legumes. *In* "Post Harvest Biology and Biotechnology" (H. O. Hultin and M. Milner, eds.). Food and Nutrition Press, Westport, Connecticut.

Rockland, L. B., and Metzler, E. A. (1967). Quick-cooking lima and other dry beans. *Food Technol. 21* (3A), 26A.

Rockland, L. B., and Nishi, S. K. (1979). Tropical grain legumes. *In* "Tropical Foods: Chemistry and Nutrition" (G. E. Inglett and G. Charalambous, eds.), Vol. 2, p. 547. Academic Press, New York.

Rockland, L. B., and Radke, T. M. (1981). Legume protein quality. *Food Technol. 35*, 427.

Rockland, L. B., Jones, F. T., and Hahn, D. M. (1977). Light and scanning electron microscope studies on dry beans: Extracellular gelatinizat on of lima bean starch in water and a mixed salt solution. *J. Food Sci. 42*(5), 1204.

CHAPTER 9

MINOR COMPONENTS OF RICE: CHANGES DURING STORAGE

ROBERT R. MOD
ROBERT L. ORY
Southern Regional Research Center
United States Department of Agriculture
New Orleans, Louisiana

I. INTRODUCTION

Measurement of amylose:amylopectin ratios is, and has been, the preferred method for predicting desirable cooking properties of rice (1-5). However, this ratio does not always correlate completely with observed differences in cooking behavior of rice. Varieties in the same amylose class may differ in quality of the processed rice, suggesting that factors other than amylose content may be important (6,7). The increase in water absorption and decrease in extractable solids during cooking of milled rice were similar regardless of the form in which the rice was stored or the amylose content of the rice (8).

Rice hemicelluloses have received little attention in earlier studies (9-12), but studies have increased as a result of their importance in cell wall structure and as a dietary fiber source (13-21). Ferulic acid and lignin in rice have received very little attention in the earlier studies (22,23), but because of possible crosslinking of ferulic acid by ester linkage to hemicellulose in the cell walls, interest has increased (24). These minor components of rice may affect the cooking properties and the changes rice undergoes with aging.

Composition and structure have also been investigated in other cereal hemicelluloses: wheat flour pentosans (25-30), barley (31,32), and oats (33-35). Wheat endosperm cell walls contain soluble (36) and insoluble (37) arabinoxylans containing small amounts of ferulic acid bound by ester linkage to the pentosans. Markwalder and Neukom (38) showed the presence of ester-bound diferulic acid in water-insoluble pentosans of wheat endosperm.

Rice undergoes several changes during the storage or aging process that modify the cooking, processing, eating, and nutritional qualities and affect the commercial value of the grain. Freshly harvested milled rice becomes a pasty gruel upon cooking, but after storage for several months, the tendency of the kernels to disrupt and stick together is greatly reduced. Metcalf *et al*. (28) reported that lipids also affected the gelling characteristics of wheat starches. Mod (39) has found ferulic acid associated with lipids from rice flour.

This chapter covers factors that can affect cooking properties and quality of aging and fresh rice, such as chemical composition and chemical reactions of the minor components of rice that are associated with quality, nutritional value, and cooking properties.

III. HEMICELLULOSES

A. Chemical Composition

The composition of rice hemicelluloses from a long-grain (Starbonnet) rice grown in three areas and a medium-grain (Calrose) rice grown in California are shown in Tables I-III.

All of the rices contained the same sugars: rhamnose, arabinose, xylose, mannose, galactose, and glucose. Shibuya and Iwasaki (13) detected a seventh sugar, fucose, in milled rice. This was also found by Pascual and Juliano (40) in rice bran cell walls. Shibuya and Iwasaki (13) reported that the hexuronic acid was galacturonic acid, whereas Pascual and Juliano (40) reported it as a mixture of galacturonic and glucuronic acids.

The amounts of protein associated with these sugars in the hemicelluloses are not as uniform. The protein content of water-soluble (WS) bran hemicellulose is much higher than that of WS endosperm hemicellulose, and the amino acids are atypical of those in seed storage proteins (e.g., low arginine and glutamic acid, high hydroxyproline, alanine, and serine; see Table II). Amino acids of the alkali-soluble (AS) hemicelluloses do not show the high arginine of storage proteins. Glutamic acid is high and hydroxyproline, alanine, and serine are lower than that in WS hemicelluloses (see Table III). Yamagishi *et al.* (41) also found large amounts of hydroxyproline in WS rice bran hemicellulose, present as O-α-L-arabinofuranoxyl-hydroxyproline. No one has been able to confirm a role for hydroxyproline in crosslinking of rice hemicelluloses or on aging properties, but the presence of additional hydroxyproline provides binding sites for hydrogen bonding or ester linkages.

There is a growing interest in dietary fiber contents of foods, especially cereal grains, and on the effects of aging or storage of plant foods on fiber quality. Crude (CF) and neutral detergent fiber (NDF) contents of intact rice bran and endosperm and brown rice are shown in Table IV. There are no essential differences in the endosperm, but bran samples are quite different.

B. Mineral-Hemicellulose Interactions

Interactions of hemicelluloses with minerals will affect their availability from foods that contain the minerals. Dietary fibers also act as ion exchange resins and can produce undesirable effects by reducing mineral absorption (42-45). Both WS and AS hemicelluloses bound maximum amounts of Cu, Fe, and Zn within 30 min *in vitro* (17) (Table V), but digestive tract enzymes released significant amounts of the bound minerals to make them potentially available for absorption (Table VI). Protein such as is present in brown rice affects the interaction of minerals with rice hemicellulose *in vitro* (17). Increased protein content had an inverse effect on release of copper by trypsin, but with hemicellulase and pepsin, it increased the release of iron and zinc (Table VII). *In vitro* tests on interactions of Ca, Mg, and Mn with rice hemicelluloses (18) yielded similar results. Both AS and WS hemicelluloses bound about 100% of the Ca and Mg and 70% of the Mn, but amounts released by enzymes varied. Protein generally decreased the amounts of minerals bound and increased the release by enzyme treatment. Phytic acid, present in the bran layer of brown rice, also affects availability of minerals and mineral-hemicellulose interactions. Table VIII illustrates the effect of phytic acid on binding of Ca and Mg to rice AS hemicellulose. As phytate concentration increases, binding to hemicellulose decreases. Rice hemicelluloses have many sites for hydrogen binding or crosslinking--sites that can be affected by oxidation-reduction during storage or aging and ultimately affect rice cooking properties.

III. FERULIC ACID

A. Association with Hemicelluloses

Ferulic acid is present in cereal grains in both free and combined forms. Fausch *et al.* (36) found it in wheat hemicellulose, Markwalder (38) found it in pentosans of wheat endosperm, and Mod *et al.* (17) found ferulic acid in rice bran and endosperm hemicelluloses. Neukom (46) showed that wheat flour pentosans underwent oxidative gelation to form diferulic acid from two ferulic acid residues, to cause crosslinking of the pentosans. This crosslinking tends to have a stabilizing effect on structure of the large polysaccharides that may ultimately affect cooking properties of the wheat or rice.

B. Effects on Viscosity of Rice Flour

Extraction of hemicellulose from rice flour changed the viscosity of the flour, as measured in the Brabender Amylograph-Viscograph (17). Hemicellulose-free rice flour had different viscosity from that of intact rice flour. Addition of WS endosperm hemicellulose back to long- (Starbonnet) and medium (Calrose)-grain rices had differing effects (Figs. 1-3). Peak viscosity (Fig. 1), viscosity after 20 min hold (Fig. 2), and viscosity cooled from 94 to 50°C (Fig. 3) decreased for long-grain rice flour but increased for medium-grain rice. The summary of these results (Table IX) also shows that gelatinization time is unaffected by addition of hemicellulose back to the rice flour. Viscosity of the waxy variety, Waxy Pirmi, was similar to the long-grain Starbonnet. Medium-grain Nortai showed little or no effect with added hemicellulose (Figs. 4-6). These results, summarized in Table X, are lower than corresponding values for Starbonnet and Calrose. Deobald (47) stated that waxy-rice flour has less than 0.5% amylose in the starch, has a lower peak viscosity than some of the short-grain rices, and has practically no setback (cooled to 50°C) viscosity.

After identifying ferulic acid with rice hemicellulose, the arabinogalactan fraction of rice was purified by the procedure of Neukom and Markwalder (26) for wheat. Addition of the ferulic acid-polysaccharide complex to intact rice flour lowered the entire heating and cooling curves by 185 B.U. (Table XI). Thus the ferulic acid-polysaccharide complex appears to be responsible for the viscosity reduction caused by addition of WS bran hemicellulose to intact rice flour. Although it was not found with the WS endosperm hemicellulose, a possible role of the ferulic acid-polysaccharide complex in cooking properties of rice cannot be ruled out. Ferulic acid is associated with the insoluble endosperm hemicellulose and may be responsible for the insolubility of the pentosans. When oxidized, diferulic acid can be formed by oxidative coupling of two ferulic acids to cause crosslinking of the pentosans. This phenomenon could be responsible for the observed changes in rice during aging, particularly in eating qualities.

C. Changes in Stored Rice

Total phenolic compounds in rice are in constant flux during aging (48). The total phenolics decreased in both long- and medium-grain rices up to 6 months' storage, then appeared to increase slightly after 12 months (Table XII). The lower amount of *cis*- and *trans*-ferulic acid in brown rice compared to that in milled rice may be attributed to higher amounts of *cis*- and *trans*-p-coumaric acid in the total phenolics of brown rice. During this period, lignin content of the rice grain also changed (Table XIII).

Carbohydrate esters of ferulic acid fluoresce blue under UV light, and upon treatment with ammonia vapor they turn green, suggesting attachment to polysaccharide rather than to lignin (49). Insoluble rice endosperm hemicellulose fluoresced similarly under UV light (48), providing evidence for the attachment of ferulic acid to polysaccharide in rice rather than to lignin. Fresh rice endosperm cell walls fluoresce light purple under UV light, possibly because of crosslinking of adjacent polysaccharide chains by oxidative coupling of two ferulic acid molecules, as proposed by Neukom and Markwalder (26). Hardley and Jones (50) reported that diferulic acid fluoresces purple under UV light. Such crosslinking could explain the physicochemical changes observed in rice after long periods of storage or aging and could evolve by the following reactions:

1. Arabinoxylan-ferulic acid crosslinking via formation of diferulic acid
2. Arabinoxylan-ferulic acid crosslinking with cysteine in the protein moiety of an adjacent chain
3. Starch-protein interactions in the cells

TABLE I

Sugar Composition of Rice Hemicelluloses

variety	composition (weight %) Rham	Arab	Xyl	Man	Gal	Glu	protein, N x 5.95	uronic acid	Arab/Xyl ratio
A. Alkali-Soluble Rice-Bran Hemicelluloses									
Starbonnet (La.)	2.6	34	30	tr	5	4	9.58	10	1.1/1
Starbonnet (Ark.)	2.0	24	27	0.9	10	tr	15.89	12	0.9/1
Starbonnet (Tex.)	1.7	24	32	1.0	9	2	18.45	12	0.8/1
Calrose (Cal.)	1.3	22	29	0.7	8	4	26.42	9	0.8/1
B. Alkali-Soluble Rice Endosperm Hemicelluloses									
Starbonnet (La.)	1.1	32	24	1	5	18	9.46	10	1.3/1
Starbonnet (Ark.)	3.0	13	10	3	5	18	27.91	10	1.3/1
Starbonnet (Tex.)	1.2	22	12	10	3	15	26.78	10	1.8/1
Calrose (Cal.)	3.0	17	17	3	6	34	10.83	10	1/1

(Table continued on next page)

TABLE I (continued)

C. Water-Soluble Rice Bran Hemicelluloses									
Starbonnet (La.)	Trace	38.7	13.7	0.2	30.4	0.1	5.23	11	2.8:1
Starbonnet (Ark.)	0.1	38.6	7.1	0.1	31.4	3.2	5.40	12	5.4:1
Starbonnet (Tex.)	0.3	41.8	11.5	2.0	31.1	2.4	4.80	6	3.6:1
Calrose (Cal.)	Trace	38.5	8.7	1.5	28.2	3.3	4.20	11	4.4:1

D. Water-Soluble Rice-Endosperm Hemicelluloses									
Starbonnet (La.)	0.7	8.3	2.4	0.7	11.0	51	3.09	6	4/1
Starbonnet (Ark.)	0.6	9.0	1.4	8.4	1.9	47	2.38	8	6/1
Starbonnet (Tex.)	1.0	7.4	1.1	1.6	11.8	54	3.57	9	7/1
Calrose (Cal.)	0.7	3.4	0.5	3.4	5.0	55	1.67	9	7/1

TABLE II

Amino Acid Composition of Water-Soluble Rice Hemicelluloses[a]

A. Rice Bran

Amino Acid	Variety			
	Starbonnet La.	Starbonnet Ark.	Starbonnet Tex.	Calrose Cal.
Ala	15.6	14.9	15.3	13.6
Val	4.4	6.1	5.1	6.8
Gly	3.3	2.7	5.1	5.1
Ile	1.1	0.7	1.7	1.7
Leu	2.2	2.0	1.7	1.7
Pro	4.1	3.4	3.4	3.4
Thr	6.7	6.8	6.8	6.8
Ser	11.1	11.5	11.9	10.2
Met	1.1	0.7	1.6	1.7
Hypro	28.9	28.4	27.1	23.7
Phe	1.1	1.7	1.7	1.4
Asp	4.4	4.7	4.7	5.1
Glu	8.9	10.1	8.5	10.2
Tyr	1.1	0.7	1.7	1.7
Lys	3.3	2.7	1.7	3.4
His	3.3	2.7	1.7	3.4
Arg	1.1	2.7	1.7	3.4

(Table continued next page)

TABLE II (continued)

B. Rice Endosperm				
Ala	9.4	8.0	9.5	7.9
Val	7.3	7.1	4.2	4.4
Gly	6.3	6.3	4.2	7.0
Ile	3.1	3.6	2.1	2.6
Leu	6.3	5.4	4.2	5.3
Pro	5.2	4.5	3.2	4.4
Thr	5.2	5.4	5.3	5.3
Ser	9.4	8.0	11.6	9.6
Met	1.0	1.0	1.1	0.4
Hypro	9.4	6.3	8.4	5.3
Phe	4.2	4.5	2.1	4.4
Asp	7.3	11.6	13.7	13.2
Glu	8.3	12.5	13.7	14.0
Tyr	3.1	4.5	3.2	4.4
Lys	4.2	6.3	4.2	6.1
His	3.1	1.8	6.3	1.8
Arg	7.3	3.6	4.2	3.5

a/ Weight % of recovered amino acids.

TABLE III

Amino Acid Composition of Alkali-Soluble Rice Hemicelluloses[a].

A. Rice Bran

Amino Acid	Variety			
	Starbonnet La.	Starbonnet Ark.	Starbonnet Tex.	Calrose Cal.
Ala	5.3	8.7	6.9	7.1
Val	4.7	5.1	5.1	5.6
Gly	5.6	6.3	6.3	6.1
Ile	3.0	3.0	3.4	3.8
Leu	7.0	9.0	8.8	9.3
Pro	5.3	7.5	6.4	7.3
Thr	3.3	3.6	3.4	4.5
Ser	5.0	5.7	5.4	5.5
Met	1.7	1.2	1.2	1.8
Hypro	0.7	—	—	0.5
Phe	6.0	6.0	5.8	6.0
Asp	12.3	14.7	14.1	13.8
Glu	22.3	11.4	13.9	11.6
Tyr	2.0	3.3	2.4	2.2
Lys	6.6	5.7	5.4	5.3
His	3.3	3.9	4.4	3.0
Arg	6.0	4.2	7.1	6.3

(Table continued on next page)

TABLE III (continued)

B. Rice Endosperm				
Ala	6.9	6.6	6.6	6.0
Val	5.3	6.8	6.1	6.3
Gly	6.3	6.1	5.5	6.3
Ile	2.9	3.9	2.9	3.8
Leu	6.9	7.6	7.4	6.6
Pro	4.8	4.7	4.0	4.7
Thr	3.4	3.9	3.2	3.4
Ser	5.6	5.3	5.3	4.7
Met	1.9	0.4	1.3	0.9
Hypro	1.1	—	—	0.3
Phe	4.8	4.5	4.2	4.1
Asp	13.0	12.9	14.2	14.1
Glu	21.2	21.1	22.2	23.8
Tyr	2.9	3.2	3.4	2.8
Lys	6.1	6.6	6.4	6.3
His	1.9	1.6	1.7	1.3
Arg	5.3	3.9	5.7	4.1

a/ Weight % of recovered amino acids.

TABLE IV

Dietary Fiber Contents of Rice Bran, Endosperm, and Brown Rice.

Variety	Bran		Endosperm		Brown Rice	
	% CF	% NDF	% CF	% NDF	% CF	% NDF
Starb. (La.)[a]	9.2	29.7	0.3	2.9	1.1	6.0
Starb. (Ark.)	11.7	34.2	0.3	2.7	---	---
Starb. (Tex.)	10.3	29.8	---	---	---	---
Calrose (Cal.)	21.1	44.7	0.3	2.4	1.1	3.8
Nato (Tex.)	15.3	33.0	---	---	---	---
Brazos (Tex.)	11.5	28.7	---	---	---	---
Soft Winter Wheat	---	---	---	---	3.2	18.3

a/ Starb. = Starbonnet; Soft Winter Wheat = sample obtained from U. S. Grain Inspection Service for comparison.

TABLE V

Binding of Minerals by Rice Hemicelluloses Before and After Enzyme Treatment *In Vitro*.

	Initial Amt. Bound (μg/g)	Amounts Still Bound After Treatment (μg/g): Hemicellulase	Pepsin	Trypsin
A. Copper:				
		Alkali-Soluble Hemicellulose		
	9,949±120	7,323±203	6,294±171	4,100±82
		Water-Soluble Hemicellulose		
	7,179±136	5,698±153	3,943±174	4,206±191
B. Iron:				
		Alkali-Soluble Hemicellulose		
	5,088±68	5,164±125	5,045±244	4,726±37
		Water-Soluble Hemicellulose		
	5,047	3,207	2,024	3,184
C. Zinc:				
		Alkali-Soluble Hemicellulose		
	9,399±70	3,409±106	8,630±264	9,356±336
		Water-Soluble Hemicellulose		
	3,000	245	810	2,378

TABLE VI

Release of Minerals from Rice Hemicelluloses by Enzyme Treatment *In Vitro*.

Mineral/Hemicellulose Complex	% Released by Enzymes[a]		
	Hemicellulase	Pepsin	Trypsin
Cu-alk. sol.	26	37	59
Cu-water sol.	21	45	41
Fe-alk. sol.	0	0	7
Fe-water sol.	36	60	37
Zn-alk. sol	64	8	0
Zn-water sol.	92	73	21

[a] Values calculated from data in Table V.

TABLE VII

Effect of Protein on Release of Minerals from WS Rice Hemicellulose by Enzymes *In Vitro*.

Protein Content %	% Released by Enzymes:		
	Hemicellulase	Pepsin	Trypsin
A. Copper:			
9.8	63	40	49
22.8	82	44	45
37.7	21	45	41
B. Iron:			
9.8	11	6	0
22.8	24	18	0
37.7	37	60	37
C. Zinc:			
9.8	74	34	38
22.8	88	71	61
37.7	92	73	21

TABLE VIII

Effect of Phytic Acid on Binding of Ca and Mg by Rice Alkali-Soluble Hemicellulose.

Phytic Acid % (W:W)	Ca Bound (μg/g)	Mg Bound (μg/g)
0	9,294	6,868
33.3	7,503	5,949
50.0	5,525	5,222

TABLE IX

Effect of Hemicellulose on Hemicellulose-Free Rice Flour

Variety and Additives	Gelatinization		Temp. at Peak Viscosity (°C)	Amylograph Viscosity		
	Temp. (°C)	Time (Min.)		Peak (B.U.)	After 20 Min. at 94°C (B.U.)	Cooled To 50°C (B.U.)
Starbonnet						
Hemicellulose-Free	66	18	91.5	620	265	520
10mg " added	65	18	92.0	610	255	500
30mg " added	65	18	91.5	580	215	470
40mg " added	65	18	92.0	600	215	450
Calrose						
Hemicellulose-Free	60	20	90.0	515	220	440
10mg " added	60	20	91.0	520	220	440
20mg " added	60	20	90.0	520	235	460
40mg " added	60	20	90.0	580	235	465

TABLE X

Effect of Hemicellulose on Hemicellulose-Free Rice Flour

Variety and Additives	Gelatinization		Temp. at Peak Viscosity (°C)	Amylograph Viscosity		
	Temp. (°C)	Time (Min.)		Peak (B.U.)	After 20 Min. at 94°C (B.U.)	Cooled To 50°C (B.U.)
Nortai						
Hemicellulose-Free	57	22	90	320	90	210
20mg " added	57	22	90	310	92	200
40mg " added	57	22	90	310	90	215
Waxy Pirmi						
Hemicellulose-Free	54	8	66	500	190	275
10mg " added	54	8	66	475	175	260
20mg " added	54	8	66	435	140	210
40mg " added	54	8	66	410	130	160

TABLE XI

Effect of Hemicellulose and Ferulic Acid-Polysaccharide on Viscosity of Intact Starbonnet Rice Flour.

Fraction Added	Amylograph Viscosity Values		
	Peak	After 20 Min. Hold at 94°C.	Setback at 50°C.
	(B.U.)	(B.U.)	(B.U.)
Intact (control)	530	300	570
500 mg. Bran Hemicellulose	420	260	500
90 mg. Endos. Hemicellulose	500	290	550
Intact (control)	500	300	570
100 mg. Free Polysaccharide	470	300	560
200 mg. Ferulic Acid-Polysaccharide	315	170	400

TABLE XII

Changes in Total Phenolics Content of Rice During Storage

Source/Variety	Months in Storage		
	0	6	12
	(μg/g)	(μg/g)	(μg/g)
Brown Rice:			
Labelle (long grain)	425	304	400
Pecos (medium grain)	294	350	400
Milled Endosperm:			
Labelle	149	120	227
Pecos	76	135	102

Changes in *cis*/*trans* Ferulic Acid Content of Total Phenolics

Source/Variety	0	6	12
	(%)	(%)	(%)
Brown Rice:			
Labelle	86.3	77.2	78.9
Pecos	80.1	72.4	73.7
Milled Endosperm:			
Labelle	95.1	—	83.9
Pecos	95.7	—	83.3

TABLE XIII

Changes in Lignin Content of Rice During Storage.

Source/Variety	Months in Storage (µg/g)		
	0	6	12
Bran:			
Labelle (long grain)	3.0	2.6	3.6
Pecos (medium grain)	3.8	3.3	5.3
Milled Endosperm:			
Labelle	0	0.3	0
Pecos	0	0.1	0

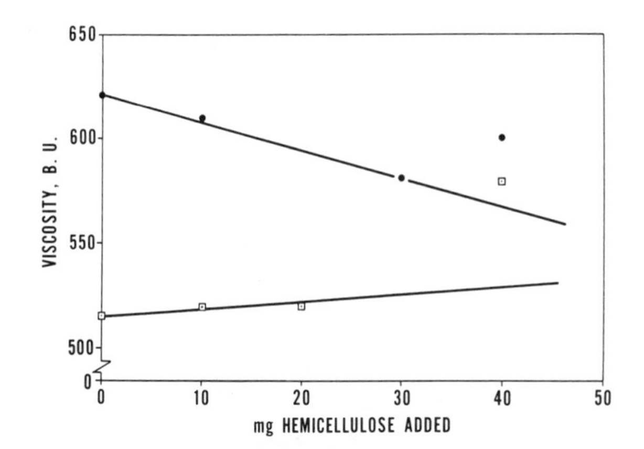

Fig. 1. Effect of added rice hemicellulose on peak viscosity of rice flours: Starbonnet, long grain (●); Calrose, medium grain (▣).

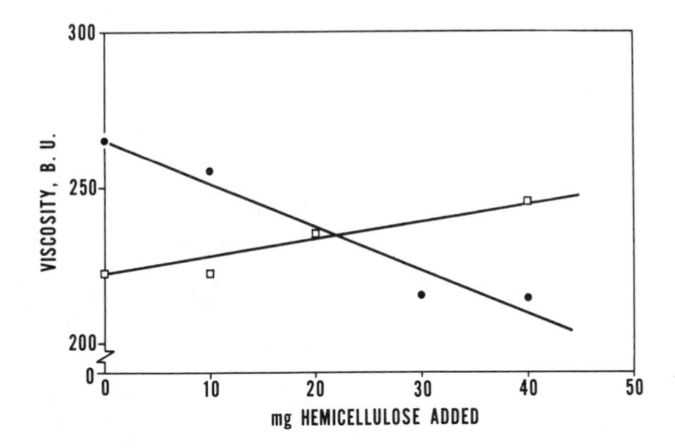

Fig. 2. Effect of added rice hemicellulose on viscosity after 20 min hold of rice flours at 94°C (same samples as Fig. 1).

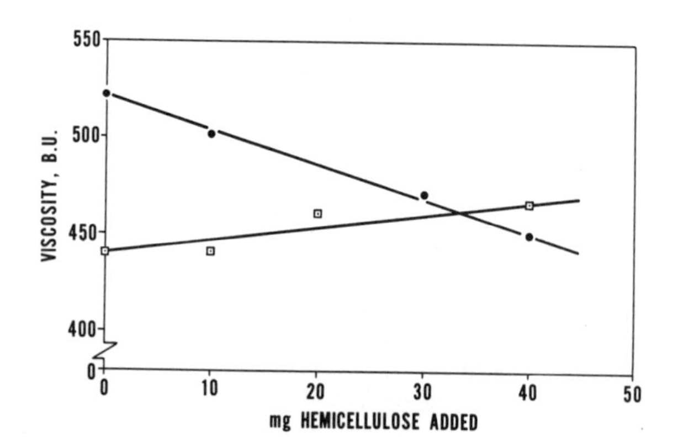

Fig. 3. Effect of added rice hemicellulose on viscosity cooled to 50°C (setback viscosity) of rice flours (same samples as Fig. 1).

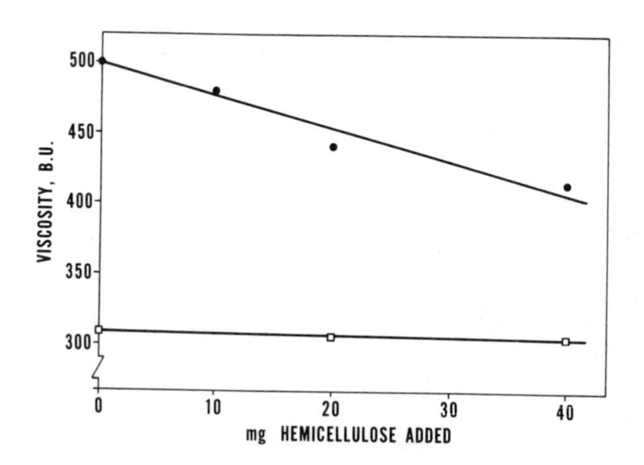

Fig. 4. Effect of added rice hemicellulose on peak viscosity of rice flours: Waxy Pirmi, waxy rice (●); Nortai, medium grain (□).

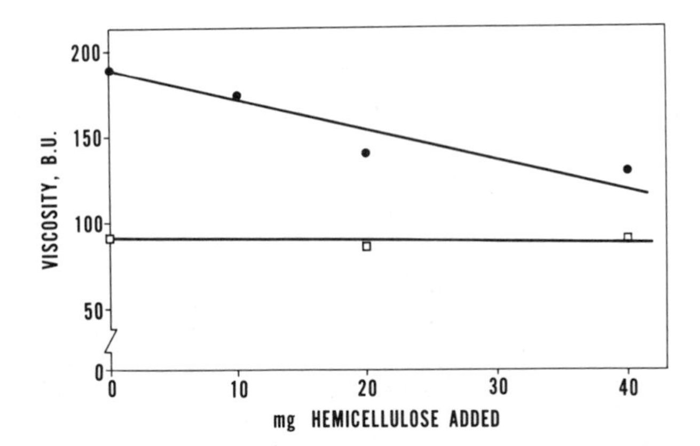

Fig. 5. Effect of added rice hemicellulose on viscosity after 20 min hold of rice flours at 94°C (same samples as Fig. 4).

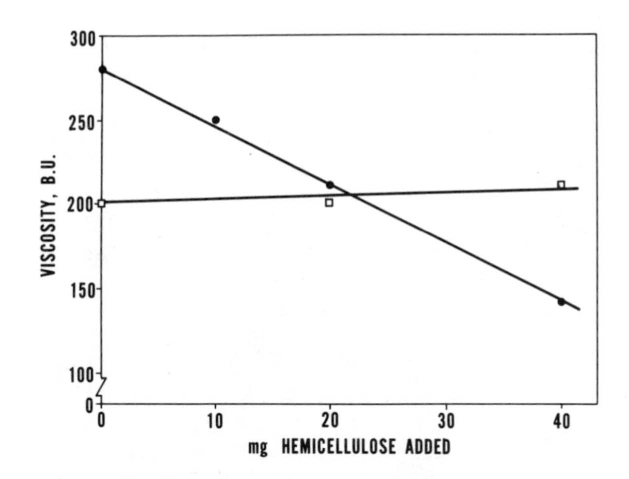

Fig. 6. Effect of added rice hemicellulose on viscosity cooled to 50°C (setback viscosity) of rice flours (same samples as Fig. 4).

REFERENCES

1. Rao, B. S., Vasudeva Murthy, A. R., and Subrahmanya, R. S. (1952). *Proc. Indian Acad. Sci., Sect. B 36B,* 70.
2. Halick, J. V., and Keneaster, K. K. (1956). *Cereal Chem. 33,* 315.
3. Williams, V. R., Wu, W. T., Tsai, H. Y., and Bates, H. G. (1958). *J. Agric. Food Chem. 6,* 47.
4. Juliano, B. O., Onate, L. V., and Del Mundo, A. M. (1965). *Food Technol. 19,* 1006.
5. Bollich, C. N., and Webb, B. D. (1973). *Cereal Chem. 50,* 637.
6. Palmiano, E. P., and Juliano, B. O. (1972). *Agric. Biol. Chem. 36*(1), 157.
7. Primo, E., Cases, A., Barber, S., and Barber, C. (1962). *Rev. Agroquím. Tecnol. Aliment. 2*(4), 343.
8. Villareal, R. M., Resurrection, A. P., Suzuki, L. B., and Juliano, B. O. (1976). *Staerke 28,* 88.
9. Bevenue, A., and Williams, K. T. (1956). *J. Agric. Food Chem. 4,* 1014.
10. Matsuo, Y., and Manba, A. (1958). *Hakko Kogaku Zasshi 36,* 190.
11. Gremli, H., and Juliano, B. O. (1970). *Carbohydr. Res. 12,* 273.
12. Cartano, A. V., and Juliano, B. O. (1970). *J. Agric. Food Chem. 18,* 40.
13. Shibuya, N., and Iwasaki, T. (1978). *Agric. Biol. Chem. 5,* 201.
14. Shibuya, N., Misaki, A., and Iwasaki, T. (1983). *Agric. Biol. Chem. 47*(10), 2223.
15. Mod, R. R., Conkerton, E. J., Ory, R. L., and Normand, F. L. (1978). *J. Agric. Food Chem. 26,* 1031.
16. Mod, R. R., Conkerton, E. J., Ory, R. L., and Normand, F. L. (1979). *Cereal Chem. 56,* 356.
17. Mod, R. R., Ory, R. L., Morris, N. M., and Normand, F. L. (1981). *J. Agric. Food Chem. 29,* 449.
18. Mod, R. R., Ory, R. L., Morris, N. M., and Normand, F. L. (1982). *Cereal Chem. 59,* 538.
19. Normand, F. L., Ory, R. L., and Mod, R. R. (1979). *In* "Dietary Fibers: Chemistry and Nutrition" (G. E. Inglett and I. Falkehag, eds.), p. 203. Academic Press, New York.
20. Normand, F. L., Ory, R. L., and Mod, R. R. (1981). *J. Food Sci. 46(4), 1159.*
21. Maningat, C. C., and Juliano, B. O. (1982). *Phytochemistry 21*(10), 2509.
22. Leonzio, M. (1965). *Riso 16,* 331.

23. Leonzio, M. (1967). *Riso 16*, 313.
24. Mod, R. R., Normand, F. L., Ory, R. L., and Conkerton, E. J. (1981). *J. Food Sci. 46*, 571.
25. Neukom, H., Providoli, L., Gremli, H., and Hui, P. A. (1967). *Cereal Chem. 44*, 238.
26. Neukom, H., and Markwalder, H. V. (1975). *Carbohydr. Res. 39*, 387.
27. Cole, E. W. (1967). *Cereal Chem. 40*, 411.
28. Metcalf, D. G., D'Appolonia, B. L., and Gilles, K. A. (1968). *Cereal Chem. 45*, 539.
29. Fincher, G. B., Sawyer, W. H., and Stone, B. A. (1974). *Biochem. J. 139*, 535.
30. Perlin, A. S. (1951). *Cereal Chem. 28*, 382.
31. Aspinall, G. O., and Ferrier, R. J. (1958). *J. Chem. Soc.*, p. 638.
32. Igarashi, O., and Sakurai, Y. (1966). *Agric. Biol. Chem. 30*, 642.
33. Morris, D. L. (1942). *J. Biol. Chem. 142*, 881.
34. Peat, S., Whelan, W. J., and Roberts, J. G. (1957). *J. Chem. Soc.*, p. 3916.
35. Preece, I. A., and Hobkirk, R. (1953). *J. Inst. Brew. 59*, 385.
36. Fausch, H., Kuendig, W., and Neukom, H. (1963). *Nature (London) 199*, 287.
37. Geissmann, T., and Neukom, H. (1973). *Cereal Chem. 50*, 414.
38. Markwalder, H. V., and Neukom, H. (1976). *Phytochemistry 15*, 836.
39. Mod, R. R. (1984). Unpublished data.
40. Pascual, C. G., and Juliano, B. O. (1983). *Phytochemistry 22*, 151.
41. Yamagishi, T., Matsuda, K., and Watanabe, T. (1976). *Carbohydr. Res. 50*, 63.
42. Branch, W. J., Southgate, D. A. T., and James, W. P. T. (1975). *Proc. Nutr. Soc. 34*, A120.
43. James, W. P. T., Branch, W. J., and Southgate, D. A. T. (1978). *Lancet 1*, 938.
44. Reinhold, J. G., Faradji, B., Aradi, P., and Tsmail-Beigi, F. (1976). *J. Nutr. 106*, 493.
45. McHale, A., Kies, C., and Fox, H. M. (1979). *J. Food Sci. 44*, 1412.
46. Neukom, H. (1980). *In* "Autoxidation in Food and Biological Systems" (M. G. Simic and M. Karel, eds.), p. 249.
47. Deobald, H. J. (1972). *In* "Rice Chemistry and Technology" (D. F. Houston, ed.), p. 264. Am. Assoc. Cereal Chem., St. Paul, Minnesota.
48. Mod, R. R., Conkerton, E. J., Chapital, D. C., and Yatsu, L. Y. Unpublished data.

49. Harris, P., and Hartley, R. (1976). *Nature (London) 259,* 508.
50. Hartley, R. D., and Jones, E. C. (1975). *Phytochemistry 15,* 1157.

CHAPTER 10

CHEMICAL AND MICROBIOLOGICAL STABILITY OF SHOYU (FERMENTED SOY SAUCE)

TAMOTSU YOKOTSUKA
Kikkoman Corporation
Noda-shi, Chiba-ken, Japan

Handbook of Food and Beverage Stability: Chemical, Biochemical, Microbiological, and Nutritional Aspects

I. INTRODUCTION AND HISTORY

The amino acid-containing seasonings around the world are divided into two categories, the chemical and the fermentative; the first originated in Europe a few hundred years ago, the second in China 3000 years ago. Of the fermentative soy sauce there are two principal varieties in the Orient, the Chinese and the Japanese. Chinese soy sauce is made largely from soybeans with only a small amount of wheat flour as raw materials; Japanese soy sauce is made from approximately equal parts of soybeans and wheat kernels. Production of the former has basically followed the old method developed in China, while production of latter was developed in Japan in the late seventeenth century based on technology imported from China and Korea. The history of fermented soybean foods in China and Japan is roughly indicated in Table I (35 Years History of Noda Shoyu Co. Ltd., 1955; The History of Kikkoman, 1977; Sakaguchi, 1979; Wang and Fuang, 1981; Bo, 1982).

The name of chiang-yu appeared in China for the first time in the book "Ben-Chao-Gong-Mu" in 1590, while in Japan, shoyu appeared for the first time in the book "Ekirinbon-Setsuyoshu," written in 1598. It must be noted that chiang-yu and shoyu in these books were distinctly different products, although the Chinese characters for

both products were the same; 醤
"Ben-Chao-Gong-Mu" described the preparation of chiang-yu as mixing boiled soybeans with raw barley flour in the ratio 3:2, pressing the mixture into cakes, leaving the cakes in a room until they became covered with yellow molds, mixing the cakes with salt water, and aging the mixture under sun. The liquid obtained after pressing the mash was chiang-yu. We can see the distinct similarity in this chiang-yu to the soy sauce now popular in the Southeast Asian countries, except that in these countries, the koji (mold culture) is cultured with the mixture of soybeans and wheat flour in a ratio from 10:1 to 10:2 in a granular form, without being formed into balls or cakes. In the northern part of China and Korea, the mixture of cooked soybeans and wheat flour is still pressed into bricks or balls and then cultured with molds to make koji.

In Japan, in the production of a kind of soy sauce known as tamari, cooked soybeans and a small amount of wheat or barley flour are extruded into small pellets of 15 to 20 mm diameter and then cultured with molds. The koji is mixed with a small amount of salt water to make a rather hard mash. This mash is so hard that it cannot be mixed or agitated; rather the liquid part of the mash separated in the bottom layer is siphoned up and sprayed on the surface of the mash. The liquid part of an aged mash is drained off from the bottom of the keg, and the residue is usually extracted with cool or boiled salt water to get a second batch of tamari. The consumption of tamari has declined greatly, so that it now represents less than 4% of the total production of shoyu. The original appeal of tamari shoyu was that it is rich in the dilicious taste derived from a high content of amino acids. The highest total nitrogen content of the best quality of tamari was near 3%. The major reasons for the retarded consumption of tamari in recent years in Japan is that it is too expensive, lacks good volatile flavor (especially that produced by yeast fermentation), and sometimes exhibits an inferior volatile flavor caused by bacterial contamination. The traditional centers of tamari production have been in central Japan, near the city of Nagoya.

The technology of preparing shoyu made great progress and the production of shoyu increased during the Edo period, 300-400 years ago. Many books were published at that time regarding shoyu preparation, and the manufacturing procedure of koikuchi-shoyu, now the representative Japanese shoyu, was established about 300 years ago. The center of industrial production of koikuchi shoyu was in Noda and Choshi, both in Chiba Prefecture, near Tokyo. Shoyu production in these cities began in 1561 and 1616, respectively, and more than

20,000 kiloliters of shoyu were produced in the two cities in 1 year.

The light-colored shoyu, usukuchi, was invented in Tatsuno, western Japan, in 1666. Wheat kernels slightly exceed soybeans in the mixture from which usukuchi is prepared, and more water is used in usukuchi than in koikuchi production. Both kinds of shoyu, koikuchi and usukuchi, were differentiated from tamari, which was made almost exclusively from soybeans. The production of tamari was industrialized also in the late seventeenth century.

II. SHOYU IN JAPAN (Yokotsuka, 1961, 1980, 1981a,b,c, 1983)

A. Kinds of Shoyu

The Japanese Agricultural Standard (JAS) classifies shoyu into five kinds, with three production methods and three grades. The names of five kinds of shoyu and their amount of production are listed in Table II. Koikuchi means dark in color, usukuchi light in color; both are made from approximately equal parts of soybeans and wheat kernels. More than 85% of shoyu is koikuchi. Shiro-shoyu is made mostly from wheat kernels with a very small amount of soybeans and is very light in color. On the contrary, saishikomo-shoyu is prepared from koji and unpasteurized shoyu instead of salt water. Therefore, it is very dark in color and rich in solids content. The current annual production of shoyu averages 1.2 million kiloliters.

The results of chemical analysis of different kinds of shoyu obtained on the Japanese market are given in Table III. Koikuchi and usukuchi are rich in alcohol, but tamari initially has a higher nitrogen content but only trace amounts of alcohol, unless alcohol is added later. Similar differences are observed between the Japanese koikuchi and usukuchi, and the soybean sauce and fish sauce produced in the Southeast Asian countries (Table IV).

On the other hand, the soy sauce produced in the Orient may be categorized as follows on the basis of its contents of alcohol and lactic acid:

1. Low alcohol (trace) and low lactic acid (trace): protein chemical hydrolysate type
2. Lower alcohol (<0.5%) and higher lactic acid (0.5-1.2%): tamari type
3. Higher alcohol (1-3%) and higher lactic acid (0.8-1.5%): koikuchi and usukuchi type

The following features distinguish genuine koikuchi shoyu from other similar products in the Orient:

1. Wheat kernels are used in equal or greater amounts than soybeans as raw materials.
2. Defatted soybean grits are now most often used; sometimes, however, the soybean component is made up of either whole soybeans or defatted soybean grits mixed with a small amount of whole soybeans.
3. Koji molds, *Aspergillus oryzae* or *A. sojae*, are cultured on the whole mixture of raw materials in granular form.
4. The salty mash, composed of koji and a considerable amount of salt water, is subjected to vigorous lactic and yeast fermentations at <30°C.
5. After pressing the aged salty mash to produce raw shoyu, the pressing residue is usually not washed with salt water to obtain a second batch.
6. Raw shoyu is pasteurized at 70 to 80°C, followed by gradual cooling.

The resulting chemical composition of koikuchi shoyu can be described as follows:

1. The large amount of carbohydrates present in the raw materials results in a final product rich in both glucose and alcohol.

2. Koji molds grow better on the finely crushed soybeans and wheat kernels than they do on either whole soybeans covered with wheat flour or a mixture of soybeans and wheat flour made into balls, pellets, or cakes. This is because the greater surface area of raw materials allows greater enzymatic activity in the koji.

3. The protein involved in raw materials is highly degraded because of (1) the high enzymatic activity of koji, (2) the improved autoclaving of soybeans and wheat kernels, and (3), the improved conditions of salty mash fermentation. The resulting product contains more total nitrogen and amino nitrogen, especially free glutamic acid.

4. Such volatile flavor-producing compounds as *p*-alkylphenol and guaiacol derivatives, produced from wheat kernels by the mold and yeast fermentations, constitute an important part of the volatile flavor of koikuchi.

5. The raw shoyu is rich in alcohol and lactic acid, as well as many other aromatic flavor-producing compounds produced by vigorous lactic and yeast fermentations.

6. The final product exhibits a dark reddish brown color and strong "brown" flavor resulting from pasteurization

carried out at a fairly high temperature and followed by gradual cooling.

The typical chemical composition of good-quality genuine fermented koikuchi shoyu is 1.5-1.8% (grams per 100 ml) total nitrogen, 3-5% reducing sugar (mainly glucose), 2-3% ethanol, 1-1.5% polyalcohol (primarily glycerol), 1-2% organic acid (predominantly lactic acid), and 16-18% sodium chloride (NaCl), and the pH is 4.7-4.8. To be palatable, shoyu should consist of one-half of the nitrogenous compounds as free amino acids and >10% of the nitrogenous compounds as free glutamic acid.

The JAS recognizes the three following production methods for shoyu: (1) genuine fermented or honjozo, (2) semichemical or semifermented, or shinshiki-jozo, and (3) amino acid solution-mixed.

According to JAS, in the preparation of genuine fermented shoyu, >80% of heat-treated soybeans and wheat kernels should be cultured with koji molds to make koji, and the koji then mixed with salt water to make mash or moromi. Moromi should be fermented with lactobacilli and yeasts, and then well aged. The total amount of shoyu checked by JAS in 1981 was about 1.076 million kiloliters, of which 72% was genuine fermented, 24% was semichemical or shinshiki jozo, and the remaining 4% was amino acid solution-mixed.

The protein chemical hydrolysate (HVP), which is prepared by degrading defatted soybean meal with concentrated HCl at >100°C, is not regarded as shoyu by the JAS. The merits of HVP are its higher content of tasty amino acid (e.g., glutamic and aspartic acids) and its moderate price; on the other hand, the demerits are the lack of flavorful metabolites of lactic and yeast fermentations, and too high a content of volatile sulfur and carbonyl compounds, which worsen the odor of the product and render it unstable upon blending with fermented shoyu and subsequent storage.

The JAS establishes three grades for each variety of shoyu: special, upper, and standard. The grade is determined by organoleptic evaluation, total nitrogen content, soluble solids without NaCl content, and alcohol content. Only high-quality shoyu made by fermentation can qualify for the special grade. The official standard shoyu used as the control for the organoleptic evaluation of special grade is issued annually by the Central Inspection Association of Shoyu. The analytical standard of JAS for special grade of koikuchi shoyu is as follows: >1.5% total nitrogen, 16% extract, and >0.8% alcohol. About 60% of Japanese shoyu was special grade in 1981. Blending of fermented shoyu with chemical hydrolysate is permitted only for upper and standard grades, to the extent that the characteristic flavor of fermented shoyu is not spoiled, usually no more than a 50% mixture. It is estimated that there are <3000 shoyu manufacturers in Japan. The five major producers account for 50% of the total production, and the 50 producers next in line account for 25%. The biggest manufacturer's share in the market is ~30%.

B. Manufacture of Koikuchi Shoyu

The flow sheet of koikuchi fermentation given in Fig. 1 outlines five steps: treatment of raw materials, koji making, mash controlling, mash pressing, and refining.

Whole soybeans or, more commonly, defatted soybean grits are moistened and cooked with steam under pressure. Wheat kernels, the other half of the raw materials, are roasted at 170 to 180°C for a few minutes, then coarsely crushed into four or five pieces each. The soybean-wheat kernel mixture is then inoculated with a pure culture of *Aspergillus oryzae* or *A. sojae* and spread out to a depth of 30 to 40 cm on a large perforated stainless-steel plate having either a rectangular shape measuring 5 × 12 m, or a doughnut shape with a diameter of 15 to 30 m. The raw materials are aerated for 2 or 3 days with temperature- and moisture-controlled air, which comes up from bottom holes through the materials to give the proper conditions for mold cultivation and enzyme formation. The temperature of the materials is kept at ~30°C, and the 40-43% moisture level of the materials at the beginning of cultivation decreases to 25 to 30% after 2 or 3 days. This allows the mold to grow throughout the mass and provides the enzyme necessary to hydrolyze the protein, starch, and other constituents of the raw materials. This mold-cultured material is called koji. The koji is mixed with salt water of a 22-23% salt content to a final volume of 120 to 130% of that of the raw materials. The mash, or moromi, is transferred to the deep fermentation tanks, formerly wooden kegs holding about 5 to 10 kiloliters or concrete tanks holding 10-20 kiloliters. These tanks are now being replaced by resin-coated iron tanks of 50 to 300 kiloliters. The moromi is held for 4 to 8 months, depending on its temperature, and occasionally agitated with compressed air to mix the dissolving contents uniformly and to promote microbial growth. During the fermentation period, the enzymes from koji mold hydrolyze most of the protein to amino acids and low molecular weight peptides. Around 20% of the starch is consumed by the mold during koji cultivation, but almost all of the remaining starch is converted to simple sugars; in turn, more than half of these simple sugars are fermented to lactic acid and alcohol by lactobacilli and yeasts, respectively. The pH drops from an initial value of 6.5 to 7.0 to 4.7 to 4.9. The lactic acid fermentation in the beginning stage is gradually replaced by yeast fermentation. Pure cultured *Pediococcus halophylus* and *Saccharomyces rouxii* are sometimes added to the mash. The salt concentration of the mash stabilizes at around 17 to 18% (w/v) after 1 or 2 months.

An aged shoyu mash is filtered under high hydraulic pressure through cloth. The pressure for pressing is increased in two or three steps, sometimes reaching 100 kg/cm^2 in the final stage, which decreases the moisture content of the press-cake to <25%. A diaphram-type pressing machine is often employed for shoyu mash filtration, resulting in a press-cake moisture content of >30%. The press-cake is used to feed cows and ducks.

The liquid component obtained by pressing of the mash is clarified and then pasteurized at 70 to 80°C and stored in a semiclosed tank. The clear middle layer is bottled or canned, or sometimes spray-dried. The oil layer separated from the heated shoyu consists of free higher fatty acids and their ethyl esters derived from the yeast metabolism of soybean and wheat oils, and it is sometimes utilized as an antifreezing agent for paint.

III. STABILITY OF COLOR OF SHOYU

Shoyu color ranges from yellow to dark reddish-brown depending on the varieties of the product. The color results not from the true solution of pigments but from the colloidal dispersion of socalled melanoidin compounds, which change the tones from reddish brown to yellow upon dilution with water. Shoyu color is generally produced by oxidative and nonoxidative browning reactions, which generally occur between amino compounds and sugars. The intensity, tones, and stability of the color of shoyu to heat and oxidation are dependent on the proportions of raw materials used (especially the mixing ratio of soybeans and wheat kernels), the forms of soybeans (whole or defatted), the forms of wheat (whole kernels or wheat bran), the kinds of koji mold, the conditions of koji preparation, the temperature and the duration of fermentation of salty mash or moromi, the kinds of lactic bacteria and the degree of yeast fermentation of salty mash, and the conditions of pasteurization of raw shoyu.

The mechanism of the formation of color during such processes as treatment of raw materials, koji cultivation, salty mash fermentation, pressing of mash, and pasteurization of raw shoyu, is for the most part the result of nonenzymatic, nonoxidative, and heat-dependent chemical reactions. This color formation during mash fermentation is predominantly the result of the so-called Maillard reaction between amino compounds and sugars, and that during pasteurization is mostly the result of Strecker degradation along with the Maillard reaction. This color formation is always accompanied by the formation of the desired shoyu flavor.

On the other hand, the color increase or darkening of shoyu upon marketing occurs during transportation, storage, and consumption; the extent of color change depends on such factors as the size and gas permeability of the containers, and the length and temperature of storage and consumption. This color darkening greatly reduces the flavor quality and chiefly results from nonenzymatic and oxidative chemical reactions, which are accelerated by temperature elevation and are catalyzed by such heavy metals as iron and copper. The color of koikuchi and usukuchi shoyu packed in tightly sealed glass bottles is relatively stable but darkens soon after opening the seals. Shoyu packed in 1-liter polyethylene-terephthalate (PET) bottles or in tin cans of the same volume darkens upon exposure to air twice as quickly as shoyu packed in glass bottles of the same volume.

The color density of the salty mash or moromi of koikuchi shoyu when the enzymatic degradation of raw materials and lactic and yeast fermentations are nearly complete (usually 3-4 months from the beginning) approximately doubles by the end of aging (6 or 8 months from the beginning). Furthermore, the color density of the liquid part of this aged mash again doubles by the pasteurization steps, which is usually carried at about 70 to 80°C followed by gradual cooling.

The consumption rate in Japan for a 1- or 2-liter bottle of koikuchi shoyu is 2-4 weeks for an average family consisting of two to four persons. When the color density of koikuchi shoyu more than doubles, mainly as a result of oxidative darkening upon use, the flavor deterioration becomes noticeable to the ordinary consumer. This occurs after ∿35 days' storage at 28°C, for example. The browning velocity of shoyu product sealed from the air increases about 10-fold upon exposure to the air at the same room temperatures. The susceptibility to oxidative color darkening of koikuchi shoyu is two to three times that of usukuchi shoyu.

In general, raw materials containing less pentose and pentosan, salty mash fermentation with greater reducing power of lactobacilli, sufficient fermentation of salty mash with yeasts, pasteurization of raw shoyu accompanied by less browning reactions, lighter color, lower concentrations of total nitrogen and sugars, lower pH values, lower rH values, lower content of iron and copper, packaging in glass bottles instead of plastic ones, and lower storage temperature, all tend to help stabilize the product to oxidation.

A. Color Compounds of Shoyu

Kurono and Katsume (1927) characterized the color of shoyu as consisting of melanoidin pigments mainly consisting of two compounds: $C_{27}H_{17}N_3O_{13}$ and $C_{27}H_{15}N_3O_{12}$. Mitsui and Kusabe (1957) also isolated two kinds of shoyu pigment, one of which was considered to be the same as that isolated by Kurono.

Omata *et al.* (1955b) isolated the pigmenting substances of shoyu as two fractions, acidic and basic, by column and paper chromatography, and then spectrophotometrically determined their increase during the brewing of mash and the storage of shoyu. The acidic fraction increased more than the basic fraction.

Hashiba (1971) isolated the browning compounds in shoyu by gel filtration with Sephadex G-25 into three peaks, PI, PII, and PIII, according to the elution. PI increased during oxidative storage, while PIII increased remarkably in the pasteurization process. The increase of PI gave the dark-brown color, while that of PIII gave the red tone of the color of shoyu, respectively. Hashiba (1973a) purified the melanoidin produced during storage of shoyu at 37°C for 50 days by dialysis against running water, DEAE-cellulose chromatography, and Sephadex G-100 gel filtration until a single band appeared on disk electrophoresis. The color of melanoidin thus obtained did not increase by heating or by oxidation. In addition, it had no effect on the browning of shoyu upon heating and oxidation, and it considerably reduced ferricyanide and consumed oxygen. Upon hydrolysis, the melanoidin liberated sugars such as glucose, xylose, galactose, and arabinose, and all kinds of amino acids found in shoyu.

Motai *et al.* (1972) and Motai and Inoue (1974) fractionated the color material of shoyu into eight color components by DEAE-cellulose chromatography with stepwise elution. The color tone of each peak became darker, and the E_{450} and molecular weight became higher according to the order of elution. Bright tone of color components increased when shoyu was heated, while dark tone of color components increased when it was oxidized. On the other hand, the melanoidin pigments prepared by heating an aqueous solution of glycine and xylose at 100°C for 2 hr were chromatographically fractionated into eight color components. The fractionated color components from shoyu and those from the glycine-xylose model system exhibited the corresponding behaviors both for heating and oxidation. According to spectral measurements, elemental analysis, and amino acid analysis, the color components appeared to be very similar to each other in their chemical structures, having stepwise different molecular weights. The infrared absorption spectra of the eight peaks were of the same patterns, which suggested that they were those of melanoidins. Thus the color of shoyu has been recognized to consist of at least eight kinds of melanoidin pigments of different degrees of polymerization.

B. Measurement of Shoyu Color

The color of shoyu represented by the CIE system was reported by Omata and Ueno (1953a) to be a dominant wavelength of 590 to 620 nm, an excitation purity of 86-88%, and a luminous transmittance of 0.14 to 0.17. Umeda and Saito (1956) determined the color of several shoyu samples prepared from different aging periods of mash according to the CIE system, and a set of color standards for shoyu was prepared from chemical pigments so as to correspond to the above color differences, which consisted of 30 degrees of aqueous color solution of the same visual distance packed in glass test tubes. The method for determining the color intensity of a shoyu sample by comparison to the above color standard is very simple and convenient when the shoyu has the same color tone, having been separated from the same kinds of mash or just after pasteurization; however, where different color tones exist because of oxidation during storage or different production methods, shoyu color is difficult to determine by this method.

This color standard was originally prepared for the color determination of koikuchi shoyu, but its transmittances were not always identical with those of koikuchi shoyu, when determined at extremely short or long wavelengths; thus a special light source is needed for the actual color determination. Furthermore, this color standard could not be applied for the determination of the very light color of usukuchi or shiro shoyu, whose color tones differ immensely from those of koikuchi shoyu.

Okuhara *et al*. (1977) prepared an improved set of color standards for shoyu based precisely upon the spectrophotometric transmittances determined with koikuchi, usukuchi, and shiro shoyu as indicated in Table V. The new color standards consist of two sets of color tubes, one for koikuchi and usukuchi shoyu consisting of 27 color tubes, and the other for shiro shoyu consisting of 16. The new standards exhibit spectrophotometric transmittances almost identical to those of the above three kinds of shoyu covering the wide ranges of wavelength, so that ordinary white light can be used as the light source for the color determination.

According to Okuhara (1973), the photometric absorption of shoyu color has no peak and is greater in the short-wavelength range than in the long-wavelength range; therefore, the color tone of diluted shoyu is pale yellow but shifts from orange to dark red in accordance with increase in concentration.

The spectrophotometric absorption of stepwise-diluted raw shoyu, pasteurized shoyu, and pasteurized oxidized shoyu are indicated in Table VI, VII, and VIII, respectively. The x and y values of a 40% concentration of raw shoyu (40 ml raw shoyu + 60 ml water), 20% concentration of pasteurized shoyu, and 20% pasteurized oxidized shoyu were similar to each other, but their Y values (luminosity) were 28.3, 31.6, and 24.1%, respectively.

According to Motai (1976), a linear relationship was found between the logarithm of absorbance (log A) and the wavelength (450-650 nm) in the color distribution of both shoyu and melanoidin prepared from a model system, although a similar phenomenon is generally recognized for the color of whisky, cola drinks, beer, caramel, and miso. There was no change in 10 g A per 100 nm (designated as ΔA) of each of eight pigments fractionated from shoyu or melanoidin of the model system, when each of them was heated or oxidized. According to 10 g A per 100 nm in this case can be used as the parameter to indicate the color tone of shoyu. It has been generally recognized that the reddish color increases upon heating of shoyu, while in oxidative browning the color of shoyu darkens (Okuhara *et al.*, 1969). Motai (1976) observed three types of browning of shoyu or increase of color density (E_{450}), as shown in Fig. 2:

- Type a. The color tone darkens along with the increase of ΔA, which occurs in nonenzymatic oxidative browning during the storage of shoyu in open air.
- Type b. The color tone is unchanged along with the unchanged ΔA, which occurs in the heat-dependent browning of shoyu bottled or canned with a seal.
- Type c. The color tone becomes bright along with the increase of ΔA, which occurs in the heat-

dependent browning of shoyu during aging of salty mash and pasteurization of raw shoyu.

The ΔA values of koikuchi and usukuchi shoyu on the shelf were reported to be 0.63-0.70 and 0.56-0.60, respectively. The smaller ΔA value of usukuchi shoyu is due to the shorter aging period of mash and to the shorter pasteurization time in the preparation of usukuchi shoyu as compared to those of koikuchi shoyu.

C. Color Darkening during Storage and Consumption of Shoyu

Sawa *et al*. (1978) reported on the color increase of commercial shoyu during storage on the shelf, in warehouses, and during consumption at home. They expressed the color gradation of shoyu by its optical density (OD) spectrophotometrically determined at 530 and 610 nm. The OD values determined at 610 nm were adjusted for 530 nm. The temperatures in the warehouses ranged between 17 and 40°C in summer and between -7 and +15°C in winter depending on the room temperature and facilities. The OD values of usukuchi shoyu packed in 2-liter glass bottles and 1-liter PET bottles increased by 13 to 17% and 26 to 36%, respectively, from June to September when stored in these warehouses. The OD value of usukuchi shoyu packed in 2-liter glass bottles and stored for 1 year in a rather large warehouse increased by 30%, mostly during the 6 months including summer.

The home temperature ranges during consumption of shoyu were 20-34°C in summer, 10-25°C in spring and autumn, and 5-15°C in winter. The increase in OD values of koikuchi shoyu was twice that of usukuchi shoyu when packed in different sized packages and stored at 28, 22, and 15°C; this is summarized in Table IX. No detectable flavor deterioration was observed in the above experiments except for the shoyu packed in 1-liter plastic bottles, in which the color darkening proceeded relatively quickly, and the flavor deterioration began to be recognized when the original OD value more than doubled. The same observation was made for shoyu packed in 2-liter glass bottles when stored for 35 days at 28°C.

Sawa and Nishikawa (1981) compared the color stability of commercial shoyu from various manufacturers while it was consumed at home. The increasing ratio of OD values of usukuchi and koikuchi shoyu packed in 2-liter glass bottles differed very little for the same kind of shoyu, and ranged between 80 and 108% for usukuchi and 90 and 110% for koikuchi, respectively, as shown in Table X.

Almost no differences in color tone were observed for the different brands of shoyu; they consisted of 76 to 79% at OD_{430} nm (yellow), 16 to 18% at OD_{530} nm (red), and 5 to 6% at OD_{610} nm (blue).

In accordance with the daily removal of shoyu from the bottles little by little, the OD_{430} nm of the shoyu remaining in bottles decreased; on the other hand, OD_{530} nm and OD_{610} nm increased, the greater increase being OD_{610} nm.

The $\Delta\alpha$, which is the ratio between OD_{530} nm and OD_{610} nm of commercial shoyu packed in 2-liter glass bottles, ranged between 3.28 and 3.55; this ratio changed to 2.28 to 2.49 after 45 days of daily consumption at home, indicating the decrease of red tone and the increase of blue tone. The $\Delta\alpha$ values of both commercial bottled shoyu after 1 to 2 years' storage and shoyu that was darkened nonoxidatively at 65°C were nearly unchanged.

The OD increases during consumption of shoyu produced by methods with different total nitrogen contents, and packed in 18-liter cans, are indicated in Table XI. The OD increase of shoyu packed in 36-liter polyethylene bottles and that for shoyu kept in larger tanks during transportation and consumption were also reported by the same authors.

D. Browning Mechanisms of Shoyu

1. Color Formation during Brewing and Pasteurization

About 50% of the color of koikuchi shoyu is produced during fermentation and aging of mash, and the remaining 50% during pasteurization; most of the pigmentation is considered to result from the heat-dependent browning of the Maillard reaction between amino compounds and sugars.

One example of color formation in the course of shoyu preparation is shown in Table XII (Motai, 1976).

Moriguchi and Ohara (1961) recognized that when soybeans mixed with 0.8% $K_2S_2O_3$ were cooked with steam and then subjected to enzymatic digestion according to the usual method of usukuchi shoyu preparation, the color of shoyu obtained was lighter by 37% as compared to the control.

According to Burton *et al.* (1963), almost all carbonyl compounds react with amino radicals, and α-ketoaldehydes (pyruvic aldehyde, 3-deoxyosones), diketons (diacetyl), and α-β-unsaturated aldehydes (crotonaldehyde, furfurals) are above all reactive among them. α-Hydroxyaldehydes change into α-β-unsaturated aldehydes after dehydration and react with amino compounds. Reducing sugars react with amino radicals and produce α-ketoaldehydes (3-deoxyosones) and α-β-unsaturated carbonyls (unsaturated osones), and take part in the browning reactions. Xylose is 10 times more reactive with amino radicals than glucose. The content of hexoses in shoyu is 6-10 times that of pentoses, but it has been pointed out that the sugars that mainly take part in the browning of shoyu are pentoses such as xylose and arabinose (Kato and Sakurai, 1962; Kamata and Sakurai, 1964; Shikata *et al.*, 1971a). The decreases of sugar content of salty shoyu mash over 180 days' fermentation are indicated in Table XIII; note that the steepest decrease is observed with xylose (Okuhara *et al.*, 1975).

The color intensity of unpasteurized (or raw) shoyu and its reducing sugar content are related with a correlation coefficient of .795, significant at the 1% level, and the color intensity of pasteurized shoyu and its reducing sugar content are related with a correlation coefficient of .887, significant at the 1% level (Okuhara *et al.*, 1969). When the glucose, xylose, arabinose, or galactose was added to raw shoyu and heated for 5 hr at 80°C, remarkable browning was seen with xylose but almost none with glucose, although glucose is the major sugar contained in shoyu. Okuhara *et al.* (1969) mixed three kinds of shoyu (A, B, and C) with 0.025, 0.5, 0.75, 1.0, 1.25, 1.5, 1.75, and 2.0% xylose, respectively, and heated the mixtures at 80°C for 5 hr. Their results are shown in Fig. 3. The color increase and the pentose consumption showed a linear relationship, but the color increases per 1 mg consumption of xylose of the three kinds of shoyu differed from each other. Then a mixture of xylose, glucose, glycine, lactic acid, and 18% salt water in the average concentrations of shoyu was heated after adjusting its pH to 4.8, but the color increase was much smaller than that observed upon heating shoyu or HVP. The direct participation of xylose in the heat-dependent browning of shoyu was calculated to be only 10-20%.

The neutral fraction containing sugars was separated from shoyu and was added to the original shoyu and heated for 5 hr at 80°C. The amount of browning in this case corresponded to only 12% of that of the control shoyu heated in the same way. When the same neutral fraction was added to a peptide solution prepared enzymatically by hydrolyzing soybean proteins, and heated in the same way as above, the amount of browning observed was only minimal (Okuhara *et al.*, 1975). These experiments indicate only a small amount of direct participation of the neutral fraction of shoyu in the heat-dependent browning reaction.

According to Okuhara *et al.* (1969), shoyu produced in such a way that sugars in the salty mash were not so much consumed by yeast fermentation or by other reactions has a higher velocity of browning by heat. According to Okuhara *et al.* (1975), not only glucose but also xylose is consumed by the yeast fermentation of shoyu mash. This in turn not only hinders the formation from these sugars of intermediate compounds of heat-dependent browning of shoyu, but also results in decreased amounts of these intermediate compounds once produced from these sugars, perhaps as a result of reduced yeast fermentation. These findings explain shoyu obtained from mash that is well fermented with yeast is of lighter color and at the same time has a slower browning velocity by heat. Ordinary koikuchi shoyu mash prepared on a laboratory scale was divided into two portions. Yeast starter was added to one portion and allowed to ferment for about 4 months; the other portion was stored for the same period of time with addition of some toluene in order to stop the yeast fermentation. The results are indicated in Fig. 4. About 2 mg/ml more xylose was consumed in the yeast-fermented mash than in the nonfermented mash. When 8 mg/ml of xylose was added to the fermented mash, the heat-induced browning velocity of both fermented and nonfermented mashes became the same; this suggests that the yeast fermentation not only reduced the xylose content of mash but also suppressed the browning reaction of intermediate compounds.

Among the amino compounds that react with sugars in the so-called Maillard reaction, the peptides are more reactive than the free amino acids as indicated in Table XIV (Okuhara *et al.*, 1975). From these results, it was presumed that the precursors of the heat-dependent browning reactions of shoyu are compounds produced from sugars and amino compounds by Maillard reactions during storage of salty mash, and that formation of these compounds is hindered by the yeast fermentations.

Okuhara and Saito (1970) recognized a slight effect on the depression of heat-dependent browning of shoyu decolorized by heating with such reducing agents as ascorbic acid or cystine, or such metals as Zn, Al, Fe, Na, and Mg, or by electrolysis.

It is known that the color of shoyu is also dependent on the kinds of lactobacilli or *Pediococcus halophylus* in the shoyu salty mash (Fujimoto *et al.*, 1980). Kanbe and Uchida (1984) divided the lactobacilli in shoyu salty mash into two groups according to their reducing potency. The rH value of the salty mash naturally inoculated with lactobacilli as the control decreased to 7.5 around the time of their maximum growth, that is about 50 days from the mash making, while the salty mash inoculated with 1×10^4/g of *P. halophylus* No. 34, which was isolated as having the strong reducing potency, showed an rH value of 6.0 at the peak of their growth. The color of salty mash inoculated with No. 34 after 180 days of storage was more than 35% lighter than that of the control mash. Raw shoyu separated from these two kinds of salty mash was pasteurized under the same conditions, and the heat-dependent and oxidative browning rates were determined on the same basis of the contents of NaCl 17.2% and TN 1.57% of shoyu. The rates of heat-dependent browning and oxidative browning of the shoyu prepared from the test mash were 24 and 18% less than those of the shoyu from the control mash, respectively. It was also observed with the test shoyu that the reducing power for potassium ferricyanide and the contents of hydroxymethyl furfural, reductones, and 3-deoxyosones, all of which belong to the so-called intermediate browning compounds, were smaller than those of the control shoyu.

Okuhara *et al.* (1970) statistically analyzed the relation between the components of raw shoyu and its browning rate. The correlation and multiregression models were calculated as indicated in Table XV. To obtain the various concentrations of shoyu components, in proportion I the weight of koji per unit amount of raw materials was varied, and in the proportion II the concentration of salt in mash was varied. The components that correlated significantly with the browning rate were the concentration of koji in mash (CK), the weight of koji from a unit amount of raw materials (WK), the amount of shoyu obtained from a unit amount of raw materials (L), nonamino nitrogenous compounds (TN - FN), titratable acidity (TA), reducing sugar of shoyu (RS), and reducing power of raw shoyu [R(N)]--above all TN - FN and RS. These correlations alone seem to indicate that the browning pigment is produced from TN - FN and RS. Okuhara *et al.* (1971) also suggested that TN - FN might significantly correlate with the browning of shoyu on the basis of the following experiments:

1. The browning of shoyu is much faster than that of the solution of sugar and amino acids.
2. The browning of shoyu is dependent on the degree of yeast fermentation and the amount of TN - FN of mash.
3. The browning of a solution of xylose and glycine is accelerated by adding a small amount of shoyu.
4. The browning of a solution of xylose and glycine is accelerated by adding the product of soybean protein digested by the crude extract of shoyu koji as the enzymatic source.

Shikata *et al.* (1971b) digested the raw materials of shoyu with an enzyme mixture including cellulase, diastase, and protease, and observed that the color degree of the digested solution correlated closely with its formol nitrogen (FN) × pentose percentage, as shown in Fig. 5.

Motai *et al.* (1975) brewed shoyu by varying the ratio of the concentration of soybeans to wheat kernels as raw materials; the extent of browning of shoyu by heat increased with increasing soybean concentrations. The color tone of shoyu caused by heating lightened in accordance with increase in the soybean concentration, whereas the color tone of shoyu produced from wheat kernels alone exhibited the darkest tone of all. These workers isolated the amino fractions and the sugar fractions from the product of enzymatic degradation of defatted soybean meal and that of wheat kernels, respectively, and determined the contributions of these fractions to the heat-dependent browning of shoyu. It was suggested that the contributions of each amino fraction of soybeans and wheat kernels were 75 and 25%, respectively, and those of each sugar fraction were 44 and 56%, respectively. From these data the contributions of soybeans and of wheat kernels to the heat-dependent color formation of shoyu were calculated to be 60 and 40%, respectively.

Omata and Nakagawa (1955a,b) recognized that some ether-soluble carbonyl compounds of shoyu, including furfural and acetaldehyde, darken its color. Kato (1958, 1959) concluded that aromatic amine-*N*-xylosides decompose to form red pigments of melanoidin by weak acid catalysis, and furfural is not an intermediate in melanoidin formation. Kato (1960) isolated 3-deoxyxylosone or 3-deoxy-D-glucosone from the above reaction mixture as bis(2,4-dinitro)phenylhydrazones, and pointed out their significance in the browning reaction as the intermediate in melanoidin formation. Kato *et al.* (1961) identified 3-deoxy-D-glucosone also in fermented shoyu, amounting to 8 mg% in koikuchi shoyu, 3 mg% in usukuchi shoyu, and 17 mg% in tamari shoyu, respectively; however, it was not found in HVP. The content of 3-deoxy-D-glucosone in fermented shoyu is much higher than that of furfural, which was reported to be 0.2-0.7 mg% by Omata and Nakagawa (1955b). Moreover, it was pointed out that 3-deoxy-D-glucosone is more reactive with amino acids than furfural. The amount of 3-deoxy-D-glucosone formed during the preparation of shoyu is shown in Table XVI (Kato *et al.*, 1961).

3-Deoxypentosone was also isolated from pasteurized shoyu but in much smaller amounts than 3-deoxy-D-glucosone; this was attributed to its instability. When pasteurized shoyu was mixed with xylose and kept at 37°C, a rapid increase of 3-deoxypentosone was observed. When pasteurized shoyu was kept at 40°C or heated at 80°C, the hexose and pentose content of shoyu decreased along with the increase of color degree determined by absorbance at 470 nm, and 3-deoxy-D-glucosone gradually increased and then decreased after reaching a peak. The same result was obtained when xylose was added to shoyu. This reaction was accelerated by oxygen or air, but the reaction proceeded also in the absence of air (Kato and Sukarai, 1962).

From these results, the reaction mechanism of color increase of pasteurized shoyu was presumed to proceed as follows: aldose → 3-deoxyosone → color pigments.

When the chemically synthesized 3-deoxyosones, including 3-deoxypentosone, 3-deoxy-D-glucosone, and 3-deoxygalactosone, were added to shoyu, large amounts of browning color substances were produced that were water soluble, but those produced when furfural was added to shoyu were barely soluble in water and in small amounts. It was considered that in the aminocarbonyl reactions the majority of 3-deoxyosones are utilized by the excess amount of amino acids instead of converting into hydroxymethylfurfural (HMF), resulting in the formation of water-soluble pigments (Kato and Sakurai, 1963). These are shown in Fig. 6.

Okuhara *et al.* (1969) reported the increase of absorbance determined at 500 nm when raw shoyu of the koikuchi type was heated at 40, 50, 60, 70, 80, and 90°C, as indicated in Fig. 7. The heat-dependent browning velocity of raw shoyu increased twofold in accordance with a temperature elevation of 10°C above room temperature, but threefold at temperatures >60°C.

Motai (1976) observed a linear increase of color intensity in accordance with the elevation of temperature during pasteurization of shoyu, described by the equation:

$$D = \alpha \times 10^{-0.040t}$$

where D = time to attain a definite color degree, α = a constant, and t = temperature of pasteurization.

The α value varies with the concentrations of pentose and total nitrogen of shoyu, and generally the higher the (pentose × total nitrogen), the lower the α value. A 2.5- or 3-fold increase in browning speed was observed for the ordinary koikuchi shoyu in accordance with a 10°C temperature elevation within the range of 50 to 90°C (Motai, 1976; Onishi, 1970).

Generally the higher the pH the greater the browning reaction, but within the average range of pH values of shoyu (4.6-4.9), there are no practical differences in the amounts of heat-dependent browning.

2. Color Formation on the Shelf and after Opening the Seal

Omata and Ueno (1953b) pointed out that the color change of pasteurized shoyu during storage is not due to the enzymatic reactions but is greatly affected by the action of air and less so by temperature and light. When the color of shoyu darkens with aeration, its transmittance curve does not change remarkably.

When pasteurized shoyu is stored in a sealed glass bottle or can, color intensity increases as the result of browning by heat without changing the ΔA value. The increase in color intensity of shoyu stored in the open air is much greater than in sealed containers, as shown in Figs. 8 (Okuhara *et al.*, 1969) 9 (Hashiba, 1977). This is caused by nonenzymatic oxidative browning, in which the ΔA value decreases in accordance with the increase of the ratio between the surface area and the volume of the shoyu. These effects are shown in Fig. 10 (Motai and Inoue, 1974).

The influence of iron on the browning reaction of shoyu is well known. Furuta and Ohara (1954) added 30-60 ppm Fe^{3+} to shoyu mash and observed a 12-20% increase in browning just after the addition as compared to the control, and a 21-30% increase after storage for 20 days at 30°C. The effect of adding Fe^{3+} to heated shoyu was almost indifferent to the temperature of heating at 80 and 100°C, and was not so remarkable as that in mash.

According to Hashimoto *et al.* (1970), the average content of iron in shoyu is 20-30 ppm, and that is reasonably calculated to be derived from the raw materials; soybeans, wheat, salt, and water. Most of the iron in raw and heated shoyu was recognized to be in the form of Fe^{2+}. The addition of tannic acid or potassium ferricyanide was effective in removing 60-70% of the iron in shoyu without affecting its organoleptic quality. When these chemicals were added to shoyu during heating, ∿93% of the iron was removed. The influence of Fe^{2+} on the color increase of shoyu during storage was smaller than that of Fe^{3+} and Cu^{2+}. However, the rate of color increase during the storage of shoyu containing 7-10 ppm Fe that had been reduced by the above procedure was about equal to that of ordinary shoyu, containing 20-23 ppm Fe. The oxidative increase in color of shoyu containing 2 ppm Fe was slightly lower than that of untreated shoyu containing 32 ppm Fe.

Hashiba *et al.* (1970) reported that Fe^{2+} and Mn^{2+} contribute to the increase in color intensity of shoyu with oxidation, while the other trace metal ions (Cu^{2+}, Zn^{2+}, Co^{2+}, and Cd^{2+} did not. Iron ion (Fe^{2+}, Fe^{3+}) had no effect on the color increase of shoyu upon heating.

The reduction in content of such heavy metals as Fe and Mn in shoyu to <1 ppm did not affect the heat-dependent browning of shoyu, but decreased the oxidative browning to one-third that of the untreated shoyu, as indicated in Table XVII (Hashiba and Abe, 1984).

Shoyu samples were divided into cation (C) and acidic-neutral (A + N) fractions by Amberlite 120 (H^+ type), and the browning of each fraction was compared. In oxidative browning, the participation of C fraction and iron was considerable but not that of the A + N fraction. On the other hand, heat-dependent browning involved no participation of iron and high participation of the A + N fraction. It was suggested that some unknown compounds present in the C fraction react strongly with iron in oxidative browning (Hashiba and Abe, 1984).

Okuhara *et al.* (1972) heated raw shoyu samples at 60, 70, and 80°C, respectively, to obtain different color densities. Then these pasteurized samples were shaken in air for 24 hr at 40°C, and the increment in OD at 600 nm was determined The amount of oxidative browning of shoyu thus determined with both raw and pasteurized shoyu was correlated only with the initial color intensity of shoyu and was unaffected by the heating temperature, as shown in Fig. 11. It was also observed that pasteurized shoyu having the same color intensity, having been prepared by heating raw shoyu at various temperatures and times for various lengths of time, browned to the same color densities by oxidation--despite the varying contents of reducing sugars, peptides, amino acids, and other compounds--if the shoyu is made from the same raw materials and the concentrations of both total nitrogen and sodium chloride are adjusted to be the same.

The effective participation of reductones in oxidative browning reactions was pointed out by Hodge (1953). The nonenzymatic oxidative browning of shoyu during storage has been attributed to the participation of such intermediates of the Maillard reaction as reductones, Amadori rearrangement compounds, and melanoidins. The ascorbic acid added to shoyu, which is among the reductones, changes into dehydroascorbic acid by oxidation, which reacts with amino acids to deepen the color of shoyu (Omata *et al.*, 1955a).

Okuhara *et al.* (1972) found that the oxidative browning and the formation of reductones in shoyu during heating were nearly proportional to the initial color intensity. The oxidative browning took place simultaneously with the consumption of reductones during oxidation, and the reductone consumption correlated with the volume of air used for the oxidation of shoyu with a logarithmic curve. The oxidative browning pigment appeared to be formed by the aminocarbonyl reaction of the oxide of reductones, and not by the reaction of the dehydro compound of reductones. On the other hand, the mechanism of oxidative browning of shoyu was analyzed by multiple-correlation analysis, and the result indicated that the oxidative browning was mainly correlated with both the oxidation of reductones formed by the heat treatment of shoyu and the Baumé or the amount of soluble solid in the shoyu.

Hashiba (1973b) separated shoyu into three fractions--a cation fraction, a neutral fraction, and an anion fraction--using ion Amberlite 120 (H^+ type). When these three fractions were stored separately, only the cation fraction darkened considerably. When they were combined and stored, the color of the mixture increased at nearly the same rate as that of the original shoyu. The effects of the anion fraction containing organic acids and the ashed cation fraction on the overall browning of shoyu were calculated to be 10-20% and 20%, respectively. The sum of the contribution rates of the anion fraction, the neutral fraction, the amino acids, and the ashed cation fraction in the browning of shoyu was calculated to be ∿40%. Compounds responsible for the remaining 60% were considered to exist in the cation fraction. It was suggested that such compounds had strong reducing power and oxygen uptake ability.

Hashiba (1974) prepared a simulated shoyu, which was an amino acid solution containing glucose (5%), xylose (1%), NaCl (17%), and lactic acid (2%), and they adjusted the final pH to 5.0. This sugar-amino acid model system was stored for aging for 3 months at 30°C under anaerobic or aerobic conditions, and subsequently for 2 weeks more at 37°C under aerobic conditions to examine the oxidative browning. The oxidative browning of the model systems increased with increase in aging period, and the model systems aged under anaerobic conditions darkened less than those aged under aerobic conditions. Adding 40 ppm Fe^{2+} to the model system, which is the average content of Fe^{2+} in shoyu, accelerated the above oxidative browning. An Amadori product, 1-deoxy-1-glycine-D-fructose, was isolated from the aged glucose-glycine model system, and it caused a marked increase in the rate of the oxidative browning. Hashiba (1975) also isolated an Amadori compound (1-deoxy-1-diglycine-D-fructose) from a glucose-diglycine model system, and this Amadori compound promoted the oxidative browning of the aqueous solution of glucose and diglycine, which furthermore was hastened by 30 to 40 times in the presence of Fe^{2+}. In the browning reaction between glucose and triglycine, similar intermediates were also detected.

The Amadori compounds composed of aromatic or heterocyclic amino acids, such as fructose-triglycine, fructose-phenylalanine, fructose-histidine, and fructose-tryptophan, were especially reactive in oxidative browning, and this type of browning was synergistically accelerated by the presence of Fe^{2+} and Mn^{2+}. Oxygen was thought to accelerate the breakdown of Amadori compounds to liberate amino acids and glucose (Hashiba, 1976). The Amadori compounds derived from pentose (e.g., xylose-glycine) browned more rapidly than those from hexose (e.g., glucose-glycine). In the reaction between glucose and seven peptides, the liberation of C-terminal amino acids by the cleavage of peptide bonds was observed. It was suggested that the amino acids were liberated from the peptide in Amadori compounds, because the peptide bond in Amadori compounds was found to be more labile than that of the free peptide (Hashiba *et al.*, 1977). Amadori compounds were isolated from shoyu by ion exchange chromatography, gel filtration, and paper chromatography (Hashiba, 1978). Five compounds were identified, and their contents in shoyu were estimated to be as follows: fructose-glycine (∼0.2 m*M*), fructose-alanine (∼0.3 m*M*), fructose-valine (∼1.2 m*M*), fructose-isoleucine (∼1.3 m*M*), and fructose-leucine (∼1.5 m*M*). The Amadori compounds exhibited remarkable browning in the presence of oxygen and iron, but darkened very little without these two components. Amino acids promoted the oxidative browning of the Amadori compounds. These are shown in Table XVIII (Hashiba and Abe, 1984). Amadori compounds from pentose or peptides were considered to be so unstable that they would have been decomposed while passing through the resin, and thus they were not isolated from shoyu.

Such Amadori compounds as fructose-glysine (F-Gly) and fructose-alanine (F-Ala) reacted with iron and produced red pigments, from which the colorless compounds as indicated in Fig. 12 were separated (Hashiba and Abe, 1984).

According to Motai and Inoue (1974b), the color compounds of shoyu consist of different degrees of polymerized melanoidins, and the oxidative color increase of shoyu through the polymerization occurs according to the equation

$$E = K \times M^{\alpha}$$

where E = color intensity determined by $E(1\%, 1\text{ cm})_{450}$, M = molecular weight, and K and α are constants.

The value is almost completely independent of the time of heating and the kinds of sugar in the melanoidin reaction, but it is dependent on the kinds of amino compounds and above all their molecular weights. The values of K and α were calculated with shoyu as compared with miso and various melanoidins, as shown in Table XIX. The table suggests that the melanoidins of shoyu originated from di- or tripeptides.

Hashiba (1973c), treated raw shoyu by ultrafiltration and found that its color intensity was reduced to one-tenth to two-fifths of the initial color intensity, and the color increase of the treated shoyu on heating was one-third to one-half that of untreated shoyu. No sedimentation was found in the course of heating raw shoyu treated with ultrafiltration. When pasteurized shoyu was subjected to ultrafiltration, the color increase was depressed to about one-half that of untreated shoyu. The substances related to the browning of shoyu (e.g., Fe^{2+}, 3-deoxyglucosone, hydroxymethylfurfural, reductone, carbonyl compounds, and ferricyanide-reducing compounds) were removed by ultrafiltration, and 5-7% of the total nitrogen and 20% of the reducing sugar in shoyu were also lost during the same procedure.

Okuhara *et al.* (1975) fractionated raw shoyu by stepwise ultrafiltration to determine the molecular size of the major intermediate compounds both by heat and oxidation. A 350-ml sample of raw shoyu was filtered through an ultrafiltration paper, cutting the molecular weight to 10,000 and producing 50 ml of filtrate. The residual shoyu was treated likewise with ultrafiltration paper, cutting the molecular weight of 10,000 to 5,000, 5000 to 1000, 1000 to 500, and <500, successively, obtaining 50 ml of filtrate in each case, with the exception of the filtration through MW 500, in which 100 ml of filtrate were obtained. The total nitrogen content of each shoyu filtrate was adjusted to 1.547% (w/v), and OD values were determined. These shoyu samples were then heated at 80°C for 5 hr, and the increase of OD values at 500 nm, or OD-I was determined and compared to that of the control. On the other hand, each filtrate was shaken with air for 24 hr at 30°C to produce oxidative browning, and the OD-II was determined and compared to that of the control. These results are shown in Table XX. As indicated in the table, the intermediate compounds of the browning reaction in shoyu seem to have a wide range of molecular size, but the majority seem to fall between 1000 and 5000 MW, and the participation of the compounds having <500 MW in the overall browning of shoyu is about 50%.

According to Hashiba (1981a,b), the participation of peptides in browning during the aging of shoyu mash was remarkable, but amino acids participated more than peptides in the oxidative browning process. The contributions of pentose and hexose to oxidative browning of shoyu were estimated to be 75 and 25%, respectively.

Temperature elevation up to 65°C promotes the oxidative browning of shoyu, above which browning stops (Motai, 1976). Lactic and citric acids both promote oxidative browning of shoyu (Hashiba, 1973), but phosphoric acid has no effect on this process, although it promotes the heat-dependent browning of the Maillard reaction (Kato, 1956).

IV. TURBIDITY FORMATION DURING STORAGE AND CONSUMPTION OF SHOYU

A. General Aspects of Turbidity and Sediment Formation during Storage of Commercial Shoyu

Commercial shoyu only rarely becomes turbid or forms sedimentation on the bottom or floating materials on the surface during storage or when used for cooking.

The liquid components, press-filtered from a well-aged salty mash or from raw shoyu, is stored for sedimentation over a period of several days, after which the precipitate on the bottom and the floating material on the surface are separated out by decantation or by filtration accomplished with aids such as kieselguhr or silica-containing earth. Such clarified raw shoyu is seldom put on the commercial market as it is, because it usually contains easily heat-coagulable materials composed mainly of proteins and is thus very unsuitable for commercial purposes. Although it is possible to remove the heat-coagulable material during salty mash fermentation or by molecular-size filtration of raw shoyu, raw shoyu prepared by such procedures is not yet found on the market.

Raw shoyu is usually pasteurized as the final step in manufacture for several reasons previously mentioned; among them is the absolute removal of heat-coagulable materials, an important process but very difficult because of the complex mechanisms involved. The turbidity and sediment formation that occur during pasteurization of shoyu are not dependent simply on the heating temperature but on the amounts and kinds of heat-coagulable protein and other material dissolved in raw shoyu, the temperature and duration heating, the content of total nitrogen and salt and the pH of shoyu, as well as the content of HVP mixed with fermented raw shoyu. This sediment formation occurs gradually, taking several days.

Therefore, if raw shoyu is pasteurized at 70 or 80°C as usual, and the transparent heated shoyu is taken out immediately or separated from coagula by decantation or filtration after 1 or 2 days' storage in a tank, and then bottled or canned to make the final product, the clear shoyu product sometimes becomes turbid or forms sediment on the bottom of the container during several days or several months of storage, and the commercial value of the products is seriously damaged.

The possibility of secondary sediment formation in commercial bottled shoyu was tested for 60 brands of koikuchi shoyu and 10 brands of usukuchi shoyu that were purchased on the market (Hashimoto *et al.*, 1970c; H. Hashimoto, unpublished, 1975; Yokotsuka *et al.*, 1971a). The shoyu samples to be tested were kept at 30, 37, 40, and 50°C for 1 to 3 days in a test tube to observe at regular intervals the turbidity or flocculation formation. The potential to produce turbidity or sediment during storage over long periods of time at room temperature was observed for 40 koikuchi and 4 usukuchi brands. Hashimoto *et al.* (1970b) reported that higher total nitrogen content and lower pH value tend to increase the turbidity of shoyu during storage. The greater the amount of HVP mixed with fermented shoyu, the longer the product could be stored without turbidity formations. At storage temperatures $<30^{\circ}C$, turbidity formation apparently decreased, and it absolutely stopped at $<5^{\circ}C$. When raw shoyu was heated at $\geq 50^{\circ}C$, secondary sedimentation occurred in the final product during storage, but it must be noted that raw shoyu heated at 70°C produced secondary sediment upon bottled storage to a degree much less than that heated at 60 or 80°C. A high concentration of NaCl inhibited the separation of the precipitate. The sediment formed most readily upon heating raw shoyu with a pH of 4.6 to 4.7, but the time required for sediment formation in heat-processed shoyu was shorter when the pH was lower than when it was higher. The turbidity mentioned above is caused mainly by heat-coagulable native proteins derived from the raw materials (mainly soybeans) and/or enzymes of koji molds, but other rare causes are known, such as crystallization of barely soluble amino acids (e.g., tyrosine and phenylalanine) or barely soluble chemical preservations added extraneously to shoyu (e.g., butyl-*p*-hydroxybenzoate or calcium salts of oxalic or citric acids, or calcium or magnesium salts of organic phosphoric acids), or polymerized Maillard pigments of shoyu produced mainly by oxidation.

B. Coagulation and Sediment Formation during Pasteurization Process of Normally Brewed Raw Shoyu

1. Chemical Properties of Heat-Coagulable Material in Shoyu

According to Hashimoto *et al.* (1971), the heat-coagulated material produced by heating raw shoyu is equivalent to 10% in volume and 0.025-0.05% in weight of shoyu, and 0.2-0.4% of the total nitrogen content. The heat-coagulated material of raw shoyu was classified according to the pasteurization process used. The first sediment of shoyu, which was formed by heating at 85°C for 30 min, was composed of 81.6% protein, 15.2% carbohydrate, and 2.3% ash, and the yield was 2 mg/liter. The secondary sediment, formed after separating the first sediment by further treatment of shoyu at 55°C for 4 days, was composed of 89.2% protein, 9.6% carbohydrate, and 1.2% ash that was mostly phosphate, and the yield was 252 mg/liter. The secondary sediment was solubilized by hydrolysis with pronase-P (the purified protease from *Streptomyces griseus*) at pH 6.5, liberating peptides and amino acids; however, the sediment was not hydrolyzed by the same enzyme in the presence of 2 mol NaCl at pH 5.0. The amino acid composition of the heat-coagulated material of shoyu is completely different from that of soybeans or wheat kernels with regard to the aspartic acid:glutamic acid ratio or the proline:leucine ratio. The major ingredient of heat-coagulated material in shoyu was recognized by immunological identification to be undenatured proteins derived from the enzymes produced by koji molds and not from the raw materials. The precursor of heat-coagulable material was purified according to the procedures described in Fig. 13. The elution curve of the most purified protein fraction separated from protease (D in Fig. 13, bottom line) coincided with both the absorption curve at 280 nm and that of amylase activity. The most purified fraction became turbid upon heating at 85°C for 30 min, but the turbidity was less than that of fraction C (Fig. 13), which contained protease, under the same heating conditions, suggesting that the protease might be promoting the turbidity formation.

2. Mechanisms of Heat Coagulation and Sediment Formation

According to Hashimoto and Yokosuka (1972, 1974), the time required for sediment formation upon heating raw shoyu is not proportional to the temperature of heating, and sediment forms more slowly in shoyu heated at 70°C than that at 60 or 80°C. These facts may help explain why the heat-induced sediment formation in shoyu is not merely caused by heat coagulation of undenatured protein (Fig. 14). The addition of a small amount of raw shoyu to the shoyu heated at 85°C for 20 min promoted the formation of heat coagula, indicating that raw shoyu contains some heat coagulation-promoting factors. The addition of three kinds of protease from koji molds (i.e., alkaline protease and neutral proteases I and II) purified from raw shoyu also greatly reduced the time of coagulation. Of the three, alkaline protease was the strongest catalyst for coagula formation in shoyu heated at

40°C. The heat coagulation-promoting activities of these proteases in heated shoyu disappeared on addition of potato inhibitor or EDTA. The sediment-forming activity disappeared when the pH of the shoyu was adjusted to <3.0 or >9.8. The activity was the most stable at around pH 5.8, where about 55% of the original proteolytic activity remained after incubation at 40°C for 20 hr. Proteolytic activity began decreasing with incubation at 45°C for 10 min; it continued decreasing as the temperature increased and was completely lost after heating at 75°C for 10 min. However, activity was again observed after heating above 80°C for 10 min (see Fig. 15).

According to Sekine (1972), the heat-resistant neutral protease II isolated from koji molds, *Aspergillus sojae*, specifically loses its activity at around 75°C, presumably because of autolysis. Therefore, it may be considered that the minimal coagula formation in shoyu heated at 75°C is caused by the inactivation of non-heat-resistant alkaline protease and neutral protease I, and heat-resistant neutral protease II. In the coagula formation of shoyu heated at more than 75°C the thermostable protease seems to play an important role, its activity withstanding heating to 120°C for 10 min.

The influence of temperature on the amount of coagulation during heating shoyu is shown in Fig. 16 (Hashimoto *et al.*, 1970b). The optimal temperature for maximum coagulation of shoyu is $\sim$60°C.

For industrial pasteurization, raw shoyu is heated at $\sim$80°C and then rapidly or gradually cooled to 60°C, continuing with gradual cooling at room temperature. Higher temperature and longer heating time of raw shoyu is preferable to sediment out the maximum amount of coagulable material. This leads to a final product with thoroughly stabilized clarity. At the same time, however, too much heating, especially above 70°C, results in undesirably dark color and "brown" flavor.

Several technological innovations have allowed the following changes in shoyu manufacture: (1) greater protease activity retention in raw shoyu than previously as the result of an increased protease activity of koji, due to both improvements in the technology of koji making and the reduction in the fermentation period of salty mash, (2) a reduction in the salt content of salty mash, (3) a reduction in the amount of chemical reactions during pasteurization of shoyu so as to prepare a lighter colored and less heat-flavored product preferred by the consumer, and (4) a reduction in chemical changes during heat treatment of raw materials by HTST method. All of these relate to the increase of heat-coagulable substances in raw shoyu and insufficient precipitation of coagulated precursors of sedimentation of heated shoyu.

Nakadai (1984) analyzed the relationship between the activities of various koji enzymes and the amount of coagula formed during the heating phase of shoyu production by stepwise multiple-regression analysis on data obtained from brewing trials with 83 different strains of koji molds. Contributing proportions to the amount of coagula were as follows: alkaline plus neutral protease activity (31.9%), α-amylase activity (15.3%), pectin-lyase activity (14.4%), acidic protease activity (8.4%), chitinase activity (8.2%), xylanase activity (2.9%), and α-galactosidase (2.3%).

3. Sediment Acceleration during Pasteurization Process

When raw shoyu is heated at 70 to 80°C and then cooled either rapidly or slowly, the suspension of coagulated substances formed sometimes does not sediment over several days of storage, which hinders the process of obtaining a good yield of clarified pasteurized shoyu by decantation or by filtration.

It seems reasonable to consider the pasteurization process in shoyu manufacture as comprising two stages: (1) to denature and coagulate the raw proteins of koji molds, resulting in the formation of an invisible and/or visible suspension of flocculant precipitate in heated shoyu, and (2) to associate the denatured protein molecule to each other through hydrophobic bond by the action of both heat and proteases, forming particles large enough to sediment on the bottom of the container (Hashimoto and Yokotsuka, 1974).

Motai *et al.* (1981) investigated the contribution of various enzymes of koji molds to the sedimentation of coagula during the shoyu pasteurization process by measuring the amount, density, and particle sizes of the coagula. The higher the density of the sediment and the molecular weight of proteins, the faster the sediment precipitated. The sediment tended to precipitate upon increasing the number of particles ranging in size from 2.0 to 5.04 nm. During the pasteurization process, particles of 5.04 to 12.7 nm were at first rapidly degraded to 0.79 to 1.28 nm in size and then gradually restored to 2.0 to 5.04 nm by coagulation. The density also increased with longer holding time, while the higher molecular weight protein in the sediment decreased. Thus it was shown that precipitability of the sediment improved gradually with increase in its density and particle sizes during the pasteurization process.

According to Motai *et al.* (1983), when the amount of α-amylase increased, the amount of coagula in shoyu also increased, but there was no effect on density and particle size. Upon heating acidic protease-free shoyu, obtained from raw shoyu by passing through a pepstatin-Sepharose column, the amount of coagula and the rate of coagulation were reduced in comparison with shoyu containing acidic protease. There was little change in density distribution in the absence of acidic protease. Alkaline protease increased the amount and density of the coagula and enhanced the early stage of coagulation, but retarded the final stage of coagulation. These findings indicate that the retardation of coagulation at the final stage was not due to proteolytic action, and that α-amylase had no effect, acidic protease had some promoting effect, and alkaline protease had a remarkable retarding effect on the sedimentation of coagula in shoyu.

4. Prevention of Sediment Formation in Marketing Shoyu

Commercial shoyu ideally should be guaranteed to retain its clarity for at least 6 months of storage on the shelf at average temperatures. Clarity stability can be checked by noting the absence of flocculation formation when the shoyu is kept at 55°C for 4 days, or when shoyu mixed with 0.05% pronase is kept at 55°C for 4 hr (H. Hashimoto, unpublished, 1975).

Several methods have been considered to prepare the marketing shoyu with stabilized clarity.

1. Removing the precursors of heat-coagulable material from raw shoyu by ultrafiltration: When compounds of >10,000 MW are removed from raw shoyu through ultrafiltration, no coagula formation is observed during the pasteurization process; as already mentioned, this procedure also helps to improve the color stability but, on the other hand, is always accompanied by some loss of total nitrogen.

2. Removing the precursors of heat-coagulable material in salty shoyu mash by heating: Generally speaking, the protease activity of salty shoyu mash decreases as the fermentation period lengthens, as the salt and alcohol contents increase, and as the pH values decrease. Noda *et al.* (1982) heated a salty shoyu mash at 40°C for 120 hr (I), at 45°C for 72 hr (II), or at 50°C for 24 hr (III), and then measured by filtration through paper the amounts of sediment formed in the pasteurized shoyu after heating at 65°C for 120 min and then at 85°C for 60 min. Sediment decreased by 60, 84, and 99%, respectively, as compared to that of control shoyu, which was obtained from non-heat-treated salty mash by the usual

method. This procedure increased the color intensity of shoyu by 15 to 20%. When each filtrate of pasteurized shoyu was stored at 45°C for 15 days, secondary sediment formed after 14 days for (I), after 9 days for (II), but none for (III).

Arai (1984) heated salty shoyu mash at 85°C for 30-50 min, or 88°C for 10 to 30 min, or 90°C for 5 to 20 min, or 94°C for 0 to 5 min, following each treatment with rapid cooling to <50°C within 30 min. The clarified shoyu, press-filtered from these heat-treated salty mashes, was pasteurized at 105°C for 30 sec and then filtered with the aid of kieselguhr. This refined shoyu did not form secondary sediment after storage at 45°C for 10 days.

3. Utilizing the protease activities contained in raw shoyu to accelerate sedimentation during the pasteurization process: Adding 0.15% raw shoyu (w/v) to shoyu that had been heated at 85°C for 30 min and then cooled to 40°C, remarkably reduced the time needed to complete sedimentation to 10 hr as compared to 45 hr where no raw shoyu was added.

During the usual pasteurization process of heating shoyu at ∿80°C, most of the proteases present in raw shoyu, other than the heat-resistant neutral protease II, are inactivated and not utilized to accelerate sedimentation. A pasteurization method has been developed that effectively accelerates sediment formation: raw shoyu is heated and kept at 60°C, and then heated at 80°C for 30 min (Yokotsuka *et al.*, 1971a). Alternatively, raw shoyu may be heated at 80 to 85°C for ∿30 min and then cooled rapidly to 60 to 50°C; with this procedure the thermostable neutral protease accounts for most of the acceleration of sediment formation.

H. Hashimoto (unpublished, 1975) added raw shoyu in varying proportions to heated shoyu, which had been prepared by heating at 85°C for 30 min. The blended shoyu was incubated at 60°C, and sediment was found to form most rapidly when the ratio of raw to treated shoyu was 1:1. This indicated that heating shoyu at >80°C facilitates sediment formation more than heating at 60°C. A recycling pasteurization procedure was developed (Yokotsuka *et al.*, 1971c; H. Hashimoto, unpublished, 1975), in which part of a batch of raw shoyu to be pasteurized is recycled through a heat exchanger where it is heated at 80°C, to the remainder of the batch until the temperature becomes ∿60°C. Then the entire batch is heated at 80°C for 30 min, at 115°C for 5 sec, in order to give an adequate color and heat flavor to the product and to sterilize the spore-forming bacteria. This method of pasteurization effectively utilizes not only thermolabile and thermostable proteases but also the easily coagulable nature of heated shoyu at 80°C, to accelerate sediment formation during pasteurization.

Heating shoyu at 85°C for 30 min and then cooling and holding at 70 to 75°C produced no observable sediment formation. This is because the neutral protease II is specifically inactivated at this temperature. However, shoyu prepared in this manner formed a small amount of sediment after >6 months of storage at room temperature.

4. Adding commercially available proteases from various sources and culture extract of *Aspergillus* molds to heated shoyu to accelerate sediment formation (Hashimoto and Yokotsuka, 1973): This accelerating activity was well correlated with the protease activity measured at pH 5.0, the pH of shoyu. Thermostable acid protease was proposed to be more effective in accelerating sediment formation. In view of this point, a thermophilic fungus, *Penicillium duponti* K1014, was selected because of its high level of productivity of thermostable acid protease. The enzyme was found to be most active at pH 4.6 and 75-80°C for sediment formation of shoyu.

5. Inactivating the sediment-promoting proteases remaining in refined shoyu (i.e., clarified shoyu separated from heat coagula by decantation after 4 to 5 days' storage of pasteurized shoyu) with aminocarboxylic acids such as EDTA, HEDTA, and DTPA (Hashimoto *et al.*, 1970d): These chelating agents completely inhibit the formation of secondary sediment of shoyu even with a final concentration as low as 0.1% without affecting the organoleptic quality of shoyu. Shoyu treated with this procedure did not form secondary sediment after storage for 3 months at 30°C. However, DEG and NTA were not effective.

6. Removing the precursors of sediment and sediment-promoting enzymes remaining in refined shoyu by absorbents and precipitants (Hashimoto *et al.*, 1970c; Yokotsuka *et al.*, 1971b): It was observed that treating shoyu with bentonite, acid clay, or active carbon had a strong effect on the prevention of sediment in refined shoyu. The greatest effect was expected when treatment with these absorbents was carried out at >80°C, little effect was observed at <70°C. Acid clay absorbed part of the nitrogenous compounds and color of shoyu but absorbed hardly any free amino acids. A very small amount of Al was eluted from acid clay into the shoyu. Refined shoyu treated with 1.0% acid clay, bentonite, or active carbon formed no secondary sediment after storage for 3 months at either 30 or 37°C.

C. Turbidity Caused by Insufficient Heat-Denaturation of Soybean Protein as Raw Material

When the native protein in soybeans is not sufficiently denatured by heat treatment because of either insufficient temperature and heating time, or nonuniform heating, the undenatured raw protein cannot be fully hydrolyzed by the enzymes of koji molds, with the result that highly salt water-soluble protein that is not coagulable during the heat of pasteurization goes into the final product. Such shoyu becomes turbid when diluted with water or and then heated. This phenomenon occurs in the preparation of Japanese clear soup, for example.

Although the turbidity problem is prevented by uniformly denaturing the protein in raw materials by heat treatment, overheating of soybeans reduces their enzymatic digestibility, thus decreasing product yield. The conditions for heating soybeans as raw materials are indicated in Fig. 17 (Yokotsuka *et al.*, 1966; Yasuda *et al.*, 1972). The potential for such turbidity formation is usually checked by diluting shoyu 6-10 times with water and then boiling for 3 min. Other tests consist of adding such solvents as methanol, ethanol, propanol, acetone, or trichloroacetic acid, adjusting the shoyu pH to 3.5, or subjecting shoyu to high-pressure liquid chromatography that reveals soybean protein as a separate peak from those of other raw proteins derived from wheat kernels and koji enzymes (Sekine, 1984).

Sasaki *et al.* (1975, 1979) prepared experimental shoyu containing raw protein derived from soybeans insufficiently denatured by heat, and separated the turbidity-causing material from the shoyu by Sephadex G-75 column chromatography in a yield of 8%. This material constituted 71% protein and 18% carbohydrates on a dry basis. The protein fraction of the material was found to contain more glutamic acid, proline, and lysine than the original soybean protein. The average molecular weight of the protein was found to be 170,000. This material produced turbidity in shoyu when diluted and then heated at pH 3, whereas the isoelectric precipitation of soybean protein produced turbidity at pH 4.5. The enzymes of koji molds digested this material after heat treatment more slowly than soybean protein. In these aspects, this turbidity-causing protein was greatly different from soybean protein.

Tamura and Aiba (1982) also studied the turbidity-causing material in shoyu upon dilution and heating. They prepared experimental shoyu from soybean meal that was insufficiently heat denatured by mixing with 130% volume of water and then autoclaving at an atmospheric pressure of 5 kg/cm for 30 sec. The turbidity-causing material was first separated from the shoyu by dialysis against water at 8°C for 96 hr. The material obtained was then separated from acidic polysaccharides through stepwise precipitation with decreasing pH values, and further purified through gel filtration. The purified material was judged to be almost pure protein from the electrophoresis pattern; it had a total nitrogen content of 16.49% and was of MW 210,000. It contained greater amounts of hydrophobic amino acids such as glycine, alanine, valine, isoleucine, and leucine than did soybean protein. This purified material produced the maximum turbidity in shoyu at pH 5.0-5.6. This protein was presumed to be similar to glycinin-T, and the intermediate enzymatic product of undenatured 11S globulin, which is less digestible by the enzymes of koji molds than 7S protein in the presence of high salt concentrations.

Turbidity-causing material in shoyu can be enzymatically hydrolyzed by processing longer or by reducing the salt concentration of mash by dilution with water. Turbidity from raw materials derives mainly from soybeans and not from wheat, because the wheat protein is easily digested by the enzymes of koji molds.

V. STABILITY OF FLAVOR COMPONENTS OF SHOYU FOR HEATING AND FOR OXIDATION

A. Formation of "Fire Flavor" in Shoyu in the Course of Pasteurization

Pasteurization of raw shoyu obtained from a thoroughly aged mash is carried out with several purposes: (1) removal of living cells of molds, yeasts, lactobacilli, and sometimes of heat-resistant bacterial spores, (2) inactivation of microbial enzymes, sometimes including phosphatases that degrade 5'-nucleotides added to shoyu as flavor enhancers, or esterases that degrade *p*-hydroxy-alkyl ester added as a preservative, (3) removal of heat-coagulable substances such as the native protein of enzymes derived from koji molds, (4) clarification of shoyu by floating the suspended oil particles up to the surface and by precipitating the suspended solid particles to the bottom, (5) increase of the color intensity as the result of the so-called browning reaction, (6) formation of the appropriate amount of "brown flavor" or "fire flavor" in Japanese, (7) removal of unpleasant volatile flavor components such as hydrogen sulfide that were produced during fermentation, and (8) increase in the resistance of the product to film-forming yeast, lactobacilli, and other harmful microbes. In the pasteurization of usukuchi shoyu, major efforts are directed to avoiding an increase in the color intensity while at the same time achieving the other aims described above as much as possible.

The industrial pasteuriqation of shoyu is generally conducted at around 75 to 85°C, followed by gradual cooling in the case of koikuchi shoyu, or rapid cooling in the case of usukuchi shoyu. However, higher temperatures than the above sometimes become necessary, especially when bacterial spores must be killed--by so-called HTST heating at 115°C for 5 sec, for example.

The quality of "fire flavor" of shoyu produced by pasteurization is naturally dependent on the temperature and duration of heating, but it is generally believed that holding for an adequate length of time at between 70 and 80°C is necessary for the formation of good "fire flavor," and this flavor formation goes along with the formation of an adequate color intensity of the product.

The sum of chemical changes that take place during the industrial pasteurization of koikuchi shoyu having a total nitrogen content of ∿1.5% is generally equivalent to that of heating shoyu at 80°C for 3 to 5 hr; in the 1950s, by contrast, it was 85°C for 5 hr, when the total nitrogen content of the highest quality koikuchi shoyu was ∿1.2%. When raw koikuchi shoyu containing 1.2% total nitrogen was heated at 85°C and immediately poured into open wooden kegs holding 5-10 kiloliters and subjected to natural cooling, the increasing ratio of titratable acidity, and the decreasing ratios of reducing sugar and of amino-nitrogen contents on the fourth day of storage were 43, 19, and 5%, respectively. The increasing or decreasing ratios of shoyu ingredients determined every 5 hr while heating at 85°C for 35 hr are indicated in Fig. 18 (Yokotsuka and Takimoto, 1956). One of the distinct chemical changes occurring during heating of shoyu is the increase in several kinds of aldehydes, attributable to Strecker degradation, in which amino acids change into aldehydes having one less carbon molecule. The total aldehyde contents of raw and pasteurized shoyu were reported to be 0.003 and 0.006%, respectively (Yamada, 1928). Another remarkable increase of carbonyl compounds during heating of shoyu is that of furfural and 5-hydroxymethylfurfural, which is caused by aminocarbonyl reaction between sugars and amino compounds. The α-diketon compounds in raw shoyu amount to 0.1 mg%, and reach 0.3 mg% after pasteurization; these are composed of diacetyl, acetylpropionyl, and acetylbutyryl in the ratio of 100:20:5. In addition, glyoxals are present in an amount that is equivalent to 20% of that of α-diketon compounds (Asao and Yokotsuka, 1963).

The content of free phenolic compounds (calculated as 4-ethyl guaiacol) increased from 0.017 to 0.048% after 5 hr heating of koikuchi shoyu at 80°C, which was determined by the Folin-Ciocaltou method as for the ether extract of shoyu. At the same time, a gradual increase in the conjugated form of the phenolic fraction was also recognized. Vanillic acid, ethyl-vanillate, 4-ethyl guaiacol, and their phenol esters combined with acetic acid or benzoic acid were presumed to be the major components in the above fraction (Yokotsuka and Takimoto, 1958).

An increase in the content of ether-soluble sulfur-containing compounds is observed during heating of shoyu, which is composed of dimethyl sulfide, mercaptan, and presumably mercaptals. The content and the increase of these sulfur-containing compounds during heating of fermented shoyu are much smaller than those with HVP. One example of quantitative analysis of flavor constituents of koikuchi shoyu is indicated in Table XXI (Sasaki *et al.*, 1980). According to Nunomura *et al.* (1983), acetoin, furfuryl alcohol, 4-hydroxy-5-methyl-3(2H)-furanone (HMMF), acetaldehyde, propanal, acetone, 2-methyl propanal, dimethyl sulfide, ethylene sulfide and 3-(methylthio)propanal increased in amount, and methanol, 1-propanol, 2-phenyl ethanol, and 4-hydroxy-5-ethyl(or methyl)-2-methyl(or ethyl)-3(2H)-furanone (HEMF) decreased in amount in the course of heating shoyu at 100°C. Acetoin, furfuryl alcohol, HMMF, propanal, 2-methyl propanal, 3-methyl butanal, dimethyl sulfide, ethylene sulfide, and 3-(methylthio)propanal increased after heating to 100°C over 5 min, but they increased no further with further elevation of temperature, thereby with a decrease of 1-propanol.

The changes in contents of organic acids and of amino acids of koikuchi shoyu during heating at 80°C for 5 hr are indicated in Table XXII and Table XXIII, respectively (Ueda and Tanitaka, 1977).

B. Stability of Shoyu Flavor during Storage on the Shelf

It is generally observed that flavor deterioration occurs in parallel with the oxidative browning of shoyu. Shoyu flavor deterioration is more closely related to the increase of darkening or the decrease of ΔA value caused by oxidative browning than to the increase of color intensity or the increase of red color caused by heat-induced browning. Changes in volatile flavor components, especially a decrease in ethyl acetate and an increase in acetaldehyde, were observed to accompany the oxidative browning of shoyu (Onishi, 1970).

Tsukiyama *et al.* (1981) reported on the oxidative and nonoxidative quality changes of shoyu during storage as shown in Tables XXIV-XXVI. The most remarkable changes were the increase in aldehydes, 2,3-diacetone, and phenylcyanide-reducing power, the decrease in arginine and histidine by oxidative browning, the increase in titratable acidity, volatile acids, reducing power, carbonyl compounds except for acetone, and an unknown compound X, and the decrease in pH, formol nitrogen, and amino acids brought about by heat-dependent nonoxidative browning during storage.

C. Stability to Evaporation of Flavor Components of Shoyu

The results of a quantitative analysis of the headspace gas from koikuchi shoyu that had been evaporated by passing helium gas through shoyu at 20°C, is indicated in Table XXVII (Sasaki and Nunomura, 1979). The odor units of 6 of 14 constituents of this headspace gas were calculated as shown in Table XXVIII, which suggested that the volatile flavor of newly pasteurized koikuchi shoyu is mainly expressed by the lower boiling aliphatic aldehydes and ethanol. However, judging from the very low threshold values of compounds containing less sulfur and free phenolic compounds, it may be reasonably presumed that they also significantly contribute to the flavor quality of pasteurized shoyu. The time dependence of rapid loss of four major volatile flavor constituents of shoyu is indicated in Fig. 19. Results were determined by pouring 10 ml shoyu into a flat dish at 23°C (Sasaki and Nomura, 1981).

VI. EFFECTS OF CONTAINERS ON QUALITY OF SHOYU

About half the shoyu produced in Japan is consumed at home and is packed in 1-liter plastic bottles or 2-liter glass bottles. The other half, which is consumed institutionally and industrially, is usually packed in 18- or 180-liter steel cans, or conveyed in bulk and stored in larger size tanks holding 1-10 kiliiters. Wooden kegs of Japan cedar (*Cryptomeria japonica*), containing 14.4-18 liters, were the major marketing containers of shoyu since the seventeenth century until round 1950, although china bottles containing less than 2 liters were also generally used. The industrial use of 1.8-liter and later 2.0-liter (1914) glass bottles for shoyu began around 1910. Around 1950 Japan began using steel cans containing 0.5 or 1 gallon for exporting shoyu. Around 1965 the use of wooden kegs as containers for shoyu was totally discontinued; at about the same time, 1-liter PVC bottles became popular as disposable shoyu containers. The glass bottle sealed with a crown cap remains the best shoyu container as compared with other materials with regard to (1) water loss via evaporation through container materials, (2) browning caused by the action of oxygen penetrating the container materials, (3) loss of flavorful compounds by evaporation through or by absorption into container materials, (4) flavor deterioration caused by compounds migrating from container materials or by oxygen penetrating container materials, and (5) loss of chemical constituents other than flavorful compounds through absorption by container materials. Cedar kegs are the worst containers with regard to the above considerations; for example, it absorbs the *p*-hydroxybutylbenzoate that is added as a preservative, thus rendering the product susceptible to contamination with film-forming yeast, thereby greatly reducing the commercial value. On the other hand, the terpenelike flavor derived from cedar was formerly flavored by consumers.

The drawbacks of glass bottles as shoyu containers are its heavy weight and mechanical weakness; however, fortified glass prepared by such methods as tempering or ion exchange has improved upon these drawbacks to a great extent.

With regard to the carcinogenicity of PVC monomer (VCM) remaining in PVC bottles, for a while bottles made of several types of multilayer plastics such as PPE (polypropylene-eval or conjugated ethylene-vinyl alcohol) or PPN (polypropylene-nylon) were used as shoyu containers in Japan, combining the advantages of reduction in the water permeability of polyethylene or polypropylene, with the very low oxygen permeability of eval or nylon. These multilayer bottles have been replaced by PET bottles since 1978, however, because of the latter's many advantages as described below. On the other hand, the PVC bottles are again being widely used, subject to strict governmental regulations of VCM content of <1 ppm in Japan. The biaxial stretching blow manufacturing method (BO-PET), a great technological improvement over the direct-blow method (CD-PET), has rapidly increased the popularity of PVC bottles as shoyu containers of 500 to 1000 ml volume. Glass bottles still have some advantages over PET bottles in that shoyu in glass bottles undergoes less weight loss, however, and flavor deterioration during storage. Nevertheless, PET bottles are replacing glass bottles in the 500- to 1000-ml volume range because of their many advantages including lighter weight, safety, transparency, the acceptable level of small changes in color, flavor, and weight loss, and finally the great environmental advantage that it is easily incinerated, producing only carbon dioxide and water. The degree of stability of shoyu when packed in 1-liter PVC and PET bottles as compared with that in 360-ml glass bottles is shown in Table XXIX, and the result of sensory evaluations in Table XXX (Saiki, 1979).

When steel cans are used as containers for shoyu, the volume of headspace, the nature of resin coating the inside of cans, and the elution of iron into the shoyu become issues. Tin-free steel cans (TFS) exhibit less iron elution than do ordinary steel cans, perhaps because of the better adherence of resin to the surface of TFS. Urea resin is more often used than phenol resin as the inside coating resin of steel cans used for shoyu. Replacement of headspace gas in the containers with nitrogen gas created a minimal effect on the keeping quality of shoyu (Saiki, 1979, 1979).

A paper carton box made from multiple layers of polyethylene (PE)-paper-PE-aluminum foil-PE-PE proved very effective for keeping quality when the shoyu was packed without headspace and when precautions were taken to avoid pinholes in the aluminum foil, according to the author's experience.

Saiki (1978) found nearly no effect of sunlight on the color change of koikuchi shoyu but did observe a certain amount of color darkening in usukuchi shoyu packed in a colorless glass bottle as compared to a PVC bottle, especially colored PVC, because the PVC cuts out light of up to 370 nm wavelength.

VII. MICROBIAL STABILITY OF SHOYU

The microbes that can grow in good-quality fermented shoyu containing 15-20% salt are limited to such yeasts as *Saccharomyces*, *Pichia*, and *Torulopsis*, and no serious health hazards involving microbial contamination of shoyu have ever been reported. However, consumer demand has led to a decrease in the salt content of shoyu. At the same time, chemical preservatives are generally not used for health reasons; moreover, modern consumers prefer the reduced color intensity and the less intense "heat flavor," both of which are produced by the pasteurization of shoyu at reduced temperatures. All of these changes have increased the microbial susceptibility of shoyu, as well as the potential for contamination by film-forming yeasts and/or various harmful lactobacilli or other bacteria.

A. Film-Forming Yeasts

Some film-forming yeasts belonging to *Zygosaccharomyces rouxii* (Barnett *et al.*, 1983), the same as *Saccharomyces rouxii* (Lodder and Kregar-van Rij, 1952), or *Z. japonicus* and *Z. sulsus*, as formerly classified in Japan, are the most common microbes that are sometimes found in a good-quality koikuchi shoyu, which is composed of >17% salt, >1.5% total nitrogen, >2.0% ethanol, 4.7-4.8 pH, and ∿0.85 water activity, and is pasteurized at >70°C followed by gradual cooling. *Pichia anomala* (Lodder and Kregar-van Rij, 1952) are also found as film-forming yeasts in the same shoyu, but non-film-forming yeasts belonging to *Z. major* and *Z. soya* (Barnett *et al.*, 1983) and *Candida versatillis* and *C. etchellsii* (Barnett *et al.*, 1983; same as *Torulopsis versatillis* and *T. etchellsii*, Lodder and Kregar-van Rij, 1952) are seldom found in ordinary commercial shoyu; although these yeasts are the dominant ones in the salty shoyu mash fermentation, they are killed during pasteurization. Almost the same comments can be made about good-quality usukuchi shoyu of more than 1.15% total nitrogen and more than 19% salt content. Good-quality koikuchi and usukuchi shoyu, as mentioned above, are initially resistant to contamination with these film-forming yeasts, but if diluted with water, the chemical components of shoyu become easily so contaminated. Once commercial shoyu is contaminated with these yeasts, the white film that forms on the surface greatly damages its commercial value; also, the color fades somewhat, the reducing sugar content remarkably decreases, the glutamic acid content also decreases to some extent, a bitter and sour taste becomes noticeable, and the characteristic volatile flavor of koikuchi shoyu is replaced by other odors and finally by a foul odor.

According to Imabara and Nakahama (1968), when film-forming *Zygosaccharomyces* yeasts grow and form film on the surface of shoyu, more benzaldehyde and benzoic acid and less phenylethyl alcohol are produced from phenylalanine than those metabolized by non-film-forming *Zygosaccharomyces* shoyu yeasts. The same authors (1971) also determined the volatile flavor constituents produced by *Hansenula anomala* and *Pichia membrafaciens* in the film-forming conditions to be ethyl acetate and fusel oil, respectively, which amounts were remarkably more than those produced under non-film-forming conditions.

B. Film-Forming Yeast-Static Activity of Shoyu

The yeast-static activity and its increase in the course of pasteurization of koikuchi shoyu were reported by Yokotsuka (1954). Removal of the acidic fraction of shoyu by treatment with ion exchange resin greatly reduces the yeast-static activity, but this activity returns to pretreatment levels when the shoyu is heated at 90°C for 3 hr. The activity to prevent the growth of film-forming yeast on the surface of shoyu was determined for 35 volatile flavor components either isolated from or possibly present in koikuchi shoyu. The minimum effective concentrations of these compounds in preventing the growth of these yeasts on the surface of shoyu diluted with twice the volume of water were determined, and the results are shown in Table XXXI.

Most of the yeast-static activity in the ether extract of shoyu has been found to exist in the acidic fraction, which contains organic acids and free phenolic compounds. During the course of pasteurization of koikuchi shoyu held at 80°C for 5 hr or at 90°C for 1 hr, the titratable acidity of shoyu increased by some 25%, most of which was found to be caused by the increase of acetic acid, and the free phenolic content increased from 0.014 to 0.048% calculated at 4-ethyl guaiacol. Hanaoka (1976) reported on the formation of yeast-static activity upon heating the solution of M/10 amino acid and M/5 xylose in the water extract of shoyu koji at 120°C for 60 min. A correlation between the increase of yeast-static activity and the increase of color intensity of the solution was observed with some exceptions, as indicated in Fig. 20.

C. Fortification of Yeast-Static Activity of Shoyu

To prevent with certainty the contamination of film-forming yeasts derived from air or from containers, benzoic acid or butyl-*p*-hydroxybenzoate is legally added in Japan. The upper limits for the allowable amounts of these chemical preservatives as stipulated by the Japanese government are 0.6 g/kg for benzoic acid and 0.25 g/liter for butyl-*p*-hydroxybenzoate in the case of shoyu. Because of the insolubility of butyl-*p*-hydroxybenzoate, sodium benzoate has become the most popular chemical preservative used for shoyu; however, consumer preference has of late led to the increase of ethanol content in shoyu by fermentation or by external addition with or without aseptic bottling of shoyu. The extent of yeast-static activity of shoyu to be fortified with preservatives should be determined from the initial individual yeast-static activity of shoyu; this is dependent on the type of shoyu, the concentrations of solids (including total nitrogen, reducing sugar, salt, and ethanol), the pH values, the kinds of yeast, and the method to be used for determination of yeast-static activity.

Hanaoka (1964, 1976) divided shoyu of an average koikuchi type (total nitrogen 1.5%, sodium chloride 17.0%, ethanol 1.5%, reducing sugar 3.0%, and pH 4.8) into two fractions, ether soluble and nonsoluble; each fraction exhibited a film-forming yeast-static activity equivalent to 100 to 200 ppm benzoic acid. Adding 2% sodium chloride to the shoyu increased the yeast-static activity equivalent to 110 ppm benzoic acid within shoyu NaCl concentrations between 16 and 22%, and adding 1% ethanol increased the yeast-static activity equivalent to 120 ppm benzoic acid within shoyu ethanol concentrations between 0.5 and 3.5%.

On the other hand, Takakusa *et al.* (1975) reported that the yeast-static activity fortified by adding 300 ppm sodium benzoate to shoyu was equivalent to that obtained by adding 10 ppm butyl-*p*-hydroxybenzoate or 1.5% ethanol.

D. Harmful Lactobacilli and Spores of *Bacillus subtilis*

Due to consumer concern about excess dietary salt intake, the low-salt containing or salt-reduced shoyu is becoming more and more common on the market. When the salt concentration of shoyu falls below 15%, contamination not only by film-forming yeasts but by some lactobacilli such as *Lactobacillus plantarium* occurs and seriously damages the commercial product; as a result, an increase of organic acids and a decrease of pH are observed, along with the rapid disappearance of glutamic acid and aspartic acid through decarboxylation, which sometimes causes breakage of glass or plastic containers with tight seals (see Fig. 21; Hanaoka, 1976). In some of these cases, butyl-*p*-hydroxybenzoate is more effective than benzoic acid as a preservative within the legally permitted concentrations; in addition, it is often necessary to add >5% ethanol. The heat-resistant spores of *Bacillus subtilis* remaining in shoyu that has been pasteurized at <100°C sometimes germinate when the shoyu is diluted with water in some processed foods, thus spoiling the commercial products. Pasteurization using the HTST heating method is presently the most common method for avoiding this problem.

E. Bactericidal Action of Shoyu

Ujiie and Yokoyama (1956) reported on the bactericidal action of koikuchi shoyu (Kikkoman brand) for nine kinds of intestinal pathogenic bacteria including *Escherichia communis, Shigella flexneri, Vibrio cholerate-Inaba, Salmonella typhi-Shikata,* and *Bacillus subtilis* (B31); this activity was attributed to the acidity, high osmotic pressure, and some chemical components contained in the shoyu. These findings are shown in Table XXXII. Yamamoto *et al.* (1978) reported that *Escherichia coli* and *Staphylococcus aureus* were easily killed in 18% salt solution dissolved with 50 ppm butyl-*p*-hydroxybenzoate; thus the salt solution synergistically enhanced the bactericidal action of butyl-*p*-hydroxybenzoate. The time needed to kill all of the above two kinds of bacteria in shoyu was dependent on the initial number of cells; 4-6 hr for 10^3/ml, 24-48 hr for 10^5/ml, and 5-7 days for 10^7/ml. This bactericidal action is predominantly due to the salt concentrations, supplemented by pH value and alcohol concentrations; there is also some influence from the total nitrogen concentrations and other factors. G. Sakaguchi, I. Onishi, and K. Shinagawa (1975) tested the fate of staphylococci during incubation in 9% salt and 17% salt (w/v) fermented koikuchi shoyu, respectively, to which were added no chemical preservatives. The staphylococci (10^6 cells/ml) in 17% salt shoyu were killed almost completely within 3 hr; 90% of the cells were killed in 9% salt shoyu within the same period of time, whereas the time needed for 90% death in 17% salt shoyu was only 13-14 min. These experimental results indicate that the contribution of salt content to the death of staphylococci is great. But the times necessary to kill 90% of staphylococci in 10 and 17% salt-containing phosphate buffer of pH 4.7 were 980-1440 min and 460-530 min, respectively. These results suggest that components other than salt participate in the bactericidal action of shoyu. The behavior of *Clostridium botulinum* in shoyu was also studied. The strains of type A remained without germination for 3 months at 30°C, and the strains of type B decreased slightly in number in the same period of time.

TABLE I.
History of Fermented Soybean Foods

China	Japan
Shu-ching (700BC)	
Chu (koji)	
Chiang (fish, bird, flesh)	Manyo-Shu (350-759)
Chi-Min-Yao-Shu (532-549)	hishio (chiang) (fish, meat, soybeans)
Chu (crushed wheat, wheat flour made into balls or cakes, or cooked rice.	
Chiang (soybeans, or wheat)	
Shi, Shi-tche	
	Koma-hishio and Miso
Tang-dynasty (618-906)	Taiho-Law (701)
	Soybean hishio, Miso,
	Kuki (Shi)
Ben-Chao-Gong-Mu (1590)	Ekinrinbon-Setsuyoshu (1598)
Chiang-yu	Shoyu (Chiang-yu)
	Honcho-Shokkan (1692)
	Shoyu, Miso, Tamari.

Note: Industrial production of shoyu in Japan; Noda (1561), Choshi (1516), Tatsuno (1666), and that of miso, Sendai (1645).
Export of shoyu from Japan (1668).
Visit of C.Thunberg to Japan from Sweden (1774)

TABLE II.

Production Amount & Kinds of Shoyu in Japan

Total production	1,187,148 kl	
Total sales	1.184,306 "	
	(Bureau of Foods, Japan)	
Koikuchi	902,862 kl	84.4 %
Usukuchi	138,261 "	12.9 "
Tamari	20.885 "	2.0 "
Saishikomi	3,130 "	0.3 "
Shiro	5,042 "	0.5 "
Total	1,070,180 kl	100.0 %

(authorized by Japanese Agricultural Standard)

Supplied by Japan Shoyu Inspection Association, 1982.

TABLE III.

Chemical Analysis of Genuine Fermented and Special Grade of Shoyu in Japanese Market (January, 1984).

Kinds of shoyu		Be	NaCl	TN	FN	RS	Alc	pH	Ex	Col
Koikuchi	I(6)	22.2	17.0	1.57	0.91	3.59	2.00	4.89	19.5	12
"	II(3)	23.0	16.9	1.68	0.94	4.48	2.21	4.97	21.6	12
"	III(1)	23.7	16.4	1.98	1.11	5.48	2.47	4.87	24.5	14
"	IV(5)	19.7	13.3	1.55	0.90	4.04	3.16	4.90	21.1	11
"	V(2)	16.5	8.7	1.54	0.89	4.28	3.67	4.34	21.4	13
Usukuchi	I(5)	22.2	19.4	1.21	0.72	4.49	2.30	4.91	16.2	28
"	II(1)	22.4	18.0	1.48	0.85	4.87	2.65	4.86	18.7	28
"	IV(1)	19.3	14.7	1.19	0.73	4.82	3.79	5.00	18.0	28
Tamari (2)		23.1	17.5	1.84	1.03	2.74	2.53	4.95	21.2	7
Saishikomi (1)		28.9	13.1	1.96	0.94	12.2	1.46	4.82	38.5	2
Shiro (1)		25.3	17.9	0.51	0.28	17.2	0.08	4.74	21.2	46

TN; total nitrogen%(g/100ml), FN; Formol nitrogen, RS; reducing sugar, Alc; alcohol, Ex; extract without salt, Col; number of color standard issued by the Japan Shoyu Inspection Association, the smaller the darker. TN of three kinds of Special Grade of Koikuchi, Tamari and Saishikomi; I(ordinary), more than 1.50, II(super), more than 1.65, III(ultra super, more than 1.80. Ex of three kinds of Special Grade of Usukuchi and Shiro; I(ordinary), more than 14, II(super), more than 15, ultra super, more than 17. Salt content; IV(usujio), less by 20%, V(genen), less by 50%, than the standard, which is Koikuchi 17.5%, Usukuchi 19.9%, Tamari 17.9%, Saishikomi 15.6 %, and Shiro 17.9%, as determined by JAS.
The kinds of Special Grade are determined in the bylaws of Japan Shoyu Inspection Association.
The too high content of alcohol (underlined) may be added on after fermentation.

Arranged from J. Japan Soy Sauce Research Inst. 10, p68(1984).

TABLE IV.

Typical Soysauces and Fishsauces in the World

No. Type	Be	NaCl %(w/v)	Total nitrogen %(w/v)	Reducing sugar %(w/v)	Alcohol %(v/v)	Color intensity
1. Koikuchi (Japan)	23.0	16.9	1.68	4.48	2.21	++
2. Usukuchi (Japan)	22.2	19.4	1.21	4.49	3.67	+
3. Soysauce (HongKong, Lo Chau)	16.9*	26.4	1.31	3.32	0	+++
4. Soysauce (HongKong, Sang Chau)	12.7*	24.7	0.78	7.51	0	+++
5. Soysauce** (Taiwan)	25.0	15.5	1.62	6.53	1.53	+++
6. Soysauce** (Singapore)	31.9	20.9	2.62	8.15	0.23	+++
7. Tamari (Japan)	23.3	19.0	1.70	0.86	0.09	+++
8. Soysauce (Indonesia, Kecap Manis)	----	8.90	0.55	32.3	0.06	++++
9. Soysauce*** (USA)	24.4	20.9	1.35	1.76	0.01	+++
10. Fishsauce** (Philippines, patis)	8.73*	28.1	1.53	1.01	0.11	+
11. Fishsauce (Japan, Shottsuru)	6.79*	25.5	1.17	0.60	----	+

From Shinshi (1983)(nos 3 and 4), Shinshi (1981)(Nos 10 and 11), and Kikkoman Corporation (1984), (Nos. 1, 2, 5-9).

* Extract without salt, **Average figures of three kinds, *** Average figures of two kinds.

The Nos 1,2, and 7 are the genuine fermented products, but the HVP contents of other Nos are not known.

TABLE V.
Spectrophotometric Transmittance of Shoyu

nm	Koikuchi	Usukuchi	Shiro	nm	Koikuchi	Usukuchi	Shiro
400	0.7	1.0	2.1	560	58.2	78.7	74.1
10	0.8	1.2	3.8	70	62.7	82.0	76.0
20	1.0	2.0	6.6	80	67.0	84.7	78.5
30	1.5	3.7	11.0	90	70.8	87.1	80.6
40	2.5	6.8	15.4	600	73.9	89.1	82.3
50	3.5	11.4	20.0	10	77.2	90.7	83.7
60	6.3	17.0	25.3	20	79.9	92.2	85.2
70	8.9	24.0	31.0	30	82.6	93.3	86.7
80	11.4	30.5	36.6	40	84.1	94.3	86.9
90	16.4	40.0	43.2	50	85.6	96.2	87.4
500	22.0	47.9	49.7	60	88.0	95.4	88.4
10	27.4	55.7	55.4	70	89.7	96.2	89.5
20	34.3	61.8	59.3	80	90.5	96.7	90.4
30	40.8	67.2	64.3	90	92.0	97.1	91.2
40	47.3	71.7	68.1	700	93.2	97.5	92.4
50	53.0	75.0	70.6	10			

Note: Measured by Hitachi Model 181

Fom Okuhara et al. (1977).

TABLE VI.
Color of Raw Shoyu

Concentration of Shoyu	Y	x	y	Dominant Wave Length (nm)
0.1	66.9	0.453	0.434	583
0.2	49.6	0.505	0.450	583
0.3	36.8	0.547	0.428	588.3
0.4	28.3	0.568	0.419	590.7
0.5	23.5	0.586	0.402	593.8
0.6	18.6	0.615	0.372	600
0.7	15.7	0.617	0.374	599.5
0.8	12.6	0.639	0.363	604.9
0.9	11.3	0.647	0.350	605.5
1.0	10.4	0.649	0.348	606

Measured according to the CIE System

$$x = \frac{X}{X + Y + Z}, \quad y = \frac{Y}{X + Y + Z}$$

From Okuhara et al. (1977).

TABLE VII.
Color of Pasteurized Shoyu

Concentration of shoyu	Y	x	y
0.1	53.3	0.489	0.415
0.2	31.6	0.569	0.416
0.3	21.3	0.596	0.396
0.4	15.1	0.629	0.365
0.5	10.8	0.648	0.336

Measured according to the CIE system.

From Okuhara (1973).

TABLE VIII.
Color of Oxidized Shoyu

Concentration of shoyu	Y	x	y
0.1	45.3	o.521	0.442
0.2	24.1	0.572	0.416
0.3	13.0	0.630	0.366
0.4	6.8	0.678	0.303

Measured according to the CIE system.

From Okuhara (1973).

TABLE IX.

Coefficient of Color Increase of Shoyu while Using.

Temp. C.	Koikuchi			Usukuchi		
	2L,G. 45days	1L,P. 24days	500ml,P. 12days	2L,G. 45days	1L,P. 24days	500ml,P. 12days
28	2.84	1.92	1.48	2.34	1.96	1.40
22	1.95	1.39	1.34	1.91	1.59	1.35
15	1.39	1.27	1.21	1.64	1.45	1.28

A 40 ml shoyu was taken out every day for use.

G : glass bottle, P : plastic bottle.

From Sawa et al. (1978).

TABLE X.

Comparison of Color Darkening of Marketing Shoyu during Consumption.

Kinds		OD just after opening the seal	Darkening coefficient (%) Days of storage					
			0	10	20	30	40	45
Usukuchi shoyu	A	0.294	0	9	29	56	84	108
	B	0.406	0	17	36	61	80	96
	C	0.344	0	9	29	52	75	96
	D	0.312	0	12	27	51	75	90
	E	0.350	0	8	26	46	62	81
	F	0.287	0	9	20	36	53	64
Koikuchi shoyu	G	0.475	0	16	38	46	90	111
	H	0.446	0	14	40	63	87	108
	I	0.459	0	13	31	55	86	109
	J	0.465	0	21	38	59	83	104
	K	0.447	0	17	39	60	84	102
	L	0.501	0	17	34	52	75	91

A 200 ml shoyu sample was removed on every 5th day. The color increase was measured at 530 nm (Usukuchi) and 610 nm (Koikuchi) at 22 C in a 3 mm cell.
From Sawa and Nishikawa (1981).

TABLE XI.

Color Darkening of Various Kinds of Shoyu during Consumption

Kinds of shoyu	Temp. C	Darkening Coefficient (%)								
		Days of storage								
		0	2	4	6	8	10	12	14	16
A	15	0	0.9	3.1	4.3	5.9	7.8	10.5	16.0	24.0
A	22	0	1.2	3.4	6.8	11.6	13.2	19.1	23.9	33.0
A	28	0	1.9	5.8	7.9	13.5	17.2	22.0	32.0	45.3
B	15	0	2.9	3.2	5.8	8.5	9.6	11.7	18.1	24.0
B	22	0	2.3	2.1	7.0	11.0	12.3	15.2	21.0	28.4
B	28	0	2.9	4.9	7.0	9.6	14.9	22.6	26.0	39.8
C	15	0	0.4	1.2	3.1	5.5	8.3	11.2	14.1	18.7
C	22	0	0.6	2.1	4.5	8.2	12.3	16.4	20.9	26.0
C	28	0	1.2	3.1	5.5	9.6	15.0	20.9	28.2	39.3
D	15	0	1.8	3.2	4.5	7.2	10.0	12.7	16.9	24.4
D	22	0	2.4	3.6	4.8	8.1	11.4	15.1	21.1	31.0
D	28	0	2.7	4.4	6.0	9.3	13.0	18.1	26.5	38.2

Two liters shoyu were removed on every second day from an 18 liter can.

A: Genuine fermented Usukuchi, TN 1.24%
B: Genuine fermented Usukuchi, TN 1.17%
C: Semi-fermented Usukuchi, TN 0.98%
D: Genuine fermented Koikuchi, TN 1.53%

From Sawa and Nishikawa (1981).

TABLE XII.
The Color Formation of Shoyu during Preparation

Period of mash fermentation (months)	ΔA	Color degree (E_{450})	Percent of color formation
1*	0.45	2.48	13.7
2*	6.08	6.08	33.7
3*	8.54	8.54	47.2
6** (pasteurized)	18.07	18.07	100.0

* The color of liquid part of mash was determined.
** The press-filtrate of mash was pasteurized and the color was determined.

From Motai (1976).

TABLE XIII.
Sugar Contents in Shoyu Mash Fermentation

Days**	Glucose*	Galactose*	Xylose*	Arabinose*
7	63.9	8.8	8.7	5.4
14	72.7	7.1	8.1	5.3
30	70.3	7.6	6.5	5.2
60	30.2	7.9	3.0	3.9
90	20.0	6.2	1.6	2.7
120	23.9	5.6	1.4	2.5
150	21.4	5.6	1.4	2.3
180	26.3	5.4	1.2	1.9

* mg/ml

** days of storage of salty shoyu mash

From Okuhara et al. (1975).

TABLE XIV.
Browning of Peptide Solution

Peptide of amino acid	OD 500nm
Glycine	0.1442
Glycylglycine	1.503
Glycylglycylglycine	0.598
Glycylglycylglycylglycine	0.632
α-Alanine	0.107
α-Alanylalanine	0.606
Peptide solution	3.156

Peptide or amino acid was dissolved in 0.2M acetic buffer solution (pH 4.8) containing 1% xylose so as to bring the total nitrogen content of the solution to 0.7%, and the solution was heated at 80°C for 5 hrs. The peptide solution was prepared to be TN-FN 0.7%.

From Okuhara et al. (1975).

TABLE XV.

Components, Correlated Significantly to Browning of Shoyu

Dependent	C(HF)	C(OXH)		C(HN)		C(N)	
Population	II	I	II	I	II	I	II
Variable							
CK			**	*	**		**
WK	*		**	*	**	**	**
L			*		**		**
TN-FN		*	*	**(++)	**	**(+)	**(++)
$FN\text{-}NH_3\text{-}N$		-*	**	(+)			*
$NH_3\text{-}N$				*			
TA	(-)			**(+)	**	**	**(+)
RS		*	*	**	**(++)	**(++)	**
Alcohol				-**		-**	
Org.Acid		-**		-*		-**	
pH						-**(--)	
R(N)		-**		**	**	**	**
R(NH)				*	*		
Red(N)			**(+)				
Red(NH)					**		
NaCl		(-)		*			
C(N)		**(++)	**	**			
C(H)			*(-)		**		
C(HN)	*	(-)					

From Okuhara et al (1970).

*Linear correlation is significant at 5% level. ** Linear correlation is significant at 1% level. (+) or (-) Partial correlation is significant at 5% level. (++) or (--) Partial correlation is significant at 1% level. + Positive correlation. - Negative correlation. Symbols of variable are ; CK The ratio of koji to the mash water. WK The weight of koji (kg) per material unit. L Volume of shoyu per material unit. TN-FN Subtracted Formol nitrogen (%) from total nitrogen (%). $FN\text{-}NH_3\text{-}N$ Subtracted ammonium nitrogen from Formol nitrogen. $NH_3\text{-}N$ Ammonium nitrogen(%). TA Titratable acidity (meq/10ml). RS Residual reducing sugar(%). Alcohol %(V/V). Org.Acid Organic acid (meq/ml). pH of raw shoyu. R(N) Reducing power of raw shoyu. R(HN) Subtracted RN from reducing power of pasteurized shoyu. Red(N) Reduction (μg/ml) of raw shoyu. Red(HN) Subtracted Red(N) from reductone(μg/ml) of pasteurized shoyu. NaCl %(W/V). C(N) Color intensity of raw shoyu. C(H) Color intensity of pasteurized shoyu. C(HN) Subtracted C(N) from C(H). C(OXH) Subtracted C(H) from color intensity of oxidized shoyu. C(HF) Subtracted color intensity from C(H).

TABLE XVI.

The Formation of 3-deoxy-D-glucosone (3DG) during the Preparation of Shoyu .

Samples	3DG mg%
Koji (72 hours cultivation)	0
Liquid part of 3 days mash	0.9
" 6 "	3.2
" 10 "	10.0
" 33 "	14.0
Raw shoyu after 4 months	20.0

From Kato et al. (1961).

TABLE XVII.

Heat-dependent and Oxidative Browning of Shoyu, from which Heavy Metals are removed .

	TN %	FN %	RS %	TA	Fe++ ppm	△ E350 Heat-dependent browning*	Oxidative browning**
Original shoyu	1.52	0.88	4.2	14.7	34	0.365(100%)	0.360(100%)
Heavy-metals removed shoyu***	1.48	0.87	4.1	14.0	1	0.345(95%)	0.120(33%)

* heated at 80 C for 4 hrs.

** stored at 37 C for 5 days.

*** Chelate resin, Dowex A-1 treated.

From Hashiba and Abe (1984).

TABLE XVIII.

Oxidative Browning of Amadori Compounds isolated from Shoyu.

Amadori compounds*	Browning after 14 days at 37 C ($\Delta \underline{E}$ 550)		Contents ofAmadori compounds in pasteurized shoyu. mmol/liter
	Mon-oxida-tive	Oxidative	
F-Gly	0.012	0.266	0.2
F-Ala	0.009	0.360	0.3
F-Val	0.009	0.360	1.2
F-iso Leu	0.006	0.400	1.3
F-Leu	0.010	0.310	1.5
Mixture of aminoacids and sugar**	0.000	0.003	---

* 0.1M Amadori compounds were added to amino acid mixture.

** 0.1M glucose was added to amino acid mixture.

From Hashiba and Abe (1984).

TABLE XIX.

k and α Values of Miso Color compared with those of Shoyu Color and various Melanoidins

Melanoidin	k	α
Miso	4.57×10^{-4}	1.32
Shoyu	4.47×10^{-4}	1.30
Gly-xylose system	2.73	0.29
Lys- " "	1.45	0.39
Glu- " "	0.30	0.56
Gly_2- " "	0.11	0.70
Gly-Leu- " "	0.0115	0.95
Gly_3- " "	2.70×10^{-4}	1.45

From Motai and Inoue (1974b).

TABLE XX.

Heat-dependent and Oxidative Browning of Each Fraction of Shoyu divided by Ultrafiltration.

Heat-dependent browning

M.W. of fraction	Ratio of concentration	TN %	TN 0.154			Increase X TN ratio***
			OD	ΔOD-I	Increase**	
--10000	6	1.967	0.846	0.294	0.93	0.99
10000-5000	5	1.827	0.469	0.320	1.13	1.12
5000-1000	4	2.150	1.488	0.415	1.31	1.52
1000-500	3	1.568	0.137	0.269	0.85	0.72
500--	1	1.547	0.094	0.203	0.64	0.54
Control		1.845	0.530	0.316	1	1

Oxidative browning

M.W. of fraction	ΔOD-II	Increase	Increase X TN ratio
--10000	0.186	1.50	0.99
10000-5000	0.075	0.60	0.59
5000-1000	0.165	1.33	1.55
1000-500	0.038	0.31	0.17
500--	0.035	0.28	0.23
Control	0.124	1	1

* OD of fraction - OD of control
** ΔOD-I of fraction / ΔOD-I of control
*** TN of fraction / TN of control
From Okuhara et al. (1975)

TABLE XXI.

Results of Quantitative Analysis of Flavor Components in Koikuchi Shoyu.

Ethanol	31,501,10	Furfuryl alcohol	11.93
Lactic acid	14,346,57	Isoamyl alcohol	10.01
Glycerol	10,208,95	Acetoin	9.78
Acetic acid	2,107,74	n-Butyl alcohol	8.69
HMMF	256,36	HDMF	4.83
2,3-Butanediol	238,59	Acetaldehyde	4.63
Isovaleraldehyde	233,10	2-Phenylethanol	4.28
HEMF	232,04	n-Propyl alcohol	3.96
Methanol	62,37	Acetone	3.88
Acetol	24,60	Methionol	3.65
Ethyl lactate	24,29	2-Acetylpyrrole	2.86
2,6-Dimethoxyphenol	16,21	4-Ethylguaiacol	2.77
Ethyl acetate	15,13	Ethyl formate	2.63
Isobutylaldehyde	14,64	Gamma-butyrolactone	2.02
Methyl acetate	13,84	4-Ethylphenol	Trace
Isobutyl alcohol	11,96		

From Sasaki et al. (1978).

TABLE XXII.

Quantitative Changes of Organic Acids in Koikuchi Shoyu when heated at 80 C for 5 Hours.

	Before heating mg/100ml	After heating mg/100ml	Increasing ratio
Acetic acid.	134.5	10.4	104.4
Pyruvic acid	2.3	3.2	139.1
Formic acid	10.8	16.7	154.6
Lactic acid	660.3	662.7	100.3
Succinic acid	20.6	20.2	98.0
Pyroglutamic acid	230.1	250.0	108.6
Glycolic acid	19.8	20.1	101.5
Malic acid	43.2	44.6	103.2
Citric acid	59.9	60.6	101.1
Total organic acid	1181.5	1218.5	103.1

From Ueda and Tanitaka (1977).

TABLE XXIII.

Quantitative Changes of Amino Acids in Koikuchi Shoyu when heated at 80 C for 5 Hours.

Amino acid	Before heating mg/100ml	After heating mg/100ml	Difference %
Tyrosine	60	55	91.6
Phenylalanine	290	284	97.9
Lysine	431	418	97.0
Histidine	71	70	98.6
Arginine	199	161	80.9
Aspartic acid	458	442	96.5
Threonine	270	263	97.4
Serine	365	358	98.1
Glutamic acid	1129	1069	94.7
Proline	369	363	98.4
Glycine	211	202	95.7
Alanine	358	356	99.4
cystine	trace	trace	----
Valine	339	339	100.0
Methionine	99	97	98.0
Isoleucine	311	308	99.0
Leucine	476	472	99.1
Total	5436	5257	96.7

From Ueda and Tanitaka (1977).

TABLE XXIV.

The Chemical Analysis of the Shoyu used for the Storage Test.

	NaCl%	TN%	pH	Alc%	RS%	FN%	Glut%	AcI*	AcII**	OD
Cont.	19.2	1.18	4.82	2.45	5.18	0.63	1.29	8.13	7.50	0.280
A	19.3	1.18	4.71	2.24	5.21	0.62	1.18	8.88	7.36	1.117
B	19.2	1.18	4.65	2.30	5.18	0.62	1.25	9.10	7.20	0.600
C	19.2	1.18	4.76	2.42	5.14	0.62	1.14	8.60	7.36	0.395
D	19.2	1.18	4.50	2.40	4.76	0.60	0.98	9.97	7.58	1.200

A : Stored at 30C for 100 days oxidatively without shaking.

B : Stored at 30C for 16 days oxidatively with shaking.

C : Stored at 30C for 100 days non-oxidatively.

D : Stored at 95C for 3 hours non-oxidatively.

% : w/v--NaCl, TN, RS, FN, and Glut. v/v--Alc.

Ac I : ml of 0.1N NaOH to titrate 10ml shoyu at pH 7.0

AcII : ml of 0.1N NaOH to titrate 10ml shoyu from pH 7.0 to pH 8.3.

OD : determined at 530 nm in 3mm cell.

From Tsukiyama et al. (1981).

TABLE XXV.

Amino Acid Composition of Shoyu after Storage.

Amino acid	Initial mg/100ml	Oxidative storage 30C,100days	Non-oxidative storage 30C, 100days	Non-oxidative storage 95C, 3 hours
NH_3	145	152	153	142
Lys	280	295	297	215
His	100	89	86	97
Arg	316	284	299	257
Asp	315	310	300	290
Thr	152	142	142	140
Ser	209	202	199	191
Glu	1,287	1.177	1,140	924
Pro	264	263	267	273
Gly	174	173	173	162
Ala	224	233	227	209
Cys	17	11	14	3
Val	249	262	254	243
Met	128	114	115	95
I-Leu	252	254	244	241
Leu	410	402	397	402
Tyr	48	51	50	54
Phe	248	254	240	236
Total	4,813	4,658	4,597	4,174

From Tsukiyama et al. (1981).

TABLE XXVI.

The Quantitative Changes of Lower Boiling Constituents of Shoyu during Storage.

Compounds	Initial	Oxi-dative storage 30 C 100days	Non-oxi-dative storage 30 C 100days	Effect of tempera-ture*	Effect of oxygen**
Acetaldehyde	3.5	11.25	4.5	28.6	193
iso-Butyl aldehyde	0.64	2.13	1.93	202	31.2
Acetone	0.88	7.98	1.13	28.4	778
Ethyl acetate	6.49	5.00	7.82	20.5	-43.5
iso-Valer aldehyde	1.34	4.21	4.30	220	-6.7
n-Propyl alcohol	3.00	2.52	3.16	5.3	-21.3
iso-Butyl alcohol	15.9	12.4	16.6	4.4	-26.4
n-Butyl alcohol	5.52	4.23	5.61	1.6	-25.0
iso-Amyl alcohol	14.8	11.0	14.8	0	-25.7

* ((non-oxidative storage - initial) / initial) X 100.

** ((oxidative storage-non-oxidative storage) / initial X 100.

From Tsukiyama et al. (1981).

TABLE XXVII.

Quantitative Analysis of Headspace Gas from Shoyu.

Compounds	Concentrations ppm ($\bar{x}$, n=10)	Coefficient of variation (%)
Methanol	9.45	4.43
Acetaldehyde	3.76	9.58
Ethanol	5,605.18	3.50
Propionaldehyde	1.70	8.52
Acetone	2.09	3.75
Ethyl formate	1.66	3.02
n-Propyl alc.	0.82	5.64
Isobutyraldehyde	6.38	3.16
Ethyl acetate	33.41	1.83
Isobutyl alc.	3.79	1.75
n-Butyl alc.	0.69	10.75
Isovaleraldehyde	8.17	2.88
2,3-Pentenedione	0.76	8.25
Isoamyl alc.	2.36	9.38

From Sasaki and Nunomura (1979)

TABLE XXVIII.

Odor Units of 6 Components of Headspace Gas from Shoyu.

Compounds	Concent-rations ppm	Threshold ppm in water	Odor units	Relative odor units(%)
Ethanol	5,605,18	1.83×10^{-1}	30,629.40	33.05
Ethyl acetate	32.41	6.0×10^{-1}	55.68	0.06
isovaleraldehyde	8.17	1.5×10^{-4}	54,466.67	58.77
Isobutyraldehyde	6.38	9.0×10^{-4}	7,088.89	7.65
Acetaldehyde	3.76	1.5×10^{-2}	250.67	0.27
Propionaldehyde	1.70	9.5×10^{-2}	178.95	0.19

From Sasaki and Nunomura (1981)

TABLE XXIX.

The Stability of Koikuchi Shoyu packed in PVC, PET and Glass Bottles during Storage at 30 C.

Materials	BO-PET***	PVC	Glass
Volume, ml.	1000	1000	150
Weight of plastic, g.	35	39-40	230
Loss of weight after 100 days, g.	3.364	2.702	-----
Color number** (9)	6-5	5	8-7
OD at 500 nm* after 100 days **(8.04)	3.003	3.125	2.283
OD at 600 nm* after 100 days **(1.950)	0.785	0.839	0.532

* revised for 10 mm thickness of shoyu

**( ) ; starting figures

*** BO-PET: PET bottles made by biaxial stretching blowing method.

From M.Saiki (1979)

TABLE XXX.

The Sensory Evaluation of Shoyu packed in PVC, PET and Glass Bottles storaged at 30 C for 4 Weeks.

		PET (1000ml)	Glass (350ml)
When compared to PVC bottles	Mean	0.2273	0.50000
	T-O value	2.0174	3.1691
	Significance	non-discriminating	discriminating at 1%
When compared to glass bottles	Mean	2.2273	
	T-O value	1.0000	
	Significance	non-discriminating	

Panel member 22, Scoring method.

From M.Saiki (1979)

TABLE XXXI.

The Minimum Effective Concentrations of Flavor Substances found in Shoyu to prevent the Growth of Harmful Yeasts (Film-forming Yeasts).

Minimum effective concentrations	Flavorous substances found in shoyu	Flavorous substances probably present in shoyu
0.01%		isobutyl benzoate
0.02%	1-hydroxy-3-methoxy-4-ethylbenzene caproic acid	
0.04%	ethylvanillate isoamyl acetate	propyl benzoate
0.06%	benzoic acid benzaldehyde ethylisovalerianate	benzacetal
0.1%	isovaleraldehyde isovaleracetal vanillic acid	vanillin
0.1%	1-benzoyloxy-2-methoxy-2-methoxybenzene tyrosol acetic acid propionic acid isobutyric acid isovaleric acid ethyl acetate ethyl lactate ethyl palmitate ethyl oleate ethyl linolate ethylmercaptane isoamyl alcohol butanol ethanol(less than 1%)	isobutylacetal

From Yokotsuka (1954).

TABLE XXXII.

The Bactericidal Nature of Fermented Shoyu for the Intestinal Pathogenic Bacteria

Bacteria	Sample No.1		Sample No.2	
	Initial counts	Time for total death	Initial counts	Time for total death
Co	10X2.6^{8}	48 hrs	10X2.6^{8}	6 hrs
	10X4.8^{2}	24 hrs	10X4.9^{2}	6 hrs
Sh	10X2.7^{8}	72 hrs	10X2.7^{8}	48 hrs
	10X5.2^{2}	24 hrs	10X5.2^{2}	6 hrs
T	10X3.4^{8}	48 hrs	10X3.4^{8}	6 hrs
	10X4.8^{2}	24 hrs	10X4.3^{2}	6 hrs
PA	10X1.6^{8}	24 hrs	10X1.6^{8}	24 hrs
	10X4.8^{2}	6 hrs	10X4.7^{2}	6 hrs
PB	10X2.2^{8}	168 hrs	10X2.2^{8}	144 hrs
	10X5.2^{2}	120 hrs	10X5.0^{2}	72 hrs
E	10X2.3^{8}	72 hrs	10X2.3^{8}	48 hrs
	10X4.4^{2}	24 hrs	10X4.4^{2}	24 hrs
Ch	10X1.7^{8}	1 hr	10X1.7^{8}	10 min
	10X5.4^{2}	10 min	10X5.4^{2}	5 min

Co, E. communis; Sh, Shigella flexneri; Ch, Vibrio cholerate-Inaba; T, Salmonella typhi-Shikata; PA, Salmonella paratyphi (B29); PB, Salmonella paratyphi-Toyota; E, Salmonella enteritidis (B24).
Sample No. 1 was unpasteurized shoyu, and Sample No. 2 was pasteurized at 75°C and then sodium benzoate was added to make the solution 0.07% shoyu. Each sample contained some 10^5 per ml B. subtilis, but no other bacteria were found (nowadays these are totally killed by HTST heating). The surface of No. 1 shoyu was covered with the film of film-forming yeasts some days after opening.
The chemical analysis of the shoyu used were:

	NaCl	TN	RS	Alc	pH
No.1	18.0%	1.44%	2.80%	1.80%	4.6
No.2	18.5	1.45	2.50	1.20	4.6

By a preliminary test, it was confirmed that the shoyu diluted more than ten times with water did not prevent the growth of these bacteria.
From Ujiie and Koyama (1956).

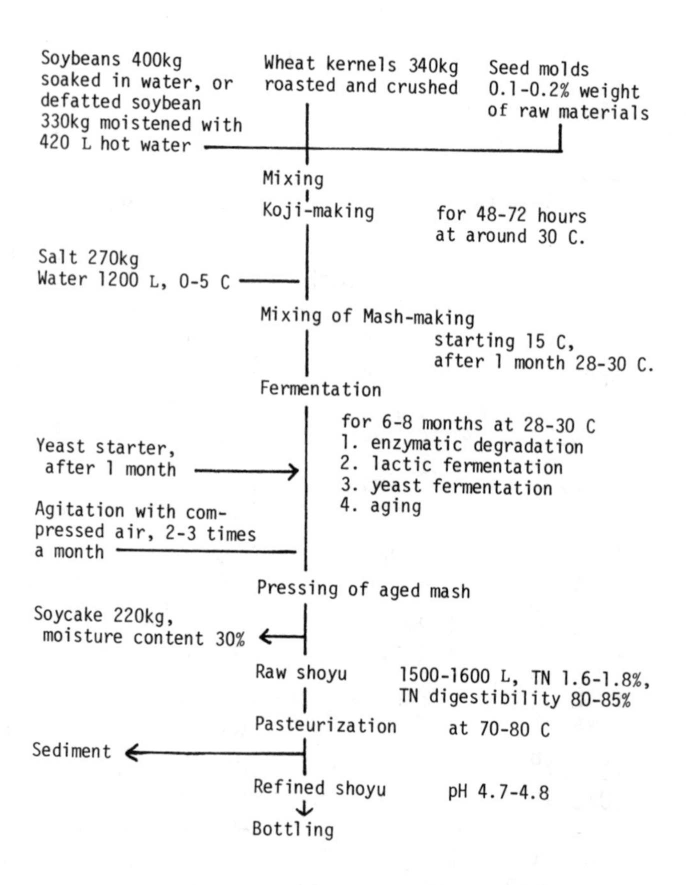

Fig. 1. Flow sheet showing process of koikuchi shoyu fermentation.

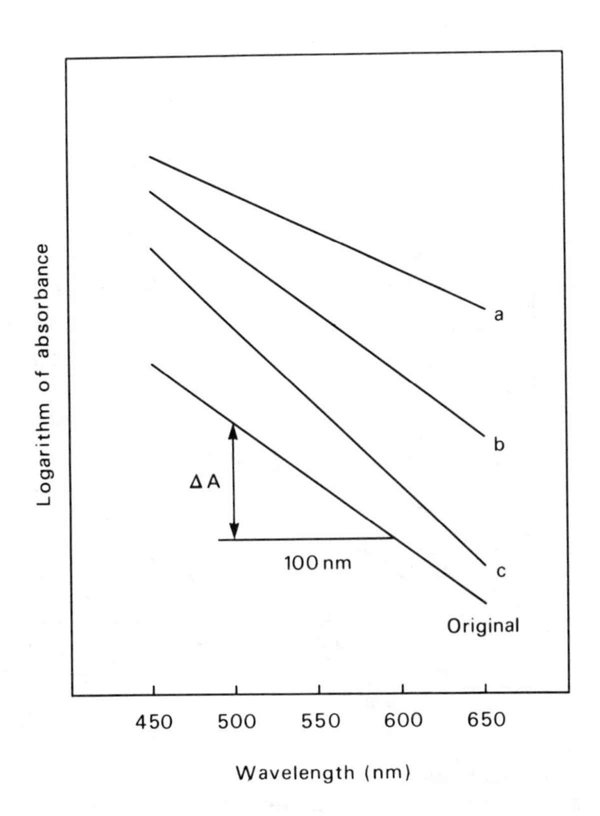

Fig. 2. Change in ΔA during the browning of shoyu (types a-c, see text) (from Motai *et al.*, 1972).

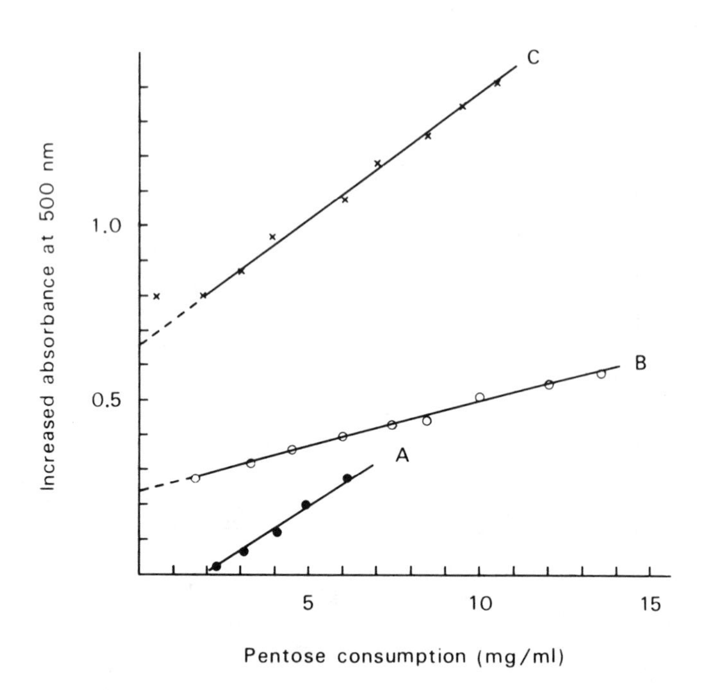

Fig. 3. Relation between xylose (pentose) consumption and pigment formation in three kinds of shoyu, A-C; see text (from Okuhara *et al.*, 1969).

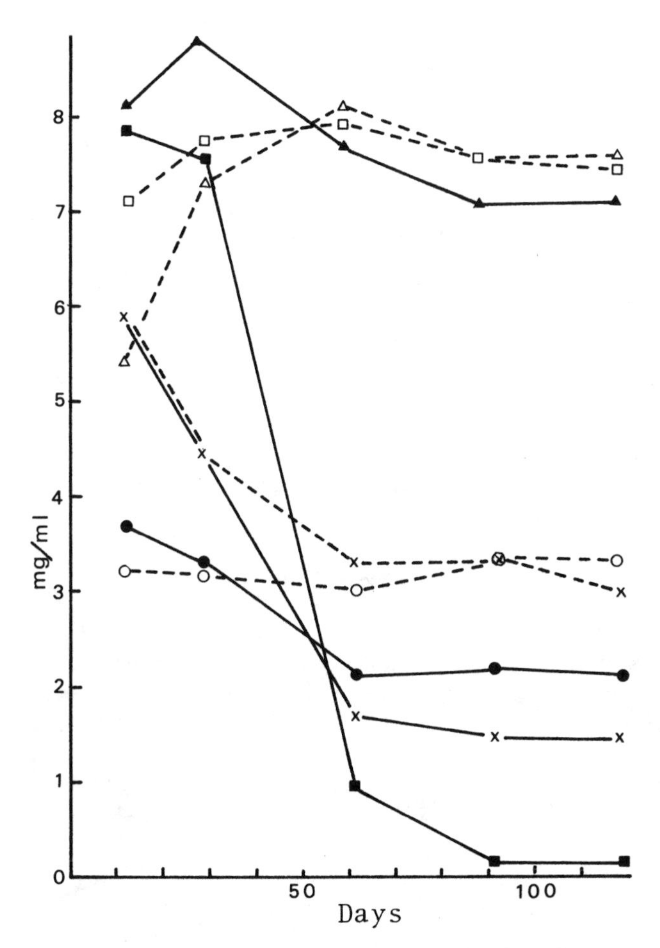

Fig. 4. Consumption of sugars by yeasts in shoyu fermentation. ——— fermented, ----- non-fermented. □■ glucose, △▲ galactose, xx xylose, ○● arabinose. Scale of glucose is reduced to one-tenth (from Okuhara *et al.*, 1975).

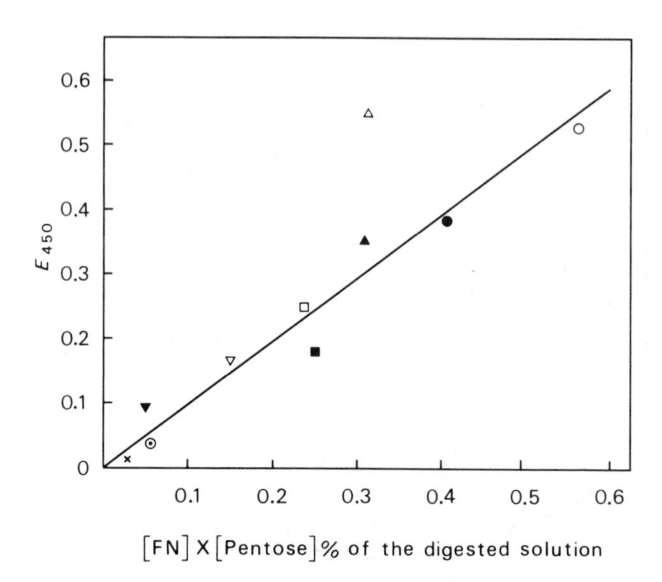

Fig. 5. The relation between FN × pentose percentage and color degree. × Polished rice; ⊙ corn; ▼ corn gluten; ▽ domestic wheat; ■ wheat gluten; △ wheat bran; ○ defatted soybean; ● whole soybean; □ dehulled domestic wheat; ▲ imported wheat (from Shikata *et al.*, 1971b).

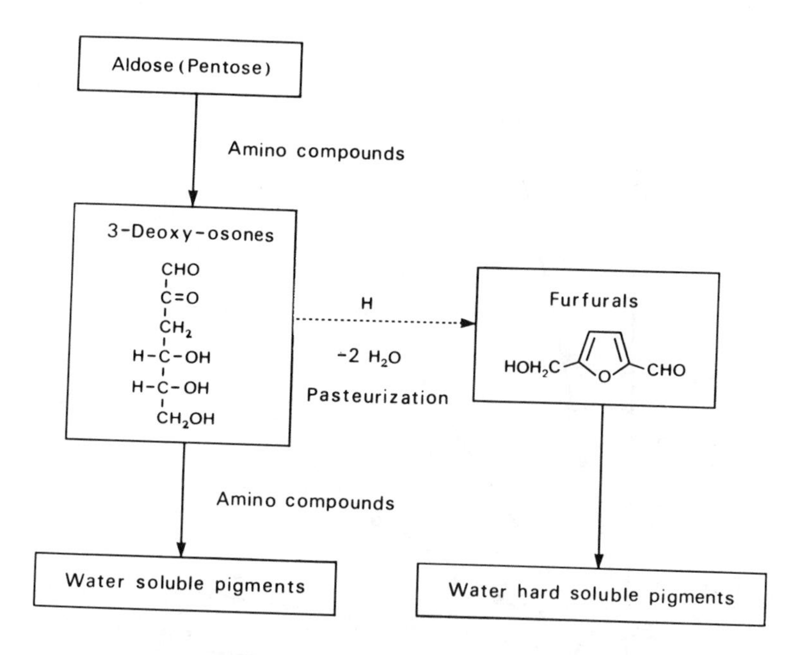

Fig. 6. The mechanisms of heat-dependent color formation of shoyu (from Kato and Sakurai, 1963).

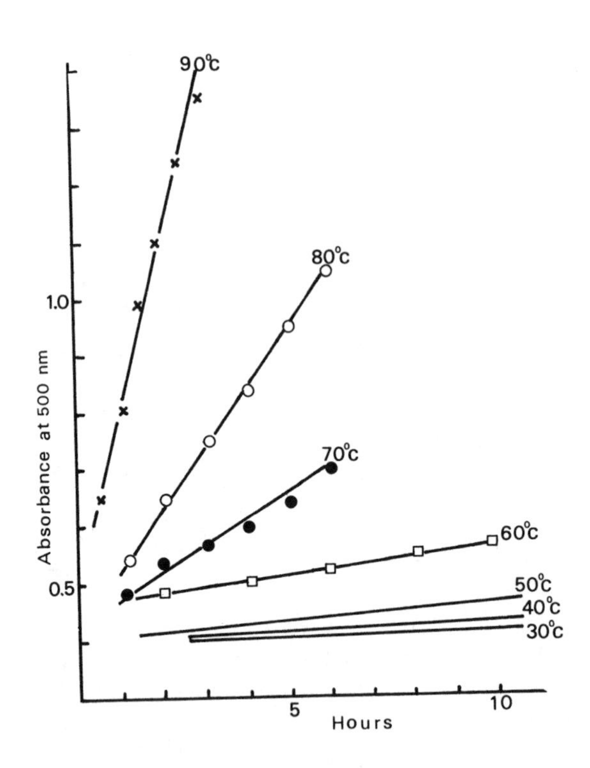

Fig. 7. Effect of temperature on velocity of pigment formation of raw soy sauce (from Okuhara *et al.*, 1969).

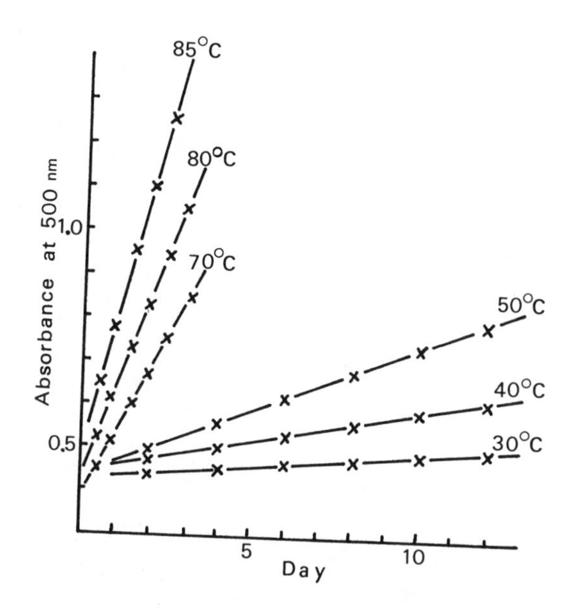

Fig. 8. Oxidative browning of shoyu. Shoyu was heated at various temperatures under aerated conditions (from Okuhara *et al.*, 1969).

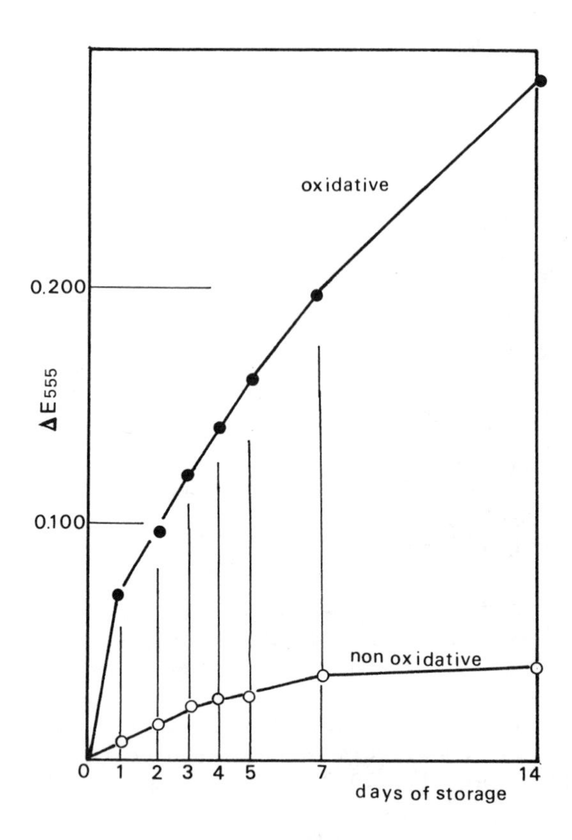

Fig. 9. Effect of oxygen on the browning of shoyu. Shoyu was stored at 37°C, and the color of shoyu diluted 20 times with water was determined (from Hashiba, 1977).

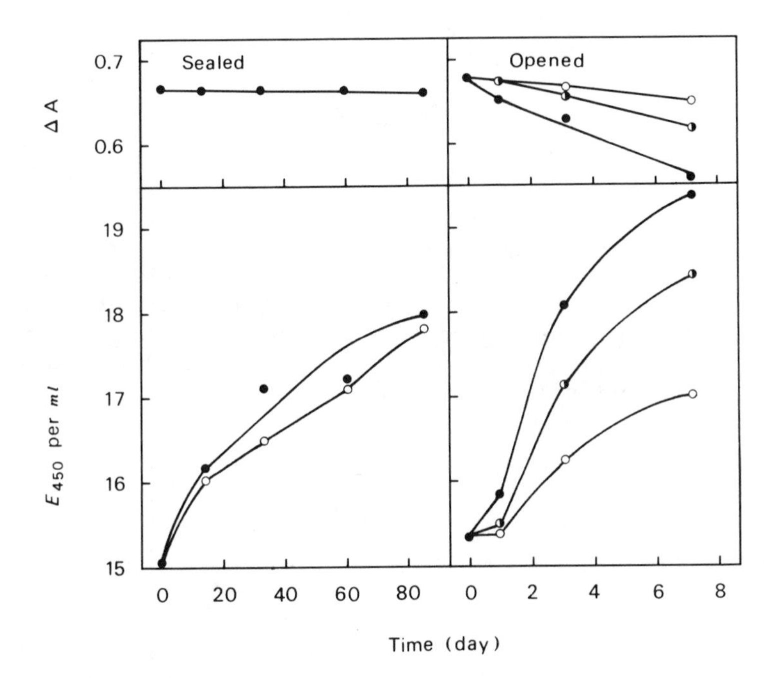

Fig. 10. Change in color of shoyu on storage at 30°C. Sealed: ● 18-liter can; ○ 2-liter bottle. Opened: ● radius (r) = 6.5 cm; ◑ r = 4.5 cm; ○ r = 2.75 cm. For the opened condition, 300 ml of shoyu was placed in beakers with the same height but different surface areas, covered with cellophane film, and stored at 30°C (from Motai and Inoue, 1974).

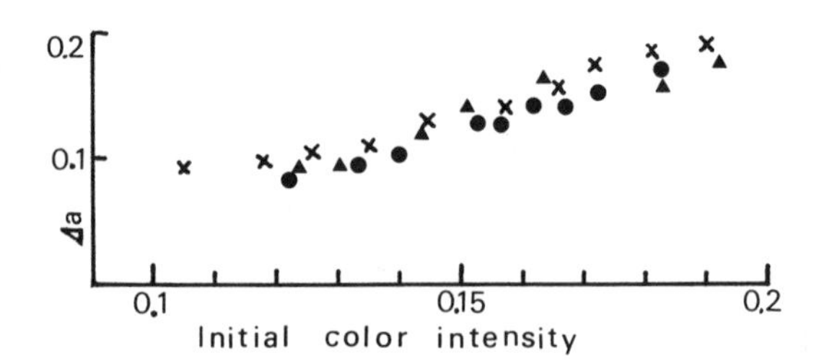

Fig. 11. Relation between oxidative browning and initial color intensity of shoyu. Pasteurizations were carried out at 80°C (×), 70°C (▲), and 60°C (●). The pasteurized shoyu was oxidized by shaking at 40°C for 24 hr. Δ*a*, Increment in OD at 600 nm by oxidation (from Okuhara *et al.*, 1977).

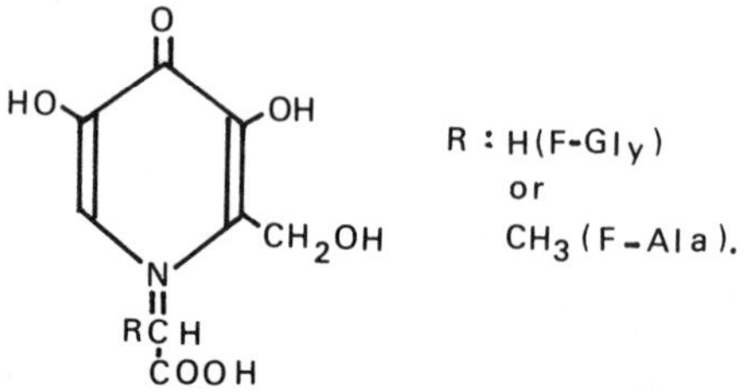

HO OH
CH_2OH
N
RCH
COOH
O

R : H(F-Gly)
or
CH_3 (F-Ala).

Fig. 12. Colorless compounds separated from red pigments that were produced by oxidative browning between F-Gly or F-Ala and iron (from Hashiba and Abe, 1984).

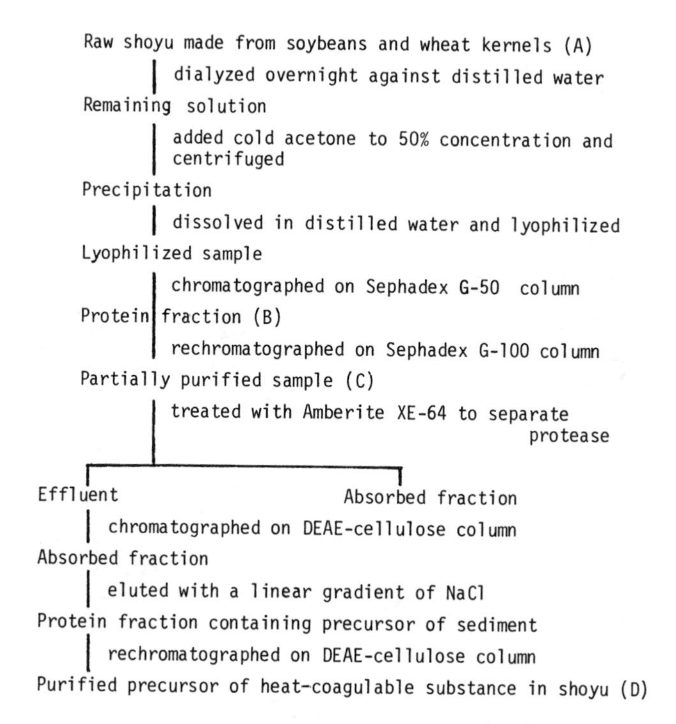

Fig. 13. Purification of heat-coagulable material in raw shoyu (from Hashimoto *et al.*, 1971).

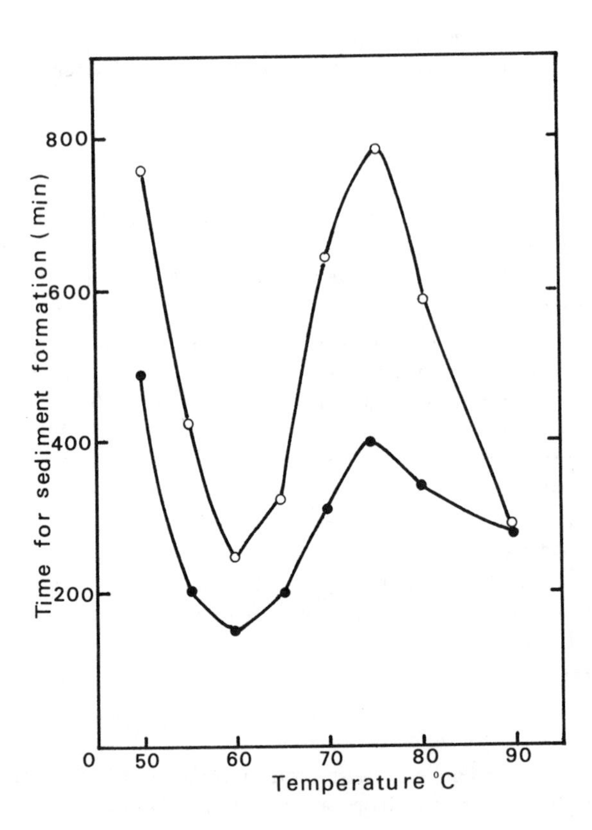

Fig. 14. Effect of heating temperature on the time required for sediment formation in shoyu. Ten milliliters of each sample of raw shoyu were pipetted into duplicate 10-ml glass-stoppered test tubes and heated to various temperatures, at which the times required for sediment formation were measured. ● Raw shoyu made from soybeans; o raw shoyu made from defatted soybean grits (from Hashimoto and Yokotsuka, 1974).

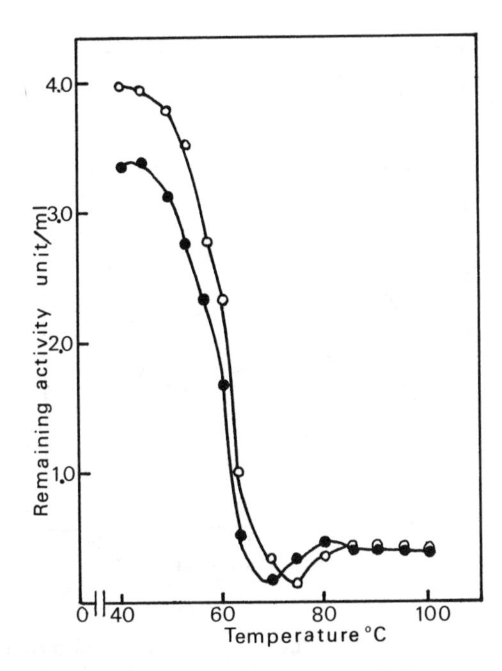

Fig. 15. Effect of heating temperature on the stability of protease in raw shoyu. Raw shoyu (1 ml) was incubated for 10 min at various temperatures, and the residual activity was measured at pH 5.0 by using milk casein (pH 5.0) containing 21% NaCl as substrate. o Raw shoyu made from defatted soybean grits (pH 4.8); ● raw shoyu made from soybeans (pH 4.7) (from Hashimoto and Yokotsuka, 1974).

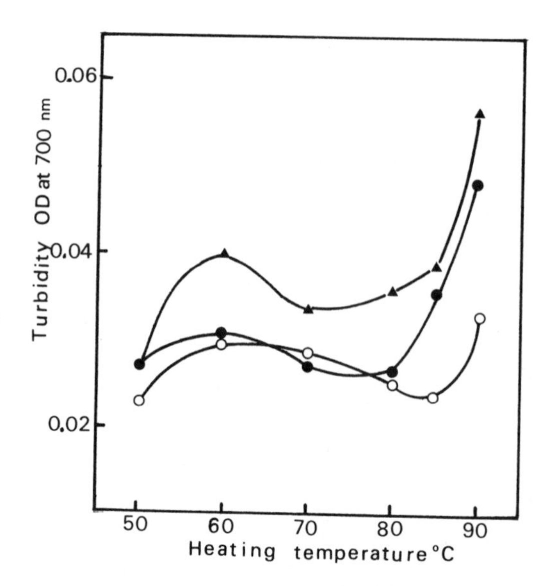

Fig. 16. Effect of heating temperature of shoyu on turbidity appearing during heating. Heating time: o 0.5 hr; ● 1.0 hr; ▲ 3.0 hr. The turbidity during heating was measured at 700 nm according to the text (from Hashimoto *et al.*, 1970).

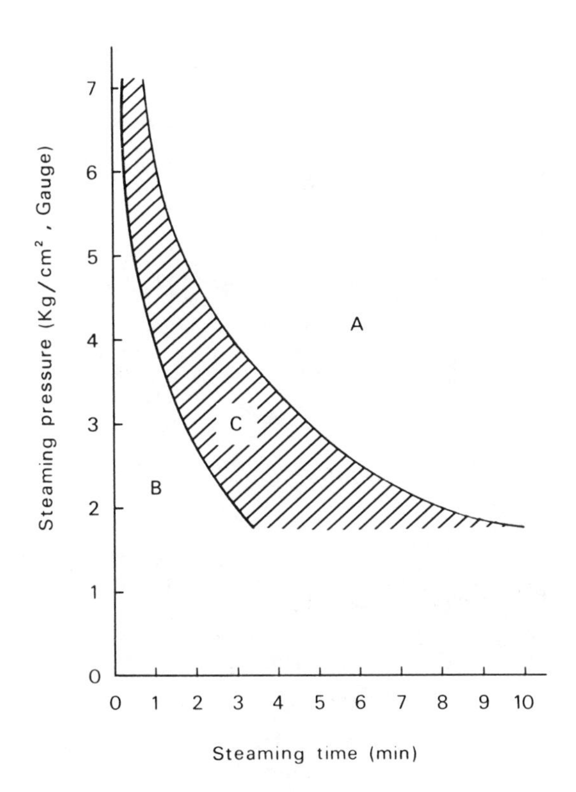

Fig. 17. Denaturation of soybean protein by steaming at 130% moisture. (A) Overdenaturation region; (B) underdenaturation region; (C) proper denaturation region for shoyu production (from Yokotsuka *et al.*, 1966).

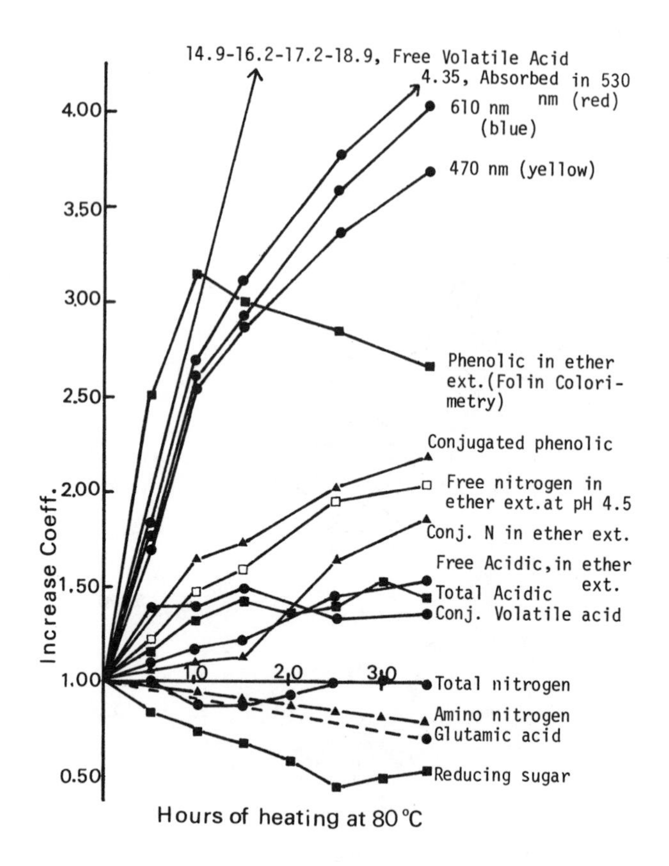

Fig. 18. Increase or decrease coefficients of ingredients in shoyu caused by heating (from Yokotsuka and Takimoto, 1956).

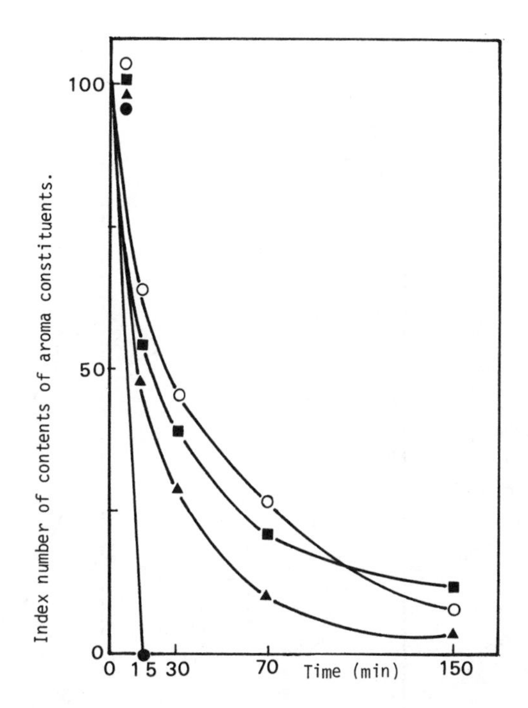

Fig. 19. Time dependence of loss of aroma constituents from the headspace. o Ethanol; ■ isovaleraldehyde; ▲ ethyl acetate; ● propionaldehyde (from Sasaki and Nunomura, 1981).

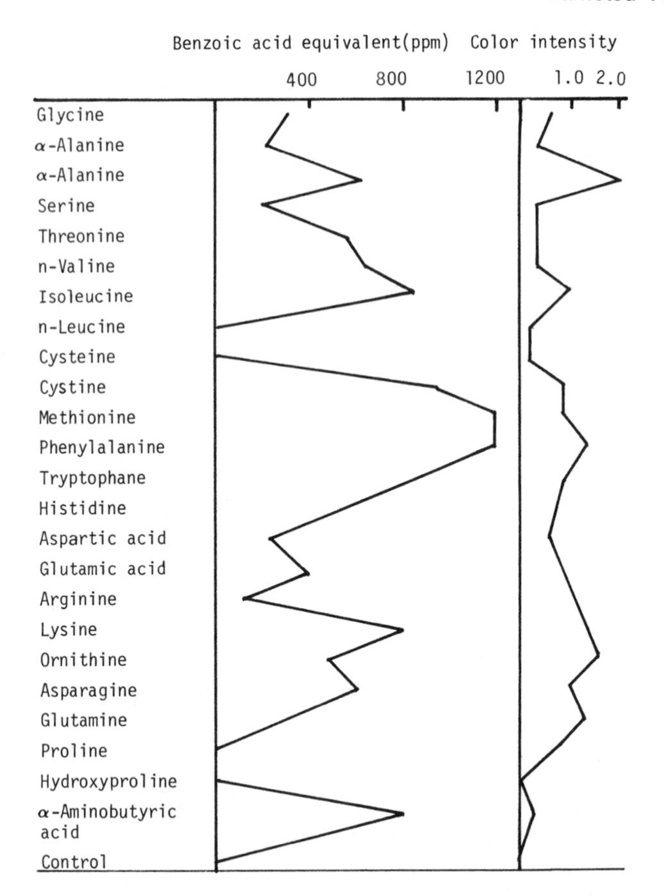

Fig. 20. Yeast-static activity of the mixture of M/10 amino acid and M/5 xylose heated at 120°C for 60 min, and the color intensity of the heated mixture (from Hanaoka, 1976).

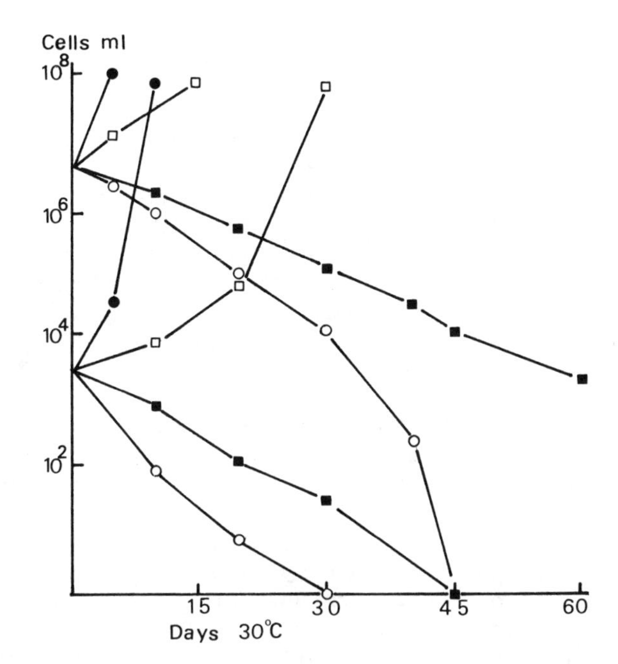

Fig. 21. Relation between the growth of *Lactobacillus plantarium* and the salt (NaCl) content of shoyu with the following chemical components: TN 1.4%, ethanol 1.5%, RS 3.0%, pH 4.8. NaCl content: ● 9%; □ 12%; ■ 15%; o 18% (from Hanaoka, 1976).

REFERENCES

Arai, T. (1984). Production method of shoyu. Japanese Patent Sho-59-45852, Kikkoman Corporation.

Asao, Y., and Yokotsuka, T. (1963). Studies on flavorous substances in shoyu. Part 24. α-Diketon compound, 4. *J. Agric. Chem. Soc. Jpn.* *37*(10), 569-575.

Barnett, J. A., Payne, R. W., and Yarrow, D. (1983). "Yeasts, Characteristics and Identification." Cambridge Univ. Press, London and New York.

Bo, T.-A. (1982). Origin of chiang and chiang-yu and their technology of production. *J. Brew. Soc. Jpn.* *77*(6), 365-444.

Burton, H. S., McWeeny, R. M., Lundin, R. E., Guadugni, D. G., and Ling, L. C. (1963). Characterization of an important aroma component of bell peppers. *Chem. Ind. (London)*, p. 490.

Fujimoto, H., Aiba, T., and Goan, M. (1980). Studies on lactic acid bacteria in shoyu mash. Part 3. Effect of temperature on the growth and fermentation of lactic acid bacteria. *J. Jpn. Soy Sauce Res. Inst.* *6*(1), 5-9.

Furuta, T., and Okara, K. (1954). The effect of iron-salts on the color tone and intensity of shoyu. *Seas. Sci.* *2*(2), 6-12.

Hanaoka, Y. (1964). Studies on preservation of shoyu. II. On the antifungal action of shoyu. *J. Ferment. Technol.* *42*(9), 553-564.

Hanaoka, Y. (1976). Microbial quality change of shoyu and its prevention. *J. Jpn. Soy Sauce Res. Inst.* *2*(5), 221-226.

Hashiba, H. (1971). Studies on browning compound in shoyu (soy sauce). *J. Agric. Chem. Soc. Jpn.* *45*(1), 29-35.

Hashiba, H. (1973a). Studies on melanoidin from soy sauce. *J. Agric. Chem. Soc. Jpn.* *47*(11), 727-732.

Hashiba, H. (1973b). Non-enzymatic browning of soy sauce. Use of ion exchange resins to identify types of compounds involved in oxidative browning. *Agric. Biol. Chem.* *37*(4), 871-877.

Hashiba, H. (1973c). Non-enzymatic browning of soy sauce decolorized by ultrafiltration. *J. Agric. Chem. Soc. Jpn.* *47*(4), 269-291.

Hashiba, H. (1974). Effect of aging on the oxidative browning of sugar-amino acid model systems. *Agric. Biol. Chem.* *38*(3), 551-555.

Hashiba, H. (1975). A glucose-diglycine condensation product participating in oxygen-dependent browning. *J. Agric. Food Chem.* *23*, 539-542.

Hashiba, H. (1976). Participation of Amadori rearrangement products and carbonyl compounds in oxygen-dependent browning of soy sauce. *J. Agric. Food Chem.* *24*(1), 70-73.
Hashiba, H. (1977). Oxidative browning of shoyu. *Kagaku to Seibutsu* *15*(3), 156-158.
Hashiba, H. (1978). Isolation and identification of Amadori compounds from soy sauce. *Agric. Biol. Chem.* *42*(4), 763-768.
Hashiba, H. (1981a). Oxidative browning of soy sauce. III. Effect of organic acids, peptides and pentose. *J. Jpn. Soy Sauce Res. Inst.* *7*(1), 19-23.
Hashiba, H. (1981b). IV. Oxidative browning of Amadori compounds. V. Isolation and identification of Amadori compounds from soy sauce. *J. Jpn. Soy Sauce Res. Inst.* *7*(3), 116-120, 121-124.
Hashiba, H., and Abe, K. (1984). Oxidative browning and its control with the emphasis on shoyu. *Proc. Annu. Meet. Agric. Chem. Soc. Jpn.* April 4, p. 708.
Hashiba, H., Koshiyama, I., Sakaguchi, K., and Iguchi, N. (1970). Effects of iron and some trace metal ions in shoyu on the increase of color of shoyu. *J. Agric. Chem. Soc. Jpn.* *44*(7), 312-316.
Hashiba, H., Koshiyama, I., and Fukushima, D. (1977). Oxidative browning of Amadori compounds from amino acids and peptides. *In* "Protein Crosslinking-B. Nutritional and Medical Consequences" (M. Friedman, ed.), pp. 419-448. Plenum, New York.
Hashimoto, H., and Yokotsuka, T. (1972). Studies on sediment of shoyu. IV. Sediment forming activity in fermented raw shoyu. *J. Ferment. Technol.* *50*(4), 257-263.
Hashimoto, H., and Yokotsuka, T. (1973). The significant effect of thermostable acid protease on sediment formation of shoyu. *J. Ferment. Technol.* *51*(9), 661-669.
Hashimoto, H., and Yokotsuka, T. (1974). Mechanisms of sediment formation during the heating of raw shoyu. *J. Ferment. Technol.* *52*(5), 328-334.
Hashimoto, H., Yoshida, H., and Yokotsuka, T. (1970a). Removal of iron from soy sauce. *J. Ferment. Technol.* *48*(1), 50-55.
Hashimoto, H., Yoshida, H., and Yokotsuka, T. (1970b). Studies on sediment of shoyu. I. Factors influencing the formation of secondary sediment of shoyu. *J. Ferment. Technol.* *48*(8), 493-500.
Hashimoto, H., Yoshida, H., and Yokotsuka, T. (1970c). Studies on sediment of shoyu. II. Effect of chelating agents and surfactants on the prevention of secondary sediment of shoyu. *J. Ferment. Technol.* *48*(8), 501-505.

Hashimoto, H., Yoshida, H., and Yokotsuka, T. (1970d). Studies on sediment of shoyu. III. Effect of absorbants and precipitants of secondary sediment of shoyu. *J. Ferment. Technol. 48*(10), 622-627.

Hashimoto, H., Yoshida, H., and Yokotsuka, T. (1971). Studies on sediment of shoyu. V. Composition of sediment formed by heating raw shoyu and isolation of precursors of sediment. *J. Ferment. Technol. 49*(7), 642-649.

Hodge, J. E. (1953). Chemistry of browning reactions in model system. *J. Agric. Food Chem. 1*, 928-943.

Imabara, H., and Nakahama, T. (1968). Studies on pellicle producing soy sauce yeast. VIII. On the flavor substances. *J. Ferment. Technol. 46*, 876-884.

Imabara, H., and Nakahama, T. (1971). Studies on pellicle producing yeast. X. Relationship between the production of aroma causing substances and the characteristics of pellicle-formation in Hansenula and Pichia yeasts. *J. Ferment. Technol. 49*(8), 661-667.

Kamata, H., and Sakurai, Y. (1964). Color formation of soy sauce. *Seas. Sci. 11*(2), 21-26.

Kanbe, C., and Uchida, K. (1984). Selection of shoyu-lactobacilli having strong reducing power and their application for the stabilization of shoyu color. *Proc. Symp. Brew. Jpn. Assoc. Brew. 16th, 1984*, pp. 37-40.

Kato, H. (1956). Studies on browning reactions between sugars and amino compounds. I. pH dependency of browning interactions between various kinds of reducing sugar and amino compound. *Bull. Agric. Chem. Soc. Jpn. 20*, 273-278.

Kato, H. (1958). Studies on browning reactions between sugars and amino acids. III. Identification of red pigments and furfural produced from aromatic amino-N-xylosides. *Bull. Agric. Chem. Soc. Jpn. 22*(2), 85-91.

Kato, H. (1959). Studies on the browning reactions between sugars and amino acids. IV. Identification of 5-hydroxy-hexosides and its role in the browning reaction. *Bull. Agric. Chem. Soc. Jpn. 23*(6), 551-554.

Kato, H. (1960). Studies on browning reactions between sugars and amino acids. V. Isolation and characterization of new carbonyl compounds, 3-deoxy-osones formed from N-glycosides and their significance for browning reaction. *Bull. Agric. Chem. Soc. Jpn. 24*, 1-12.

Kato, H., and Sakurai, Y. (1962). Studies on the mechanisms of browning of soybean products. II. The role of 3-deoxy-osones, hexose and pentose in the color increase of pasteurized shoyu. *J. Agric. Chem. Soc. Jpn. 36*, 131-137.

Kato, H., and Sakurai, Y. (1963). Studies on the mechanisms of browning of soybean products. III. Meaning of 3-deoxy-osones and furfurals in the browning of pasteurized shoyu. *J. Agric. Chem. Soc. Jpn. 37*, 423-425.
Kato, H., Yamada, Y., Isaka, K., and Sakurai, Y. (1961). Studies on the mechanisms of browning of soybean products. I. Isolation and identification of 3-deoxy-D-glucosone for shoyu and miso. *J. Agric. Chem. Soc. Jpn. 35*, 412-415.
Kurono, K., and Katsume, H. (1927). Chemical composition of shoyu pigments. *J. Agric. Chem. Soc. Jpn. 3*, 594-613.
Lodder, J., and Kregar van Rij, N. J. W. (1952). "The Yeasts: A Taxonomic Study." North-Holland Publ., Amsterdam.
Mitsui, M., and Kusabe, E. (1957). Studies on shoyu color. II. *J. Brew. Soc. Jpn. 52*, 816-822.
Moriguchi, S., and Ohara, H. (1961). Studies on the effect of sulfite applied to the treatment of raw material for light-colored shoyu. 1. Determination of the kind and amount of sulfite for use, and small scale brewing test therewith. *J. Ferment. Technol. 39*, 283-288.
Motai, H. (1976). Browning of shoyu. *Nippon Shokuhin Kogyo Gakkaishi 23*(8), 372-384.
Motai, H., and Inoue, S. (1974a). Conversion of color components of melanoidin produced from glycine-xylose system. *Agric. Biol. Chem. 38*(2), 233-239.
Motai, H., and Inoue, S. (1974b). Oxidative browning in color of shoyu. Studies on color of shoyu. Part 2. *J. Agric. Chem. Soc. Jpn. 48*, 329-336.
Motai, H., Inoue, S., and Nishizawa, Y. (1972). Studies on color of shoyu. Part 1. Characteristics of shoyu color and separation of color components. *J. Agric. Chem. Soc. Jpn. 46*(12), 631-637.
Motai, H., Inoue, S., and Hanaoka, Y. (1975). The contribution of soybean and wheat on the color formation of shoyu. *Nippon Shokuhin Kogyo Gakkaishi 22*, 494-500.
Motai, H., Ishiyama, T., and Matsumoto, E. (1981). Physicochemical analysis of coagulation and precipitation of shoyu sediment formed by heating. *J. Agric. Chem. Soc. Jpn. 55*(4), 325-331.
Motai, H., Hayashi, K., Ishiyama, T., and Sonehara, T. (1983). Role of main enzyme proteins of koji-mold *Aspergillus* in sedimentation of coagula in soy sauce during pasteurization. *J. Agric. Chem. Soc. Jpn. 57*(1), 27-36.
Nakadai, T. (1984). Statistical estimation of key-enzymes determining the amount of coagula during the heating of soy sauce production. *J. Agric. Chem. Soc. Jpn. 58*(6), 533-557.

Noda, Y., Mine, H., and Nakano, M. (1982). Effect of keeping warm of raw shoyu or shoyu mash on the prevention of sediment produced by heating. *J. Jpn. Soy Sauce Res. Inst.* *8*(4), 176-181.

Nunomura, N., Mori, S., Sasaki, M., and Motai, H. (1983). Studies on flavor compounds of shoyu. Part 12. The effect of heating on the flavor compounds. *Proc. Annu. Meet.--Agric. Chem. Soc. Jpn.* , p. 237.

Okuhara, A. (1973). Browning of shoyu. *Kagaku to Seibutsu* *10*(6), 383-390.

Okuhara, A., and Saito, N. (1970). Color of soy sauce. III. Effect of reducing reagents on brewing of soy sauce. Color of soy sauce. IV. Decolorization and depression of browning by reduction. *J. Ferment. Technol.* *49*(3), 177-189, 190-201.

Okuhara, A., Nakajima, T., Tanaka, T., Saito, N., and Yokotsuka, T. (1969). Color of soy sauce. II. Stability of color. *J. Ferment. Technol.* *43*(1), 57-68.

Okuhara, A., Tanaka, T., Saito, N., and Yokotsuka, T. (1970). Color of soy sauce. V. Application of multiple correlation analysis on browning mechanisms. *J. Ferment. Technol.* *48*(4), 228-236.

Okuhara, A., Saito, N., and Yokotsuka, T. (1971). Color of soy sauce. VI. The effect of peptides on browning. *J. Ferment. Technol.* *49*(3), 272-287.

Okuhara, A., Saito, N., and Yokotsuka, T. (1972). Color of soy sauce. VII. Oxidative browning. *J. Ferment. Technol.* *50*(4), 264-272.

Okuhara, A., Saeki, M., and Sasaki, T. (1975). Browning of soy sauce. *J. Jpn. Soy Sauce Res. Inst.* *1*(4), 185-189.

Okuhara, A., Nakajima, T., Mori, S., and Nagahori, T. (1977). Studies on color standard of shoyu. *J. Jpn. Soy Sauce Res. Inst.* *3*(5), 194-197.

Omata, S., and Nakagawa, Y. (1955a). On the color of soy. Part 3. On the substances increasing color intensity of soy sauce. *J. Agric. Chem. Soc. Jpn.* *29*, 165-168.

Omata, S., and Nakagawa, Y. (1955b). On the color of soy. Part 4. On the effect of furfural on the color of soy sauce. *J. Agric. Chem. Soc. Jpn.* *29*, 215-219.

Omata, S., and Ueno, T. (1953a). On the color of shoyu. Part 1. Color of shoyu and its characteristics for the active charcol. *J. Agric. Chem. Soc. Jpn.* *27*, 570-575.

Omata, S., and Ueno, T. (1953b). On the color of shoyu. Part 2. Factors causing the color change during storage. *J. Agric. Chem. Soc. Jpn.* *27*, 575-580.

Omata, S., Ueno, T., and Nakagawa, Y. (1955a). On the color of soy sauce. Part 5. Effect of reductones on the color of soy sauce. *J. Agric. Chem. Soc. Jpn.* *29*, 251-256.

Omata, S., Ueno, T., and Nakagawa, Y. (1955b). On the color of soy sauce. Part 6. On the soy sauce pigment. *J. Agric. Chem. Soc. Jpn. 29*, 256-260.
Onishi, T. (1970). Changes of aroma and color during process of soy sauce. *Seas. Sci. 17*(4), 1-5, 47-53.
Saiki, M. (1978). Plastics as the material for the containers of liquid seasonings. *Shokuhin Kogyo 11*(2), 27-36.
Saiki, M. (1979). Container for shoyu. *J. Ferment. Assoc. 74*(6), 365-371.
Sakaguchi, K. (1979). "Search for the Route of Shoyu," Vol. 1, pp. 252-266.
Sasaki, M., and Nunomura, N. (1979). Studies on flavor compounds of shoyu. VII. Lower boiling fractions. *Proc. Annu. Meet.--Agric. Chem. Soc. Jpn.*, p. 31.
Sasaki, M., and Nunomura, N. (1981). Flavor components of top note of shoyu. *J. Chem. Soc. Jpn.* No 5, pp. 736-745.
Sasaki, M., Nunomura, N., and Yokotsuka, T. (1980). Studies on flavor compounds of shoyu. IX. Quantitative analysis of shoyu flavor. *Proc. Annu. Meet.--Agric. Chem. Soc. Jpn.*, p. 282.
Sasaki, S., Katsuta, S., Uchida, K., Chiba, H., and Yoshino, H. (1975). The studies on "N" materials in soy sauce. Part I. *J. Jpn. Soy Sauce Res. Inst. 1*(2), 97-103.
Sasaki, S., Uchida, K., and Yoshino, H. (1979). The studies on "N" substances in soy sauce. Part II. Composition and properties of "N" materials in soy sauce. *J. Agric. Chem. Soc. Jpn. 53*(8), 255-260.
Sawa, T., and Nishikawa, M. (1981). Studies on variation of quality in soy sauce. Part 2. Coloration of soy sauce in large vessels on various conditions. *Jpn. Soy Sauce Res. Inst.* 7(5), 212-215.
Sawa, T., Yamasaki, S., Tanaka, Y., and Nishikawa, M. (1978). Studies on variation of quality in soy sauce. Part 1. Coloration of soy sauce during the storage and spending on various conditions. *J. Jpn. Soy Sauce Res. Inst. 4* (3), 97-100.
Sekine, H. (1972). Neutral protease I and II of *Aspergillus sojae*. Isolation in homogeneous form, and some enzymatic properties. *Agric. Biol. Chem. 36*(2), 198-206, 207-216.
Shikata, H., Okuno, T., and Moriguchi, S. (1971a). Color of Usukuchi soy sauce. I. Browning reaction by reducing sugars and amino acids. *Seas. Sci. 18*(4), 28-39.
Shikata, H., Omori, K., Okuno, T., and Moriguchi, S. (1971b). Color of Usukuchi soy sauce. II. Browning reaction of soy sauce mash. *Seas. Sci. 18*(5), 149-162.

Takakusa, H., Shikata, H., Omori, K., Nishiyama, T., and Moriguchi, S. (1975). Studies on behavior of soy sauce and antiseptic effect of soy sauce. *J. Jpn. Soy Sauce Res. Inst. 1*(4), 190-194.

Tamura, J., and Aiba, T. (1982). The studies on N-material in soy sauce. Part 1. Purification and some properties of N-material. *J. Jpn. Soy Sauce Res. Inst. 8*(5), 196-201.

Tsukiyama, R., Sasabe, K., and Date, S. (1981). Studies on the variation of quality of soy sauce. Part 3. Changes of several components during storage of light color soy sauce. *J. Jpn. Soy Sauce Res. Inst.* 7(5), 216-220.

Ueda, R., and Tanitaka, S. (1977). The quantitative changes of organic acid and amino acid in the course of pasteurization of shoyu. *J. Jpn. Soy Sauce Res. Inst. 3*(1), 19-23.

Ujiie, F., and Yokoyama, K. (1956). The bactericidal nature of fermented shoyu for the intestinal pathogenic bacteria. *J. Jpn. Food Hyg. Assoc. 6*(4), 1.

Umeda, I., and Saito, N. (1956). On the standard color of shoyu. *Seas. Sci. 1*(1), 29-40.

Wang, H. L., and Fuang, S. F. (1981). History of Chinese fermented foods. *Misc. Publ.--U.S., Dept. Agric. FL-MS-33.*

Yamamoto, Y., Moriya, T., and Tsujihara, T. (1978). Studies on sterilize ability in seasoning foods. Part 1. Sterilize ability of compositions of soy sauce. No. 1. *J. Jpn. Soy Sauce Res. Inst. 4*(3), 101-104.

Yasuda, A., Mogi, K., and Yokotsuka, T. (1972). Studies on the cooking method of proteinous materials for brewing. Part 1. High temperature and short cooking method. *Seas. Sci. 20*(7), 20-24.

Yokotsuka, T. (1954). Studies on flavorous substances in shoyu. Part 13. On the yeast-static substance in brewed soy and their precursor in the heated soy. *J. Agric. Chem. Soc. Jpn. 28*, 114-118.

Yokotsuka, T. (1961). Aroma and flavor of Japanese soy sauce. *Adv. Food Res. 10*, 75-134.

Yokotsuka, T. (1980). Recent advances in shoyu research. *Proc. Int. Symp. Recent Adv. Food Sci. Technol. 1980*, pp. 1-22.

Yokotsuka, T. (1981a). Recent advances in shoyu research. *In* "Quality of Foods and Beverages: Chemistry and Technology" (G. Charalambous and G. Inglett, eds.), Vol. 2, pp. 171-196. Academic Press, New York.

Yokotsuka, T. (1981b). Application of proteinous fermented foods. *In* "Traditional Food Fermentation as Industrial

Resources in ASCA Countries" (S. Saono, F. G. Winarno, and D. Karyadi, eds.), pp. 145-180. Indonesian Institute of Science, Jakarta.

Yokotsuka, T. (1981c). Risks of mycotoxin in fermented foods. *In* "Advances in Biotechnology" (M. Moo-Young, ed.), Vol. 2, pp. 461-466. Pergamon, Toronto.

Yokotsuka, T. (1983). Scale up of traditional fermentation technology. *Korean J. Appl. Microbiol. Bioeng.* *11*(4), 353-371.

Yokotsuka, T., and Takimoto, K. (1956). Studies on flavorous substances in shoyu. Part 14. Flavorous substances in heated shoyu. 1. *J. Agric. Chem. Soc. Jpn.* *30*(2), 66-71.

Yokotsuka, T., and Takimoto, K. (1958). Studies of flavorous substances in shoyu. Part 15. Flavor substances in heated shoyu. 2. *J. Agric. Chem. Soc. Jpn.* *32*(1), 23-26.

Yokotsuka, T., Mogi, K., Fukushima, D., and Yasuda, A. (1966). Dealing method of proteinous raw materials for brewing. Japanese Patent 929910, Kikkoman Shoyu Co., Ltd.

Yokotsuka, T., Yoshida, H., and Hashimoto, H. (1971a). Production method of shoyu. Japanese Patent, announced Nov. 17, 1971.

Yokotsuka, T., Yoshida, H., Saito, N., Okuhara, A., Hashimoto, H., and Nakajima, T. (1971b). Production method of shoyu. Japanese Patent 628109, Kikkoman Corporation.

Yokotsuka, T., Yoshida, H., and Hashimoto, H. (1971c). Production method of shoyu. Japanese Patent 646002, Kikkoman Corporation.

CHAPTER 11

SHELF-LIFE OF MILK

WILLIAM W. MENZ
Winston-Salem, North Carolina

Handbook of Food and Beverage
Stability: Chemical, Biochemical,
Microbiological, and Nutritional Aspects

I. INTRODUCTION

Milk and milk products have been a most important part of the North American diet since 1611, when the first cows were brought to Virginia. The dairy industry has grown so that today it provides U.S. consumers with a wide variety of milk and dairy products.

Milk from all kinds of animals is the basis for a standard food in different parts of the world. In the United States, however, the cow alone furnishes almost all of the milk consumed. Some goats' milk is being produced and consumed, but this chapter considers cows' milk alone.

Before delving into the subject of the shelf-life of milk, let us look at some statistics about the considerable annual consumption of milk and dairy products in the United States.

A. U.S. Milk Production

In 1983, the latest year for which complete figures are available from the U.S. Department of Agriculture (USDA), a record of 137.6 billion lb of milk was produced and marketed by U.S. dairy farmers, with an additional 2.3 billion lb of milk used on the farm (1). Milk production of the individual cow has increased over the years, resulting in an increase per cow from 3000 lb annually in 1835, to an average of ∼12,000 lb/year in 1983. Top producers have given up to 28,000 lb/year in a number of cases.

Production and consumption figures in the United States are given in pounds by the USDA, and all figures for milk and dairy products quoted in this chapter follow this example.

Table I gives total U.S. milk and dairy products sales for the period 1976-1983.

B. Conversion of Milk into Various Products

One quart of milk weighs 2.15 lb, a gallon 8.6 lb. Specific gravity of milk at 16°C is 1.032. The quantity of milk used to produce 1 lb of a specific product depends mainly on the initial fat content of the milk, and this fat content varies with different cattle breeds, with different locations of the country, and, to some extent, with the season. Table II gives an average by which the industry is guided (2).

At the outset it should be mentioned that milk is among the most perishable of all foods because the excellent nutritive composition of fluid milk provides an ideal medium for bacterial growth. It may undergo many flavor changes unless constantly protected against contamination and adverse environmental conditions.

Let us identify many of the dairy products discussed in the following sections.

1. Fluid Milk

Milk is defined by the Code of Federal Regulations (CFR), Title 21, as the lacteal secretion, practically free from colostrum, obtained by the complete milking of one or more cows. Depending on the milk fat content, the fresh fluid product is "milk," "low-fat milk," or "skim milk," and is so labeled. Milk (i.e., whole milk) must have a minimum of 3.25% milk fat and not less than 8.25% solids not fat (3,4). Table III (5) gives the general composition of milk.

Low-fat milk, from which a sufficient amount of milk fat has been removed, has a milk fat content ranging from 2% down to 0.5% and must have not less than 8.25% milk solids not fat (3,4). Skim milk has a fat content of <0.5%, and the milk solids not fat must equal at least 8.25% (3,4).

Milk solids not fat include protein, carbohydrate, water-soluble vitamins, and minerals, whereas the total solids content of milk includes all the constituents of milk except water.

All fluid milk products shipped in interstate commerce are pasteurized, ultrapasteurized, or ultrahigh-temperature (UHT) processed before being sold for beverage use. Declaration of the term "pasteurized" on the container is optional, whereas the term ultrapasteurized or UHT is mandatory when applicable. Some states permit the sale of "certified raw milk" within the state's borders. Pasteurization is the heating of raw milk at sufficiently high temperature for a specific length of time to destroy pathogenic bacteria. Pasteurization also destroys yeast, molds, and 95-99% of all other nonpathogenic bacteria. It inactivates most enzymes that might cause spoilage through the development of off-flavors. Thus pasteurization makes milk bacteriologically safe and increases its keeping quality (6,7).

2. Flavored Milk

Flavors, with or without coloring and nutritive sweeteners, may be added to a specific standardized milk. The caloric and carbohydrate value will vary according to the milk fat content and the amount and composition of the added ingredients. Protein, mineral, and vitamin content of flavored milk will approximate that of the milk to which flavoring has been added (3,4).

3. Evaporated Milk

Evaporated milk is made by preheating milk to stabilize proteins, concentrating it in vacuum pans at 50 to 55°C to remove about 60% of the water. Then the product is homogenized and standardized to the required percentages of components, and vitamins are added. Finally it is sealed into containers and sterilized by heating. Currently the product is UHT treated and aseptically packaged. Federal standards require that the milk fat and the total milk solids be at least 7.5 and 25%, respectively. Evaporated milk must have vitamin D added to provide 25 IU/fluid oz. The addition of vitamin A is optional (3,4).

4. Sweetened Condensed Milk

Sweetened condensed milk is similar to evaporated milk except that a safe and suitable sweetener such as sucrose is added. The amount of sweetener added (40-45%) is sufficient to prevent spoilage.

Federal standards require that condensed milk contain not less than 8% by weight of milk fat and not less than 28% by weight of total milk solids (4,5).

5. Dry Whole Milk

Dry milk is the product resulting from the removal of water from milk, and it contains not less than 26% milk fat and not more than 4% moisture (3,4).

6. Nonfat Dry Milk

Nonfat dry milk is the product resulting from the removal of fat and water from milk, and it contains the lactose, milk proteins, and milk minerals in the same relative proportions as in the fresh milk from which it is made. The fat content is not over 1.5% by weight unless otherwise indicated, and it contains not over 5% by weight of moisture (3,4).

7. Dry Buttermilk

Dry buttermilk is the product resulting from the removal of water from liquid buttermilk derived in the manufacture of butter. It contains not less than 4.5% milk fat and not more than 5% moisture (3,4).

8. Butter

In most countries, including the United States, butter must legally contain at most 16% by weight of water and at least 80% by weight of milk fat. Butter may either be salted or unsalted; salted butter contains about 0.6 to 2% salt.

9. Cream Products

Cream is the liquid milk product, high in fat, separated from milk and adjusted through the addition of various milk products and containing not less than 18% milk fat. Several cream products identified by federal standards are as follows (3,4,6,7):

Half-and-half: not less than 10.5% but less than 18% milk fat
Light cream: not less than 18% but less than 30% milk fat
Light whipping cream: not less than 30% but less than 36% milk fat
Heavy cream: not less than 36% milk fat

All cream products are pasteurized or ultrapasteurized and homogenized.

Dry cream is obtained by removal of water only from pasteurized milk and/or cream; homogenization is optional. It contains not less than 40% but less than 75% milk fat and not more than 5% moisture.

10. Raw Certified Milk

The certification of raw milk originated in 1893 when dairy facilities had not yet achieved the sanitary excellence that now exists in the United States. Some very few states still allow the intrastate sale of certified raw milk, the production of which is under strict control of the respective state milk control agencies.

11. Acidophilus Milk

Skim, low-fat, or whole milk products containing *Lactobacillus acidophilus* (>500 million units/ml) have been prepared and are available on the market. The culture remains inactive within milk during storage at refrigeration temperature but is activated upon consumption through the heat generated by the intestinal tract. The natural flavor and the consistency of the milk to which the culture was added are unchanged (8).

12. Acidified and Cultured Milk and Cream Products (9,10)

These products are known as sour cream or cultured sour cream, acidified sour cream, sour half-and-half or cultured sour half-and-half, acidified sour half-and-half, and the various types of yogurt including regular, low fat, and nonfat.

a. Sour Cream and Cultured Sour Cream. Sour cream contains not less than 18% milk fat, pasteurized or ultrapasteurized, homogenized cream cultured with *Streptococcus lactis* at 22°C until the acidity is $\geq 0.5\%$, calculated as lactic acid. Sour cream is usually a straight fermentation product, although rennet extract may be added to produce a thicker product.

b. Acidified Sour Cream. This product is like regular sour cream except that it is soured with safe and suitable acidifiers, with or without the addition of lactic acid-producing bacteria.

c. Sour Half-and-Half and Cultured Product. Containing not less than 10.5% milk fat but less than 18%, as well as lactic acid-producing bacteria, this product has a titratable acidity, expressed as lactic acid, of not less than 0.5%.

d. Acidified Sour Half-and-Half. Here half-and-half is soured with a suitable and safe acidifier. The product has the same milk fat content and titratable acidity as sour half-and-half.

e. Yogurts. Yogurt can be manufactured from fresh whole, low-fat, or skim milk. Milk solids not fat are often added to the product. Fermentation is usually accomplished by a one-to-one mixed culture of *Lactobacillus bulgaricus* and *Streptococcus thermophilus*. The milk is pasteurized or ultrapasteurized, and homogenization is optional. The federal standards specify that yogurt contain not less than 3.25% milk fat, low-fat yogurt 0.5-2%, and nonfat yogurt not more than 0.5%. All three types must contain not less than 8.25% milk solids not fat and have a titratable acidity of not less than 0.9%. The three styles of yogurt are (1) sundae style, with fruit at the bottom of the container, (2) blended style (Swiss, French, etc.), with fruit or flavors blended throughout the yogurt, and (3) unflavored.

13. Frozen Dairy Products (11,12)

The principal frozen dairy products are ice cream, frozen custard, ice milk, sherberts, water ices, and mellorine. These products may be hard or soft frozen; hard-frozen products are packaged in bulk in different-sized containers or molded into various novelties, while soft-frozen products are served directly from the freezer.

The Food and Drug Administration (FDA) in Title 21, Part 135 of the CFR, defines ice cream as a food produced by freezing, while stirring, a pasteurized mix consisting of a number of optional but specific dairy ingredients, along with certain optional caseinates used only under specific conditions, a nutritive carbohydrate sweetener, stabilizing and emulsifying agents, and flavoring agents. The inclusion of whey solids is also permitted. Federal standard requirements for frozen products are given in Table IV (13). The various amounts of components outlined in the table for a given frozen product (e.g., ice cream, ice milk) indicates its quality rating, from regular to premium product.

There are no federal standards for frozen yogurt. In composition frozen yogurts compare most closely to ice milk.

Another product for which there is currently no federal standard is a freezer-dispensed milk shake. It is made from an ice milk-type mix of slightly different composition.

14. Whey

Whey is the watery by-product in cheese manufacture and represents ~90% of the original milk volume used in cheese manufacture. It contains over half of the original milk nutrients. The >43 billion lb of whey produced in 1983 (6.5% average solids content) represent almost 3 billion lb of nutrients, primarily lactose and protein. Composition of whey is given in Table V.

15. Lactose

Lactose is the predominant carbohydrate in milk and whey, and accounts for about 54% of the total solids-not-fat content of milk and more than 70% of the solids content of whey.

16. Casein

Casein, the major protein of milk, can be isolated from skimmed cows' milk by a variety of commercial processes and finds considerable application in the manufacture of not only foods, including imitation food products and animal feed, but also in nonfood industrial applications. The casein portion constitutes $\sim$3% of fluid milk and is distinguished from the "whey proteins" (0.6%) by their insolubility and tendency to precipitate and coagulate at the isoelectric point (pH 4.5).

II. PHYSICAL PROPERTIES OF MILK

Before proceeding with the shelf-life characteristics of fluid milk and some related dairy products, it is important to survey some of the general physical properties of fluid milk. Table VI outlines these properties.

III. PROTECTING THE QUALITY OF MILK

Of utmost importance in protecting the quality of milk and dairy products is the fact that the base material--namely, the milk furnished by the cow--is a quality product, free from impurities and excessive bacterial contamination. Milk and dairy products are among the most perishable foods because of their excellent nutritive composition and fluid form. Cows' milk provides an ideal medium for bacterial and enzyme growth. If it is not kept according to scientifically determined measures against contamination and adverse environmental conditions, milk will undergo many undesirable flavor changes shortly after production; these changes will eventually cause the milk to become "rancid." The many nutrients in milk, beneficial to human growth and health, must be retained throughout the production, processing, transportation, and distribution of the product.

Dairy and public health technology have progressed to such an extent that consumers can today rely on the safety, nutritional value, and aesthetic appeal of milk and dairy products, even if the point of consumption is hundreds and thousands of miles away from the point of production. Among the most important reasons for this reliability are the many federal, state, and local regulations controlling the production and manufacturing of milk and dairy products.

A. Grade A Pasteurized Milk Ordinance (PMO)

The most important aspect of this regulation is that it is the most effective tool in protecting the quality of the total U.S. milk supply and is based on recommendations made by the FDA, which is part of the U.S. Public Health Service. The PMO stipulates the necessary steps in producing, processing, and transporting the U.S. milk supply and involves complete cooperation on the part of sanitary and regulatory officials at various levels of government and all segments of the dairy industry, including educational and research organizations.

In many cases the requirements of state and local governments are even more stringent than the guidelines of the PMO, which in itself assures a milk considered by many to be the most quality controlled in the world.

To attain this high quality, sanitary practices as outlined in the PMO include the following:

1. Inspection and sanitary control of farms and milk plants
2. Examination and testing of herds to eliminate bovine diseases related to public health, such as mastitis and brucellosis
3. Regular instruction and necessary sanitary practices for personnel engaged in all phases of handling milk
4. Proper processing of milk such as pasteurization, ultrapasteurization, and UHT treatment
5. Laboratory examination and monitoring of milk for impurities such as pesticides and antibiotics

Grade A milk is obtained, following the above outline, from dairy farms that conform to all requirements of the PMO or its equivalent as enforced by local and state authorities. The milk must be obtained from healthy cows tested and found to be free of disease and disease-producing organisms. The raw milk, immediately after being obtained from the cow, must be cooled to 5°C and maintained at no higher temperature from the completion of milking until processing at a dairy plant. At the dairy plant it is analyzed to ensure that it contains not more than a specified bacterial count. Many dairy processing plants check the milk before unloading for the presence of antibiotics and to certify that antibiotics, if present, do not exceed the actionable level of 0.01 IU/ml or 6 parts per billion (ppb) when considering penicillin contamination, based on sodium benzylpenicillin, which has an activity of 1670 IU/mg. Converted into parts per billion, the above ratio would be equal to 6 ppb. The FDA stipulates that the tolerance for antibiotics in milk is zero, but considering that analytical methods are not able to handle zero degree of contamination, the authorities are willing to tolerate an actionable level of 0.01 IU/ml. Incidentally, few methods are available that are sensitive enough to measure such low levels; the only completely dependable method are the Delvotest P, which takes about 2.5 hr to complete, or the Charm test (by Penicillin Assays, Inc.), which can be completed within 15 min. Both tests can measure at a level <0.01 IU/ml, namely at 0.005 IU/ml.

The reason for considering in some detail the antibiotic contamination of milk is that the presence of antibiotics at levels higher than 0.01 IU/ml has a definite effect on the shelf-life of milk and particularly on the making of cultured dairy products by delaying acid production and leading to abnormal fermentation in the cultured dairy product. This abnormal acid production in itself has no particular effect on the keeping quality of the product, but it might cause some flavor deterioration.

A number of our own research investigations established approximate levels of antibiotics in milk for the inhibition of various culture organisms. The concentration of penicillin required to inhibit pure and mixed cultures is somewhere between 0.002 and about 0.6 IU/ml, a considerable spread depending on the type of culture used. It can be concluded from this that in cheese making a penicillin concentration of 0.002 IU/ml in milk will delay acid production and lead to abnormal fermentation in the cheese.

Another of our research investigations has shown that penicillin concentrations of between 0.05 and 0.06 IU/ml inhibit culture growth. While these figures are somewhat higher than those previously reported, the point is still that penicillin concentrations react differently to the particular starter culture utilized in cheese making.

As previously mentioned, antibiotics have no known effect on the keeping quality of milk, but in some instances, they do contribute to the development of off-flavors. The zero-tolerance level was set by the FDA in 1956 in order not to stimulate sensitivity reactions in consumers. Little information is available on how daily intake of small amounts of penicillin might affect the population. A small proportion of the population is susceptible to developing penicillin sensitivity, which can range from mildly transient to potentially serious. However, there are no known cases of persons being affected by consuming milk that might be contaminated. Furthermore, there is no evidence that those not prone to developing sensitivity would be made sensitive by consuming milk containing penicillin.

Of greater importance is the possibility of certain infectious nondairy bacteria developing resistance to penicillin and becoming refractive to penicillin-containing milk, as well as the possibility that dairy foods made from such milk may sometimes contain antibiotic-resistant strains of infectious bacteria.

The presence of other unintentional microconstituents in fluid milk has caused some problems for the dairy industry and public health officials. Radioactive materials, in addition to antibiotics and pesticides, are being detected with ever-increasing precision.

Radioactivity from environmental contamination may reach the population through food supply such as dairy products. The risk of contamination is very minimal and does not warrant any changes in food technology or in food habits. Milk is being used as an analytical agent for obtaining information on current radionuclide concentrations, because milk is convenient to analyze and is produced under many different environmental conditions.

B. Pasteurizing, Ultrapasteurizing, and UHT Processing (4,6,7,17-19)

The milk produced in the United States is considered the world over to be the product containing the least bacteria. Nevertheless, even under the best sanitary conditions, disease-producing organisms may enter the raw milk production process from environmental sources. The PMO details procedures for the proper pasteurization of milk and milk products to safeguard consumer health and to extend the shelf-life of the product. Therefore, pasteurization is mandatory by law for all grade A milk and milk products moved in interstate commerce for retail sale.

Pasteurization is the heating of milk in specially designed equipment at a sufficiently high temperature for a specific length of time to destroy pathogenic bacteria. Pasteurization also destroys yeasts, molds, and 95-99% of all nonpathogenic bacteria. It inactivates most enzymes that might cause spoilage through the development of off-flavors. Therefore, pasteurization makes milk bacteriologically safe and increases its keeping quality without changing the nutritive value to any great degree. None of the main constituents (e.g., proteins, fat, carbohydrate, and minerals) are affected, nor are most of the vitamins, except for a slight loss of vitamin K and about 10% losses for thiamine and vitamin B_{12}. Vitamins A and E are subject to oxidative deterioration but appear to be quite stable in the fluid products. Ascorbic acid in milk may be destroyed depending on the time-temperature relationship during pasteurization.

Ultrapasteurization, a fairly new industrial process, involves thermal processing above 138°C for at least 2-4 sec, depending on the degree of heat involved. The higher the temperature the shorter the holding time. This process assures the destruction of all microorganisms with the exception of some heat-resistant, nonpathogenic spores. After packaging it extends the shelf-life of the product thus packaged under refrigeration without changing its nutritive value. Ultrapasteurization might reduce the whipping property of whipped cream, but the cream can be whipped with great success if a cold and narrow-bottomed container is used (20).

Ultrahigh-temperature (UHT) processing of milk is identical to ultrapasteurization; at its conclusion the milk is immediately packaged in a container that has been sterilized with a hydrogen peroxide spray (19). Fluid products so processed can be kept unrefrigerated for at least 3 months. With complete control over the processing of this type of fluid product, the nutritive value of the milk should not change and it should taste like regular pasteurized milk. Table VII outlines the thermal processing of milk and the common time-temperature ratios.

C. Homogenization (6,7,15)

The purpose of homogenization is to impart into the fluid product a uniform composition and palatability without altering the composition. During homogenization the fat globules are mechanically reduced in size and dispersed in a fine emulsion throughout the fluid milk. The fat globules do not adhere to each other or coalesce and therefore do not rise and form a creamy layer. This process is an optional process, but all milk sold in the United States is homogenized.

The homogenizing process involves pumping the milk under pressure of about 2000 to 2500 pounds per square inch (psi) through the miniscule orfices of a homogenizer. The fat globules are thus reduced from 2 to 6 μm to <2 μm. All homogenized milk must be pasteurized to reduce and inactivate the enzyme lipase, which can cause hydrolytic rancidity of the milk, that is, a considerable deterioration of the product. After homogenization the milk should be kept at all times at refrigeration temperature.

The process of homogenization has a definite influence on the shelf-life of milk in that it improves the keeping quality of the product as well as its nutritive value. It imparts a uniform consistency within the milk and eliminates separation of the cream from the remainder of the fluid body; the curd tension is reduced so that a softer curd is being digested, which aids the digestive process, and there is also less susceptibility to oxidized flavor induced by copper contamination. Aside from the whiter color of the fluid product, the milk proteins are more readily coagulated by heat or acid treatment, so that care must be taken to avoid curdling; finally, homogenization produces increased susceptibility to off-flavors induced by sun or fluorescent light.

All of the processes described in this section lead to a product that must be handled with care. Caution must be exercised to handle the product properly both during processing and at home. The most important factor is the application of the correct temperature when handling and storing the product.

Table VIII gives the approximate storage life of milk products at specific temperatures (4,6,7,17,23)

D. Care of Purchased Milk

Proper handling of the dairy product by the processor is a must, and pull dating, which is mandatory for many states, is designed to guarantee a product with a normal shelf-life. To preserve the quality of milk the consumer also has some definite obligations.

1. Pull Date

Pull date or open dating is the sometimes mandatory but often voluntary inclusion of a date on the milk or dairy product carton, to indicate to the consumer when the milk or dairy product should be withdrawn from sales. It is used by the dairy industry to reflect the age of the individual products on sale. It does not indicate, as is so often mistakenly assumed, the shelf-life of the dairy products. Generally, depending on storage condition in the dairy case of the market and other localities selling dairy items, and subsequent care at home, the products will remain fresh and usable for a few days after expiration of the pull date. Regulations on the affixing of a pull date vary among states and municipalities, and no set pattern is uniformly in force.

In states where enforcement is not mandatory, pull dates of 2 weeks can be observed. This period is sufficient to ensure that products are fresh and pleasant-tasting. In many of these states a milk commission sets the rules and regulations, and usually its members are dairy plant managers.

2. Handling of Dairy Products (6,7,15,17,24,25)

Since dairy products are highly perishable, care must be taken to observe certain practices to preserve quality. The most important practices, often not followed, are as follows:

1. Use containers that protect milk from exposure to sunlight or bright daylight, and most importantly the fluorescent light within a display dairy case. The effect of fluorescent light on milk and some dairy products can contribute to the development of off-flavor, as well as reductions in riboflavin, ascorbic acid, and vitamin B_6 content.
2. Store milk and dairy products in refrigerated temperature of 5°C as soon as possible after purchase.
3. Keep containers for milk and dairy products closed to prevent absorption of other food flavors in the refrigerator. The milk is still fresh even if some other food flavor has been absorbed, as long as all handling directions are followed.
4. Use the dairy products in the order they were purchased, and serve cold.
5. Return the container to the refrigerator as soon as possible to prevent bacterial growth. If dairy products or milk are exposed to temperatures of 5°C for any length of time, the shelf-life has been greatly reduced.

6. In the case of milk and other fluid dairy products, never return unused portions to the original containers.

IV. SHELF-LIFE OF MILK

Almost everyone considers the shelf-life of dairy products as the length of time after processing that the products will retain their quality. However, the shelf-life of milk is greatly influenced by the handling of the product after it is taken from the animal, and even though shelf-life is technically the period between processing and the time at which milk and dairy products become unacceptable to the consumer, the early preprocessing period is equally important. When speaking of the "keeping" quality of milk and dairy products one refers to shelf-life, a term that really has no adequate working definition. Consumable life might be a more descriptive term than shelf-life, but there is still no adequate means of measuring it objectively, much less of predicting it. Since the shelf-life of cheese and related products is discussed in detail in Chapter 2 of this book, we concentrate here on the shelf-life of milk and dairy products other than cheese.

The fluid dairy products--namely the various types of milk--are the most important when considering shelf-life. Many of the well-known dairy products are subject to various manufacturing processes that in themselves add considerably to the shelf-life of the product. Our discussion is directed chiefly toward the fluid product.

Fresh cultured items such as buttermilk, sour cream, yogurt, and cottage cheese have a longer shelf-life than milk if kept at the correct refrigeration temperature because of their acidity. Some manufacturers of these products add small amounts of sorbate (0.1%) which inhibits yeast and mold formation.

The dairy farmer's contribution to the overall shelf-life of milk entails a high-quality milking operation in which recommended practices are followed. One example is the time of feeding as well as the type of feed, both of which have a significant influence on the amount of feed flavor found in milk. In the design of the milking system, including the milk transfer arrangements, it is important to avoid extensive agitation to guard against chemical breakdown of the milk fat. Cleanliness in the production of milk and an efficient cooling system will yield a product of extremely low bacterial count and few flavor defects due to bacterial action.

Dairy farm operations in the United States are considered the most sanitary in the world. The permissible bacteria count is set at 100,000 units/ml, and hardly any U.S. dairy operation even comes close to this count. Only when milking cattle are diseased, often unknown to the farmer, is this figure surpassed; many U.S. operations can boast a count that is only 5% of the permissible 100,000 units/ml.

As mentioned before, the most important factor to consider besides cleanliness is that strict attention be paid to temperature control of the milk from cow to final consumption. Raw milk must be kept at $\leq 5^{\circ}C$ without interruption until processing.

Storage temperature is critical in the shelf-life of milk. Table IX shows the effect of storage temperature on the shelf-life of milk and fluid milk products (26).

On the average it can be assumed that a 5° increase in storage temperature will reduce the shelf-life of milk by 50%. To be explicit, a carton of milk having a shelf-life of 10 days if kept at 40°F, will spoil in 5 days if kept at 45°F or 2.5 days if kept at 50°F.

The consistently sanitary and refrigerated handling of milk during all stages of processing has greatly reduced the influence on spoilage of acid-producing bacteria, but as these methods have improved, spoilage by psychrotrophic bacteria has become important.

It must be recognized that as the technology of processing of milk has improved over the years and made conditions unfavorable for the growth of one group of bacteria, the proliferation of another group might be favored. Psychrotrophic gram-negative populations proliferate in milk held at near 0°C (27), and the most commonly isolated bacteria from processed milk are thermoduric psychrotrophs (28).

A. Keeping Quality of Raw Milk

It is the initial bacterial content of raw milk, collected after 2 to 3 days of storage on the farm, that determines the extent to which quality is maintained during subsequent storage for up to 13 days. When raw milk collected on the farm is cooled to $<5^{\circ}C$ within 15 min of milking, the number of bacteria do not increase if this temperature is maintained for 72 hr. However, in milk that is not cooled immediately after milking and kept up to 1 or 2 hr above the 5°C mark, the number of bacteria do increase even when the milk is kept at $<5^{\circ}C$ for 72 hr (29). Thus the longer the delay before cooling, the more effective is the period of preincubation in promoting bacterial growth. The same investigator also concluded that the increase in the bacterial number is greater when milk inoculated with gram-negative psychrotrophic bacilli is pre-incubated before storage at 5°C than when there is no prein-cubation period. Coliform growth is depressed at low temperatures, while the number of bacteria and the number of psychrotrophs increases. Higher temperatures (i.e., 3-5°C) are favorable to coliform growth (30). Similarly, raw milk stored at >5°C develops a rancid flavor, associated with an increase of fatty acids after 3 to 4 days; storing milk at 10°C causes rancidity after 1 day of storage (31).

B. Keeping Quality of Pasteurized Milk

The keeping quality of cold-stored pasteurized milk is limited by the degree of recontamination with psychrotrophic gram-negative bacteria. If the pasteurized milk is made from a raw product of good microbiological and organoleptic quality, stored in the absence of light, the flavor of the milk will remain quite stable until the bacterial count exceeds 5×10^6 to 20×10^6 cells/ml (32). Until this point the milk has a pleasant taste; therefore, the microbiological keeping time can be defined as the time elapsing from pasteurization and filling until the above defined limit. The keeping quality of the fluid milk depends on (1) the bacterial number of different groups of bacteria present after pasteurization, (2) a possible lag phase, (3) the growth rate, and (4) the bacterial count giving rise to rejection of the product by those in charge of handling or regulating the product.

Each different type of bacteria coexisting in pasteurized milk grows at its own rate. These growth rates can be plotted according to Arrhenius plots (33), which deal with the growth rates of pure cultures in pasteurized milk at various temperatures. In an Arrhenius plot the logarithm of the growth rate is plotted against the reciprocal of the absolute temperature. Proceeding from certain initial numbers of bacteria after pasteurization and filling, it is possible to calculate the keeping quality at various temperatures of storage if the Arrhenius plots are known. The same authors describe a system whereby the effect of varying temperature on the bacterial growth rate can be computed. Such computations can be made equally well using as a basis either the Arrhenius calculations or the even better fitting linear relation between the root of the growth rate and the absolute temperature, which was developed by Ratkowsky and others (34).

Mistry and Kosikowski (35) have developed a time-temperature indicator strip to be used in estimating how much of the shelf-life of the milk product has elapsed or even whether it has ended.

It should be pointed out that when milk is homogenized the growth rates of the bacterial species present in the pasteurized milk do not change. Yet a slightly shorter shelf-life may result from homogenization, in that a slight delay does occur in the flattening of the logarithmic growth phase, compared to that of nonhomogenized milk, during which most bacteria are concentrated in the cream layer (36). A detrimental effect of high bacterial load in raw milk on the keeping quality of pasteurized milk has been described (37). These findings are based on the decline of flavor in relation to the taste of the freshly pasteurized milk.

Pasteurized milk keeps from two to three times longer than does raw milk of similar quality and bacterial count (38). This extended keeping quality is applicable if the product is stored at 0°C. Pasteurized milk kept at that temperature had a shelf-life of from 8 to 12 weeks, and its spoilage, marked by proteolysis, was of bacterial origin. Most of the bacteria were gram-negative pseudomonads. The bacteria that are heat shocked during pasteurization have a longer lag phase of growth, and the higher the storage temperature after the pasteurization the shorter the lag phase--$\sim$7 days at 7°C storage temperature, and longer at 4.5°C (39).

C. Bacterial Spoilage

Bacterial spoilage of pasteurized milk occurs when psychrotrophs contaminate the milk after pasteurization and when thermoduric psychrotrophs survive pasteurization. If unpasteurized milk has a high bacterial count, so will the pasteurized product, and spoilage will occur sooner, according to a number of authors (32,40,41). However, other workers have indicated that the amount of bacteria found in raw milk has no effect on the keeping quality of the pasteurized milk (42-44). The total bacterial count as a function of time and storage was found to be unrelated to the initial bacterial count (45). This means that the colony count in fresh pasteurized milk is of little value in predicting the shelf-life of milk. Some of these authors state that it is not the number of bacteria but their types that determine the shelf-life of milk.

There is very little agreement among authors regarding the sources of bacterial contamination. The basic conclusion to be drawn is that various parameters such as pH, free fatty acidity, and standard plate count coincide most frequently with the occurrence of off-flavors.

Some (46,48) have reported that bacteriological tests such as standard plate count and coliform counts before and after pasteurization are a poor indicator of the keeping quality of milk. On the other hand, flavor scores before and after pasteurization were significantly correlated with keeping quality, but they accounted for only a small proportion of the quality and thus did not measure all the factors determining keeping quality. By all indications the deterioration of pasteurized milk during storage is due to microbial influences, but it may not be due to the growth of bacteria alone. Keeping quality is also influenced by the process of filling and by the type of filling equipment used. The bacterial count was usually higher after 7 to 14 days with piston fillers than with rotary fillers (48).

D. Stored Milk Deterioration

When no postpasteurization contamination is present, off-flavor usually develops in milk after storage for 2 weeks at 7°C. If the bacterial count in the raw milk was high before pasteurization, an off-flavor--usually bitterness--develops in the pasteurized product (49). This effect is observed even with a low bacterial count of the pasteurized milk and with or without postpasteurization contamination.

How does the consumer judge milk as good or bad? Good-tasting milk is characterized as having a pleasing, slightly sweet taste immediately on taking the sample into the mouth, and no unpleasant aftertaste. Such good-tasting milk should be recognized and be used as a standard in judging off-flavors. Good milk quality and flavor begins with the correct feeding of the cow, even before the first drop of milk is produced. The feeding program must restrict the intake of weeds and other roughages that might impart an off-flavor to the milk. The cow's environment must be well maintained during feeding and milking, and the cow itself must at all times be kept in the most sanitary condition. There is no use in considering the shelf-life of milk if the conditions for producing a high-quality product are not met during the initial stages of production.

V. DEFECTS IN SHELF-LIFE OF MILK

Off-flavors are the most obvious shelf-life defects, but how are they identified? The most common off-flavors found in milk fall into three basic categories: absorbed, bacterial, and chemical. Absorbed flavors are identified as feedy, barny, cowy, unclean, and weedy. Bacterial-caused flavors are acid, malty, unclean, and putrid. Chemical flavors are cowy, salty, oxidized, and sunlight caused.

Absorbed flavor defects can occur at any point in the life of the milk. They can develop before, during, and after the milking process, and as a result of improper refrigeration (i.e., leaving milk uncovered in the refrigerator). Bacterial flavor defects usually result from contact of the milk with improperly washed or improperly sanitized milking equipment, external contamination from dirty teats, and improper cooling. Finally, chemical defects can occur both before and after milking. Cowy flavors may result when the animal is suffering from disease. Genetic makeup, type of feed, or stage of lactation can cause certain cows to produce milk that is susceptible to rancidity and oxidation. On the other hand, rancid or oxidized defects can also be induced by poor handling techniques or faults in the equipment that handles the milk after milking. Foreign off-flavors can be caused by medications, a reaction to pesticides, antibiotics, or any number of contaminants.

The classification system given in Table X is the one used in the United States. It identifies off-flavors in milk on the basis of their causes.

A. Lipolysis

Chemical reactions are the major cause of off-flavor. Milk undergoes a number of subtle chemical and physical changes detrimental to the manufacture of a high-quality product. Among these reactions, lipolysis, which can be defined as the enzymatic hydrolysis of milk fat, is perhaps of the greatest economic significance. The accumulation of the reaction products, especially the free fatty acids, is responsible for the common off-flavor developed in milk, frequently referred to as rancidity or, more correctly, hydrolytic rancidity. Lipolysis results from the enzymatic splitting of the milk fat leading to the accumulation of free fatty acids, many of which possess this rancid flavor. In normal milk the fat is surrounded by a membrane and is therefore largely protected from attack by the lipolytic enzymes. If, however, the milk is subjected to unusual handling, such as excessive agitation, the fat globule may be ruptured and a new membrane formed consisting of casein and associated lipolytic enzymes. Under these conditions the enzymes have access to the fat, which is then hydrolyzed, leading to free fatty acid production and rancidity.

The question arises which enzymes are primarily responsible for lipolysis in milk. Since the actual demonstration of lipolysis in milk in the mid-1930s, many have suggested that this defect might be due to the presence in milk of more than one enzyme. Phase distribution studies of lipolytic activity led to the view that in fresh milk there were two lipases, one acid and one alkaline. In addition, it has been well established that psychrotrophic bacteria in milk can produce lipolytic enzymes that show high thermal stability. These psychrotrophs are able to provoke lipolytic processes even during refrigeration of milk. In contrast to milk, the original enzymes of which are labile to heat, microbial lipases of psychrotrophic bacteria are extremely stable to heat. While the organisms themselves are destroyed by pasteurization, the lipolytic enzymes of bacterial species frequently found in milk are not or are only slightly deactivated at these or higher temperatures. This is one reason the quality of pasteurized milk products may be affected by the biochemical activity of lipolytic microorganisms. Fatty acids set free by lipolysis are those that produce a rancid flavor in milk. These are the even-numbered fatty acids having 4-12 carbon atoms.

B. Psychrotrophs

The use of the term psychrotrophs in this context should be explained. When applied to bacteria it describes microorganisms capable of growth, ever so slowly, at 8°C, irrespective of their ideal growth temperature. The term psychrophilic is more familiar, but it is restrictive in that it describes organisms having an optimum growth temperature in the range of 10 to 16°C. This writer has used the term spore formers, that is, bacteria that survive high-temperature treatments and are capable of multiplying at refrigeration temperatures of 4°C.

C. Oxidation

Oxidized flavor is characterized by a cardboard or tallow-like taste, which is sharp when the sample is held in the mouth but disappears almost completely upon swallowing. The off-flavor can develop upon storage; it is more prevalent in pasteurized creamlike milk, skim milk, and fluid cream products, but can develop in homogenized milk during periods of storage. Copper, iron, rust, excessive chlorine, sunlight, and exposure of milk to bright fluorescent light in a dairy case are known factors contributing to the oxidized flavor. The problem also varies with the feed type, with the time of the year, and among individual cows.

Oxidized flavor may take several days to a week to develop to the point of "consumer complaints." Because it may develop long after the milk is secreted from the cow, it seems logical to seek one single contributing factor. This factor is storage of milk. The dairy industry all over the world is making a great effort to prevent off-flavor in milk, but the same preventative measures are not always followed in the home or in the store. For example, our research has focused on the effect of fluorescent light, as encountered in a dairy display case, on the flavor and selected nutrients of homogenized milk held in conventional containers. Homogenized milk was packaged in three different types of half-gallon containers: unprinted fiberboard, blow-molded plastic, and clear flint glass. The milk was held in a sliding-door display case with fluorescent light exposure of 100 fc for 144 hr. The fiberboard container afforded protection from light-activated off-flavor up to 48 hr whereas milk in plastic and glass containers developed an off-flavor only 12 hr after exposure. The milk kept in plastic or glass half-gallon containers exhibited no differences in organoleptic quality. Similarly, riboflavin destruction in plastic and glass was not significantly different and amounted to about 10 to 17% loss following 72 hr of exposure. No significant loss of riboflavin could be demonstrated in the milk held in fiberboard as compared to the control. Ascorbic acid losses were evident in all milk samples independent of type of container material; however, losses of vitamin C from milk held in plastic and glass were much more rapid than in milk held in fiberboard. Exposure to light of milk in all three containers tested had no effect on the amino acid composition as compared to the control milk held in the dark. These results reinforce present thinking that protection of milk from light during marketing is necessary to ensure flavor quality and to a lesser extent nutrient value, thereby obtaining the optimum shelf-life of the product (51).

D. Proteolysis

In regard to proteolytic activity, heat-stable proteases have been isolated from psychrotrophic bacteria in raw milk (52,53). It should be pointed out that much of the work on lipolysis and proteolytic activity has been done under the auspices of the Dairy Research Foundation, of which this author was the founder and first administrator. In many cases the information reported in this chapter is based on basic research results done for the foundation by various land grant universities throughout the United States. Therefore, in some cases the information reported is from unpublished reports of recent date, and the information will remain unpublished until patent protection has been obtained. The above authors have done much work on proteolytic activities in milk and on the ultrapasteurization and aseptic packaging of fluid milk products. The foundation, through its diligent pursuit of the sterile milk question, is perhaps very much responsible for having guided this program through basic and applied research and can claim that the existence of sterile milk products in the U.S. market is a direct result of the foundation's insistence on such research.

Protease in UHT milk causes much gelation and is most active at a pH of 6 or 7. As early as 1975 it was discovered (52) that the protease produced by a strain of gram-negative psychrotrophs, present at between 1000 and 10,000/ml in sterile milk stored for 2 days at 4°C, was enough to promote spoilage of the milk in less than 3 days at 4°C. The same authors devised a system to inactivate the protease in UHT-treated milk. Crude preparations of protease in a simulated milk ultrafiltrate were 98% inactivated when heated to 55°C for 10 min, while only 70% were inactivated after 60 min of heating at the same temperature.

The occurrence of bitter flavor in milk seems to be associated with the presence of heat-stable protease. While appreciable amounts of lipases can be inactivated by heating milk to 64°C for 10 sec (54), heat treatment that would inactivate protease would also denature the proteins. Until the proteases have been more fully investigated, any type of inactivation of these enzymes seems unlikely for the time being. It may be possible to employ a protease inhibitor if one can be economically found.

VI. RESEARCH EFFORTS TO COMBAT SHELF-LIFE DEFECTS

The competitive inhibition of one group of bacteria by another group to improve the shelf-life of raw milk has been investigated (55). The authors suggested that the development of oxidized off-flavors in milk may be prevented by decreasing the oxidation-reduction potential of the raw milk by controlled inoculation with growing bacteria. The bacteria would be destroyed during the pasteurization process.

Superoxide dismutase is a naturally occurring enzyme in milk that is inactivated during the pasteurization process. It has been shown that this enzyme possesses antioxidant characteristics. Research sponsired by the Dairy Research Foundation has shown that the addition of minute quantities of this enzyme might be beneficial to the shelf-life of milk, but additional work needs to be done to prove its value (56).

Also considered has been the antibacterial action of the lactoperoxidase system in milk as well as the use of lysozyme to inhibit bacterial growth. This means that it might be possible to improve the keeping quality of milk by adding catalase-negative bacteria or immobilized glucose oxidase to provide peroxide to couple the system. Lysozyme is found in high concentrations in human milk, and to simulate this, it has been added to cows' milk. Crude lysozyme has been prepared from the spleen of cattle and may provide an economical source of the inhibitor (57).

A. On-Farm Processing

One method being investigated by some food science departments of universities at the request of the Dairy Research Foundation is the on-farm processing of milk and the subsequent concentration of the product through the use of ultrafiltration or reverse osmosis. It has been demonstrated that subpasteurization of raw milk at 74°C for 10 sec causes 98% inactivation of milk lipase and limits growth of bacteria, as compared to those in unheated raw milk (58). For these studies the milk was stored after subpasteurization in a bulk tank at 1°C. After 1 week the milk was pasteurized, and after storage for 14 days at 4°C the milk compared favorably with milk pasteurized immediately after milking and similarly stored. Using a subpasteurization temperature of 65°C for 15 sec had no effect on the whey proteins; they were not denatured and the heat coagulation time was unaffected (59).

Pasteurization shortly after milking and subsequent transportation of milk in sealed containers for further processing at dairy plants is another possibility. Also of potential value is concentrating the milk and removing some of the water so that greater amounts of the actual milk components can be transported; the ramification of such a system go beyond simply extending the shelf-life of milk.

B. Packaging

In a handbook that provides only limited space, many of the factors influencing shelf-life of milk cannot be elaborated on. One of the more important issues is the handling and packaging of fluid milk products. A major source of postpasteurization contamination of milk is in filling the containers at the packaging machine. If care is taken in that phase, the shelf-life of milk can be considerably extended. One of the most important aspects of sterile milk production is the use of sterile packages, for example the packaging of UHT-treated milk. A number of manufacturers have introduced an aseptic package that may also be suitable for packaging pasteurized milk.

Reducing the oxygen levels in milk is very important (60); it has been shown that a reduction from 9 to 12 ppm to 1-3 ppm at 3°C slowed the growth of certain bacteria by 63%. The shelf-life of milk may be improved through aeration of the product during pasteurization and packaging, and through packaging the product using materials of low oxygen permeability. Some (61) have proved that the infusion of carbon dioxide into milk might inhibit the growth of common spoilage organisms, but this might contribute to the survival of psychrotrophs able to withstand low oxygen tension.

As mentioned previously, the shelf-life of milk bottled in clear containers is reduced with exposure to fluorescent light as encountered in dairy display cases in the market. Exposure of milk to fluorescent lights in the presence of oxygen causes oxidative degradation of both the protein and lipids, and also decreases the vitamin contents with increased exposure time. No difference is observed for milk stored either in fiberboard or blow-molded cartons and thus protected from exposure to fluorescent light (62).

C. Sterile Dairy Products

A relatively new process that is in common use in European countries but much less in the United States, is the ultrapasteurization of milk and dairy products followed by aseptic packaging of the product. This consists of heating milk and dairy products at or above 138°C for 4 sec or less, the length of exposure depending on the processing temperature used. Research at North Carolina State University, the U.S. pioneer on this subject, has shown that a temperature of 149°C for 2 sec produced a product free of objectionable off-flavors.

Ultrapasteurization ensures the complete destruction of all microorganisms with the possible exception of some non-pathogenic, highly heat-resistant spores. This process, with subsequent aseptic packaging, extends the shelf-life of the product in an unrefrigerated atmosphere to at least 3 months. This author has tasted some UHT-treated, sterile-packaged milk at a South German Research Institute that was kept unrefrigerated for 28 months with no negligible loss of a pleasing flavor. Ultrapasteurization doesnot reduce the nutritive value of the dairy product so treated, since the levels of most nutrients in milk are not significantly affected by UHT processing and extended storage. Upon FDA approval of the use of hydrogen peroxide as a sterilizing agent for aseptic packages for UHT-treated dairy products, the U.S. production of UHT-treated milk, with subsequent sterile packaging, became possible. A number of U.S. dairies are using this process to produce a shelf-stable product, but it is doubtful that fluid milk products treated by ultrapasteurization and aseptic packaging will be a consumer success. With our extended system of refrigeration, this more expensive product, will not sell well. Specialty products such as flavored milks and cream items are much more suitable for this process, and it will be these specialty products that will be increasingly prepared by this sterilization process. Presently 50% of U.S. cream products are ultrapasteurized, but they are not packaged in what would be considered a completely sterile container.

Research is presently being carried out to determine the advantages of direct versus indirect superheating of milk (63). So far the work has shown that there is not much difference in flavor or shelf-life span of ultrapasteurized milk using either of the heating processes. With the indirect unit, the product is heated via a heat-conducting surface, so the milk is separated from the heating medium. After reaching the desired processing temperature, the product is maintained at this temperature during the short holding time, after which immediate cooling occurs in a separate heat exchanger, where product and cooling medium again are kept separated by a heat-conducting barrier.

In the direct UHT system, the milk is mixed with saturated steam under pressure. This permits rapid heating of the product as the steam condenses. Again the product is held at the desired processing temperature for a given hold time. Cooling is carried out in the direct system by a vacuum that removes the water added from condensing steam and flash-cools the product.

Holding times for UHT processing are relatively short in comparison to the holding times needed in conventional canning processes. The holding times, usually only a few seconds, are determined together with the correct temperature to be applied to ensure complete destruction of spoilage and disease-causing microbial spores while minimizing undesirable physical, chemical, and biological transformations that can occur within the product.

The desirable changes that occur during the UHT process are inactivation of biologically viable materials such as enzymes, microorganisms, and their spores. Undesirable changes include some losses of quality such as taste, color, and some nutrients.

The aseptic filling process is a very important part of the milk's sterility. In most cases the containers are formed from a roll that has been presterilized; as they are shaped the containers are again sterilized with hydrogen peroxide spray and heat treatment. After filling they are heat sealed. A number of container materials are available that meet the aseptic requirements, including plastic films, plastic beakers, plastic film sachets, and blow-molded thermoplastic bottles.

Quality control methods so far have not been standardized. Detection of contaminated containers after incubation is simple and many methods have been published, but the interpretation of results is left to the processor, and it has been the practice for processors to develop their own quality control programs.

D. Shelf-Life Prediction of Refrigerated Milk

Muir and Phillips (64) have examined the quantitative relation between raw milk quality and product quality. They present comprehensive new information on the distribution of the growth rate of psychrotrophic bacteria in raw milk. The results of these different concepts have been incorporated into model calculations that explore the likely "safe" shelf-life of raw milk under various conditions. These calculations provide a useful guideline for the safe storage of raw milk under practical conditions.

VII. SHELF-LIFE OF OTHER DAIRY PRODUCTS

The quality of the raw milk precursor is the most important determinant in the shelf-life of dairy products. Most of these products fall into the same "perishability" category as milk, meaning that these products have about the same distribution time as the fluid milk product itself.

Thus chocolate milk, half-and-half cream, skim milk, low-fat milk, light cream, and whipping cream must be treated exactly like whole milk to obtain maximum shelf-life.

Fresh cultured dairy products such as buttermilk, sour cream, and yogurt must also be kept at $\leq 5°C$; however, these products, because of their acidity, have a longer shelf-life than the above fluid milk products, but only if properly refrigerated. Some brands of cultured dairy products gain additional shelf-life by incorporating a small amount (0.1%) of sorbate, a mold and yeast inhibitor.

A. Butter

A considerable proportion of the raw milk produced in the United States is used for making butter. Recall that 21.6 lb of milk are required to make 1 lb of butter. This product is manufactured from pasteurized cream; churned in clean, sanitary equipment and washed with pure water, it contains very few microorganisms. Since most butter is slightly salted, it possesses better keeping quality, as related to freedom from flavor deterioration caused by microbial action, than unsalted sweet butter.

Table XI (65) compares the total counts of molds, yeasts, and bacteria per gram of refrigerated, lightly salted versus unsalted butter.

Overall proper storage conditions are essential for maximum shelf-life. Most butter currently manufactured passes through distribution channels rather quickly. A temperature of 0 to 5°C is considered satisfactory for short-term storage. If butter is to be stored for a long period, a temperature of -18 to -23°C should be used. Unsalted butter should always be stored in the freezer compartment before selling to consumers. Butter should also always be stored in an environment free of odor, since it freely absorbs odors from the atmosphere or from odorous materials with which it might come into contact.

B. Frozen Dairy Products

The principal frozen dairy product is ice cream, which along with frozen custard, ice milk, sherberts, ices, and mellorine makes up the bulk of frozen dairy products production. Good-quality criteria of ice cream and other frozen dairy products are flavor, body and texture, appearance, melting characteristics, and freedom from excessive bacteria. Good manufacturing practices also demand freedom from any extraneous matter. These criteria apply equally to all types of manufactured products on sale, both commercial and premium. The criteria for a premium rating are a product's higher fat and/or nonfat milk solids content, lower overrun with a weight higher than the minimum of 4.5 lb/gallon, and high-quality flavoring. Flavor is largely a summation of the flavor contribution of each of the raw materials, although off-flavor may result from bad manufacturing processes or prolonged storage of the ice cream, particularly if nuts are included. The object of manufacturing good ice cream is to produce a smooth-textured product with good bite-resistance and to be able to maintain these qualities despite temperature fluctuations and/or long storage.

When ice cream and related products reach the consumer, their texture should be smooth and velvety. The ice crystals must be extremely small; the air cells must also be small and evenly distributed. There should be no separate fat particles, and the product should be free from crystallized sugars, which, if present, impart a sandy, undesirable texture. The mouth-feel upon consuming should not be excessively cold, dry, or astringent. If all of these organoleptic requirements have been met through good manufacturing practices, then it should be possible to keep the ice cream in frozen storage of -25 to -30°C for $\geq$12 months without introducing any flavor defects.

Freeze Concentration of Milk

Milk may freeze accidentally or freezing may be utilized as a specific means for preserving milk and dairy products. The fluid product is not usually frozen other than accidentally, because freezing might result in various physical defects including (1) destabilization of the fat globule membrane, resulting in the development of separate fat particles and free fat, (2) a change in the colloidal calcium phosphate, and finally (3) the destabilization of the protein system.

Considerable research has been directed toward making freeze concentration applicable to fluid milk and whey products. In particular the work at the University of

Wisconsin (66-68) has shown that if, prior to freezing, fluid milk is separated by ultrafiltration into its main components (i.e., the protein fraction and carbohydrate fraction), which are then separately frozen, none of the above defects occur. It has even been shown that the separate fraction can be packaged into one common carton for storage, and both fractions can be reconstituted as one.

This research has found additional application, in that freeze concentration can be used for the isolation of casein. This is not pertinent to the subject at hand, but it should be of interest to the food industry.

The freeze concentration of milk and whey is being pursued from a commercial standpoint by one U.S. company with the aim of demonstrating the feasibility of concentrating dairy products by freeze crystallization and introducing this technology to the dairy industry.

XI. SUMMARY

Much of the work described in this chapter is being carried out under the auspices of the U.S. Dairy Farmer. The Dairy Farmer, through its extensive financial contribution to dairy research through grants to land grant universities, is helping the dairy industry to produce a dairy product of superior quality with long-lasting shelf-life and excellent stability.

Today's consumer determines the acceptability of fluid milk by flavor and shelf-life. A uniformly good taste and acceptable keeping quality are essential for the dairy industry to maintain its good sales position.

Dairy farmers and processors are the most important link in maintaining these criteria for having an excellent dairy product in the market. To obtain these criteria was relatively easy when milk was produced, processed, and delivered to the consumer within a very short time such as 24 hr. Currently milk is usually collected from the farm every other day and is normally processed in plants 5 days a week.

The shift from home delivery to store purchase of milk and dairy products makes excellent sanitation a must. Store sales of fluid milk have lengthened the time between processing and consumption. They have also introduced another group of persons handling the product, which in itself has a definite influence on the shelf-life of the product.

Since we believe that product temperature is the most important factor influencing the shelf-life of milk, care must be taken that milk and dairy products are displayed in the dairy case using the best manufacturing practices recommended. Product display temperature should be uniformly maintained in the storage case as well as in the refrigerator at home at <5°C, since as a rule of thumb, for every 5° rise in temperature, the shelf-life is reduced by 50%. It should be kept in mind that freshly produced products must give the industry an excellent reputation, as the consumers use the products many days later. Every effort should be made to assure a 14-day shelf-life under good conditions for fluid milk products.

TABLE I. Yearly Milk and Dairy Products Sales in the United States 1976-1983[a]

Year	Fluid milk[b]	Butter	Cheese[c]	Condensed and evaporated milk	Frozen products[d]
1976	55,960	932	4381	1829	6000
1977	55,825	860	4434	1769	6067
1978	55,965	894	4712	1657	6068
1979	55,835	921	4824	1646	5960
1980	55,773	894	4827	1586	6006
1981	55,444	880	5035	1640	6083
1982[e]	54,785	891	5175	1608	6115
1983[e]	54,830	897	5144	1621	6331

	Whole-milk powder	Nonfat dry milk powder	Buttermilk powder	Whey powder
1976	33	743	38	527
1977	34	798	59	530
1978	56	640	50	543
1979	66	692	42	597
1980	63	642	42	604
1981	39	559	45	629
1982[e]	34	569	40	669
1983[e]	43	627	44	722[f]

[a] In millions of pounds. Reference (1).
[b] Excludes milk used on farms; includes fluid whole milk, cream mixtures, low-fat milk, skim milk, buttermilk, flavored milk drinks, and yogurt.
[c] Includes whole and part skim-milk cheese and cottage cheese.
[d] Includes ice cream, ice milk, sherbert, novelty products, and mellorine.
[e] Preliminary.
[f]

TABLE II. Conversion Factors for Processing Whole Milk into Various Products

Product	Amount of whole milk required to produce 1 lb of product (lbs)
Butter	21.2
Cheese	10.0
Evaporated milk	2.1
Condensed milk	2.3
Whole-milk powder	7.4
Powdered cream	13.5
Ice cream[a]	12.0
Cottage cheese	6.25[b]
Nonfat dry milk	11.0[b]

[a]One gallon.
[b]Skim milk.

TABLE III. Composition of Milk[a]

Components	Whole-milk content	
	1 cup - 244 g	100 g
Water (g)	214.70	87.99
Energy (kcal)	150.0	61.0
kJ (kcal × 4.184)	627.0	257.0
Protein N × 6.38 (g)	8.03	3.29
Lipids, total (g)	8.15	3.34
Saturated (g)	5.07	2.08
Monounsaturated (g)	2.35	0.96
Polyunsaturated (g)	0.30	0.12
Cholesterol (mg)	33.0	14.0
Carbohydrates (g)	11.37	4.66
Ash (g)	1.76	0.72
Minerals (mg)		
Calcium	291.0	119.0
Iron	0.12	0.05
Magnesium	33.0	13.0
Phosphorus	228.0	93.0
Potassium	370.0	152.0
Sodium	120.0	49.0
Zinc	0.93	0.38
Vitamins		
Ascorbic acid (mg)	2.29	0.94
Thiamine (mg)	0.093	0.038
Riboflavin (mg)	0.395	0.162
Niacin (mg)	0.205	0.084
Pantothenic acid (mg)	0.766	0.314
Vitamin B_6 (mg)	0.102	0.042
Folacin (μg)	12.0	5.0
Vitamin B_{12} (μg)	0.871	0.357
Vitamin A (RE)	76.0	31.0
Vitamin A (IU)	307.0	126.0

[a]Reference (5).

TABLE IV. Federal Standards for Frozen Dairy Products[a]

Product	Lb/gallon[b]	Lb total solids/gallon[b]	Milk fat (%)	Total nonfat milk solids[c] (%)	Whey solids[d] (%)	Total milk solids[b,c] (%)
Ice cream	4.5	1.6	10[b]-14	6[b]-10	1.5-2.5	20
Bulky flavored ice cream	4.5	1.6	8[b]	8	2.0	16
Ice milk	4.5	1.3	2[b]-7	4[b]-9	1.0-2.25	11
Bulky flavored ice milk	4.5	1.3	2	7[b]	1.75	9
Frozen custard	4.5	1.6	10[b]	10	2.5	20
Mellorine	4.5	1.6	--			6% Vegetable or animal fat
Sherbets	6.0	--	1 -2	1[b]		5
Water ices	6.0	--	--	--	--	--

[a]Reference (13).
[b]Minimum.
[c]Caseinates cannot be used to satisfy any part of the total milk solids requirements.
[d]Solids from either concentrated or dried whey.

TABLE V. Composition of Whey[a]

Component	Amount per 100 g of whey
Water	93.1 g
Food energy	26 kcal
Protein	0.9 g
Fat	0.3 g
Lactose (carbohydrate)	5.1 g
Ash	0.6 g
Calcium	51 mg
Phosphorus	52 mg
Iron	0.1 mg
Sodium	--
Potassium	--
Vitamin A	10 IU

[a]Reference (14).

TABLE VI. Physical Properties of Fluid Milk[a]

Property	Value
Titratable acidity	0.16 ± 0.02%
pH	6.6 ± 0.2 at 25°C
Surface tension	55.3 dynes
Specific gravity	1.032 ± 0.004
Freezing point	-0.5400°C
Boiling point	100.17°C
Specific heat at:	
0°C	0.920
15°C	0.938
40°C	0.930
Coefficient of expansion at:	
10°C	0.9975
15.6°C	0.9985
21.1°C	1.000
Viscosity	1.6314 cP
Electrical conductivity	45-48 × 10^{-4} mho
Osmolality[b]	275 Osm/kg

[a]References (6,15).
[b]Source: The Doyle Pharmaceutical Co., Minneapolis, Minnesota. Osmolality of a solution is based on the number of particles in a solution; the greater the number of particles, the higher the osmolality.

TABLE VII. Time-Temperature Ratios for Thermal Processing of Milk[a]

Temperature (°C)	Time	Industrial nomenclature
63.0	30.0 min	Low time low temperature (LTLT)
71.5	15.0 sec	High temperature short time (HTST)
60.5	1.0 sec	High temperature shorter time (HTST)
100.0	0.01 sec	Higher-heat, shorter time (HHST)
138.0	2.0-5.0 sec	Longer shelf-life; ultrapasteurization
138 to 150°C	2.0-5.0 sec	Ultrahigh temperature (UHT)[b]

[a]References (4,21,22).
[b]This category gives a product that does not need to be refrigerated until opened; all others need to be refrigerated after processing.

TABLE VIII. Storage Life of Various Milk Products at Specific Temperatures

Product	Approximate storage life
All fluid milk products	8-20 days, <4°C
Sterile milk products	4 months at 21°C 12 months at 4°C
Frozen milk products	12 months at -23°C
Evaporated milk	1 month at 32°C 12-24 months at 21°C 24 months at 4°C
Concentrated milk	≥2 weeks at 1.5°C
Concentrated frozen milk	6 months at -26°C
Sweetened condensed milk	3 months at 32°C 9-24 months at 21°C 24 months at 4°C
Nonfat dry milk, extra grade (in moisture-proof pack)	6 months at 32°C 16-24 months at 21°C 24 months at 4°C
Dry whole milk, extra grade (gas pack. max. oxygen 2%)	6 months at 32°C 12 months at 21°C 24 months at 4°C
Buttermilk	2-3 weeks at 4°C
Sour Cream	3-4 weeks at 4°C
Yogurt	3-6 weeks at 4°C
Eggnog	1-2 weeks at 4°C
Ultrapasteurized cream	6-8 weeks at 4°C
UHT-pasteurized milk	3 months at room temperature

TABLE IX. Relation of Storage Temperature to Shelf-Life of Milk

Temperature (°F)	Shelf-life (days)
32	24.0
40	10.0
45	5.0
50	2.0
60	1.0
70	0.5
80	0.5

TABLE X. Off-Flavors in Milk and Their Causes

Causes	Descriptive or Associated Terms
Heated	Cooked, caramelized, scorched
Irradiated	Light, sunlight, activated
Lipolyzed	Rancid, butyric, bitter, goaty
Microbial	Acid, bitter, fruity, malty, putrid, unclean
Oxidized	Papery, cardboard, metallic, oily, fishy
Stored	Lacks freshness, stale, coconut
Transmitted	Reed, weed absorbed, cowy, barny
Lacks fine flavor	Flat, watery
Miscellaneous	Astringent, foreign, medicinal, chalky, salty

TABLE XI. Total Counts of Molds, Yeasts, and Bacteria per Gram in Salted and Unsalted Butter[a]

	Salted			Unsalted		
Butter Age	Molds	Yeasts	Bacteria	Molds	Yeasts	Bacteria
Fresh	19	295	260,000	25	169	230,000
1 month old	25	59	77,000	340	210	2,100,000

[a]Reference (65).

REFERENCES

1. U.S. Department of Agriculture, Economic Research Service (1984). "Dairy Outlook and Situation." U.S. Govt. Printing Office, Washington, D.C.
2. Menz, W. W. (1982). *In* "Handbook of Processing and Utilization in Agriculture" (I. Wolff, ed.), Vol. I, pp. 283-290. CRC Press, Boca Raton, Florida.
3. U.S. Department of Agriculture, Agricultural Marketing Service (1980). "Federal and State Standards for the Composition of Milk Products (and Certain Nonmilk Products)," Agric. Handb. No. 51. U.S. Govt. Printing Office, Washington, D.C.
4. U.S. Department of Health and Human Services, Food and Drug Administration (1980). "Milk and Cream," Code of Federal Regulations, Foods and Drugs, Title 21 (Part 131) (Part 135). Revised as of April 1, 1981. U.S. Govt. Printing Office, Washington, D.C.
5. U.S. Department of Agriculture, Agriculture Research Service (1976). "Composition of Foods, Dairy and Egg Products, Raw, Processed, Prepared," Agric. Handb. No. 8-1. U.S. Govt. Printing Office, Washington, D.C.
6. Henderson, J. L. (1971). "The Fluid Milk Industry," 3rd ed. AVI, Westport, Connecticut.
7. Lampert, L. M. (1975). "Modern Dairy Products," 3rd ed. Chem. Publ. Co., New York.
8. Kosikowski, F. (1977). "Cheese and Fermented Milk Foods," 2nd rev. ed. F. V. Kosikowski Associates, Brooktondale, New York.
9. U.S. Department of Health and Human Services, Food and Drug Administration (1981). *Fed. Regist. 40*, 9924.
10. National Dairy Council (1972). *Dairy Counc. Dig. 43*, 19.
11. Arbuckle, W. S. (1979). "Ice Cream," 3rd ed. AVI, Westport, Connecticut.
12. Tobias, J. (1982). *In* "Handbook of Processing and Utilization in Agriculture" (I. Wolff, ed.), Vol. I, pp. 315-363. CRC Press, Boca Raton, Florida.
13. Federal Register (1977). *Fed. Regist. 42*, No. 70, 19134; *43*, No. 24, 4596; No. 88, 19384 (1978).
14. Watt, B. K., and Merrill, A. L. (1963). "Agriculture Handbook No. 8," U.S. Department of Agriculture, Agricultural Research Service, U.S. Govt. Printing Office, Washington, D.C.
15. Webb, B. H., Johnson, A. H., and Alford, J. A. (1974). "Fundamentals of Dairy Chemistry," 2nd ed. AVI, Westport, Connecticut.

16. U.S. Department of Health and Human Services (1980). "Grade 'A' Pasteurized Milk Ordinance, 1980 Recommendations." USDHHS, Washington, D.C.
17. Aggarwal, M. L. (1975). *J. Milk Food Technol. 138*, 419.
18. Arledge, W. E. (1982). *Dairy Rec. 83*, 129.
19. U.S. Department Health and Human Services, Food and Drug Administration (1981). *Fed. Regist. 46*, 2341.
20. Menz, W. W. (1977). "Unpublished Research Results," Memo. 12-7. Dairy Res. Found., Rosemont, Illinois.
21. Burton, H. (1980). *In* "Proceedings of the International Conference on UHT Processing and Aseptic Packaging of Milk and Milk products. Sponsored by Dairy Research, and North Carolina State University, Raleigh.
22. U.S. Department of Health and Human Services, Food and Drug Administration (1978). "Grade "A" Pasteurized Milk Ordinance, 1978 Recommendations." USDHHS, Washington, D.C.
23. Arnold, S., and Roberts, T. (1982). "UHT Milk: Nutrition, Safety, and Convenience," Natl. Food Rev., Spring: 2.
24. Barnard, W. E. (1974). *J. Milk Food Technol. 37*, 346.
25. Hedrick, T. I., and Glass, L. (1975). *J. Milk Food Technol. 38*, 129.
26. Kleyn, D. H. (1973). *Dairy Ice Cream Field* September, p. 36.
27. Mocquot, G., and Ducluzeau, R. (1968). *In* "Low Temperature Biology of Foodstuffs," pp. 235-250. Pergamon, Oxford.
28. Morita, R. V. (1975). *Bacteriol. Rev. 39*, 144.
29. Stadhouders, J. (1968). *Dairy Sci. Abstr. 31*, 992.
30. Antila, V. (1971). *Dairy Sci. Abstr. 33*, 5643.
31. Shiga, K., Yoshina, M., and Someya, Y. (1970). *Bull. Natl. Inst. Anim. Ind. (Chiba) 23*, 1.
32. Punch, J. D., Olson, J. C., Jr., and Thomas, E. L. (1965). *J. Dairy Sci. 48*, 1179.
33. Langeveld, L. P. M., and Cuperus, F. (1980). *Neth. Milk Dairy J. 34*, 106.
34. Ratkowsky, D. A., Olley, J., McMeekin, T. A., and Ball, A. (1982). *J. Bacteriol. 149*, 1.
35. Mistry, V. V., and Kosikowski, F. V. (1983). *J. Food Prot. 46*, 52.
36. Stadhouders, J., and Langeveld, L. P. M. (1970). Unpublished results.
37. Patel, G. B., and Blankenagel, G. (1972). *J. Milk Food Technol. 35*, 203.
38. Sherman, J. M., Cameron, G. M., and White, J. C. (1941). *J. Dairy Sci. 24*, 526.

39. Janzen, J. J., Bodine, A. B., and Bishop, J. R. (1981). *J. Food Prot. 44*, 455.
40. Ell iker, P. R., Sing, E. L., Christersen, L. J., and Sandine, W. E. (1964). *J. Milk Food Technol. 27*, 69.
41. Overcast, W. W. (1967). *Am. Dairy Rev. 17*, 203.
42. Ashton, T. R. (1950). *J. Dairy Res. 17*, 261.
43. Storgards, T. (1961). *Dairy Ind. 26*, 909.
44. Aule, O., and Storgards, T. (1962). *Proc. Int. Dairy Congr. 16th, 1962*, p. 799.
45. Watrous, J. H., Jr., Barnard, S. E., and Coleman, W. W., II (1971). *J. Milk Food Technol. 34*, 145.
46. Hankin, L., and Stephens, G. R. (1972). *J. Milk Food Technol. 35*, 574.
47. Carey, K. D., and Andersen, R. E. (1982). *J. Dairy Sci. 65*, (Suppl. I), 76.
48. Elliker, P. R. (1968). *Dairy Sci. Abstr. 31*, 995.
49. Patel, G. B., and Blankenagel, G. (1972). *J. Milk Food Technol. 35*, 203.
50. Menz, W. W. (1978). "The Flavor of Milk." Natl. Res. Found., Athens, Greece.
51. Dimick, P. S. (1973). *J. Milk Food Technol. 36*, 383.
52. Adams, D. M., Barach, J. T., and Speck, M. L. (1975). *J. Dairy Sci. 58*, 828.
53. Adams, D. M., Barach, J. T., and Speck, M. L. (1976). *J. Dairy Sci. 59*, 823.
54. Driessen, F. M., and Stadhouders, J. (1978). *Zuivelzicht 70*, 1080.
55. Solberg, P., Hadland, G., and Authen (1962). *Proc. Int. Dairy Congr. 16th, 1962*, p. 649.
56. Hicks, C. J. (1980). *J. Dairy Sci. 63*, 1199.
57. Pavlovski, P. E., and Danilov, N. S. (1979). *Food Sci. Technol. Abstr. 13*, 852.
58. Senyk, G. F., Zall, R. R., and Shipe, W. F. (1982). *J. Food Prot. 45*, 513.
59. Coghill, D. M., Mutzelburg, I. D., and Birch, S. J. (1982). *Aust. J. Dairy Technol. 37*, 48.
60. Brandt, M. J., and Ledford, R. A. (1982). *J. Food Prot. 45*, 132.
61. Shipe, W. F., Senyk, G. F., Adler, E. J., and Ledford, R. A. (1982). *J. Dairy Sci. 68*, (Suppl. I), 77.
62. Janzen, J. J., Bodine, A. B., and Bishop, J. R. (1981). *J. Food Prot. 44*, 455.
63. Swartzel, K. R. (1982). *J. Food Sci. 47*, 1886.
64. Muir, D. D., and Phillips, J. D. (1984). *Milchwissenschaft 39*, 7.
65. Kleyn, D. (1973). *Dairy Scope* 8D.
66. Lonergan, D. A., Fennema, O., and Amundson, C. H. (1981). *J. Food Sci. 46*, 1981.

67. Minson, E., Fenneman, O., and Amundson, C. H. (1981). *J. Food Sci. 46*, 1592.
68. Minson, E., Fenneman, O., and Amundson, C. H. (1981). *J. Food Sci. 46*, 1597.

CHAPTER 12

CHEMICAL CHANGES DURING STORAGE OF TEA

Tei Yamanishi
Ochanomizu University
Tokyo, Japan

I. INTRODUCTION

The characteristics of tea as a beverage are its taste, aroma, and color. These are quite different from coffee, fruit juices, and other soft drinks.

Polyphenols such as catechins and amino acids such as theanine are the main contributors to the unique taste and color of tea. The components of essential oil in fresh tea leaves and volatile compounds developed during the manufacturing process form the characteristic tea flavor.

Handbook of Food and Beverage
Stability: Chemical, Biochemical,
Microbiological, and Nutritional Aspects

Although many types of tea are available around the world, teas may be classified into three general categories according to the manufacturing process used: fermented (black tea), semifermented (oolong and pouchong), and nonfermented (green tea).

The different manufacturing conditions cause the differences in taste, aroma, and color as well as in storage stability or shelf-life of tea.

II. CHEMICAL CHANGES RESPONSIBLE FOR DETERIORATION IN QUALITY OF TEA

The important factors affecting tea quality are flavor (aroma and taste) and color of tea liquor. Deterioration in quality of tea is caused principally by chemical changes in the components that contribute to these factors.

A. Black Tea

Freshly made black tea has a raw or "green" flavor, but this rawness is replaced by a balanced astringency and flavor. However, prolonged storage leads to a deterioration in the quality of the tea.

Deterioration in black tea is caused by (1) losses of volatile components, (2) changes in catechins, amino acids, theaflavins, and other pigments, and (3) increases in undesirable "taints" such as oxidative reaction products from fatty acids, and oxidation and condensation products from soluble polyphenols such as catechins and theaflavins. These reactions are accelerated by tea moisture content, elevated temperature, and exposure to light.

1. Volatile Components

Stagg (9) reported that the volatile fraction of black tea showed an overall decline that was accelerated with moisture uptake and, to some extent, by storage at elevated temperature, as shown in Table I. The decreases are found among aliphatic aldehydes and alcohols with b.p. 108-157°C, while other components show little change.

Table II shows the change in aroma pattern of black tea during storage at different temperatures (6). An elevated temperature accelerates the variation in aroma compositions. The proportions of (*E*)-2-hexenal, benzaldehyde, pentanol, hexanol, and (*Z*)-3-hexen-1-ol, which give the fresh green note to the aroma of black tea, decrease, while those of (*E*,*E*)-2,4-heptadienal and (*E*,*E*)-2,4-decadienal increase during storage. (*E*,*E*)-2,4-Heptadienal is one of the main components of deteriorated odor of tea and is well known to be an oxidative degradation product from linolenic acid. β-Ionone and 5,6-epoxy-β-ionone are degradation products from β-carotene that are increased by storage. All the above-mentioned changes, both decreases and increases, are more pronounced at 20°C than at 5°C in Table II.

2. Polyphenolic Compounds and Amino Acids

The chief contributors to the unique taste and color of black tea are polyphenols and amino acids. Black tea contains important polyphenolic compounds in the following proportions: catechins 5%, theaflavins 1-2%, thearubigins 15%, and complex tannins (formed from polyphenols and proteins) 5%. Theaflavins are bright orange in color and their chemical structure consists of a benzotropolone nucleus, formed by oxidation of catechins by polyphenol oxidase during the fermentation stage of black tea manufacture. Thearubigins are a complex mixture of polymeric polyphenols and are dark reddish in color. The mode of formation and the nature of thearubigins has not as yet been resolved satisfactorily.

Theaflavin content (or theaflavin:thearubigin ratio) shows a high positive correlation with evaluations of color and taste (7). Theaflavins contribute to the quality and the brightness of the color of black tea liquor, but much of the color, strength (powerful tea character), and mouth-feel is due to a heterogeneous group of compounds known as the thearubigins. Theaflavins and thearubigins constitute as much as 30 to 60% of solids in a black tea infusion (12).

Deterioration in the quality of black tea is caused by losses of theaflavins; loss of its astringency follows changes in catechins, resulting in a dull, dark color and flat taste. The loss of theaflavin upon storage is accompanied by losses of amino acids, sugars, pigments (chlorophylls, carotenoids, and flavonoids), and some volatile compounds.

Figure 1 shows the changes in moisture and theaflavin content during storage of black tea in (A) a tightly closed clear glass bottle and (B) a loosely covered wooden box (13). During a period of 22 weeks, moisture content of the tea rises from 4.2 to only 5% in an airtight glass bottle, whereas it increases to 9.9% in a wooden box from which air is not excluded. The rate of reduction in theaflavins is more gradual in the airtight bottle than in the wooden box, and the final value attained after 20 weeks in the former is higher than in the latter. The changes in the quantities of epigallocatechin gallate (EGCG) and epicatechin gallate (ECG) in the same experiment are shown in Table III (13). Both gallates change irregularly during storage, and the final level of EGCG after 16 weeks of storage is not reduced, whereas that of ECG is approximately half the original value. Therefore, reduction of ECG seems to be one of the factors in the change in tea taste to "flat". Figure 2 shows a decrease of theaflavins and increase in nondialyzable pigments, which contain highly polymerized polyphenols, during storage under different conditions. Both trends of variation are accelerated by moisture uptake and temperature elevation (9).

Table IV shows the changes in free amino acid contents in black tea that is packed in different types of sachets (9). The rate of reduction of amino acids is also influenced by moisture content. Jayaratnam and Kirtisinghe (4) reported that for humidities in the range of 32% at a temperature of 20°C under light-excluding conditions, black tea could be stored for a period of 300 days without loss of tea character. Wickremasinghe (13) suggested that light accelerates photooxidation of lipid and nonenzymic browning reaction, and causes heavier deterioration in the quality of black tea.

B. Green Tea

Steaming (Japanese type) or panning (Chinese type) is the first step of green tea manufacture, in which the polyphenol oxidase and other enzymes are inactivated and the green color of tea leaves is maintained in the finished product. Also, ascorbic acid (vitamin C) in fresh tea leaves is retained in green tea at an average level of about 250 mg/100 g *sen-cha* (Japanese type) and 200 mg/100 g *kamairi-cha* (Chinese type).

Deterioration in quality of green tea during storage is recognized by the following phenomena:

1. Vitamin C content is reduced.
2. Bright green color of tea changes to olive green and then brownish green and dull color.
3. Color of tea liquor varies from bright yellow or slight greenish tone to brownish yellow.
4. Characteristic leafy and refreshing aroma of green tea changes to dull and heavy odor.

5. The well-balanced, complex taste, which consisted of umami, astringency and bitterness, changes to flat and loses its characteristic briskness.

These deteriorations in quality of green tea are accelerated by moisture, oxygen, elevated temperature, and exposure to light, much as in the case of black tea.

1. Volatile Constituents

High-quality early spring green tea is recognized by its typical fresh natural aroma. (Z)-3-Hexenyl hexanoate, the major contributor to this characteristic aroma (10), decreases upon prolonged storage. Instead, there is an increase in 2,4-heptadienal as shown in Table V (2). Decrease of (Z)-3-hexenyl hexanoate is accelerated by higher temperature. More detailed analytical data on the variations of aroma composition in high-grade and low-grade green teas during storage at 25°C are shown in Table VI (3). Here again, a decrease of (Z)-3-hexenyl hexanoate and an increase of 2,4-heptadienal are evident, as well as increases of several unsaturated carbonyl compounds. As can be seen in Table VI, carotenoid degradation products (5) such as 2,6,6-trimethyl-2-hydroxycyclohexanone, β-cyclocitral, α-ionone, β-ionone, 5,6-epoxy-β-ionone, and dihydroactinidiolide increase markedly after 4 months' storage.

The aroma of early-spring green tea is perfectly preserved only by low-temperature storage at -70°C by comparison of sensory scores for green tea stored at 5°C or at room temperature in the case of nitrogen packing.

2. Color of Green Tea and Its Liquor

The discoloration of green tea during storage is caused by changes of chlorophyll *a* and *b* into pheophytin *a* and *b*, respectively. The change in chlorophyll *a* is greater than that in chlorophyll *b* (11). Principally, chlorophylls are not soluble in water, but a very minor amount is found in the liquor from high-quality green tea (8a).

The color of green tea liquor contains various flavonoid yellowish pigments and also oxidation and condensation products of catechins. Flavonoid pigments are said to contribute ∿24% to the total color of the tea infusion (8). The oxidation and condensation products from catechins act on the color of the green tea brew to make it brownish. Since catechins are the most important components contributing to tea taste, it is obvious that the discoloration of green tea positively correlates with the change in tea taste.

At the same time, there are high correlations between the rate of conversion of chlorophyll to pheophytin and the deterioration in quality of green tea (11).

3. Moisture and Vitamin C

The content of vitamin C in green tea decreases during storage, and the rate of decrease correlates with the moisture content. Changes in moisture and vitamin C content of green tea during storage in pouches made of various materials are shown in Table VII (1). Packaged tea in the aluminum-film combination pouch shows the best quality among five different packaged tea samples. The shelf-life of the packaged green tea can be expected to be at least 3 months if green tea is packaged in a plastic film pouch, in which moisture content of <5.5% and a vitamin C residual ratio of >70% are retained in green tea (1).

There are positive correlations between decrease of vitamin C and the quality deterioration, as shown in Fig. 3 (1). When the green tea is stored in a nitrogen-packed can and kept at low temperature (5°C), vitamin C is retained almost perfectly even after 12 months' storage. If the green tea is stored in a can from which air is not excluded, and is kept at room temperature, vitamin C content decreases markedly within 3 months, as can be seen in Fig. 3.

Deterioration in quality of green tea is greatly affected by oxygen as well as moisture and elevated temperature. The storage stability of green tea is the lowest among various teas including black tea, oolong tea, and pouchong tea. To protect the quality of green tea during storage most effectively, it is necessary to use nitrogen packing or vacuum packaging.

TABLE I

Relative Composition of Volatile Fraction of Black Tea after Storage for 6 Weeks Expressed in Arbitrary Unit/Unit of Dry Matter

Atmosphere		N_2	N_2	Air	Air	Air	Air
Relative Humidity		Amb*	54%	58%	Amb*	54%	Amb*
Storage Temp.		4°C	35°C	17°C	17°C	35°C	35°C
Moisture content %	Before Storage	2.91	11.49	7.60	3.05	11.68	2.67
Hexanal	21	23	16	8	11	14	11
1-Penten-3-ol	36	39	2	40	42	2	57
E-2-Hexenal	278	174	12	143	129	0	110
Pentanol	2	45	56	55	42	41	29
Z-2-Penten-1-ol	57	36	0	32	37	0	35
Z-3-Hexen-3-ol	62	49	73	53	51	29	44
Hexanol	4	5	6	3	3	0	3
Linalool oxide(trans)	103	105	112	114	101	80	95
Linalool & Octanol	319	328	220	279	243	155	225
Phenylacetaldehyde	51	40	10	29	45	8	31
Methyl salicylate	10	30	55	28	25	43	30
Nerol & Geraniol	2	0	0	0	0	0	0
Phenethyl alcohol	2	0	0	0	0	0	0
Other 17 Compounds	2179	1985	1883	1798	1972	2096	1957
Total	3126	2859	2445	2582	2701	2468	2627

* Amb = ambient.

Analysis of volatile fraction ;

500 mg Black tea → steam distillation → Distillate → GC.
(10 μl)

TABLE II

Changes of Aroma Composition[a] of Black Tea[b] after Storage for 11 Weeks

Components	Before Storage	Storage Temp. 5°C	Storage Temp. 20°C
Carbonyl compounds			
	%	%	%
Hexanal	0.37	0.17	0.11
Nonanal	0.40	0.49	0.52
E-2-Hexenal	3.22	1.15	0.75
E-2-Octenal	0.33	0.13	0.17
E,E-2,4-Heptadienal	0.60	1.23	1.56
E,E-2,4-Decadienal	0.57	0.75	0.99
2,4,6-Decatrienal	0.58	0.51	0.91
Benzaldehyde	0.37	0.18	0.19
Phenylacetaldehyde	0.79	0.84	1.31
β-Ionone	0.65	0.92	1.64
5,6-Epoxy-β-ionone	0.19	0.24	0.33
Alcohols			
Pentanol	0.31	0.10	0.05
Hexanol	0.70	0.60	0.51
Octanol	0.31	0.44	0.55
1-Penten-3-ol	0.05	0.03	0.02
Z-2-Penten-1-ol	0.59	0.85	0.74
Z-3-Hexen-1-ol	8.87	7.41	6.10
E-2-Hexen-1-ol	1.70	1.56	1.54
Linalool	28.24	31.28	28.91
Linalool oxide I (cis, furanoid)	3.46	3.18	3.21
Linalool oxide II (trans, furanoid)	12.94	11.53	12.21
Linalool oxide III (cis, pyranoid)	0.27	0.25	0.30
Linalool oxide IV (trans, pyranoid)	1.10	0.61	0.92

TABLE II (Continued)

Components	Before Storage	Storage Temp. 5°C	20°C
	%	%	%
3,7-Dimethyl-1,5,7-octatrien-3-ol	0.36	0.42	0.42
α-Terpineol	1.00	1.15	1.20
Nerol	0.25	0.37	0.46
Geraniol	3.60	4.10	4.36
Nerolidol	1.42	2.00	2.30
Benzyl alcohol	0.39	0.27	0.81
Phenethyl alcohol	0.39	0.25	0.40
Acids			
Hexanoic acid	0.16	0.16	0.42
Octanoic acid	2.00	0.46	0.91
Nonanoic acid	0.19	0.41	0.50
Decanoic acid	0.17	0.16	0.31
Dodecanoic acid	0.49	0.41	0.66
trans-Geranic acid	0.39	0.30	0.58
Esters			
Z-3-Hexenyl hexanoate	1.34	1.35	1.30
Methyl salicylate	19.32	21.70	19.85
Methyl palmitate	0.28	0.14	0.18
Nitrogenous Compounds			
Benzyl cyanide	0.25	0.22	0.33
Indole	0.63	0.45	0.68

[a] Presented by peak area percentages of gas chromatogram. GC instrument; Shimadzu GC-7A (FID) connected with a computing integrator C-RIB, Column 30m×0.25mm fused silica WCOT, PEG 20M.

[b] Uva quality tea(Sri Lanka), manufactured on middle July, 1983, packed in the Al-laminated paper pouch, sent to Tokyo on Sept. 12, 1983 by air-freight. Storage experiment started on the next day.

TABLE III

Changes in Quantities of Epicatechin Gallate (ECG) and Epigallocatechin Gallate (EGCG) during Storage of Black Tea

No. of Weeks	mg/g dry weight of black tea			
	Bottles[a]		Box[b]	
	EGCG	ECG	EGCG	ECG
0	1.04	2.04	1.04	2.04
1	1.30	1.04	1.90	1.70
2	0.60	1.00	0.90	1.44
5	1.30	0.90	1.40	2.00
8	0.84	0.90	0.86	1.34
12	0.99	0.87	0.92	1.60
16	1.02	0.99	1.02	1.00

[a]Tightly-covered, clear glass bottle.

[b]Loosely covered wooden box.

Taste of both EGCG and ECG is sharp and strong astringency with bitterness, whereas the free catechins (EGC and EC) have mild astringency and pleasant after taste.

TABLE IV

Free Amino Acid Contents in Black Tea after Storage for 31 Weeks under the Different Conditions

Package	Can[a]	Al/PE[b]		PP[c]	WP[d]-1	WP-2
Relative Humidity(%)	Amb[e]	90[f]	Amb	90	90	90
Storage Temp. (°C)	Amb	30	Amb	30	30	30
Moisture Content(%) after 31 Weeks	5.28	5.16	5.11	13.21	13.55	20.82
Amino Acids	μmol/g dry wt of tea					
Theanine	376	326	342	149	206	0
Glutamic acid	18	16	17	7	9	0
Aspartic acid	18	16	16	7	10	0
Serine	9	8	11	3	4	0
Arginine	6	6	6	3	4	0
Phenylalanine	3	4	3	2	3	0
Threonine	2	3	2	1	1	0
Glycine, Alanine	2	2	2	1	2	0
Valine	2	3	2	1	2	0
Isoleucine	2	1	1	1	1	0
Leucine	2	3	2	1	1	0
Tyrosine	2	3	3	2	2	0

[a]Unlined seamed can. [e]Ambient.

[b]Aluminum/Polyethylene laminate sachet.

[c]Paper/Plastic laminate sachet.

[d]Wax paper sachet. [f]Stored in a humidity cabinet.

TABLE V

Changes of Aroma Pattern of Green Tea[a] during Storage at Different Temperatures

Components	Before Storage	Ratio of Peak Height[b] 5°C 2 Mon	5°C 4 Mon	25°C 2 Mon	25°C 4 Mon
1-Penten-3-ol	-	-	55	32	94
Unknown	59	33	33	16	13
Z-2-Penten-1-ol	-	-	26	15	45
Z-3-Hexen-1-ol	16	17	29	26	60
Nonanal	104	69	51	24	22
2,4-Heptadienal	-	-	17	-	16
3,5-Octadienone, Benzaldehyde	-	-	14	12	17
Linalool	100	100	100	100	100
Octanol	95	83	86	86	85
Z-3-Hexenyl hexanoate	85	68	65	46	36
Nerolidol	130	123	125	133	130

[a] Packed in aluminum laminate film pouch.

[b] Expressed by relative peak height to Linalool = 100.

TABLE VI

Variation in Aroma Composition of Green Tea during Storage[a]

Components	Ratio[b] of peak area					
	High grade of spring tea			Low grade of summer tea		
	0	2	4	0	2	4 Months
1-Penten-3-ol	0.14	0.72	1.36	0.27	5.45	9.78
Pentanol	0.47	0.65	0.67	0.47	0.99	1.12
Z-2-Penten-1-ol	0.19	0.53	1.21	0.38	5.02	8.46
6-Methyl-5-hepten-2-one } Hexanol	0.29	0.58	1.08	0.29	0.58	1.08
Z-3-Hexen-1-ol	0.22	0.78	1.47	0.32	0.83	1.04
Nonanal	0.73	0.51	0.62	0.66	1.06	1.35
Linalool oxide(cis, furanoid)	0.18	0.18	0.36	1.73	1.73	1.93
E-2,Z-4-Heptadienal } Linalool oxide(trans, furanoid)	0.37	0.65	0.90	0.84	10.13	9.03
E-2,E-4-Heptadienal	0.18	0.46	1.08	0.34	6.26	11.60
E-3,Z-5-Octadien-2-one } Benzaldehyde	0.20	0.38	0.54	0.40	1.75	2.21

TABLE VI (Continued)

Linalool	1.02	1.05	1.10	3.01	3.78	3.76
Octanol	0.99	1.12	1.35			
E-3,E-5-Octadien-2-one	-	-	-	0.60	1.26	1.26
2,6,6-Trimethyl-2-hydroxy-cyclohexanone	0.20	0.52	0.95			
3,7-Dimethyl-1,5,7-octatrien-3-ol	-	-	-	1.54	2.07	2.83
β-Cyclocitral	0.21	0.43	0.66			
Z-3-Hexenyl hexanoate	1.20	1.04	0.93	0.36	0.33	0.32
Linalool oxide(trans, pyranoid)	0.83	0.67	0.75	0.53	0.56	0.61
Geraniol	0.67	0.68	0.72	0.96	1.08	1.23
α-Ionone } Benzyl alcohol	0.88	0.96	1.43	1.24	2.46	3.41
cis-Jasmone } β-Ionone	0.66	0.85	1.19	1.45	1.58	1.78
5,6-Epoxy-β-ionone	0.36	0.59	0.70	0.35	1.07	1.28
Nerolidol	0.98	0.75	0.76	0.78	0.74	0.73
Dihydroactinidiolide	0.25	0.37	0.41	0.13	0.70	0.93
Indole	0.70	0.65	0.61	1.03	1.03	0.94

[a] Packed in the aluminum laminate pouch and stored in a dark room at 25°C .

[b] Ratio of peak area of compound/peak area of internal standard(ethyl decanoate).

TABLE VII

Changes of Moisture and Vitamin C Contents during Storage[a] of Green Tea[b] in Various Packages

Material of pouch	Moisture content (%)			Percent loss of Vitamin C		
	1	2	3	1	2	3 Months
PT[c] #300.Al[d] 9μ.PE[e] 40μ	3.5	3.9	4.0	7.4	6.7	8.8
PT #300.Al 112μ.PE40μ	3.6	3.2	3.5	1.4	5.0	8.6
PT #300.Al 15μ.PE40μ	3.2	3.2	3.2	4.6	5.6	8.7
OPP[f] 20μ.PVDC[g] 2μ coating.PE 60μ	4.4	5.4	6.0	8.1	20.5	32.1
PT 30μ.PE 13μ.P[h] 40g.PE 20μ	6.9	8.8	10.8	27.9	42.1	69.4

[a]Packed in the pouch and stored at 25°C, RH 80%.

[b]High grade spring green tea, original contents of moisture and vitamin C are 3.1 % and 372.3 mg/100g tea, respectively.

[c]Cellophane.

[d]Aluminum.

[e]Polyethylene

[f]Polypropylene.

[g]Polyvinylidene chloride.

[h]Paper.

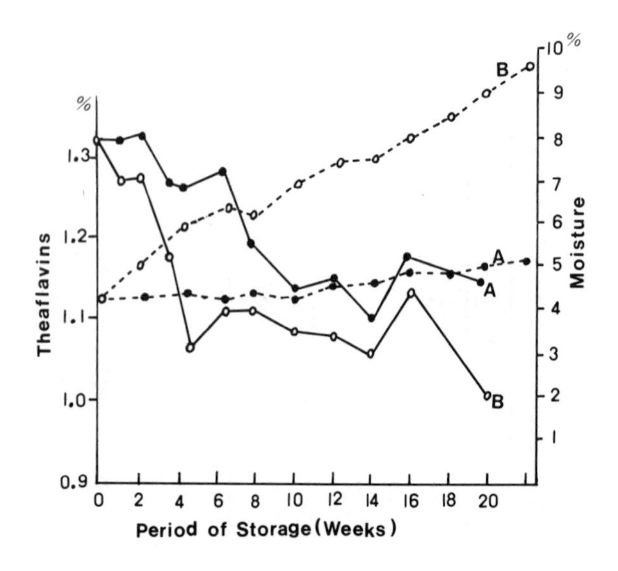

Fig. 1. Variation in theaflavin (———) and moisture contents (-----) during storage of black tea (A) in an air-tight glass bottle (●), and (B) in a loosely covered wooden box (o).

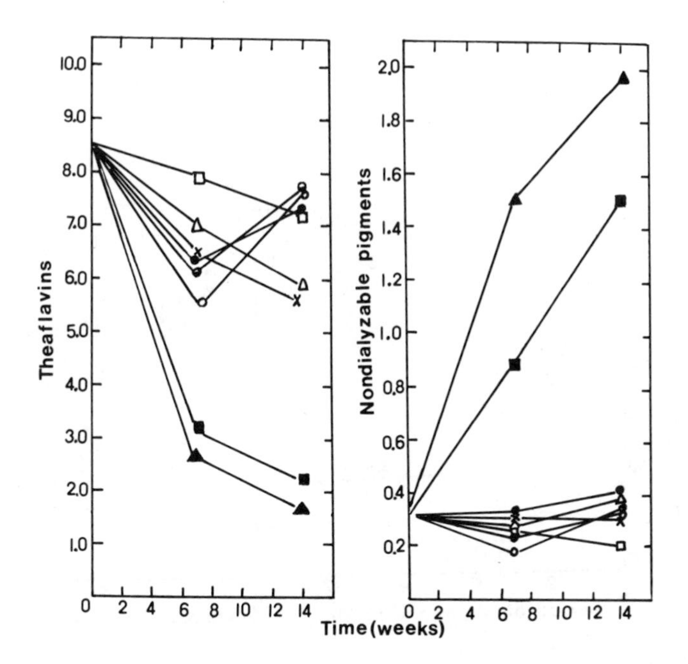

Fig. 2. Changes in theaflavins and nondialyzable pigments during storage of black tea (arbitrary units/unit dry weight).

Conditions of storage		Moisture contents after	
Temperature (°C)	RH (%)	7 weeks	14 weeks
▲ 35	75	11.10	11.88
△ 17	76	11.27	11.51
■ 35	61	8.43	8.62
□ 17	56	7.75	7.94
● 35	45	5.11	5.19
○ 17	44	5.55	5.98
× 35	0[a]	2.02	1.85
◕ 17	0	2.51	3.31

[a]Stored in a desiccator with silica gel.

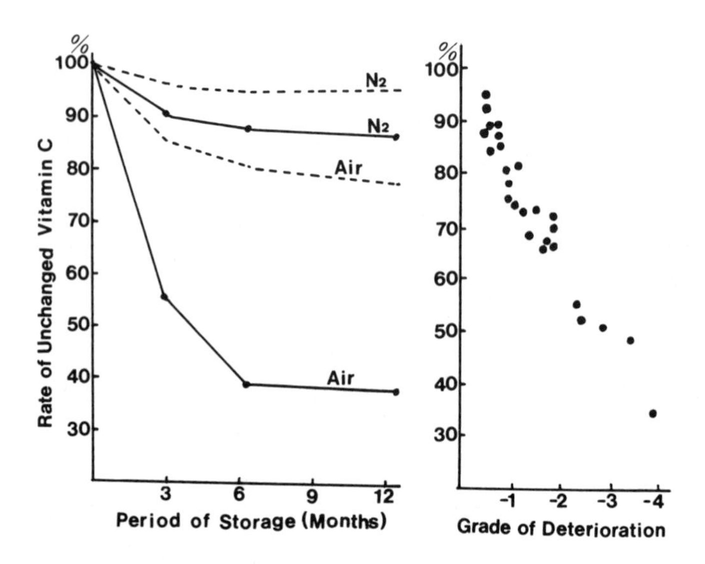

Fig. 3. Correlation between rate of unchanged vitamin C and grade of deterioration in quality of green tea. -1, Scarcely recognizable deterioration; -2, slightly; -3, considerably; -4, extremely. Storage temperature: ambient (———); 5°C (-----).

REFERENCES

1. Fukatsu, S., and Hara, T. (1971). *Study Tea 43*, 36-40.
2. Hara, T., and Kubota, E. (1979). *Nippon Shokuhin Kogyo Gakkaishi 26*, 391-395.
3. Hara, T., and Kubota, E. (1982). *Nippon Nogei Kagaku Kaishi 56*, 625-630.
4. Jayaratnam, S., and Kirtisinghe, D. (1974). *Tea Q. 44*, 170-172.
5. Kawakami, M. (1982). *Nippon Nogei Kagaku Kaishi 56*, 917-921.
6. Kobayashi, A., Kubota, K., Kihara, N., and Yamanishi, T. (1985). To be published.
7. Nakagawa, M. (1969). *Nippon Shokuhin Kogyo Gakkaishi 16*, 266-271.
8. Sakamoto, Y. (1970). *Bull. Tea Res. Stn. (Minist. Agric. For. Jpn.) 6*, 44-45.

8a. Sakamoto, Y. (1970). *Bull. Tea Res. Stn. (Minist. Agric. For. Jpn.) 6*, 47-48.

9. Stagg, G. V. (1974). *J. Sci. Food Agric. 25*, 1015-1034.
10. Takei, Y., Ishiwata, K., and Yamanishi, T. (1976). *Agric. Biol. Chem. 40*, 2152-2157.
11. Tanaka, N., and Hara, T. (1972). *Study Tea 44*, 25-30.
12. Wickremasinghe, R. L. (1978). *Monogr. Tea Prod. Ceylon 7*, 1-14.
13. Wickremasinghe, R. L., and Perera, K. P. W. C. (1972). *Tea Q. 43*, 147-152.

CHAPTER 13

COFFEE

RONALD J. CLARKE
Donnington, Chichester, Sussex, England

I. INTRODUCTION

A. Types of Coffee

Coffee, as the beverage actually consumed, is a dilute aqueous brew made by extracting roasted coffee with hot water, or by makeup (dissolution) of a preprepared extract (usually in the dry form, or "instant"), again with hot water. The term coffee, however, strictly covers also a range of transformation products beginning with the harvested coffee cherries and ending with the consumable coffee, all of which may be stored for various periods of time. The coffees to be considered here are (1) green coffee (beans), the basic raw material, normally containing ∿12% (w/w) moisture, (2) roasted coffee, (3) roasted and ground coffee, and (4) instant coffee, containing about 1-5% (w/w) moisture. These coffees are each characterized in further detail by the coffee trade, for which reference should be made to the extensive literature (4,21), and indeed for information on their manufacture. The distinctions between arabica and robusta coffee, and between wet and dry processed should be noted, together with the terms "hulled" and "hulled/cleaned" in green coffee.

B. Shelf-Life Factors

It is important in examining shelf-life data for coffee that the particular coffee in question is fully characterized, if correct conclusions are to be drawn; unfortunately, this information is not always available in the published literature. It is also important that the methods of analysis--particularly for the two main determinants of shelf-life, moisture content and headspace oxygen content--be fully understood, as well as the methods of sensory assessment.

1. Moisture Content

Methods of determination specifically for coffee, published by the International Standards Organization (or national standards bodies) are available and coming increasingly into general use (5). In practice a variety of other methods may be used, giving value of "moisture" content, either slightly higher or lower (see Table I; these methods are referred to by their numbers in subsequent tables). Most methods, of course, give arbitrary measures of water content, with respect to the true values. Nevertheless, the method of determination should be known, as small differences in actual moisture or water contents can have marked effects on shelf-life values. Most methods in use for shelf-life data are oven-based, with variants in time, temperature, pressure, and staging. Different modes of expression should also be recognized, that is, either "dry basis" or "as is" (%w/w).

Since water content is a key parameter for shelf-life, knowledge of water sorption isotherms for coffees is valuable, but studying them will not in and of itself permit assessment of shelf-life, despite the value of such concepts as the monolayer. A substantial number of such isotherms have been published in a single text by Cherife and Iglesias (12). Additional references to isotherms for green coffees are mentioned in this text.

2. Headspace Oxygen Content

The oxygen content in the headspace above a stored coffee is another important determinant in shelf-life, especially for roasted (and ground) coffee, and for soluble coffee, which again requires care in its correct and accurate method of determination, and also in the point of time at which is is actually measured (see Section III).

For shelf-life assessment purposes, for roasted coffee held within a closed impermeable package for retail sale (e.g., a vacuum can or suitable flexible laminated package), it is usually the initial O_2 content just after packing that is measured and recorded. This content will necessarily fall off with time, as the O_2 is absorbed and reacts with the coffee. The preferred mode of expression of O_2 content is in micrograms of O_2 per gram of coffee, which can, however, be calculated from the measured percentage O_2 content (% volume, of a sample of the headspace gas, by suitable metering device), gas density, the absolute pressure, and headspace/void volume within the package. This calculation can give rise to confusion, so that clarification is needed.

For example, if the package volume is, say, 517 ml for 227 g coffee (or 2.28 ml/g coffee), and the inherent density of the "solid" coffee is 1.2 g/ml, then the total headspace/void volume for gases is (2.28 - 1/1.2) = 1.45 ml/g of coffee. Suppose the measured percentage O_2 content by volume (weight) in the package, taken or converted to atmospheric pressure (1013 mb = 760 mm Hg = 30 in. Hg), is 1.0%. With an O_2 density, at a temperature of 20°C and at this atmospheric pressure, taken as 1.525 g/liter, then the actual weight of O_2 present in the package will be $1.45 \times 10^{-3} \times 1 \times 10^{-2} \times 1.425$ = 0.02 mg/g coffee or 20 μg O_2/g coffee, which is a typical figure in roast coffee packing. In this type of calculation, the voidage inside the coffee particles is sometimes ignored, but not of course the external voidage, so that a specific volume of the coffee itself is taken, rather than the "solid" coffee.

Sometimes with vacuum packs, only the vacuum applied during packing is stated. If a vacuum of 28.6 in. Hg is pulled (against 30 in. Hg atmospheric pressure), then the absolute pressure in the package is 1.4 in. Hg (equals 47.4 mb = 35 mm); and if the headspace gas is air, then the corresponding percentage O_2 content will be $1.4/30 \times 21 = 1.0\%$. We can say, therefore, that the initial O_2 content is 20 μg/g coffee as before. The partial pressure of the O_2 is, incidentally, 10 mb, another type of figure that is often quoted. If the headspace actually contains substantial quantities of carbon dioxide or other gases, then this same O_2 content is obtainable at a lower vacuum (or higher absolute pressure).

Sometimes, the package volume is not stated. As a consequence, a given percentage O_2 content figure will represent a higher true O_2 content for a vacuum-packed can than for a "hard" flexible pack containing the same weight of coffee, even at the same pressure/vacuum, on the account of the different headspace volumes present.

Radtke (17,18), has also noted that an immediately measured percentage O_2 content, in the packing of roasted and ground coffee, can rise in value if measured subsequently (say after 1 to 2 days) in a conventional vacuum pack of roasted coffee, due to slow desorption of any O_2 absorbed physically prior to packing and after or during grinding. Such a higher value, if found, represents a more true measure of the O_2 available for subsequent deterioration of the coffee. Some O_2 absorbed may have, of course, already reacted with the coffee before packing, and not be released. Some measure of this effect can be deduced in noninert gas-flushed packages by determination also of the headspace nitrogen content, with which in air, the initial O_2 content would have been associated in the ratio 21:79.

Some of these considerations apply also to instant coffee, although this product is almost invariably packed by gas displacement methods, and not under vacuum. Much instant coffee is packed, however, in ambient but dehumidified air. With green coffee, stored either in jute bags or in silos, O_2 contents are not a commercial factor. Storage in silos can be under either sealed or ventilated conditions.

3. Other Measurements

Temperature in storage, combined with moisture content and O_2 content, is of course an important factor in shelf-life. In shelf-life studies temperature is kept constant and assessment made at different temperatures. Under practical conditions, temperatures are likely to vary somewhat during expected storage life, particularly with green coffee.

Storage times are given in either days or months, or sometimes years.

4. Sensory and Quality Assessment

As with other foodstuffs and beverages, shelf-life determination of coffee in any of its forms requires a subjective assessment of "flavor quality" at any stage of storage; that is, of the consumable brew prepared from it. In general, sensory assessment is particularly subjective in the case of coffee. Flavor quality will also depend to a degree on the exact method of brew preparation (and resultant percentage solubles concentration). In assessing the quality of green coffee, the coffee beans must, of course, first be roasted (to a specified degree, usually "medium" roast) and then ground (again to a specified level) before brewing, usually by a standardized "steeping" method (with a specified water:coffee ratio). Whereas actual consumers may well make milk and sugar additions (at different levels), expert assessment will be carried out with "black" coffee. Quality may also be separately assessed by so-called aroma, especially of roasted and ground coffee, and of instant coffee, before brewing. Furthermore, there can be appearance changes during storage, especially of color (e.g., in green coffee, although this will be correlatable with flavor quality) and of caking, in the case of instant coffee.

Numerical ratings of quality are widely used, assessable from individual experts (e.g., "liquorers" in Kenya), or an average from panels composed of a number of experts (defined in different ways); and of course, from consumer test results. Flavor quality scale ratings on a 10-point scale are popular in coffee testing, for example, 9 (initial quality) as reported by Heiss *et al.* (11) for roasted and ground coffee, who also relates the values to percentage quality retentions down to 5 (as 0%) and provides qualitative statements at each level down to 1. It is therefore important to have in mind a numerical value (on whatever scale) regarding what may be regarded as acceptable, to define an allowable shelf-life. Rating scales are often determined monadically (i.e., without simultaneous assessment against a reference standard at any testing session). In following shelf-life, however, a standard should be a sample having been stored under conditions such that change is minimal, and assessed at the highest point of the scale. For coffee products, this would necessarily be storage under "deep freeze" conditions at least -20°C and preferably lower, and a nominally zero headspace O_2 content. As will be noted with roasted coffee in packages, there can be some difficulties determining what should be the "initial" sample or point of time zero; this is normally taken at the time of actual packing, although immediately after roasting (for roast whole beans, or RWB) and grinding (for roast and ground, or R and G, coffee), may well be more appropriate.

The use of rating scales needs to be allied with a statistical interpretation, chiefly by the analysis of variance among the panelists to establish significant levels, when comparing one quality level with another. Triangular testing (against the reference standard) by a panel has also been used for tests of significant change. All tests of significance, however, raise the issue of "significant to whom?", which brings us back to the characteristics of the panel judges. These issues, together with correct procedures for panel testing, are considered in a number of texts, either general or specific to coffee.

Coffee brew flavor quality is often also defined in terms of the absence or presence of particular off-flavor notes or those regarded as especially significant to coffee flavor quality (e.g., "acidity," as distinct from characteristics, often confused by consumers, of "bitterness" and "astringency"). Different coffees even in their initial condition will differ in levels to positive flavor notes, thus the need for reference comparison with zero-time samples. In ambient storage the positive flavor notes may decline in intensity, but the characteristic flavor note that eventually develops is that of "staleness." As discussed under roasted coffee, this onset can be used as a determinant of shelf-life. Flavor in the context of this chapter is the combination of odor (due to volatile components or aroma in coffee) perceived through the nasal passages during testing, plus taste (referring to the base characteristics of "acid," "sweet," "salt," and "bitter," together with various mouth-feel sensations). While changes due to storage can take place in the base characteristics, changes are primarily related to the volatile components. It is probable that this "stale" note is related to changes in these components. While coffee contains substantial amounts of relatively unstable vegetable oil, a small proportion of which enters extracts, "rancidity" of this oil is not believed to occur except under very unfavorable or lengthy storage conditions.

Quality of roasted coffee is also assessed by its headspace aroma (i.e., by "sniffing" of the dry product), which of course is entirely related to the volatile components, in particular the more volatile ones (many sulfur-containing) that enter the front nasal passages. Experience has shown that although that change can be detected in this headspace aroma, this change may not be reflected early on storage in any perceptible flavor change. Instant coffees, as will be noted, differ very considerably in initial flavor with origin, some approaching the flavor and headspace characteristics of roasted coffee, while others, especially simple spray-dried powders, will not.

Mathematical relationships between shelf-life parameters and sensory evaluation are possible, though hazardous; some tentative relationships have been noted.

5. Compositional Changes

Green coffee, roasted coffee, and instant coffee have all been noted to undergo some compositional changes during storage, and attempts have been made to correlate these with sensory assessments. None of these changes, usually small and dependent on the severity of the storage conditions, has been successfully used to replace, in any way, subjective sensory assessment of quality changes, although some interesting correlations have been found, especially with roasted coffee, which are briefly described here.

II. GREEN COFFEE

A. Direct Shelf-Life Data

There have been four major published studies: those of Stirling (22) in Kenya of a wet-processed arabica coffee, Multon (15) in France, following an earlier study of the same Angolan wet-processed arabica by Corte dos Santos (7), and Natarajan (16) in India.

Table II is taken from Stirling's graphical data, giving quality ratings after different storage times up to 1 year in a small laboratory silo at different temperatures but fixed initial moisture content.

The results of Corte dos Santos (also graphical) are set out in Table III, where quality ratings are given after closed storage in a small container at five different controlled humidities (and therefore moisture contents of the coffee) at a specific temperature over 6 months. The storage studies were actually carried out in France, with the green coffee prepared under precisely reported conditions and transported by air to Paris, with a reference standard kept in cold storage. Apart from flavor quality, increases in numbers of defective beans were also reported. The results of Multon are set out in Table IV, which extended the above work on the same coffee, to include also four different temperatures at the same different humidities. All the original sensory evaluation data, graphically presented, are not suitable for tabulation, so that a selection is taken, specifically at 75% RH. The study also included data on all the changes, including the incidence and growth of microorganisms.

Natarajan briefly reported the considerable studies on the large-scale storage of both arabica and robusta coffee, conducted in India, and "dry" and "monsoon" ambient air conditions, with particular reference to changes of appearance and color as measured by reflectance methods. His data are, however, based on a moisture method determination, involving only 4 hr oven heating at 105°C, and therefore are not detailed here.

There have been no particular reports on the effect of different O_2 contents during storage.

B. Conclusions

Wet-processed arabica coffee should keep its original quality for ≥6 months, provided that the following storage conditions are met:

1. The coffee has been dried (in its parchment form) to a moisture content of ≤11.0% (w/w) (Stirling), or 13% dry basis (=11.5% w/w, Multon); these figures probably correspond with that of Natarajan (10% w/w) by his method of determination.
2. The RHs are in the range 50-70% (Stirling) or 50-63% (Corte dos Santos, Multon).
3. The air temperature is maintained ≤20°C (Stirling) or ≤26°C (Multon).

Additional advantages will be obtained by providing ventilation (even though this may only be of ambient air) and by hulling but not cleaning it (because an additional protective layer, the silverskin, is retained). These criteria are probably similar for dry-processed arabica or robusta.

In practice, temperature and humidity conditions are likely to vary over the whole of the storage period of exported coffee in different places, making it difficult therefore to estimate shelf-life to the same final quality. Normally bagged coffee ("new crop") would not be kept in the producing country for long periods, such as 6 months. If a long period of storage is envisaged, then the coffee (if also wet processed) should be kept in its parchment or unhulled form (either in bags or silos), when for the same storage conditions, Stirling also showed that parchment coffee can be expected to maintain its original quality for at least a year. It is of interest that at one time shipments of parchment coffee were made. Stirling also commented that as a general rule, controlled ventilation is more beneficial than sealed conditions--especially at the higher temperatures--and that thermally insulated structures can offset the problem of high daytime temperatures in producing areas. Bauder (3) claimed remarkable expected advantages, even economic ones, of a system of chilled-air injection into storage buildings, largely to prevent the escalating effect of any heat generation through respiration within the coffee beans. No evidence of advantage or disadvantage has been offered on the potential value of gas-controlled storage (e.g., use of CO_2), or even "sulfiting" on a similar basis of restricting respiration.

It is the low-lying ports of export where extremes of ambient temperature and humidity can be most expected. Depending on the period of storage (which should normally be short), some changes will occur, unless more favorable storage conditions are mechanically imposed. The remaining periods of storage will be in the holds of the transporting ships and in the stores of the consuming country. In general, ventilation is recommended. A particular problem in ship transportation (including containerized shipments) is a danger from some moisture release, as a consequence of a change to a lower air humidity, rather than temperature (see sorption isotherms); such moisture may deposit locally and cause mold growth.

There are also further complications in precisely defining shelf-life for green coffee, not the least because some storage would appear to be desirable. Furthermore, some storage conditions provide flavor qualities that are desired among some eventual consumers (e.g., "aged" coffees from Venezuela and the "monsooned" coffees of India). There are also trade distinctions of "new crop" coffee (freshly harvested and processed), "second crops," and "old crop" (necessarily stored for a longer period in the producing country), which find separate acceptance.

III. ROASTED COFFEE

Roasted coffee is available commercially in two forms: roasted whole beans (RWB), and--more and more commonly--roasted and ground coffee (R and G). Consumer packages seek to provide protection against the two main deteriorative factors of moisture and O_2, by use of packaging material, impermeable to each to a greater or lesser extent, and also generally to minimize loss of coffee volatile substances, causing flavor imbalance. Packing operations are generally designed to minimize the initial O_2 content within the package, by either of two methods: (1) displacement by use of inert gases and (2) evacuation of headspace air by vacuum application (11).

The packing of roasted coffee is complicated by the release of CO_2, entrapped within the whole beans during roasting, at a rate exponentially with time, depending also on blend, degree of roast, and, importantly, grind or particle size for R and G. For this reason, the release of CO_2, if not accommodated, will cause excess pressure within a closed package of the coffee. A number of methods are available to deal with this problem: the use of a one-way valve in the package (e.g., Goglio or Hesser), and various methods of "degassing" the coffee in bulk before packing, applied for a period of time (hours, up to a day or two for R and G coffee, but longer for RWB) dependent on the amount of CO_2 needed to be released (in part, dependent on the initial vacuum or pressure in the package). The handling of the roasted coffee, and therefore opportunities for contact with O_2 and differential volatile losses prior to its actual packing, is also therefore a determinant in subsequent shelf-life of actual packages. These aspects have been studied and reported particularly at the Fraunhofer Institute of the Technological University of Munich (11).

A. Shelf-Life Data for Roasted Whole Beans

Three major sets of direct experimental data are available: two involve expert or semiexpert tasting [i.e., Cros and Vincent (8), Radtke (17), and with Piringer (18)], and another, Ernst (9), for consumer testing.

Table V shows data of Cros and Vincent for various kinds of packs, and therefore different initial O_2 contents (not stated). The choice of reference standard was regarded as important; a roasted sample left in the open 24 hr after roasting and longer before placing in storage at $-28^\circ C$ was already regarded significantly different from that left for only 7 hr.

Table VI shows the results of Radtke-Granzer taken from her graphically presented data, for RWB in an air pack.

B. Shelf-Life Data for Roasted and Ground Coffee

Tables VII and VIII show the two main sets of results graphically presented by Radtke (17) and Clinton (6), respectively, for a given moisture content and storage temperature, but different initial headspace O_2 content in closed packs. Both use 10-point scales, but numerical ratings under similar conditions are somewhat different, especially in starting values. There is continuous exponential-type change from time zero, presumably taken at the time of packing. Prepacking conditions are not given. Clinton's original paper includes graphically presented correlative data between an expert panel and a consumer panel, in order to provide additional levels of meaning for expert ratings.

Table IX gives data for the same R and G coffee and pack as in Table VIII, but with a higher moisture content.

Table X provides a compilation of temperature effects on shelf-life from different sources, which are reasonably consistent, and which includes both RWB and R and G coffees.

C. Data from Compositional Changes

A number of studies have attempted to correlate subjective quality ratings, or sensorially detectable deleterious change, with compositional changes in roasted coffee during storage, especially of selected volatile substances, determinable by gas chromatographic (GC) methods. The most explored indicator is the ratio of the contents of 2-methylfuran to butanone-2, or the so-called aroma index M/B, as examined by Reymond *et al.* (19), Kwasny and Werkhoff (14), Vitzthum and Werkhoff (25,26), and Arackal and Lehmann (2); other indicators have been examined by Radtke-Granzer and Piringer (18) and by Tressl (24), while Reymond (2) has provided a general review.

While this technique is useful for a roasted coffee of known blend and roast degree, it cannot be used directly to assess the condition of a coffee of an unknown type. An indicator such as M/B ratio is strongly dependent on blend and roast degree, though in a known manner (Kwasny and Werkhoff), and has been correlatable in a given instance with a flavor rating scale (Vitzthum and Werkhoff). The data can be used to make comparative statements of the shelf-life of a given coffee packed under different conditions (see Table XI).

C. Conclusions

There are two commercial packages of great interest in respect to their shelf-life:

1. RWB in bags (impermeable to moisture) packed in air, for a short shelf-life.
2. R and G coffee (also RWB) packed at either high-vacuum level, or gas-flushed to low O_2 content, in packages preventing ingress of either O_2 or moisture, for an extended shelf-life. In either case, an in-pack O_2 content of $\leq 1\%$ is fully feasible.

Although it is apparent that both RWB and R and G coffee deteriorate in quality continuously from the time at which the green coffee was actually roasted, the concept of Heiss (11) and Radtke (17,18) in defining three distinct quality levels during storage is most useful.

Tables XII and XIII set out shelf-life for these categories of packed products by this and other criteria.

It is clearly not possible to tabulate for all possible storage conditions, but the data, though inadequate, presented in all the tables (i.e., moisture and O_2 content, and temperature), allow some reasonable inter- or intra- extrapolations. Mathematical relationships are hazardous with subjective quality ratings; however, it is of interest that the Radtke data show a reasonable linear correlation of log (quality rating) versus time for each of the O_2 levels, and similarly for log (O_2 content) versus time at each quality level and including extrapolation to an air pack for R and G coffee. For example, the later would be expected to have a shelf-life of <u>10 to 15 days</u> to ratings 5-4, reported by M/B ratio data. In the same way, the effect of temperature is consistent with an Arrhenius-type equation relating log (rating) to the reciprocal of the absolute temperature for the same storage conditions. Some differences may well be found resulting from other differences such as blend-roast and color-grind-degassing procedures. In relation to consumer assessment, it should be noted that commercial packages of these products are not likely to be sold before 1 month after factory packing, and more likely at 3 months (6). Clinton also showed that U.S. consumers were not able to detect many changes detected by experts. In consumer use, packages will be opened and reclosed frequently; during this period the shelf-life is essentially that of an air pack. In good commercial practice for low O_2 contents, a shelf-life of 20 to 25 months for acceptable quality can be expected.

IV. INSTANT COFFEE

Although the commercial manufacture of instant coffee is now substantial in volume in the United States, Europe, and Brazil, there are only a limited number of references to studies of its stability in storage. Apart from isolated statements and general packaging-type information, the two main studies are those of Harris *et al.* (10) and Clinton (6).

Instant coffees differ very widely in character, not only because of different blends, roast degree, and extraction yield taken, but more importantly in their retained volatile component most susceptible to flavor and headspace aroma changes on storage. These differences arise, as might be expected, from different controls on drying conditions and overall processing methods that can be exercised, as described by Thijssen (23). Increased sophistication in processing is now general, compared with the early simply spray-dried powder products, so that much higher measurable volatile component contents, comparable with brewed coffee, are now found. Instant coffee is highly hygroscopic and liable to caking at about 7-8% moisture content, so that packaging to prevent moisture ingress is mandatory, on that account alone. Certain commercial instant coffees are "aromatized," by surface application of aroma-bearing coffee oils, primarily for the purpose of providing headspace aromatics for the dry product, without necessarily providing volatile components in sufficient quantity to influence "brew" flavor. Such aromatics are still strongly susceptible to change and deterioration.

A. Shelf-Life Data

Harris *et al.* conducted studies on commercial samples of a spray-dried and agglomerated and freeze-dried instant coffee (both, however, of unstated initial age and precise origin), which had been carefully repacked in impermeable packages at different headspace O_2 levels (0%, 2%, and air), and then stored at 37.8°C (i.e., so-called accelerated storage conditions) in duplicate. Their moisture contents were 4.5% (w/w) and 2.0% (w/w), respectively. Panel sensory assessments were carried out at intervals over a period of a year, on a 9-point rating scale. In general, very few significant changes were found, and none in flavor when comparing values of zero-time and final samples. However, the expert panel ratings were monadic and made little use of the spread of the scale (i.e., starting figures were only 5.8-5.5.

This information is not tabulated here, but reference should be made to the comprehensive details in the original paper.

Clinton also conducted studies on a commercial spray-dried and agglomerated coffee and a freeze-dried coffee. Table XIV shows the data for storage of these coffees at 21°C at a specific headspace O_2 content and moisture contents for each, corresponding to those used for retail sale. The rapid fall in expert panel ratings, especially for the freeze-dried sample, after 3 months is evident, although the agglomerated sample still has a rating of 5.6 even after 72 weeks. These falls are apparently exponential in character; it should be noted that although a 10-point scale was used, starting levels are some way down this scale.

Table XV shows the data for these same coffees but at higher and lower moisture contents. The decrease in quality ratings is slightly greater, the higher the moisture content (up to 5.1% for agglomerated and 3.2% for the freeze-dried coffee) for the same other storage conditions. Important aspects of this paper, however, are the correlations of expert panel ratings with consumer panel ratings, wherein quite large drops in the former are not reflected in statistically significant drops in the latter.

B. Conclusions

It is difficult to draw generalized conclusions from the range of instant coffees available, and with the very limited shelf-life data and accompanying information available.

However, many coffee extracts, simply spray-dried, are packed in jars, cans, or other packages with an air headspace (21% O_2). Provided the moisture content is maintained at less than 4-5% (w/w), it is well known that they will retain their original quality for at least 2 years at ambient conditions. With more sophisticated products (i.e., with higher retained levels of volatiles), shelf-life at still acceptable coffee quality can also be not less than ∿18 months, provided that additionally the initial O_2 headspace contents are lowered to levels <4.0% (i.e., comparable with those required for roasted coffees).

TABLE I

Methods of 'Moisture' Determination

Method Number	Method Reference	Applicable to	Procedure
1.	ISO 1446 (Guilbot)	Green	Oven heating at 48°C. Anhydrous atmosphere for 150 hours approx. under vacuum (10-20mb).
2.	EAIRO (Kenya -Wootton)	Green	Two stage oven-drying. First, at 95-100°C for one hour, then grind; second, in vacuum oven with small dry air leak, at 30 mm Hg pressure, 70°C for 22 hours.
3.	ISO 6673	Green	Oven heating of whole beans at 105 ± 1°C for 16 hours (values obtained will be 1% lower approx. than methods above).

(Table continues)

TABLE I CONTINUED

Method Number	Method Reference	Applicable to	Procedure
4.	AFNOR (V05-202)	Roasted	Air oven heating at $103^{o}C$ for 16 hours.
5.	AOAC Handbook 12th Edition	Roasted/ Instant	Vacuum oven at $100^{o}C$ for $5\frac{1}{2}$ hours.
6.	ISO 3726	Instant	Vacuum oven at $70^{o} \pm 1^{o}C$ with dry air leak, at 37.5 mm Hg, for 16 hours.

TABLE II

Quality Ratings of a Stored Green Coffee

			Parameters					
Coffee	Storage Conditions		Rating[b] following a time[c] of Storage of					
Type[a]	Temp.[d]	Other	0	1	3	6	9	12
Hulled /Cleaned	10	V[e]	4.0	4.0	5.5	5.0	5.0	5.0
	25		4.0	4.5	4.5	5.2	6.0	6.3
	35		4.0	5.1	5.0	7.2	6.8	7.0
Hulled /Cleaned	10	S[f]	4.0	4.0	4.2	4.2	4.5	4.5
	25		4.0	4.5	5.0	5.0	5.3	5.5
	35		4.0	4.7	5.5	6.3	8.0	10.0

(Table continues)

TABLE II CONTINUED

Coffee Type[a]	Storage Conditions Temp.[d]	Other	Rating[b] following a time[c] of Storage of 0	1	3	6	9	12
Hulled/ not cleaned	10	S[f]	4.5	4.0	4.5	4.0	4.8	4.1
	25		4.0	4.5	5.0	4.5	5.2	5.2
	35		4.5	5.5	6.0	6.0	7.5	10.0

Data of Stirling (1980). Intermediate temperatures not included

[a] Kenya wet-processed arabica coffee dried in parchment to 10.4 $\pm$ 0.1% $^{w}/w$ moisture content (corresponding to 55% ERH).

[b] Kenyan liquorers average rating, includes assessment of green and roasted bean appearance, 10-point scale, 1-5 acceptable, 10 rejectable.

[c] in months.

[d] in ^{o}C.

[e] Ventilated with air at a constant relative humidity of 60%.

[f] Sealed storage silo.

TABLE III

Flavour Quality Ratings of a Stored Green Coffee[a]

Parameters											
Storage Humidity[b]	EMC[c]	Rating[d] following a time[e] of Storage[f] of									
		0	15	30	45	60	75	90	120	150	180
51	11.1	3.5	-	3.0	-	3.0	-	3.5	2.5	4.5	4.0
63	13.5	4.0	-	3.5	-	3.0	-	3.5	3.5	4.0	4.0
73	16.6	5.0	5.0	4.0	4.5	4.0	4.0	-	-	-	5.0
82.5	20.3	5.0	5.5	5.0	3.0	4.5	4.5	-	-	-	6.5
92	29.3	5.0	6.0	8.0	7.5	-	-	-	-	-	-

Data of Corte dos Santos (1971)

[a] Angolan wet processed coffee.

[b] Percentage Relative Humidity, $\pm 1\%$.

[c] Corresponding equilibrium moisture content, g H_2O/100g dry coffee matter, by method 1.

[d] see Table IV.

[e] in days.

[f] Sealed storage at $26^{o}C \pm 0.5^{o}C$.

TABLE IV

Flavour Quality Ratings of a Stored Green Coffee[a]

Temperature of Storage[b]	Parameters: Rating[c] following a time[d] of Storage of									
	0	10	25	60	90	105	110	120	140	180
20	6.0	-	5.5	-	-	-	-	-	-	-
25	5.0	6.5	-	-	-	-	-	4.5	-	5.5
35	-	-	-	-	5.5	7.0	5.5	-	-	6.5

Data of Multon (1973)

[a] Angolan wet processed coffee, as in Table III.

[b] in °C.

[c] Flavour quality rating scale average (5 experts), 1 → 10 (poorest); 5.0, acceptability limit.

[d] in days at 75% R.H, equivalent to 17.0%db moisture content.

TABLE V

Flavour Quality Changes of a Stored Roasted Coffee[a]

Parameters		
Type of Pack	Time of Storage[b] for a Significant Change for, (a) at P $\gtrsim$ 95%	(b) at P $\gtrsim$ 90%
Air pack (Vrac)	40	40
Closed metallised pouch	138	138
As above, but vacuum pack (with 4 days degassing)	180	180
Vacuum pack (with one-way valve)	180	208

Data of Cros and Vincent (1980)

[a] Colombian arabica coffee, RWB, roast degree not given.

[b] days in storage at ambient temperature.

(a) Overall quality change assessed by panel triangular testing against reference sample placed in storage, 7h after roasting.

(b) Loss of aromatics by paired comparison by respondents answering correctly, to prior triangular test.

TABLE VI

Flavour Quality Ratings of a Stored Roasted Coffee[a]

Parameters	Values									
Storage Time[b]	Zero	10	13	17	21	27	35	41	59	74
Rating[c]	9.0	9.0	8.5	8.3	8.0	7.1	6.5	5.8	4.5	3.9

Data of Radtke-Granzer (1981)

[a] RWB of unstated arabica blend and roast degree.

[b] in days at 23°C in an air pack.

[c] Flavour rating scale average from panel of 8 members (two professional) on scale, 9 $\rightarrow$ 0; 8, first significant change; 5.0-5.5, second significant change; and 4, a limit of acceptability (emergence of positive staleness).

TABLE VII

Flavour Quality Ratings of a Stored Roast and Ground Coffee[a]

	Parameters										
	Rating[c] at a time[d] of Storage[e] of										
IOC[b]	Zero	2	4	6	9	10	12	14	17	20	25
0.5	9.0	-	8.5	-	6.7	-	5.5	-	4.9	4.3	4.0
1.0	9.0	-	8.0	6.5	-	-	4.5	4.0	-	-	-
3.0	9.0	7.8	6.5	5.5	4.6	-	4.0	-	-	-	-
5.0	9.0	7.0	6.0	4.2	3.5	-	-	-	-	-	-
0.0[f]	9.0	-	9.0	-	-	8.6	-	-	8.0	-	7.0

Data of Radtke-Granzer (1981)

[a] Unstated arabica blend, roast and grind degree, or moisture content ($\gtrsim$ 4%).

[b] Initial oxygen content, percentage by volume, in sealed pack.

[c] As in Table VI.

[d] in months [e] at 23°C.

[f] Reference sample packed in nitrogen, stored at -20°C.

TABLE VIII

Flavour Quality Ratings of a Stored Roast and Ground Coffee[a]

Parameters										
IOC[b]			Rating[f] at a time[g] of Storage of							
(1)[c]	(2)[d]	(3)[e]	Zero	4	6(8)	12	24	36	48	96
29	34	0.7	7.2	6.8	6.5	6.5	6.4	-	-	6.3
28	66	1.4	7.2	6.7	6.5	6.4	6.3	-	-	6.1
25	169	3.5	7.2	6.1	6.0	5.8	5.0	4.8	-	4.8
20	338	6.9	7.2	5.8	5.7	5.5	4.8	4.6	4.5	4.3

Data of Clinton (1980)

[a] Unstated blend, roast and grind degree.

[b] Initial oxygen content in pack (vacuum), either by 1, 2, or 3 below.

[c] Vacuum, inches Hg. [d] Absolute pressure in mb.

[e] Calculated $\%O_2$. [f] Rating scale, 9 → 1, expert panel.

[g] Weeks at $70^{o}F$ ($21^{o}C$) at a constant moisture content of 4.3%

TABLE IX

Flavour Quality Ratings of a Stored Roast and Ground Coffee[a]

	Parameters							
	Rating[c] at a time[d] of Storage[e] of							
IMC[b]	Zero	4	6(8)	12	24	48	60	96
4.3	7.2	6.8	6.5	6.5	6.4	-	-	6.3
5.0	7.0	6.7	6.4	6.3	6.2	-	-	6.2
5.2	6.8	6.5	6.1	6.0	-	5.2	-	5.1
6.0	6.6	6.2	5.6	5.6	-	5.1	4.9	-

Data of Clinton (1980)

[a] As in Table VIII.

[b] Initial moisture content, $\%^{w}/w$ H_2O

[c] As in Table VIII. [d] In weeks.

[e] At 29 inches Hg. initial vacuum, or 34mb total absolute pressure, or 0.7% O_2 content calculated.

TABLE X

Temperature Effects on Shelf-life of Stored Roast (and Ground) Coffee

Reference	Effect
Heiss (1977)	Each $10^{\circ}C$ temperature
Sivetz (1979)	Increases, decreases shelf-life (for same quality) by 50%, e.g. if 160 days at $-20^{\circ}C$, then 10 days at $+20^{\circ}C$. Alternatively, each $10^{\circ}C$ decrease, doubles shelf-life, e.g. from $+20^{\circ}C$ to $-20^{\circ}C$, change is $2^{40/10} = 2^4 = 16$-fold.
Ernst (1979)	From $40^{\circ}C$ to $20^{\circ}C$, 3.5 - 4.0 fold increase of shelf-life, i.e. 2^2.

TABLE XI

Comparative Effect on Shelf-life of Roasted Coffee from $^{M}/_{B}$ ratios

Parameters			
	Coffee Type[a]	Pack Type	Time[b]
1.	RWB	Air (bag)	70
2.	RWB	Vacuum	Not given
3.	RWB	As 2, but opened after 8 weeks	approx 70 additional
4.	R and G	Air	11
5.	R and G	Vacuum a) Can b) Packing	Not given
6.	R and G	As 5a, but opened/reclosed	11 20 at 4°C
7.	R and G	As 5b, but opened/reclosed	6 14 at 4°C

Data of Vitzthum et al. (1979)

[a] All coffees with initial $^{M}/_{B}$ ratio of 3.2.

[b] Time of storage, in days, at 20°C (unless other stated) to $^{M}/_{B}$ ratio of 2.3 (point of loss of coffee flavour).

TABLE XII

Summarized Assessments of Shelf-life
Roast Whole Beans in Air Packs

Ref.	Quality Criterion	Shelf-life[a]
Radtke -Granzer	'High'	28
	'Medium'	49
	'Low' (acceptable)	70
Cros et al.	Significant change from fresh sample	40
Vitzthum et al.	M/B ratio down to 2.3	70

[a] Days storage at 21-23°C, with initial moisture content below 4%.

TABLE XIII

Summarized Assessments of Shelf-life
Roast and Ground Coffee in low oxygen packs

IOC[a]	Quality Criterion	Shelf-life[b]
0.5	'High'	6
	'Medium'	12-17
	'Low' (acceptable)	20-25
1.0	'High'	4
	'Medium'	9-17
	'Low' (acceptable)	14-20

Data from Radtke-Granzer (1981)

[a] Initial oxygen content, percentage by volume.

[b] Months storage at 21-23°C, with initial moisture content below 4%.

N.B. RWB will show longer shelf-life times for the same storage conditions.

TABLE XIV

Flavour Quality Ratings of Stored Instant Coffee

	Parameters								
			Rating[d] following time of Storage[e] of						
Coffee Type[a]	MC[b]	IOC[c]	0	8	12	24	36	48	72
Spray-dried/Aggl.	4.3	4.0	7.0	6.2	6.3	5.9	-	5.9	5.6
Freeze-dried	2.5	2.0	7.2	6.4	6.4	5.6	5.1	5.0	4.8

Data of Clinton (1980)

[a] Other details of coffees not given.

[b] Moisture content, $\%^{w}/w$, by method 6.

[c] Initial oxygen content of headspace, $\%O_2$ by volume.

[d] Average rating from expert panel, on scale $9 \rightarrow 0$.

[e] Weeks at $70^{o}F$ ($21^{o}C$) storage in sealed jars.

TABLE XV

Flavour Quality Ratings of Stored Instant Coffee

			Parameters									
			Rating[d] following a Storage time[e] of									
Coffee type[a]	MC[b]	IOC[c]	0	8	12	16	24	36	40	48	64	72
F.D.	2.5	2.0	7.2	6.4	6.4	-	5.6	5.1	-	5.0	-	4.7
	1.6	2.0	7.0	6.6	-	6.2	-	-	5.3	-	5.5	-
	2.5	2.0	7.0	6.5	-	-	5.8	5.2	-	5.1	-	5.0
	3.2	2.0	7.0	6.5	-	6.0	-	-	5.1	-	4.9	-
S.D./Aggl.	4.3	4.0	7.0	6.1	6.2	-	5.9	-	-	5.9	-	6.2
	3.9	4.0	7.0	5.6	5.8	-	5.7	5.7	-	-	5.7	-
	4.4	4.0	7.0	5.7	5.9	-	5.8	5.6	-	-	5.5	-
	5.1	4.0	7.0	5.5	5.6	-	5.5	5.5	-	-	5.1	-

Data of Clinton (1980)

[a-e] As in Table XIV.

REFERENCES

1. Association Francaise de Normes (1968). French National Standard, V05-202.
2. Arackal, T., and Lehmann, G. (1979). *Chem., Mikrobiol., Technol. Lebensm. 6*, 43-47.
3. Bauder, H. J. (1973). *Proc. Colloq. Coffee, 6th, 197 ,* pp. 278-283.
4. Clarke, R. J. (1976). *Chem. Ind. (London)*, pp. 362-365.
5. Clarke, R. J. (1985). "Coffee Chemistry." Elsevier Applied Science Publ., London.
6. Clinton, W. P. (1981). *Proc. Colloq. Coffee, 9th, 1980,* pp. 273-286.
7. Corte dos Santos, A. (1973). *Proc. Colloq. Coffee, 5th, 1971,* pp. 304-315.
8. Cros, E., and Vincent, J. C. (1981). *Proc. Colloq. Coffee, 9th, 1980*, pp. 345-352; *Cafe, Cacao, The 24* (3), 203-226.
9. Ernst, U. (1979). *Chem., Mikrobiol., Technol. Lebensm. 6*, 18-24.
10. Harris, N. E., Bishov, S. J., and McBrouk, A. F. (1974). *J. Food Sci. 30*, 192-195.
11. Heiss, R., Radtke, R., and Robinson, J. (1979). *Proc. Colloq. Coffee, 8th, 1977*, pp. 163-174.
12. Iglesias, H., and Chirife, J. (1983). "Handbook of Food Isotherms," Academic Press, New York.
13. International Standards Organization (1978). "Green Coffee--Determination of Moisture Content." ISO 1446. ISO, Geneva, Switzerland.

13a. International Standards Organization (1980). "Instant Coffee--Determination of Loss of Mass at Reduced Pressure." ISO 3726. ISO, Geneva, Switzerland.

13b. International Standards Organization (1983). "Green Coffee--Determination of Loss of Mass at 105°C." ISO 6673. ISO, Geneva, Switzerland.

14. Kwasny, H., and Werkhoff, P. (1979). *Chem., Mikrobiol., Technol. Lebesm. 6*, 31-32.
15. Multon, J. L. (1974). *Proc. Colloq. Coffee, 6th, 1974*, pp. 268-277.
16. Natarajan, C. P., and Gopalakrishna Rao, N. (1970). *Proc. Colloq. Coffee, 4th, 1969*, pp. 55-58.
17. Radtke-Granger, R. (1979). *Chem. Mikrobiol., Technol. Lebesm. 6*, 36-42.
18. Radtke, R., and Piringer, O.-G. (1981). *Dtsch. Lebesm.-Rundsch. 77*(6), 203-210.
19. Reymond, D., Chavan, F., and Egli, R. H. (1965). *Food Sci. Technol., Proc. Int. Congr., 1st, 1962, Vol. 4*, pp. 595-597.

20. Reymond, D. (1983). *Proc. Colloq. Coffee, 10th, 1982*, pp. 159-176.
21. Sivetz, M., and Desrosier, N. W. (1979). "Coffee Technology," AVI Publ. Co., Westport, Connecticut.
22. Stirling, H. (1981). *Proc. Colloq. Coffee, 9th, 1980*. pp. 189-200.
23. Thijssen, H. A. C., Bomben, U. L., and Bruin, S. (1973). *Adv. Food Res. 20*, 2-111.
24. Tressl, R., and Silwar, R. (1979). *Chem., Mikrobiol., Technol. Lebensm. 6*, 52-57.
25. Vitzthum, O. G., and Werkhoff, P. (1978). *In* "Analysis of Foods and Beverages" (G. Charalambous, ed.), pp. 115-132. Academic Press, New York.
26. Vitzthum, O. G., and Werkhoff, P. (1979). *Chem., Mikrobiol., Technol. Lebensm. 6*, 25-30.
27. Wootton, A. E. (1968). *Proc. Colloq. Coffee, 3rd, 1967*, pp. 92-100.

CHAPTER 14

CITRUS FRUIT JUICES

STEVEN NAGY
RUSSELL L. ROUSEFF
Scientific Research Department
State of Florida Department of Citrus
Lake Alfred, Florida

I. INTRODUCTION

Citrus fruits are the largest and most important group of fruits used in the preparation of fruit juice beverages. The unique and distinct flavors of citrus fruits and the

Handbook of Food and Beverage
Stability: Chemical, Biochemical,
Microbiological, and Nutritional Aspects

general acceptance of these flavors by peoples worldwide have been factors contributing to the enormous growth of the citrus beverage market. In 1983 well over 3.6 billion liters of citrus fruit juices were consumed in the United States; the retail value of this enormous volume amounted to $4.4 billion. Of the total U.S. fruit juice market, the percentage market share for citrus juices amounted to 62.1% (orange juice), 7.2% (grapefruit juice), and 1.5% (lemon/lime juices) (1). Fruit nectars and apple, grape, pineapple, prune and blended fruit juices collectively, made up the remaining market share (29.2%).

Orange flavor is the most widely accepted flavor in the world fruit beverage market; grapefruit flavor is less popular than orange but more popular than lemon and lime. Lemon juice has increased in popularity because of technical advances in the production of frozen concentrate for lemonade (2).

The shelf-life of a citrus fruit beverage is primarily determined by microbial growth and by chemical changes. Although most chemical changes that occur during processing citrus juice do not lead to a reduction in shelf-life, there are some that reduce the marketability of the product by causing detrimental flavor changes, product discoloration (browning reactions), and loss of nutrients. Commercially processed citrus juice is susceptible to flavor degradation when not properly processed and when stored at warm temperatures for prolonged periods. Because of the delicate flavor quality of citrus juice, it must be kept cold (preferably at near 0°C temperature); otherwise, a disagreeable odor and off-flavor would soon develop. Changes occurring in citrus juice after processing and upon subsequent storage are of two types: (1) loss of original flavor and (2) development of flavors foreign to fresh juice (3).

The effects of processing and storage conditions on nutrient losses must be considered in any shelf-life report because of the national stress on recommended dietary allowances (RDA) values and on label declaration of nutrients in a particular food (U.S. RDA). RDA are the levels of intake of essential nutrients that are considered to meet the known nutritional needs of practically all healthy persons. No one food can meet this total requirement, but a wide variety of foods can collectively supply the RDA. Although citrus juices contain protein, fat, and carbohydrate (Table I) (4,5), nutritional labeling primarily concentrates on the contents of vitamins and minerals (Table II) (6). Most shelf-life studies of citrus juice products are concerned with the effects of processing and storage on vitamin retention, in particular vitamin C.

The type of container in which a juice is packaged can have an important bearing on the shelf-life of that product. Containers that are permeable to air will cause a faster reduction in vitamin potency than hermetically sealed, tin-coated cans. However, unlaquered tin-coated cans can, through corrosion, contribute tin, iron, and other metals to the acidic juice product. Under extreme storage temperatures (>30°C) for prolonged periods, these metals may accumulate to an extent that they exceed the safe levels established by the Codex Alimentarius Commission (7). Additionally, elevated tin levels in citrus juices have been associated with an undesirable metallic off-flavor.

II. MICROBIAL GROWTH

Pasteurization of citrus juices is intended to serve two functions: (1) inactivation of pectin enzymes responsible for gelation of citrus concentrate and clarification of single-strength juice, and (2) reduction of microbial populations. Commercial pasteurization temperatures are much higher than needed to reduce microbial populations to ineffectual levels. For orange juice a pasteurization temperature of 91°C for 7 sec is required to inactivate pectin enzymes, whereas temperatures of 74°C for 16 sec or 85°C for 1 sec will effectively reduce high microbial populations (up to 1 million organisms/ml of juice) (8). With grapefruit juice a pasteurization temperature of 99°C for 6 sec causes inactivation of pectin enzymes, whereas heat treatments of 74°C for 16 sec or 85°C for 1 sec will reduce the harmful microbial populations in grapefruit juices (9).

Pectin enzymes will not regenerate once deactivated, but this is not the case with microorganisms. Microbial populations can grow and multiply on the walls of tanks, equipment, and pipes after the point of pasteurization; additionally, airborne bacteria and yeasts common in a processing plant can also enter the product by means of open tanks, equipment, or containers during filling (8).

Single-strength citrus juice that has been commercially pasteurized and packed in hermetically sealed, tin-coated cans will generally be free of microbial spoilage during its shelf-life. However, frozen concentrate, single-strength juices reconstituted from concentrate, and chilled juices are subject to spoilage unless sanitation is carefully controlled and the product is kept near freezing. Single-strength juices reconstituted from concentrate require additional pasteurization prior to packaging, and 16 sec at 74°C is usually adequate.

There is no simple rule relating the shelf-life of a citrus juice product to a specific time period. Factors that must be considered are (1) type of citrus product (canned, chilled, concentrate), (2) plant sanitation (type of microflora present and the level of contamination; the higher the initial level the shorter the shelf-life at a given temperature), and (3) temperature during storage (warehouse, retail outlet, home).

A. Chilled Juice

Chilled juice is sold in single-strength form at the retail level and may be packed as a sterile or nonsterile product. In the sterile form, the shelf-life of the product when stored at temperatures <1.7°C (35°F) is >1 year. Chilled juice in the nonsterile form is usually packaged in paper cartons, or glass or plastic jugs. Being nonsterile, the product is subject to microbial growth and spoilage--the predominant microflora being yeasts and lactic acid bacteria. The shelf-life of a nonsterile product is related to the type of microflora present, the initial level of contamination, the temperature of storage, and the packing container (10). Citrus juice packed in glass or plastic-coated fiber packages is only "commercially sterile," which means that the organisms present are in such few numbers that growth under normal storage will not render the juice objectionable. Juice packed in glass is considered hermetically sealed, whereas fiber packages are not. Hermetically sealed containers prevent penetration of oxygen and microorganisms into the product and thus protect the product from oxidation and/or spoilage. As noted in Table III, yeasts grow at all temperatures investigated, whereas *Lactobacillus* does not grow at <10°C and *Leuconostoc* at <4.4°C. At temperatures of ≤7.2°C yeasts are the predominant flora responsible for spoilage, whereas at ≥10°C, lactic acid bacteria could outgrow the yeasts and be the cause of spoilage (10) (Table IV).

The three types of microbial spoilage that limit the shelf-life of citrus juice are (1) a fermented taste accompanied by carbon dioxide gas bubbles produced by various yeasts, (2) a buttermilk off-flavor produced by *Lactobacillus* and *Leuconostoc* bacteria, and (3) a light-colored growth on the sides of the container of the surface caused by molds (8). For a nonsterile, reconstituted juice packed in fiber packages, warehouse temperatures should range between 0 and 1.7°C (32-35°F). Storage at the retail market should also be

0-1.7°C; a 4-week shelf-life is easily attainable. Additionally, coding of the juice package at the time of packing with an expiration date is common. Processor-distributors establish an expiration code date based on time-temperature handling policies of their customers (8).

B. Frozen Concentrated Citrus Juice

The primary purpose of concentrating citrus juice is to convert single-strength juice into a product form that will be more economical to store, distribute, and sell. Modern methods of evaporative concentration involve enzyme deactivation, pasteurization, and deaeration, in addition to concentrating the juice. In the United States, the Temperature-Accelerated Short-Time Evaporator (TASTE) is the major evaporator employed to concentrate citrus juice. During the concentration of juice in a TASTE, temperatures reach ∿99°C; this temperature is sufficient to render the product commercially sterile. The product produced by the TASTE is highly concentrated (55-75°Brix) and relatively devoid of aromatic character; therefore, cutback juice, aqueous essence, and cold-pressed peel oil are added. The final product (∿42°Brix) is packed at temperatures of about -5°C and rapidly frozen at about -18°C. Commercial packs of frozen concentrated orange juice (FCOJ) have water activity values of 0.90 to 0.95, and are not microbiologically stable at temperatures of ≥4°C. Osmophilic yeasts (*Saccharomyces rouxii* and *S. uvarum*) are the primary spoilage organisms in concentrates (cans swell and burst due to gas production by yeasts). Concentrates with high Brix values (58-72°) and storage temperatures <4°C effectively reduce yeast survival (measured as colony-forming units/g concentrate) (11). Proper storage of FCOJ at refrigerated temperatures (≤0°C) in marketing chains will ensure a product shelf-life of >1 year. Extensive studies of the types of organisms viable in citrus concentrates, their thermal growth rates, and the proper handling of concentrates may be found elsewhere (12).

III. FLAVOR DEGRADATION

Perhaps the single most important factor that determines the shelf-life of a citrus juice is detrimental flavor change. Degradation of citrus flavor involves a complex group of chemical reactions differing in regard to substrate, temperature dependence of reaction rate, and conditions of acidity and moisture. Several factors have been identified as contributing to the development of off-flavors during processing and upon subsequent storage of the citrus juice product. Reduction of atmospheric oxygen during processing and packaging appears to reduce partially the short-term effects of off-flavor development. However, the major off-flavor compounds are formed by acid-catalyzed hydrolysis of flavoring oils, the formation of sulfur-containing and unidentified juice components, and products of Maillard browning. Sugars participate in browning reactions that cause darkening of the juice and also in the formation of compounds generally described as apricotlike or pineapplelike in flavor (13).

Degradation Products and Flavor Change

Prolonged storage at adverse temperatures favors formation of compounds that impart off-flavor and off-odor properties to a citrus beverage. An extensive study by Tatum *et al.* (14) identified 10 compounds in canned single-strength orange juice after 12 weeks' storage at 35°C (Table V). α-Terpineol, *cis*- and *trans*-1,8-*p*-menthanediol were derived from the essential oil fraction by acid-catalyzed hydration reactions. 2-Hydroxyacetyl furan and 3-hydroxy-2-pyrone were probably formed by acid degradation of sugars (15). Furfural was formed by degradation of ascorbic acid. 4-Vinylguaiacol, α-terpineol, and 2,5-dimethyl-4-hydroxy-3(2*H*)-furanone were judged the compounds primarily responsible for the malodorous property of temperature-aged juice. When added to freshly squeezed orange juice, 4-vinyl guaiacol imparted an old fruit or rotten flavor; α-terpineol was described as stale, musty, or piney; and 2,5-dimethyl-4-hydroxy-3(2*H*)-furanone was responsible for the pineapplelike note typical of aged juice (Table VI).

Several volatile sulfur compounds--namely, hydrogen sulfide, methanediol, sulfur dioxide, carbonyl sulfide, dimethyl sulfide, and dimethyl disulfide--have been detected in headspace gases of citrus juices (16-19). With the exception of dimethyl sulfide, these volatile sulfur compounds are not considered detrimental to flavor. Sawamura *et al.* (18) attributed the characteristic off-flavor of heated mandarin juice to dimethyl sulfide, while Shaw *et al.* (16) suggested that high levels of dimethyl sulfide may be an important contributor to off-flavor in stored, canned orange and grapefruit juices. Hydrogen sulfide, which is present in fresh citrus juice, is considerably augmented by heating and storage. It reacts with aldehydes to produce thioaldehydes; reaction with *n*-hexenal, 2-hexenal, 2-octenal, 2-nonenal, and citral (all present in citrus juice) yields compounds possessing onionlike aromas, and reaction with furfural produces a skunky aroma. Since hydrogen sulfide and furfural form during storage, it has been suggested that the skunky aroma of aged juice is due to formation of thiofurfural (20).

Diacetyl and acetylmethylcarbinol are two volatile compounds of concern to processors of frozen concentrated citrus juice. These end products of bacterial growth (*Lactobacillus* and *Leuconostoc*) bestow upon citrus concentrate an off-flavor characteristic of buttermilk. The presence of these substances in citrus juices is considered an indication of poor sanitary practices during the extracting and concentrating processes. If the diacetyl concentration of a product is $\geq$0.8 ppm, the generation rate of lactic organisms has reached an alarmingly high rate (12).

IV. SHELF-LIFE MONITORING

Various compounds have been proposed to monitor flavor change and storage abuse of processed citrus juices, namely, α-terpineol, diacetyl, 3-methyl-2-butene-1-ol, hydroxymethyl furfural, and furfural. Of these, furfural is the most widely used because (1) the furfural content of freshly processed juice is virtually zero whereas large amounts accumulate in storage-abused juice (21-23), (2) furfural can be distilled readily from a juice and enriched manyfold relative to its original level, (3) furfural is readily determined by colorimetry (24), high-performance liquid chromatography (25), and gas chromatography (23), and (4) correlations between off-flavor and furfural are excellent even though it does not contribute to the malodorous property of aged citrus juice. Maraulja *et al.* (26) showed that correlation coefficients between flavor scores and the furfural contents of canned Hamlin and Valencia orange juices, and canned grapefruit juices were highly significant (Table VII). As noted in Table VII, the rate of flavor deterioration of canned juice was dependent on both storage temperature and storage time; however, temperature was more significant than time. Nagy and Randall (27) found a similar relationship with glass-packed and canned single-strength orange juice. Furfural values within the approximate region of 50 to 70 ppb correlated with a difference in flavor in comparison to controls at the significant difference level of $p < .001$. The relationship of furfural content to storage temperature and time is shown in Fig. 1 for glass-packed, single-strength orange juice. As noted, furfural levels accumulated in a predictable manner; after 16 weeks the furfural levels in micrograms per liter of juice were 859 (30°C), 173 (21°C), 80 (16°C), 25 (10°C), and 12 (5°C). Juice stored at 30°C for 16 weeks contains about 72 times more furfural than the 5°C stored juice.

In a study conducted by Dougherty and co-workers (28), samples of canned single-strength grapefruit juice stored at 21°C (70°F), 27°C (80°F), 32°C (90°F), and 38°C (100°F) for upwards of 20 weeks were evaluated for furfural contents and flavor scores (Fig. 2). Flavor was scored by a taste panel using a 9-point hedonic scale, where 1 = dislike extremely and 9 = like extremely. Juices stored at 38°C developed distinct off-flavors within 6 weeks with a concomitant increase in furfural contents. Similar trends were found in juices stored at 32 and 27°C, but the rates of flavor deterioration and furfural buildup were slower. After 6 months, samples stored at 21°C showed a slight change in flavor and an increase in furfural, whereas control samples stored at 16°C (not shown in Fig. 2) showed virtually no change in flavor or in furfural contents.

Color is an important quality factor in the marketing of citrus juices. Change in color, primarily manifested by nonenzymatic browning, reduces consumer acceptance and, therefore is an important shelf-life determinant. Browning of single-strength juices and concentrates is quantified by several methods. Meydav *et al.* (29) measured nonenzymatic browning at 420 nm (browning index) after alcohol extraction of the brown melanoidan pigments. Kanner *et al.* (30) reported browning of orange juice concentrate as the negative change in the Hunter *L* value as measured with a tristimulus colorimeter. The *L* value is the amount of light reflected from a sample, and a more negative value indicates an absorbance increase. Browning index values (29) and tristimulus color values (30) are excellent measures of storage abuse and, additionally, are highly correlated (31).

Crandall and Graumlich (32) conducted a study to evaluate browning of orange concentrates (prepared from Florida Hamlin, Pineapple, and Valencia oranges) while stored at -22.2, -6.7, 4.4, and 26.7°C. As noted in Table VIII, products stored at 26.7°C showed increased absorbance (browning) at 420 nm after only 2-3 months' storage. These workers also reported no changes in color of any samples stored at $\leq$4.4°C for 12 months.

V. NUTRIENT LOSSES

The nutritional quality of citrus juice is essentially synonymous with vitamin content. Since consumers derive major nutritional benefits from citrus juices, factors that affect vitamin potency are of considerable importance. Modern processing techniques result in minimal loss of citrus juice vitamins (average retention is 97%), whereas the conditions of storage and the type of packaging container are the two major factors causing reductions in the vitamin potency of these nutritious beverages (33) (Table IX).

A. Vitamin C

Citrus juice products are subject to varying temperatures and storage periods during warehousing and retailing; additionally, once the product is purchased it may again be subject to different home storage conditions. Because of the nutritional importance of vitamin C in citrus juices, many studies have been conducted since the mid-1930s to quantify the loss of vitamin C during storage. Results of representative studies on orange, grapefruit, and tangerine juices are presented in Table IX. As evident, low temperatures are necessary if meaningful amounts of vitamin C are to be retained. For canned and bottled single-strength juices, storage at 21°C (this temperature may be about the average for year-round nonrefrigerated commercial storage conditions) for upwards of a year results in vitamin C retentions of >75% (34,35). Typical time-temperature profiles for vitamin C retention in commercially canned, single-strength orange juice are shown in Fig. 3. As noted, retention of vitamin C decreases as temperature and time of storage increase.

Loss of vitamin C is due to aerobic and anaerobic reactions of a nonenzymatic nature (Fig. 4) (33). Kefford and co-workers (36) described the environment of canned citrus juice as one in which there is competition for oxygen among a number of reactions, including corrosion reactions, ascorbic acid oxidation, and oxidations contributing to off-flavor and color change. Storage studies (36,37) on the loss of vitamin C in canned, single-strength orange juice have shown an initial period (about 1 to 2 weeks) of rapid loss caused by the presence of free oxygen. After oxygen was consumed, vitamin C degraded anaerobically at rates lower than by the aerobic process.

Figure 5 shows Arrhenius plots of rate constants versus the reciprocal of absolute temperature for vitamin C breakdown in canned, single-strength grapefruit juice (SSGJ) and orange juice (SSOJ) (37,38). The plot of K versus $1/T$ for SSGJ shows a linear profile for the region 10-50°C. Regression analysis of the slope yields an energy of activation (E_a) of 18.2 kcal and also shows that a temperature rise of 10°C (Q_{10}) causes an increase in the reaction rate of about 2.7. In contrast to grapefruit juice, the K versus $1/T$ plot for SSOJ shows two distinct Arrhenius profiles (two different E_a and Q_{10} values are evident). for the region 4-28°C, regression yields an E_a value of 12.8 kcal and a Q_{10} of 2.2, whereas the region 28-50°C shows an E_a of 24.5 kcal and a Q_{10} of 3.7. These different Q_{10} values agree with values reported for canned orange juice by Ross (39), who showed that between 10 and 27°C the rate of vitamin C loss doubled with each 10°C rise, whereas from 27 to 37°C the rate quadrupled.

Dehydroascorbic acid (DHA) and diketogulonic acid (DKA) are formed as breakdown products of vitamin C under aerobic conditions. DHA possesses antiscorbutic activity, whereas no activity is shown by DKA; therefore, total active vitamin C (TAVC) of citrus juices is based on the combined levels of vitamin C and DHA. Moore and co-workers (40) showed the levels of DHA in canned and bottled orange and grapefruit juices stored for 6 months at 4 and 27°C remained constant at about 1 to 2% of TAVC. Smoot and Nagy (38) showed with SSGJ stored at 10 to 50°C that the DHA (0.6-1.2 mg%) and DKA (0.3-0.7 mg%) contents remained essentially unchanged during 3 months' storage.

Effects of Container

The type of container has an important influence on the retention of vitamin C potency in a citrus beverage. Loss of vitamin C in enamel-lined cans is greater than in plain tin cans; the difference is due to residual oxygen preferentially reacting with tin in one case and with vitamin C in the other. Glass-packed juices are, in turn, inferior to canned products for vitamin C retention (40,41). SSOJ packed in glass, polyethylene, and polystyrene bottles, and wax-coated fiberboard cartons was studied by Bissett and Berry (42). Glass-packed juice lost about 10% of its initial vitamin C content after 4 months of storage at refrigerated temperatures, whereas the plastic and fiberboard-packed products lost ∿20% in only 3-4 weeks. In contrast to glass containers that are hermetically sealed, plastic bottles and fiberboard cartons are permeable to oxygen; thus, lowered vitamin retentions are expected. The retention of vitamin C in FCOJ packed in foil-lined fiberboard cartons and in polyethylene-lined fiber cylindrical cans was studied at -20.5, -6.7, and 1.1°C for a year (42). At

-20.5°C vitamin C retention was 93.5% in foil-lined cartons and 9.15% in the polyethylene-lined cans. The foil-lined carton (89% retention) proved superior to the polyethylene-lined can (44% retention) after a 1-year storage period at 1.1°C.

B. Stability of B Vitamins

B vitamins are often considered minor or trace nutrients because the amount needed in the diet to maintain good health is relatively small when compared to other nutrients. Whereas citrus juices are generally considered to be excellent sources of vitamin C, they have only recently been found to be an important source of B vitamins (43,44). Specifically only thiamine and folic acid are found in significant amounts ($\geq$10% U.S. RDA) in orange juice. However, in terms of U.S. RDA, niacin, riboflavin, vitamin B_6, and pantothenic acid all occur in quantities equal to the level of calories supplied in an average dietary intake. Limited research has been done on the determination of the stability of these vitamins (45). As seen in Tables X and XI, thiamine and pantothenic acid concentrations decrease slightly during storage. Vitamin B_6 and niacin remained relatively constant during the 4-month storage period, whereas folic acid and riboflavin appeared to increase. The apparent trend in folic acid values may be due to the instability of the assay. (All B vitamins, except thiamine, were assayed using standard microbiological procedures.) Riboflavin values on the same juices were obtained from a recently developed HPLC procedure (46). Apparently some constituent is generated during storage that interferes with the microbiological test, whereas the HPLC values show a gradual decrease with increasing storage time (Table XII). Absolute HPLC riboflavin values are about half those of the microbiological test, which could be due to the greater specificity of the HPLC test. Since the test samples were frozen for 2 years from when they were first quantified for riboflavin, some degradation of the samples may also have occurred.

TABLE I. Proximate Composition of Citrus Juices (g/100 g) (4, 5)

Juices	Water	Protein	Fat	Sugar	Acid
Orange	88.3	0.6	0.2	9.3	0.6-1.6
Tangerine	88.9	0.5	0.2	9.2	0.6-1.6
Grapefruit	90.2	0.5	0.1	6.0	1.2-2.2
Lemon	87.6	0.4	0.2	3.2	4.0-6.0

TABLE II. Average Nutrient Delivery per Serving of Reconstituted FCOJ (11.8°Brix) in Relation to U.S. RDA[a]

Nutrient	U.S. RDA (6)	Average % U.S. RDA/ 177 ml FCOJ
Vitamin A	5000 IU	1.3
Vitamin C	60 mg	120
Thiamine	1.5 mg	9.0
Riboflavin	1.7 mg	2.2
Niacin	20 mg	1.8
Calcium	1.0 g	1.7
Iron	18 mg	1.0
Vitamin B_6	2 mg	4.5
Folic acid	0.4 mg	18.7
Phosphorus	1.0 g	3.0
Magnesium	400 mg	4.5
Zinc	15 mg	0.6
Copper	2 mg	4.1
Pantothenic acid	10 mg	3.0
Vitamin D	400 IU	--
Vitamin E	30 IU	--
Vitamin B_{12}	6 mcg	--
Iodine	150 mcg	--
Biotin	0.3 mg	--

[a]FCOJ = frozen concentrated orange juice.

TABLE III. Effect of Temperature and Level of Inoculum of Test Organisms on Development of Spoilage in Orange Juice

Temp. (C)	Organism and level of inoculum					
	Yeasts		Lactobacillus		Leuconostoc	
	1/ml	1000/ml	1/ml	1000/ml	1/ml	1000/ml
	-(Days when spoilage was first detected)-					
1.7	NS[a]	21	NG[b]	NG	NG	NG
4.4	21	21	NG	NG	35	27
7.2	14	7	NG	NG	35	13
10	7	2	14	12	20	13

[a]NS = no spoilage detected after 28 days.
[b]NG = no growth.
From J. Milk Food Technol. 38: 393 (1975).

TABLE IV. Microbial Population in Orange Juice When Spoilage First Detected[a]

Temp. (C)	Yeast	Lactobacillus	Leuconostoc
	-(Microorganisms per ml)-		
1.7	280,000	NG[b]	NG
4.4	170,000	NG	22,000
7.2	270,000	NG	270,000
10.0	220,000	170,000	9,700,000

[a]Initial inoculum in juice = 100 organisms/ml.
[b]NG = No growth.
From J. Milk Food Technol. 38: 393 (1975).

TABLE V. Degradation Products in Canned Single-Strength Orange Juice After 12 Weeks Storage at 35°C[a]

Furfural	cis-1,8-p-Menthanediol
Alpha-Terpineol	trans-1,8,p-Menthanediol
3-Hydroxy-2-pyrone	4-Vinyl guaiacol
2-Hydroxyacetyl furan	Benzoic acid
2,5-Dimethyl-4-hydroxy-3(2H)-furanone	5-Hydroxymethyl furfural

[a]From J. Food Sci. 40: 707 (1975).

TABLE VI. Concentration of Degradation Products Causing Detectable Flavor Change When Added to SSOJ[a]

Compound	Conc (ppm)	Significance of difference
alpha-Terpineol	2.5	$P<0.001$
	2.0	$P<0.001$
4-Vinyl guaiacol	0.075	$P<0.001$
	0.050	$P<0.01$
2,5-Dimethyl-4-hydroxy-3(2H)-furanone	0.10	$P<0.01$
	0.05	$P<0.05$

[a]Controls stored at -18°C.
From J. Food Sci. 40: 707 (1975).

TABLE VII. Correlation Coefficients Between Furfural in Canned Citrus Juices and Flavor Scores, Storage Temperatures and Storage Times

Factor	Orange juice: Hamlin	Orange juice: Valencia	Grapefruit juice
	Correlation coefficients (r)		
Flavor	-0.709[a]	-0.801[a]	-0.692[a]
Temp.	0.680[b]	0.702[b]	0.678[b]
Time	0.380[c]	0.383[c]	0.350[c]

[a]0.1 Percent level.

[b]1 Percent level.

[c]5 Percent level of significance.

From Proc. Fla. State Hortic. Soc. 86: 270 (1973).

TABLE VIII. Development of Brown Pigments in Concentrated Orange Juice Stored at 26.7°C for Up to 12 Months[a]

Storage time (months)	Cultivar: Hamlin early	Hamlin late	Pineapple late	Valencia late
Initial	0.09	0.12	0.35	0.19
3	0.17	0.15	0.43	0.24
5	--	0.32	0.47	--
7	0.39	0.40	0.67	0.35
10	0.63	0.66	1.03	0.59
12	0.98	0.91	1.49	1.02

[a]Values are the mean absorbance at 420 nm.
From Proc. Fla. State Hortic. Soc. 95: 198 (1982).

TABLE IX. Studies on Vitamin C Retention in Processed Orange, Tangerine, and Grapefruit Juices

Product[a]	Storage Temp. (°C)	Storage Months	% Retention of vitamin C
SSOJ (canned)	9, 24, 37	12	94, 75, 17
SSOJ (canned)	10 to 26.5	24	95 to 50
SSOJ (canned)	4.5, 24.5	18	93, 60
SSOJ (canned)	1.7, 22.2, 37.8	12	100, 80, 5
SSOJ (bottled)	4.5, 24.5	18	89, 51
SSGJ (canned)	21	11	89
SSGJ (canned)	10, 20, 30, 40, 50	3	99, 97, 90, 70, 29
SSGJ (canned)	10, 18, 27	18	93, 84, 62
SSGJ (canned)	23.9	12	83
FCOJ	-20, -15, -12.2	60	100, 100, 100
FCOJ	-22, -12, -7, 0, 4	12	99, 98, 98, 97, 96
FCGJ	-22, -12, -7, 0, 4	12	98, 98, 98, 98, 97
FCTJ	-22, -12, -7, 0, 4	12	94, 94, 91, 91, 90
FCTJ	-29, -18, -12, -4	3	100, 98, 95, 98

[a]SSOJ = single-strength orange juice; SSGJ = single-strength grapefruit juice; SSTJ = single-strength tangerine juice; FCOJ = frozen concentrated orange juice; FCGJ = frozen concentrated grapefruit juice; FCTJ = frozen concentrated tangerine juice.
From J. Agric. Food Chem. 28: 8(1980).

TABLE X. Thiamine, Riboflavin and Niacin Content of Chilled Orange Juice in Glass Jars Packed at Different Months and Stored at 26.6°C (80°F)

Time of packing	Thiamine (mg/dl)			Riboflavin (mg/dl)			Niacin (mg/dl)		
	Initial	2-month	4-month	Initial	2-month	4-month	Initial	2-month	4-month
1973									
April	.072	.068	.060	.024	.034	.042	.197	.213	.216
May	.081	.065	.061	.025	.038	.049	.202	.228	.227
June	.077	.071	.063	.027	.038	.046	.206	.207	.218
July	.067	.063	.061	.026	.041	.045	.213	.215	.218
1974									
January	.069	.070	.064	.027	.038	.039	.206	.307	.294
February	.079	.072	.073	.032	.034	.044	.222	.233	.308
March	.080	.071	.070	.028	.036	.046	.338	.293	.245
Average	.075	.068	.064	.027	.037	.044	.225	.242	.245

From Proc. Int. Soc. Citric. 2:905 (1981).

TABLE XI. Folic Acid, Vitamin B_6 and Pantothenic Acid in Chilled Orange Juice Packed in Glass Jars at Different Months and Stored at 26.6°C (80°F)

Time of packing	Folic acid (mg/dl)			Vitamin B_6 (mg/dl)			Pantothenic acid (mg/dl)		
	Initial	2-month	4-month	Initial	2-month	4-month	Initial	2-month	4-month
April	.049	.047	.069	.042	.043	.046	.113	.102	.096
May	.045	.048	.059	.053	.046	.051	.147	.118	.112
June	.057	.052	.066	.056	.056	.056	.158	.141	.118
July	.050	.061	.049	.061	.060	.060	.152	.147	.124
Average	.050	.052	.061	.053	.053	.051	.142	.127	.112

From Proc. Int. Soc. Citric. 2:905 (1981).

TABLE XII. Riboflavin Storage Studies[a]

Storage time (months)	Micro (mg/100 g conc.)	HPLC (mg/100 g conc.)
0 mo.	0.084	0.041
2 mo.	0.114	0.038
4 mo.	0.112	0.037

[a]Chilled orange juice packed in glass jars and stored at 26.7°C.

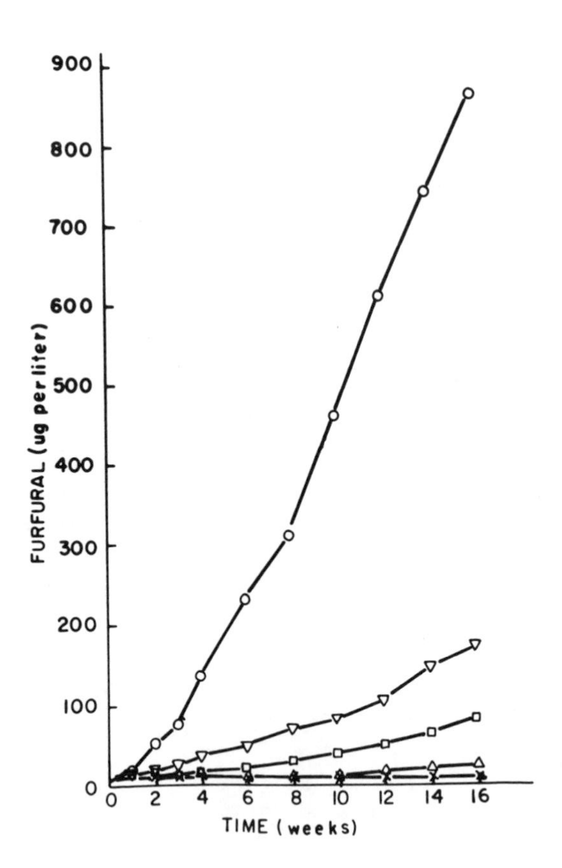

Fig. 1. Increase in furfural content in glass-packed juice over a 16-week period at 5°C (×), 10°C (Δ), 16°C (□), 21°C (∇), and 30°C (o) from reference (27).

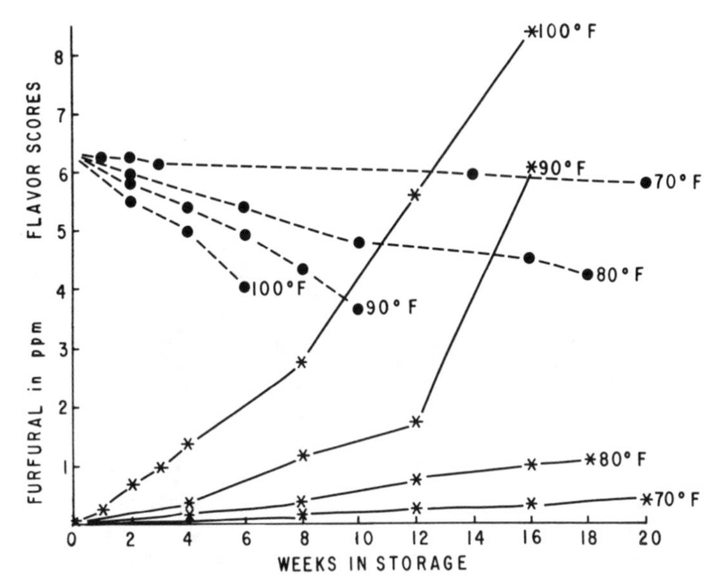

Fig. 2. Effect of storage time and temperature on the flavor score (●) and furfural content (*) of canned grapefruit juice from reference (28).

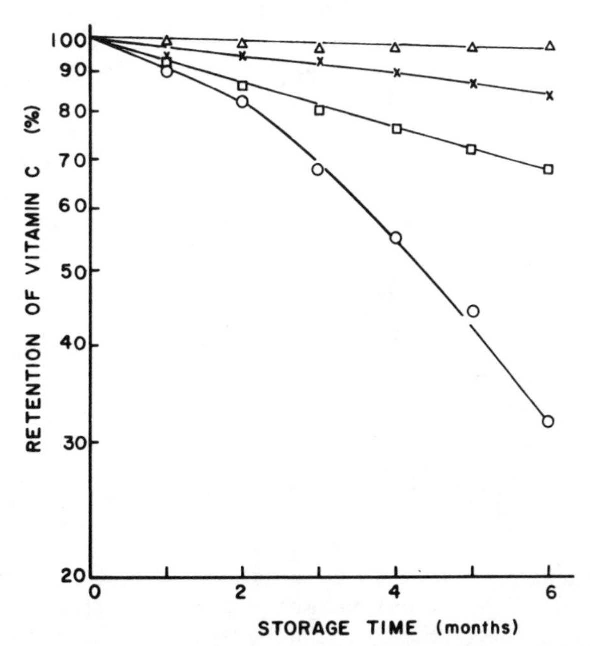

Fig. 3. Percentage vitamin C retention (logarithmic scale) versus months of storage at 4°C (Δ), 24°C (×), 32°C (□), and 37°C (o) for canned single-strength orange juice from reference (33).

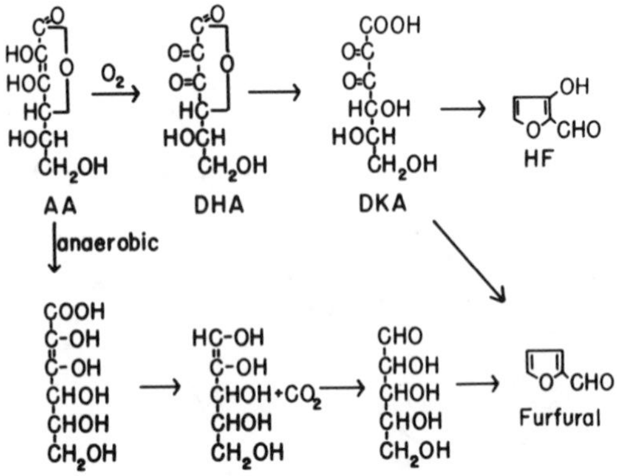

AA $\xrightarrow{O_2}$ DHA $\rightarrow$ DKA $\rightarrow$ HF

AA $\xrightarrow{\text{anaerobic}}$ $\rightarrow$ $\rightarrow$ Furfural; DKA $\rightarrow$ Furfural

Fig. 4. Possible vitamin C (ascorbic acid) degradation pathways: AA, ascorbic acid; DHA, dehydroascorbic acid; DKA, diketogulonic acid; HF, hydroxyfurfural.

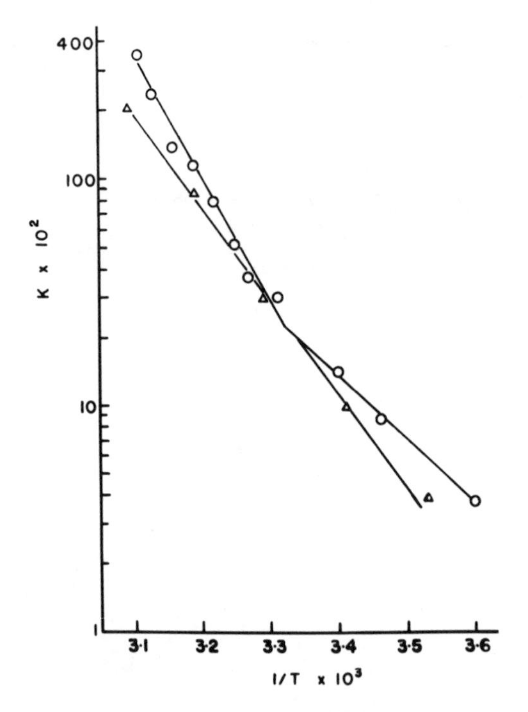

Fig. 5. Arrhenius plots of log *K* (mg of vitamin C loss/100 ml of juice per week) versus reciprocal of absolute storage temperature. The grapefruit (Δ) plot shows one linear profile, whereas the orange (o) plot shows two distinct profiles. The Arrhenius plots for orange indicate a change in kinetics at ∿28°C from reference (33). Copyright 1980 American Chemical Society.

REFERENCES

1. Anonymous (1984). *Beverage World 103* (July), 54.
2. Varsel, C. (1980). *In* "Citrus Nutrition and Quality" (S. Nagy and J. A. Attaway, eds.), p. 226. ACS Books Department, Washington, D.C.
3. Curl, A. L. (1946). *Fruit Prod. J. 25*, 356.
4. Ting, S. V. (1980). *In* "Citrus Nutrition and Quality" (S. Nagy and J. A. Attaway, eds.), p. 3. ACS Books Department, Washington, D.C.
5. Swisher, H. E., and Swisher, L. H. *In* "Citrus Science and Technology" (S. Nagy, P. E. Shaw, and M. K. Veldhuis, eds.), Vol. 2, p. 253. AVI Publ. Co., Westport, Connecticut.
6. U.S. Food and Drug Administration (1979). "Title 21 Code of Federal Regulations," Parts 100-199. U.S. Govt. Printing Office, Washington, D.C.
7. Codex Alimentarius Commission (1978). "Report of the "Twelfth Session," p. 45. FAO/WHO, Rome.
8. Carter, R. (1981). "Reconstituted Florida Orange Juice," p. 6. Florida Department of Citrus, Lakeland.
9. Carter, R. (1983). "Reconstituted Florida Grapefruit Juice." p. 7. Florida Department of Citrus.
10. Murdock, D. I., and Hatcher, S. (1975). *J. Milk Food Technol. 38*, 393.
11. Crandall, P. G., and Graumlich, T. R. (1982). *Proc. Fla. State Hortic. Soc. 95*, 198.
12. Murdock, D. I. (1977). *In* "Citrus Science and Technology" (S. Nagy, P. E. Shaw, and M. K. Veldhuis, eds.), Vol. 2, p. 445. AVI Publ. Co., Westport, Connecticut.
13. Rouseff, R. L., Nagy, S., and Attaway, J. A. (1981). *Proc. Int. Soc. Citric. 2*, 872.
14. Tatum, J. H., Nagy, S., and Berry, R. E. (1975). *J. Food Sci. 40*, 707.
15. Tatum, J. H., Shaw, P. E., and Berry, R. E. (1967). *J. Agric. Food Chem. 15*, 773.
16. Shaw, P. E., and Nagy, S. (1981). *In* "Quality of Foods and Beverages" (G. Charalambous and G. E. Inglett, eds.), Vol. 1, p. 361. Academic Press, New York.
17. Shaw, P. E., and Wilson, C. W. (1982). *J. Agric. Food Chem. 30*, 685.
18. Sawamura, M., Shimoda, M., Yonezawa, T., and Osajima, Y. (1977). *J. Agric. Chem. Soc. Jpn. 51*, 7.
19. Sawamura, M., Mitsuya, S., and Osajima, Y. (1978). *J. Agric. Chem. Soc. Jpn. 52*, 281.
20. Blair, J. S. (1964). "Citrus Station Mimeo Report," CES 65-4. Lake Alfred, Florida.

21. Kirchner, J. S., and Miller, J. M. (1957). *J. Agric. Food Chem. 5*, 283.
22. Rymal, K. S., Wolford, R. W., Ahmed, E. M., and Dennison, R. A. (1968). *Food Technol. 22*, 1592.
23. Dinsmore, H. L., and Nagy, S. (1971). *J. Agric. Food Chem. 19*, 517.
24. Dinsmore, H. L., and Nagy, S. (1974). *J. Assoc. Off. Anal. Chem. 57*, 332.
25. Marcy, J. E., and Rouseff, R. L. (1984). *J. Agric. Food Chem. 32*, 979.
26. Maraulja, M. D., Blair, J. S., Olsen, R. W., and Wenzel, F. W. (1973). *Proc. Fla. State Hortic. Soc. 86*, 270.
27. Nagy, S., and Randall, V. (1973). *J. Agric. Food Chem. 21*, 272.
28. Dougherty, M. H., Ting, S. V., Attaway, J. A., and Moore, E. L. (1977). *Proc. Fla. State Hortic. Soc. 90*, 165.
29. Meydav, S., Saguy, I., and Kopelman, I. J. (1977). *J. Agric. Food Chem. 25*, 602.
30. Kanner, J., Fishbein, J., Shalom, P., Harel, S., and Ben-Gara, I. (1982). *J. Food Sci. 47*, 429.
31. Robertson, G. L., and Reeves, M. J. (1981). *J. Food Technol. 18*, 535.
32. Crandall, P. G., and Graumlich, T. R. (1982). *Proc. Fla. State Hortic. Soc. 95*, 198.
33. Nagy, S. (1980). *J. Agric. Food Chem. 28*, 8.
34. Feaster, J. F., Tompkins, M. D., and Pearce, W. E. (1949). *Food Res. 14*, 25.
35. Freed, M., Breener, S., and Wodicka, V. O. (1949). *Food Technol. 3*, 148.
36. Kefford, J. F., McKenzie, H. A., and Thompson, P. C. O. (1959). *J. Sci. Food Agric. 10*, 51.
37. Nagy, S., and Smoot, J. M. (1977). *J. Agric. Food Chem. 25*, 135.
38. Smoot, J. M., and Nagy, S. (1980). *J. Agric. Food Chem. 28*, 417.
39. Ross, E. (1944). *Food Res. 9*, 27.
40. Moore, E. L., Wiederhold, E., and Atkins, C. D. (1944). *Fruit Prod. J. 23*, 270.
41. Edrissi, M., and Kooshkabadi, H. (1975). *Iran J. Agric. Res. 3*, 81.
42. Bissett, O. W., and Berry, R. E. (1975). *J. Food Sci. 40*, 178.
43. Streff, R. R. (1971). *Am. J. Chem. Nutr. 24*, 1390.
44. Ting, S. V., Moore, E. L., McAllister, J. W., Streiff, R. R., Hsu, J. N. C., and Hill, E. C. (1974). *Proc. Fla. State Hortic. Soc. 87*, 206.

45. Ting, S. V., and Rouseff, R. L. (1981). *Proc. Int. Soc. Citric. 2*, 905.
46. Rouseff, R. L. (1977). *Abstr. Pittsburgh Conf. Anal. Chem. Appl. Spectrosc. 28*, 173.

CHAPTER 15

SHELF-LIFE OF WINE

PASCAL RIBÉREAU-GAYON
Institut d'OEnologie
Université de Bordeaux II
Talence, France

Handbook of Food and Beverage Stability: Chemical, Biochemical, Microbiological, and Nutritional Aspects

I. INTRODUCTION

Wine occupies a special place among food products. First, there exists an exceptional hierarchy of quality for products whose main analytical parameters appear to be narrow. Furthermore, certain wines not only can be preserved for long periods but even improve with age. For some of the most prestigious wines, 10-20 years of aging are necessary to bring out the potential qualities. There are available today certain wines of the last century, aged over 100 years, the consumption of which is always remarkable.

The essential problem in storing most food products is to avoid losing the initial qualities. For wine, it is also important to know the changes and conditions responsible for quality improvement during aging, as well as those wines most susceptible to improvement during storage--that is to say, the wines capable of aging well.

II. NATURAL FACTORS OF QUALITY AND APTITUDE FOR AGING

A. Vine Plants

The fine vine plants, leading to the production of wines that age well, belong to the genus *Vitis vinifera*. At the beginning of the Quaternary Period, this species covered all of Western Europe. Varietal groups were created in the different regions, and from these have been selected the varieties cultivated in the various vineyards.

The nature of the vine plants is an essential factor in the way wines behave during aging. An important chemical change is the modification of the color and of the tannins, which also affects the taste. Combinations between anthocyanins and tannins ensure the indispensable stability of color during this change. The nature of these combinations varies with the grapes, not all of which permit the making of wines likely to age well. The Cabernet wines are high enough in tannins to require several years for refinement: prolonged aging is required to bring out all their qualities. On the other hand, Merlot wines are more mellow and may be consumed when younger.

The same consideration applies to the vinification conditions, which must be adapted to the style of each wine. Those wines that are meant to be consumed young, a few months after vinification, must be fruity and not too tannic in taste: therefore, the skins must not be macerated too long in the juice. On the other hand, the wines destined to age in the bottle for several years, or even decades, must be rich in tannins, and this is achieved by a maceration extending over 2 to 3 weeks. It is understood that grapes from the same vine plant are not necessarily adapted to the production of these two types of wines.

B. Climate

Sufficient heat is necessary for grape maturation. However, in the warmest climates a sort of uniform, "homogenized" quality results; wine aromas become heavier and lack distinctive character. For that reason, the most celebrated vineyards are not located in zones with climates that would appear to be the most favorable to viniculture. It is quite rare to find truly great vineyards in those regions that regularly experience a hot and dry climate each year during the ripening period, with watering or irrigation eventually becoming necessary.

A consequence for the great vineyards located in marginal climatic zones is the irregularity, from one year to the next, of the climatic conditions during the ripening period. Either excess humidity or drought can interfere. In this regard the type of soil in the vineyard plays an essential role in assuring a consistent water supply to the plant: it brings water during growth, and it quickly drains off excessive rainfall during ripening. Gravel and calcareous soils are particularly effective in assuring these regulatory mechanisms. In the event the natural water regulation systems prove insufficient, they are supplemented by means of artificial drainage.

The most classic regulation mechanism is that of the particularly poor gravelly soil of Médoc, in which the plant root system must penetrate to a depth of several meters. In the spring water is provided by the underground water table that rises to the surface, having been low since the end of summer. During ripening, the roots are thus relatively insensitive to the precipitation that affects the soil's superficial layers.

The regulatory phenomenon of the provisioning of water explains the existence of excellent wine-growing areas where soils apparently vary widely--and of vineyards of varying quality where soils seem to be similar.

In any case, it is well known that the quality of a vineyard soil bears no relation to the supply of minerals to the vine; the plant is undemanding and grows on poor soil.

Vitis vinifera is cultivated in the northern hemisphere between the thirty-fifth and fiftieth parallels and thus is adapted to various climates (Table I). In the colder regions, vintners choose precocious vine plants, the fruit of which can ripen before the autumn cold sets in. In warmer climates late vine plants are cultivated, because they permit economical production. It is noteworthy, however, that the different regions have selected vine plants that ripen as late as possible for the particular region. The Pinot of Bourgogne does not do as well in the warmer climate of Bordeaux. This recalls an essential concept: ripe grapes are necessary to make good wine, but a ripening that is too fast or too complete (leading to loss of aromatic elements) is undesirable.

Some vineyards have become geared to the production of wines from a single vine plant, with its own individual varietal characteristics: Pinot from Bourgogne, Riesling and Traminer from Alsace, Sauvignon from the Loire--all have acquired a fine notoriety. Numerous vineyards of more recent creation (e.g., in California, Australia, and South Africa) have also established the quality of their wines from pure vine plants. On the other hand, in other regions--Champagne and especially Bordeaux--the greatest wines originate from the association of several vines chosen deliberately in correspondence to the climate's variability from one year to the next. Depending on the meteorological conditions, for a given vineyard, each vine plant can succeed more or less. But over a long period, the judicious association of several varieties gives best quality.

Table II shows the characteristics of the principal red and white vine plants cultivated in the Bordeaux regions since 1900. The red vine plants especially are distinguished by the acidity of their fruit: for certain years in particular the malic acid content of the ripened Petit-Verdot or Malbec can be three times higher than that of Merlot. This variation in acidity certainly enters into the selection of these varieties; the Cabernets, basic vine plants because of their characteristic aroma, are complemented on the one hand by the Merlot of low acidity (which imparts suppleness in the years of insufficient maturity) and on the other hand by the Malbec and the Petit-Verdot of high acidity (which impart freshness in the years of high maturity). Today, research in supple red wines of low acidity has resulted in the almost complete disappearance of the Malbec and Petit-Verdot vine plants.

C. Ripening

During the maturation process, the evolution of the grape's composition--an essential factor in wine quality--varies considerably from one year to the next. The essential transformations that occur are the accumulation of sugars and the decrease in acidity. The ratio of sugar to acidity provides a simple means for gauging the progress ripening (Table III). At maturity, therefore, grape composition differs from year to year (Table IV). The differences occur in sugar concentration and degree of acidity, but also in the anthocyanin and tannin contents, which can vary by a factor of two. These phenolic compounds require considerable light energy for their synthesis; they are the first to be affected by cold weather--even if the sugar accumulation and decrease in acidity are normal. This explains the possibility of producing exclusively white wines in the northernmost region (Champagne, Alsace, Moselle, the Rhine), because conditions do not permit the complete maturation of red vine plants.

For example, Table V shows the harvest dates and the average grape composition for six successive vintages of the same Médoc locality. This composition is directly related to behavior upon aging. The ripest grapes are also those richest in tannin and in aroma, assuring the wine's longevity and enabling it to become refined with time.

These compositional differences reflect climatological conditions from April to September (inclusive). Generally, in the Bordeaux climate, the best ripening occurs with warm weather and little precipitation. It is possible to calculate a climatological index by subtracting the rainfall height from the daily temperature total. There is quite a good relation (Table VI) between this index and the quality of the vintage--as it appears from the taste evaluation by experts. More precise concordances could be obtained by refining the results. For example, April rainfall affects ripening less than September rainfall; the values should then be corrected by means of a coefficient that would be a function of the phenological cycle. Again, in some years ripening continues until October, and this should be taken into account in determining the climatological cycle. For that reason, Table VI does not provide a classification that corresponds to the actual quality of the late-ripening 1978 and 1979 vintages.

In any case, the conditions of the grape-s ripening and of its composition on maturity are the essential elements in the great wines' ability to age well. Among the vintage years shown in Table VI, 1959, 1961, 1975, and 1982 have produced wines that will last longest and the qualities of which will be the most likely to improve with age. The year 1961 probably was the best vintage year in the entire history of Bordeaux wine; in addition to particularly favorable climatic conditions, the ripening was facilitated by a small harvest and the grapes were small and very concentrated. Wines of this vintage are today remarkable and will probably remain so for a very long time. The vintage 1982 wines will be of similar quality and will age perfectly.

III. IN-BOTTLE AGING OF WINES

A. Aroma Changes--Development of the Bouquet

During aging in the bottle, an important development takes place in the aromatic elements, corresponding to the formation of the wine aroma called "bouquet." This change is a composite of the grape's primary and secondary aromas, the fermentation, and eventually, the wood of the oak. The chemical bases of bouquet formation remain very obscure.

Toward the end of the nineteenth century, esterification phenomena were invoked to explain this change in the aromas. Wine is an alcoholic medium containing numerous acids, among which tartaric, succinic, and lactic are the most important. In fact, upon aging increases in ethyl tartrate, ethyl succinate, and ethyl lactate are observed. However, such changes have been found in all wines, and it has been impossible to understand their different behaviors during aging. As a matter of fact, only the great wines' aroma improves with age.

Wine contains other aromatic esters, but these are of microbiological origin and are not related to aging. The ethyl esters of fatty acids have floral and fruity aromas, formed by the yeast during fermentation. They play an important role in bouquet formation, especially with white wines, and tend to decrease through hydrolysis during storage. Ethyl acetate is a sign of aroma alteration; its excessive formation results from the metabolism of acetic bacteria in poorly sealed containers.

It is well known today that bouquet development in the bottle is due to a reduction process that occurs exclusively in the absence of oxygen when the redox potential is sufficiently low. The bouquet disappears quickly or changes greatly when the wine is slightly aerated (for that reason, old wines should not be aerated ahead of time, when poured before serving). The bouquet therefore is due to oxidizable materials, forming redox systems and possessing pleasant odors only when in a reduced form. However, an initial controlled oxidation appears necessary both for the development of the bouquet and for the modification of the color and the tannins. It comes into effect during storage in wooden casks, and perhaps even in the bottle, under the influence of a few milligrams of oxygen that dissolve in the wine at filling time. This controlled aeration results in the formation of compounds the reduction of which in the bottle would favor the lowering of the redox potential. The bouquet's intensity seems related to the nature of the wine, the presence of metallic ions, the efficiency of corking, and the storage temperature. The addition of ascorbic acid to bottles of champagne at the moment of final closure improves the evolution of the bouquet by accelerating the reduction of ferric ions.

B. Color Changes

With red wines, the obvious chemical change indicative of aging is that of color. Bright red in young wine, it becomes yellower in old wine, giving rise to tones reminiscent of tile or brick--hence the expression brick-red to denote the color of wines stored in bottles for several years. In very old wines the red tone completely disappears, and yellow and brown assume predominance.

As Fig. 1 shows, in the absorption spectrum of a 1-year-old red wine, an absorption maximum of 520 nm is seen to correspond to the color red. There is also an important absorption in the ultraviolet range. Between the two, the curve shows a minimum at 420 nm, corresponding to the color yellow. For a 10-year-old wine, maximum and minimum are reduced to a shoulder, and, finally, for a 50-year-old wine the shoulder itself has practically disappeared. It is therefore possible to measure wine color by determining optical density (OD) at 520 and 420 nm. The expression $D_{520} + D_{420}$ represents color intensity; the fraction D_{420}/D_{520} expresses the shade of the coloration and corresponds quite well to the aging level of the wine.

Anthocyanins and tannins affect wine color equally. Anthocyanins are the red pigments of the grape; tannins also play an important role in the taste characteristics of the red wines. The same chemical changes that affect color during aging also exert an influence on the taste of the tannins. As a result, the general structure of the wine becomes more supple in relation to a partial precipitation of the most astringent among the phenolic compounds.

Various chemical mechanisms have been proposed for the interpretation of these reactions, including a copolymerization of anthocyanins and tannins, or a tannin polymerization. The anthocyanins disappear little by little. The pale yellow color of the tannins becomes increasingly more intense; these compounds play an essential role in the color of old wines.

Color evolution during aging is very well brought out by the figures in Table VII, in the case of wines of the same vineyard but differing in age. The anthocyanins disappear during the first years of storage. At the same time, the shade of the wine color increases, showing the greater influence of yellow (due to the tannins) relative to red (due to the anthocyanins). There is good correspondence in the case of old wines between color intensity and tannin content. The tannins are the essential constituents of the color of old wines, which contain almost no free anthocyanins. This table also shows that the older wines are much richer in total phenolic compounds and in tannins. This is certainly not due to an increase upon storage but to a greater initial content in the older vintages resulting from the conditions of viniculture and the techniques of wine making.

C. Wine Composition as a Function of Aging

Table VIII groups some other data on the composition of the great Bordeaux wines as a function of their age. The period from 1906 to 1972 is divided into five groups. From each group 11-18 samples (from three vintages) were analyzed. For each analysis and each group, the highest, lowest, and mean results are shown. The objective was to follow simultaneously the effect of aging and of the eventual changes in style that had taken place over >60 years.

1. Alcohol Content

The increase in the mean alcohol content is due not to an increase of the maximum but to a greater regularity due to a generalized enrichment in sucrose and also to better ripening of the grape at harvest time.

2. Total Acidity

The considerable decrease in acidity since the beginning of the twentieth century is related to the mastery of the malolactic fermentation technique, which has become very precise only recently. Malic acid has been completely eliminated only since 1963.

3. Volatile Acidity

Much progress has occurred in volatile acidity. Its mean value has decreased from 0.75 to 0.42 g/liter and its maximum value from 1.1 to 0.50 g/liter. This is a considerable improvement in quality; it is known that volatile acidity reflects the level of alteration due to bacteria. When the value reaches 0.6 g/liter, the characteristic sour smell and the harsh aftertaste are evident, even if the quality is not irretrievably compromised. Sulfiting is used to protect against the bacteria. Since 1963, no wine contains <50 g/liter of total sulfur dioxide at the point of bottling; this is the minimum dosage to assure adequate protection.

One reason for the increase in volatile acidity is the decomposition of residual sugars by the lactic bacteria. It is therefore important that the alcoholic fermentation be completed while leaving <2 g/liter of sugars. This result has only been achieved during the last few years and coincides with the decrease in volatile acidity.

4. Dissolved Carbon Dioxide

The amount of carbon dioxide dissolved in wine varies widely. This gas is formed during the alcoholic fermentation and has not been sufficiently eliminated before bottling. It can also result from a secondary fermentation in the bottle; this would be a real accident, however, and it almost never happens today. Carbon dioxide has an adverse effect on the taste of the aged great red wines, which must not contain more than 300 mg/liter. Beyond this amount, they appear harsh and thin, even if the characteristic tingling taste of this gas is not perceptible. This result was not always achieved between 1963 and 1972, but it has been since then.

5. Tannin Content

The results noted previously in Table VII are confirmed here. On an average, the tannin content has decreased steadily since 1906, due to (1) an increase in vine plant productivity, which gives rise to more dilute grapes, and (2) research that, through modifications in wine making, leads to more supple wines that can be consumed younger. However, it is possible that this process has gone a little too far; from 1953 to 1972, certain wines were too light for good aging. Since 1972 technical changes concerning vine cultivation and wine making have occurred: it is the exception today to find great red Bordeaux wines with a tannin index of <40. The values showing the weight of extract, that is to say total nonvolatile matter, vary like the tannin index. It is a well-known fact that the phenolic compounds represent an important part of the extract of red wines.

Finally, the values shown in Tables VII and VIII indicate that, after some years of aging, there is good correlation between the intensity of color and the tannin content, the latter being the major coloring element in old wines.

D. Factors of Aging in the Bottle

1. Temperature

The transformation of wine during aging is first dictated by temperature. It is well known that slightly higher temperatures (18-20°C) accelerate these changes, especially the formation of the bouquet. However, increased temperatures will also shorten the longevity of the wine; on the other hand, excessively low temperatures (<10°C) slow the aging process.

2. Oxygen

Another important factor is the presence of oxygen. Some oxygen is no doubt introduced during bottling and helps initiate the transformations of aging; some tests have shown that a better development of the bouquet and a faster evolution of color occur when the wine is saturated with air before bottling. However, the amount of oxygen that thereafter seeps into the bottle is by all accounts negligible. In the past it was thought that oxygen penetrated the cork of the stopper, participating in the aging process. It is now known that this is not the case. Oxygen is <u>not</u> an agent of aging. On the contrary, the presence of oxygen affects it adversely: leaking bottles contain wine of poorer taste.

3. Sulfur Dioxide

During bottling, sulfurous anhydride also exerts an influence on the wine's evolution, especially if there is a fair amount of oxygen penetration. It prevents a transitory oxidation known as "bottle sickness." It appears, however, that several months after bottling, sulfur dioxide present in the amounts normally used exerts no influence.

4. Light

Other problems that are likely to interfere with wine's normal evolution in the bottle include those relating to redox potential, the presence of inorganic ions, and especially, light.

The glass color of a wine bottle and its ability to absorb light rays affects the evolution of the wine. White wine ages more rapidly in a white bottle than it does in a colored one; the redox potential becomes lower faster and reaches a lower level in the clear glass. This could be an advantage for white wines that acquire a pleasant bouquet upon aging, but a disadvantage for those wines that need to maintain their freshness and fruitiness. Another drawback of the colorless bottle, always tied to the lowering of the redox potential, is the reduction of copper salts accompanied by the appearance of the haze known as "copper trouble." This accident is practically unknown in the case of white wines in colored bottles. Even red wines, which are less permeable and less sensitive to light, develop more harmoniously in dark-colored glass bottles.

5. Techniques for Promoting Aging

Conditions leading to faster aging of wine have been sought so as to decrease its storage time. Several different procedures have been proposed involving heat, ultraviolet irradiation, infrared irradiation, ultrasonics, even γ-ray irradiation. In practice these procedures cannot be adequately controlled; furthermore, the changes they initiate are not completely understood and do not constitute true aging.

The best-studied procedures rely on relatively violent aeration, accompanied by strong temperature variations. These are of course carried out before bottling, with the aim of reproducing in an accelerated fashion the summery conditions that age the wine and the wintery conditions that stabilize it. The violent treatments are not applicable to delicate wines; at most they make them look like old wines, but without any of the appropriate taste characteristics.

E. Conditions for Wine Storage in Bottles

Knowing the factors affecting aging, it is now possible to attempt to define the best criteria for a suitable storage location that would lead to a good development.

1. Humidity

Humidity is an essential factor--perhaps the most important one. The relative humidity in the cellar must be $\geq$50%. A humidity >50%, however, although not detrimental to a good development of the wine, will cause a fast deterioration of the labels and the packaging material. In particular, excessive humidity can favor the proliferation cf cork worms, which can appear between the stopper and the capsule. In any case, after several years the stopper no longer assures airtightness; recorking of old wines with new corks at the end of 25 years is recommended.

2. Oxygen

To minimize the introduction of oxygen, wine bottles are stored lying down so that the cork of the stopper, which is always moist, assures perfect tightness. If the bottle is stored upright, the small space between the stopper and the wine will allow oxygen penetration to become appreciable, resulting in wine deterioration. Upright storage of wine bottles must be absolutely avoided, especially during summer.

3. Temperature

Temperature must also be controlled, the extreme limits being 8 and 15°C. Below 8°C, wine development is exceedingly slow; above 15°C, development is too fast and the wine ages prematurely. It should be noted that even more than heat, it is temperature *variations* that interfere most seriously; they cause dilations and contractions in the bottle, which result in very undesirable air penetration.

4. Light

Light is also a great enemy of wine. For that reason, the use of tinted glass is recommended for the seasoning of wine; if white glass is used, one must ascertain that it will correctly absorb ultraviolet radiation. In any case, the bottles must be stored in dark cellars. Such conditions will allow the aging process to proceed more smoothly.

5. Ventilation

Cellars should be aerated to prevent bad odors. Wine is very sensitive to foreign smells, which it absorbs quite easily through the cork, according to some authors. Aeration also helps to maintain consistent humidity in cellars that tend to be either too dry or too humid.

6. Vibration

Finally, wine meant to age for long periods must be protected from vibrations such as may be caused by an underground or regular railway or by public works. Such vibrations stir up the sediment and "tire" the wine.

In any case, the conditions recommended above do not often exist in the real world of commercial distribution. The stocking of wine bottles upright, at relatively high temperatures, and exposed to the radiations of neon light fixtures is particularly contrary to reason, because these are the most unfavorable conditions possible for a good development of the wine. Rather than display the actual bottled wine in an outlet, retailers should allow prospective wine buyers to check the bottle and read the label of a model wine bottle, empty, so that they might use it as a standard for their decision.

IV. MISHAPS OF BOTTLED WINE

A. Chemical Hazes and Deposits

The chemical composition of wine is extremely complex. A total of 300 compounds have been identified, and there are probably more. It is understandable then that various chemical and physicochemical processes can bring about the formation of hazes or sediments that can change the appearance of wine, especially white wine.

1. Sediment

Traditionally the only acceptable deposits in wine are the coloring material sediments in old red wine. Technology has not yet managed to eliminate them; indeed, they have become proof of quality acquired through aging. They are related to the polymerization of the molecules of coloring matter that is involved in the modification and the refining of tannins. Upon attaining a certain size, the molecules precipitate. Precipitations that are too precocious or too heavy can be avoided by fining the wine before it is bottled. This is done by adding to the wine a protein that flocculates, carrying down the biggest molecules constituting the coloring matter.

2. Tartrate Precipitation

Another common defect is tartrate precipitation and sediment. Tartaric acid salts precipitate out at low temperatures, especially during transportation by truck and by boat under insufficiently controlled temperature conditions. Unfortunately, the preventive methods presently known are not absolutely efficient. Metatartaric acid (a less hydrous form) is a crystallization inhibitor, but its stability is not indefinite: barely a few days at slightly higher temperatures. Generally, wines are kept for several days at low temperature so that the precipitation occurs before bottling; the precipitates are then removed by filtration. However, one can never be sure of perfect stability. In fact, the solubility of tartrates is affected by variations in the wine's chemical composition, and the stability achieved at a given point is not necessarily definitive.

The point should be made that the tartrate crystals at the bottom of the bottle do not affect the wine's quality; because of their weight, they do not even interfere with the serving of the wine. Consumer education with regard to this point would prevent unjustified demands that would result in excessive and expensive treatments.

3. Oxidation Trouble

The spoiling of the color of red wines, after even limited contact with oxygen in air, is known as oxidation trouble. This defect results from the presence of an enzyme secreted by *Botrytis cinerea*, a fungus of putrefaction. This indicates that unsound grapes were used in wine making; good-quality wine need not be expected in this case, because rot causes bad odors and bad tastes. Technical means exist of inhibiting enzymes before bottling. This particular defect is thus due to sloppy techniques and is quite avoidable.

4. Defects of White Wine Clarity

Various defects of white wine clarity are known as iron trouble, copper trouble, and protein trouble. These occur under certain conditions of temperature, aeration, and lighting; they do not essentially affect wine quality. However, when they occur in the bottle, the wine must be returned to the vat, restabilized, reclarified, and rebottled. There are now excellent means for avoiding these defects, so their occurrence is a result of mistakes in manipulation. "Sunlight flavor" refers to a strange taste that appears in white wine, particularly champagne wines. This reaction is catalyzed by light and presupposes a relatively low redox potential. This produces hydrogen sulfide, which reacts with ethanol, giving rise to a mercaptan of a very offensive odor that causes the "sunlight flavor." Various additives have been tried to contain this defect; it seems that tannin might help to some extent because of its antioxidant properties. However, light is the essential factor that produces this taste defect. The wavelengths responsible are at about 370 and 440 nm. Irrespective of the luminous intensity, it is the transparency of the glass to these wavelengths that counts. In any case, to avoid this grave defect, champagne wines should be stored in the dark.

B. Microbiological Defects

1. Yeasts and Bacteria

Wine is produced using microbiological processes. Yeast, through the alcoholic fermentation, converts sugar into alcohol. Then, eventually, the lactic acid bacteria cause the decomposition of malic acid into lactic acid, a phenomenon known by the name of malolactic fermentation. By decreasing the acidity, it permits a mellowing that is good for quality; also it ensures biological stabilization, because malic acid is the most easily biodegradable component of wine.

New wine is rife with yeast and lactic and acetic acid bacteria. In principle, these should be eliminated through clarification procedures before bottling--if not completely, at least sufficiently reduced to prevent untimely developments and defects in the bottle.

Absolute sterility can be obtained, theoretically, either through a sufficiently tight filtration or by chemical treatment. In practice the complete elimination of living microorganisms is rarely achieved, however, and total stability is obtained through the use of antiseptics, of which the bisulfites are the most common. However, the activity of a given dosage of bisulfite is a function of the number of residual microorganisms. For biological stability, then, a weak microorganism population and a sufficient bisulfite dosage are necessary. If that is not the case, one can expect yeast or bacteria to develop, resulting in the appearance of suspended particles in the bottle. This is really just a defect in appearance, since quality is not impaired. In some cases, however, the growth of microorganisms can lead to a veritable decomposition of the wine. In particular, residual yeast can bring about a regular refermentation of the sweetened wines, increasing the pressure in the bottle and causing popping of the crown.

In any case, these defects are due to technical mistakes at the point of bottling and are relatively easy to correct. However, storage of wine at higher temperatures facilitates the occurrence of these defects and aggravates them.

2. Phenomenon of "Masques" in Bottles of Champagne

"Masques" are defective hazes that form in bottles of sparkling wine made by the champagne method. The secondary fermentation responsible for the froth takes place in the definitive bottle. Yeast is removed by decantation after sedimenting on contact with the stopper, the bottle being kept upside down for several weeks. The appearance of a "masque" corresponds to the formation of a yeast haze that sediments poorly.

A microbiological phenomenon contributes to this defect. The yeast employed is *Saccharomyces bayanus*, well known for its haze-forming capacity in the presence of oxygen. It is easy to see that the quantity of oxygen introduced into the bottle during filling would be sufficient to bring about the appearance of this haze.

This haze is composed of polysaccharides and some fatty acids, the presence of which make the haze adhere to the glass of the bottle. The points of adhesion are those areas that have been contaminated by lubricants used on the glass molds during the manufacture of bottles.

These yeast hazes, adhering to the bottle, cannot be eliminated during decantation; they make up the "masques" that reappear in the bottles at the point of commercialization and constitute a grave defect of presentation. These "masques" tend to be reabsorbed during storage.

C. Leaking Bottles

Several causes contribute to the appearance of "leakers": (1) cork of insufficient quality, (2) bad stopper preparation, (3) poor regulation of the jaws of the crowning machine, resulting in a crease down the length of the crown, and (4) excessive pressure inside the bottle.

Leaking may occur shortly after bottling, in which case leakage of wine is limited. Wine quality is not affected, but the leakage stains the labels.

A more serious problem may occur after many years of storage during which the stopper decomposes. This results in a substantial leaking of the wine, with correspondingly appreciable lowering of its level in the bottle. The wine is most often spoiled, and in any case the bottle becomes unsaleable. To prevent this, it is recommended that the corks of wine bottles be replaced every 25 years. On these occasions the bottles are topped up and traces of bisulfite and ascorbic acid added to guard against oxidation resulting from this manipulation.

Excessive pressure inside the bottle may also cause leakage. These appear at the point of pushing in of the crown, compressing the headspace air in the neck of the bottle. If the bottle is laid down before the cork regains its elasticity--thus assuring perfect airtightness--the excessive pressure can cause the liquid to leak, oozing between the cork and the glass. This internal pressure in the bottles is likely to increase considerably if they are dispatched rapidly in containers exposed to the sun. The temperature in the bottles can then exceed 40°C, thus it would certainly be important to ensure that wine is transported in thermally insulated containers.

Very efficient crowning methods exist that can eliminate all these pressure mishaps in the bottles. These methods rely on vacuum or, better yet, on the bursting of a carbon dioxide bubble above the wine, before penetration of the crown. Carbon dioxide being a soluble gas, its dissolution lowers the pressure a few minutes after the pushing in of the stopper. There is no further risk of leaks, and the amount of carbon dioxide injected is too small to change the wine's taste characteristics.

V. CONCLUSION

The ability of wine to age well depends on many natural factors, such as the vine plant, the locality of maturation, and the vintage.

The behavior of bottled wine (or wine in any other type of container) also depends greatly on its processing and conditioning. It is prone to many defects, hazes, and sediments, most of which are due to manipulation errors. Technology has the means to avoid them.

However, many precautions must be taken during storage to avoid the deterioration of wine: the bottles should be stored lying down, at mild temperatures, in a sufficiently humid, well-ventilated environment. These conditions apply equally to warehousing and transportation--and with even greater force in the case of wines expected to improve upon long aging.

TABLE I

CLIMATIC CONDITIONS DURING THE MONTHS OF APRIL TO SEPTEMBER IN VARIOUS WINE GROWING REGIONS (OCTOBER TO MARCH IN THE SOUTHERN HEMISPHERE)

	Total temperature (°C)	Duration of sunlight (hours)	Rainfall (mm)
Alicante, Spain	4064	1847	147
Palermo, Italy	4005	1619	138
Oran, Algéria	3908	1784	79
Jerez, Spain	3880	1930	117
Mendoza, Argentina	3909	1688	136
Patras, Greece	3811	1778	132
Perpignan, France	3691	1619	247
Florence, Italy	3659	1697	339
Adelaïde, Australia	3622	1544	177
Nimes, France	3592	1731	345
Montpellier, France	3420	1771	295
Bordeaux, France	3165	1252	358
Dijon, France	2984	1433	403
Reims, France	2782	1226	318

TABLE II

AVERAGE COMPOSITION OF MUSTS FROM BORDEAUX VINES

(Figures collected under conditions comparable for several harvests)

	Vine	Degree	Acidity (meq/L)	Tartaric Acid (meq/L)	Malic Acid (meq/L)
Red Vine-plant (cultivated on same soil)	Merlot	12°0	92	110	36
	Cabernet franc	11°3	99	102	40
	Cabernet-Sauvignon	11°4	102	111	50
	Malbec	11°0	108	98	55
	Petit-Verdot.............	12°5	120	120	60
White Vine-plant (cultivated on same soil)	Sauvignon	13°0	90	80	45
	Sémillon	12°4	80	82	32
	Muscadelle	11°2	84	90	38

TABLE III

EVOLUTION OF THE RATIO OF SUGARS TO ACIDITY DURING RIPENING (Pauillac, 1970)

	Merlot			Cabernet-Sauvignon		
	Sugars	Acidity	Ratio	Sugars	Acidity	Ratio
August 18th	72	17,6	4,0	84	17,4	4,8
August 24th	100	13,1	7,6	108	13,0	8,2
August 31st	134	9,02	14	128	9,02	14
September 7th	166	6,86	24	156	7,96	19
September 14th	172	5,68	30	156	6,67	23
September 21st	192	4,90	39	178	6,27	28
September 28th	192	5,10	37	188	5,88	32
October 5th				184	5,30	34

*** acidity expressed in grams of sulfuric acid per liter, sugar in grams per liter**

TABLE IV

COMPOSITION CLOSE TO MATURITY OF GRAPES FROM CABERNET-SAUVIGNON VINES IN TWO VINEYARDS OF THE BORDEAUX REGION - 5 SUCCESSIVE YEARS

	Year	Date	weight 100 baies	Sugar (g/L)	Acidity (meq/L)	Anthocyanins (g in 100 baies)	Tannins (g in 100 baies)
MEDOC WINES							
	1969	29 IX	85	176	126	0,25	0,36
	1970	28 IX	183	200	95	0,42	0,76
	1971	27 IX	110	185	105	0,23	0,60
	1972	9 X	101	180	150	0,22	0,41
	1973	I X	138	170	114	0,28	0,42
PREMIERES COTES BORDEAUX WINES							
	1969	29 IX	107	172	153	0,22	0,36
	1970	28 IX	115	200	123	0,41	0,72
	1971	27 IX	120	170	140	0,24	0,57
	1972	9 X	116	164	194	0,22	0,43
	1973	1 X	140	183	154	0,32	0,42

TABLE V

AVERAGE MUST COMPOSITION -HARVEST PERIOD, SAME MEDOC WINE

	MERLOT			CABERNET-SAUVIGNON		
	Harvest Period	Acidity (*)	Alcohol Degree	Harvest Period	Acidity (*)	Alcohol Degree
1977	6 to 13 X	6,7	10°2	14 to 19 X	8,3	9°4
1978	7 to 13 X	5,4	12°0	14 to 23 X	5,3	11°4
1979	1 to 7 X	4,7	12°1	8 to 15 X	5,4	11°4
1980	8 to 14 X	4,9	10°5	15 to 24 X	5,0	10°0
1981	28 IX to 14 X	4,2	10°7	5 to 13 X	4,5	10°0
1982	15 to 21 IX	4,1	12°9	22 IX to 1 X	4,1	11°5

(*) the acidity is expressed in grams of sulfuric acid per liter

TABLE VI

RELATION BETWEEN VINTAGE QUALITY AND CLIMATE CONDITIONS IN THE BORDEAUX REGION

From April to September

Year	Temperature (°C)	Rainfall (mm)	Difference	Quality
1965	3005	461	2544	Bad
1972	2900	331	2569	Indifferent
1979	2937	366	2571	Very good
1963	3010	370	2640	Bad
1968	3102	458	2644	Indifferent
1958	3142	497	2645	Indifferent
1969	3168	521	2647	Average
1977	3065	411	2654	Fairly good
1960	3130	459	2671	Average
1978	3029	326	2703	Very good
1980	3057	343	2714	Good
1971	3269	496	2773	Good
1966	3161	359	2802	Very good
1974	3124	301	2823	Good
1967	3120	278	3842	Good
1962	3123	249	2874	Good
1975	3250	362	2888	Exceptional
1983	3349	448	2901	Very good
1981	3227	289	2938	Very good
1973	3299	354	2945	Good
1970	3184	232	2952	Very good
1959	3310	331	2979	Exceptional
1964	3327	339	2988	Very good
1982	3328	289	3039	Exceptional
1961	3294	213	3081	Exceptional
1976	3386	278	3108	Very good

TABLE VII

MEASUREMENT OF PHENOLIC COMPOUNDS IN WINES OF DIFFERANT AGES – SAME CRU

(March 1964 analyses)

Year	Total Phenolics (Permanganate index)	Tannins (g/L)	Color Intensity D 420 + D 520	Inc D 420 / D 520	Anthocyanins (g/L)
1962	34	1,50	0,49	0,84	305
1961	37	1,85	0,54	0,86	165
1960	34	1,50	0,38	1,10	70
1958	48	2,35	0,42	1,05	47
1953	48	2,30	0,55	1,20	12
1947	64	3,70	0,74	1,08	21
1938	54	2,70	0,52	1,32	10
1921	68	4,40	0,80	1,32	19

TABLE VIII

ANALYSES OF VARIOUS RED WINE VINTAGES - MEDOC AND GRAVES

(Analyses of January 1976)

Number of Samples	Period		Alcoholic degree	Total Acidity (*)	Volatile Acidity (*)	Total SO_2 (mg/L)	Malic Acid (g/L)	Sugar (g/L)	CO_2 (mg/L)	Tannin (index)	Extract 100° (g/L)	Color Intensity
12	1906 to 1931	Min.	10°0	3,72	0,53	27	0	2,0	110	48	19,2	0,59
		Max.	12°3	5,59	0,90	83	3,5	4,2	230	86	43,0	1,45
		Mean	11°1	4,40	0,71	41	0,8	2,5	170	64	28,2	0,84
11	1934 to 1942	Min.	9°9	3,92	0,46	27	0	2,0	115	40	20,2	0,46
		Max.	13°1	5,39	1,10	105	3,5	3,8	335	90	35,0	1,40
		Mean	11°2	4,32	0,75	53	0,6	2,4	167	59	26,5	0,77

TABLE VIII CONTINUED

ANALYSES OF VARIOUS RED WINE VINTAGES - MEDOC AND GRAVES

17	1943 to 1952	Min	10°3	3,43	0,45	28	0	2,0	85	44	20,0	0,35
		Max.	12°7	5,29	1,05	70	2,0	3,8	735	90	35,5	1,06
		Mean	11°3	4,03	0,70	45	0,2	2,3	451	56	26,0	0,71
18	1953 to 1962	Min.	11°1	3,19	0,45	29	0	2,0	95	32	20,0	0,41
		Max.	13°3	5,29	0,72	80	1,5	2,5	585	100	27,4	1,24
		Mean	11°9	3,55	0,57	52	0,1	2,1	305	50	22,9	0,61
18	1963 to 1972	Min.	11°8	3,09	0,37	51	0	2,0	75	29	19,6	0,36
		Max.	12°7	3,53	0,50	89	0	2,5	470	60	26,6	0,64
		Mean	12°0	3,38	0,42	67	0	2,0	2,63	43	22,2	0,49

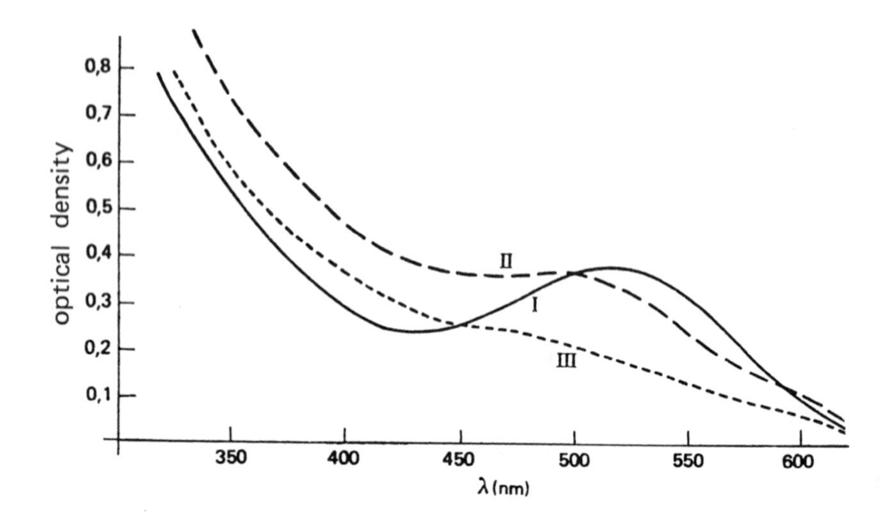

Fig. 1. Absorption spectra of three red wines of different ages: I, 1 year old; II, 10 years old; III, 50 years old.

CHAPTER 16

SAKE

TOSHITERU OHBA
MAKOTO SATO
National Research Institute of Brewing
Tokyo, Japan

Handbook of Food and Beverage
Stability: Chemical, Biochemical,
Microbiological, and Nutritional Aspects

I. OUTLINE OF SAKE BREWING

Sake, the traditional alcoholic beverage in Japan, is produced in ∿2600 breweries scattered throughout the country. Sake is usually consumed within 1 year after brewing. About 1.4 million kiloliters were consumed in 1984. Sake is consumed not only in Japan; it is also exported worldwide in annually increasing amounts. The export figure for 1983 was 1486 kiloliters. Table I shows the record of exports in recent years.

The method of sake brewing was described in detail by Kodama and Yoshizawa (9). Figure 1 outlines the sake brewing process as discussed in the remainder of this section.

A. Water

The raw materials used in sake brewing are water and rice. Since water accounts for about 80% (v/v) of the end product, the properties of the water influence the quality of sake. Colorless, tasteless, and odorless water is most suitable for sake brewing. The minerals in water are classified into effective and harmful components. Components such as potassium, magnesium, and phosphate not only serve as nutrients for *koji* mold and yeast, but also promote alcoholic fermentation (50,54). Components such as calcium and chloride serve as buffering agents for some enzymatic reactions. The harmful components are iron, manganese, and copper. Iron is the most harmful component because it gives the resulting sake an intense color and an unpleasant flavor (39). The maximum allowable iron content in the brewing water is 0.02 ppm. Therefore, appropriate treatment methods are employed to remove iron from the brewing water (51).

B. Polished Rice

The peripheral layers of brown rice contain large amounts of ash, vitamins, fats, and proteins that are undesirable for sake brewing. Brown rice is polished in the mill to remove these materials. In sake brewing, the polishing percentage rate is defined as the weight percentage of white rice obtained from brown rice. In general, cleaned rice at a 70-75% polishing rate is used, but rice at a 50% polishing rate is used in brewing superior sake. The polished rice is washed in a machine to remove the bran that has adhered to the surface of the grains, and then steeped in water. After steeping, the excess water is drained off from the grains for about 4 to 8 hr before steaming. This steeped rice is steamed in a continuous rice-steaming machine and then cooled.

C. Koji

Koji is a culture of the koji mold (*Aspergillus oryzae*) grown on and within steamed rice grains. Various enzymes such as amylases and proteases accumulate in koji.

D. Moto (Seed Mash)

Moto plays an important role as a starter of the yeast culture in carrying out the fermentation of moromi (main mash). Koji is mixed with water, steamed rice, and sake yeasts (*Saccharomyces cerevisiae*) in the prefermentation tanks. This seed mash is a concentrated culture of pure and healthy living cells of sake yeasts.

E. Moromi (Main Mash)

On the first day, a mixture of steamed rice, water, koji, and moto is put into the main fermentation tank. The fermentation process is started at low temperature (7-10°C). On the third and fourth days, additional volumes of steamed rice, water, and koji are added to the tank in order to secure the gradual growth of yeast in the mash. During the fermentation of moromi, carried out at ∿15°C, the starch in the rice is liquefied and saccharified by amylase of the koji, and this converted mixture is fermented into ethanol by the action of sake yeast. Both processes, saccharification and alcohol fermentation, advance simultaneously in a well-balanced manner. This unique, complex method, called "parallel fermentation," contributes to the high ethanol concentration, reaching nearly 20% (v/v).

F. Filtration of Mash

When the fermentation is finished, the mash is press-filtered to separate the sake from the solids. The mash residue remaining on the filtering cloth is called sake cake.

G. Fresh Sake and Filtration

The fresh sake thus obtained still contains yeast cells and turbid materials such as fiber, starch, and protein. These are allowed to settle in a vessel for 5 to 10 days at low temperature. Maintaining a low temperature is important to avoid the deterioration of sake quality due to the residual enzymes and the autolysis of yeast. After settling, the supernatant is filtered with activated carbon (200-500 g/kiloliter), cotton, and celite through a filter press. The clarified sake is blended to ensure a desirable quality, and then pasteurized for storage.

H. Pasteurization and Storage

The blended fresh sake is heated to 60 to 65°C by passing through a tube-type heat exchanger made of tinned copper, aluminum, or stainless steel. Recently, plate-type heat exchangers have become available. This pasteurization inactivates the enzymes amylase, protease, and so on, and kills the so-called hiochi bacteria, harmful microorganisms that spoil sake during storage. The history of pasteurization of sake began in the sixteenth century before Pasteur's discoveries (1868). Immediately after pasteurization, the hot sake is transferred to a sealed vessel for storage. The gradual maturation of pasteurized sake that occurs during storage is due to chemical reaction, oxidative reaction, and physicochemical changes. Usually the color of sake becomes deeper as the maturation proceeds. The maturation of flavors and the coloration are accelerated at higher temperature. Usually the rate of the changes in color and odor are rapid compared with the rate of the changes in taste. Therefore, the temperature of sake during storage should preferably be kept at 10 to 20°C, striking a balance between the maturation velocity and time of bottling.

I. Bottling

The optimum term of storage is 6-12 months depending on the storage temperature. Since the flavors and tastes differ in each tank during storage, blending is necessary to obtain a uniform and desirable quality. Since blended sake has a slight turbidity caused by proteinaceous materials, it is clarified by the fining process described later (Section II, D,2). The clarified sake is filtered with activated carbon to adjust the flavor, taste, and color. Thereafter, sake is diluted with water to the appropriate alcohol content, usually 15.0-16.5% (v/v), and filtered to remove the minute particles and microorganisms through unglazed pottery or a membrane filter having a pore size of 0.1 to 1.0 μm. The sake thus obtained is pasteurized through a plate-type heater and bottled. The bottling date is printed on the label, and then the bottled sake is transported to the wholesaler or retail shop.

An average analysis for several components of sake is given in Table II.

II. DETERIORATION AND PRESERVATION OF QUALITY

During storage and after bottling, spoilage and deterioration of sake are sometimes encountered. Changes in coloration, off-flavors, disagreeable tastes, and turbidity then occur. These are caused by protein turbidity, bacterial contamination, overaging during storage and at the shop, and sunlight exposure during transportation or window display.

A. Coloration

The color of commercial sake is slightly yellowish. At present, excessively colored sake is not accepted for sale on the market. The absorbance measured at 430 nm is 0.012-0.019 in commercial sake. A significant correlation is obtained between color and quality of sake at the 1% degree of significance ($r = -.733$, $n = 53$) (46). Since the quality of colored sake is apt to be inferior, activated carbon is used to remove the precursor of coloring compounds and colorants in sake. Sake color is classified into five categories on the basis of its cause.

1. Color Caused by Raw Materials

Flavin compounds are derived from rice, koji, and metabolic products of yeast. Riboflavin predominates, and other compounds such as flavin mononucleotide and flavin adenine dinucleotide are present in small amounts. The concentration of flavin compounds is 20-70 μg/liter in fresh sake, but they are almost completely removed by activated carbon. Flavin compounds have been found to contribute only 1-10% to the overall coloration of sake (52). Riboflavin not only acts as a photosensitizer in the coloring reaction of sake upon exposure to sunlight and increases the color intensity of sake, but also is the growth factor of hiochi bacteria (12). The quality of sake is stabilized by removing riboflavin with activated carbon.

2. Color Caused by Iron

Iron contamination occurs during the sake brewing process, so that a reddish-brown color develops. Color caused by iron is due to the formation of ferrichrysin from deferriferrichrysin (39). Ferrichrysin, one of the ferrichromes (5,6,19, 20), is one of the main components in sake coloration and is an iron complex of hexapeptide composed of 3 mol δ-*N*-acetyl-δ-*N*-hydroxyornithine, 2 mol serine, and 1 mol glycine. Ferrichrysin or deferriferrichrysin is produced by *Aspergillus oryzae* during rice koji making. Their contents in sake and in rice koji are shown in Table III.

The majority of ferrichrome compounds in rice koji are in the iron-free form (deferriferrichrysin). Most of the ferrichrysin in sake is produced from deferriferrichrysin with iron that contaminates rice grains, water, inferior activated carbon, celite, flawed enameled vessels, and bottle caps. The contribution of iron to the overall coloration of sake is 10-50%. It is difficult to remove ferrichrysin from sake unless a high concentration of activated carbon is used. Therefore, iron contamination should be prevented. Alternatively, it may be possible to utilize the koji mold mutant that does not produce deferriferrichrysin (37).

3. Color Caused by Aging

The color of sake gradually intensifies as it ages. It is considered that this colorant is a kind of melanoidin produced by the aminocarbonyl reaction (4,42). The intermediate compounds (e.g., 3-deoxyglucosone and 3-deoxypentosone) in the aminocarbonyl reaction were detected in aged sake (24,43). The contribution of aging to sake's overall coloration is 40-80% (27). The color caused by aging may be prevented by storage at low temperature and treatment with activated carbon. The temperature coefficients of components in sake concerning coloration are about 2 to 3 (26).

4. Color Caused by Sunlight Exposure

Exposure to sunlight of bottled sake during transportation or window display causes a rapid increase in the color intensity accompanied by an off-flavor composed of methyl mercaptan (CH_3SH). Three coloring reactions cause coloration, as shown in Fig. 2 (17). Reactions I and II contribute about 75% of color development. Kynurenic acid and indole acetic acid are produced from tryptophan by way of kynurenine by yeast (28). Kynurenic acid as well as flavin acts as a photosensitizer. If the light having shorter wavelengths than 380 nm is screened out, this coloration does not take place. Therefore, brown and emerald-green bottles are employed to prevent the development of color caused by sunlight exposure.

Moreover, the absorption by activated carbon of compounds involved in the coloring reaction, plus the high-vacuum atmosphere in the bottle are effective in preventing color development.

5. Color Caused by Copper

The copper content in sake is 0.02-0.4 ppm. If the copper content in bottled sake is >0.5 ppm under vacuum conditions, coloration and turbidity take place (29). The copper contamination comes from the pump and nozzle of the bottling apparatus, especially when sterilized with sodium hypochlorite solution.

B. Odor

There is a remarkable difference between the odor of fresh sake and that of aged sake. The flavor of fresh sake is fragrant owing to an odor derived mainly from esters and higher alcohols, but the taste is stimulating and rough. On the contrary, the flavor of aged sake is quiet and mellow. When the aging proceeds further, the flavor grows to resemble the odor of raochu (an alcoholic beverage in China).

The odor caused by overaging of sake, usually called hineka, is an unflavorable attribute. The following compounds were identified in aged sake by flame photometric detector-gas chromatography (FPD-GC) and gas chromatography-mass spectrometry (GC-MS): methyl mercaptan (CH_3SH), methyl sulfide, dimethyl disulfide (DMDS), 3-hydroxy-4,5-dimethyl-2(5H)-furanone (HDMF), furfural, benzaldehyde, phenyl acetic acid, *p*-hydroxybenzaldehyde, vanillin, and ethyl vanillate. It is concluded that the odor of aged sake is composed of the above compounds.

1. 3-Hydroxy-4,5-dimethyl-2(5H)-furanone

A burnt flavoring compound imparts to aged sake its characteristic and dominant odor. It is identified as 3-hydroxy-4,5-dimethyl-2(5H)-furanone (HDMF), and its structure has been confirmed by synthesis. It suggests that the compound was formed by condensation of α-ketobutyric acid with acetaldehyde, which occurred upon degradation of threonine (44). HDMF has been synthesized by Schinz *et al.* (35,36) as an α-keto-γ-lactone homolog, but had not previously been isolated as a natural product. The threshold values for HDMF were determined by the sensory evaluation test (44). HDMF was perceived as a burnt odor by all panel members. The odor units obtained from the ratio of concentration in sake to threshold value are shown in Table IV. HDMF contributes to the odor of aged sake. In other fermented foodstuffs, HDMF was found in raochu (made in Japan, more than 30 ppb) and soy sauce (more than 50 ppb) by GLC and GC-MS. HDMF has also been isolated from Sherry (3) and cane molasses (48), and was reported as the key compound of the sugary flavor. Formation and degradation mechanisms of HDMF were proposed as shown in Fig. 3.

2. Volatile Sulfur Compounds

Hydrogen sulfide (H_2S), methyl mercaptan (CH_3SH), dimethyl sulfide (DMS), and dimethyl disulfide (DMDS) were detected with FPD-GC by the headspace method in the sake brewing process (30). The main sources of sulfur compounds in brewing are thought to be the chemical degradation of sulfur-containing amino acids and the metabolism of microorganisms.

DMDS is not found in fresh sake, but in aged sake the content is 7 ppb. Since the threshold value of DMDS is 6.5 ppb, it is thought that DMDS contributes to the odors of aged sake. Also, CH_3SH content was increased by pasteurization and sunlight exposure (31). It was confirmed that DMDS and CH_3SH were formed from methionine or cysteine by degradation during pasteurization and storage of sake. The threshold values for sulfur compounds are shown in Table V (30).

Although sulfur compounds are present in a small quantity of sake, they are considered unfavorable attributes because of their low threshold values.

3. Other Odor Compounds Isolated from Aged Sake

In addition to sulfur compounds and HDMF, many compounds were isolated only from aged sake. The compounds isolated from sake are shown in Table VI.

Phenyl acetic acid was isolated from sake for the first time by Ohba (22). The formation of phenyl acetic acid is considered to proceed as shown in Fig. 4. Phenyl acetic acid was formed from phenylalanine by Strecker degradation in the process of the aminocarbonyl reaction.

Vanillin is made from ferulic acid in sake brewing. Ferulic acid in sake originates from lipid-phenol and sugar-phenol in rice grain, which are degraded by *Aspergillus oryzae* to form ferulic acid (55).

4. Dimethyl Sulfide: An Off-Flavor Compound in Sake Brewed with Old Rice

When old rice stored for ≥1 year at room temperature was used as raw material for sake brewing, an unfavorable smell, the so-called komaishu, in pasteurized and stored sake has been recognized by sensory evaluation.

The unfavorable-smelling compound was identified as DMS. The content of DMS in sake made from old rice was about 0.4-44.2 ppb (average 9.6), while only trace amounts were present in sake made from new rice (45). It has been found that DMS is formed on heat treatment of many foods and beverages and that one of the precursors is *S*-methylmethionine sulfonium salt. Likewise, it has been proven that the DMS precursor in fresh sake brewed with old rice is *S*-methylmethionine (7).

Studies have determined that a DMS precursor in old rice is glutelin containing *S*-methylmethionine (18). Brown rice samples were fumigated with methyl bromide or phostoxine and analyzed for DMS precursor. Fumigation with methyl bromide remarkably increased the DMS precursor content in the grains, but that with phostoxine had little effect, as shown in Table VII.

Fumigating wheat, soy beans, casein, and methionine with methyl bromide resulted in a remarkable increase of DMS precursor in each case. The remarkable increase in the content of DMS precursor in old rice grains is mainly ascribable to fumigation with methyl bromide, which is usually used for sterilization of the grains and as an insecticide during storage of rice (8,10).

The mechanism of the evolution of DMS in sake brewed from old rice is shown in Fig. 5. In conclusion, if sake is brewed using either new rice or old rice stored below 15°C without fumigation or fumigated with phostoxine, DMS evolution does not occur.

The threshold value of DMS in sake is about 6.2-11.6 ppb (average 8.4 ppb) (45). Although DMS is disliked as an unfavorable smell in sake, it is a favorable flavor in beer. DMS is the main sulfur compound formed during beer brewing, and the residual DMS in finished beer is considered to contribute to the flavor.

C. Taste

The coarse and stimulating taste of fresh sake disappears during storage and gradually grows smooth and harmonious. However, overaging leads to a bitter and unpleasant taste caused by the increased formation upon aging of such compounds as L-prolyl-L-leucine anhydride (PLA), methyl thioadenosine (MTA), tetrahydro-harman-3-carboxylic acid (THCA), and harman (47). The content and threshold values of bitter compounds in sake are shown in Table VIII.

PLA, a cyclic dipeptide, increases linearly with aging of sake. The mechanism of formation of PLA in sake is considered to involve the decomposition of rice protein by koji protease to a straight-chain peptide containing proline and leucine. This straight-chain peptide then cyclizes to PLA during storage. PLA was also found in other fermented foods, including sherry (1.8 ppm), raochu (9.6 ppm), soy sauce (7.8 ppm), and beer (0.7 ppm) (41).

MTA is formed from the pyrolysis of adenosyl methionine (AMe), an autolysis compound from yeast, during pasteurization and storage. AMe is not detected in beverages with a low alcohol content such as beer, wine, soy sauce, and miso (bean paste) (40).

THCA is formed by condensation of tryptophan with acetaldehyde, and is the intermediate of harman, which increases during aging and sunlight exposure of sake (32).

Taste compounds in sake are shown in Table IX. Sake containing large amounts of amino acids is apt to become inferior during storage, because amino acids are the precursor of off-flavoring compounds, colorants, and bitter compounds. The amino acid content of sake decreases markedly during storage (33). The change in content of the various amino acids in sake during storage is shown in Table X.

D. Turbidity

Turbidity is sometimes encountered in storage of bottled sake. Turbidity of sake is classified into two types: (1) protein turbidity and (2) infection by hiochi bacteria.

1. Turbidity Caused by Hiochi Bacteria

Although sake contains more than 15% ethanol, it often becomes cloudy due to the growth of specific bacteria. This phenomenon and the organisms have been called _hiochi_ and _hiochi bacteria_, respectively. Hiochi bacteria are divided into heterofermenters and homofermenters; moreover, without regard to their mevalonic acid requirement, each of these two groups is further divided into true hiochi bacilli and hiochi lactobacilli by the pH range of growth, the fermentation of sugars, and the effect of alcohol on the growth, as shown in Table XI (14).

Heterofermentation

$$C_6H_{12}O_6 \longrightarrow CH_3CH(OH)COOH + CH_3CH_2OH + CO_2$$

glucose lactic acid

Homofermentation

$$C_6H_{12}O_6 \longrightarrow 2CH_3CH(OH)COOH$$

glucose lactic acid

When sake is infected by hiochi bacteria, off-flavor and sour taste, attributable to diacetyl, acetic acid, and lactic acid, occur in addition to the turbidity.

Lactobacillus heterochiochii is the major organism detected in infected sake and is dominant among hiochi bacteria. It requires mevalonic acid as an essential growth factor (49) and has a preference for 4 to 8% alcohol as well as a tolerance for 16 to 20% alcohol.

Beer and wine, which do not contain mevalonic acid, are not infected by hiochi bacteria. Although mevalonic acid is produced in small amounts by yeast, the major portion is produced in koji by *Aspergillus oryzae*. The sanitizing of the brewing process, detection tests for bacteria (38), and perfect pasteurization (15 min at 60 to 65°C) are important to prevent the infection by hiochi bacteria (15).

Sake that becomes infected by hiochi bacteria is repasteurized at once, treated with activated carbon, and deacidified with alkali or ion exchange resins. In the worst case, the infected sake is discarded.

2. Protein Turbidity

Fresh sake is pasteurized at 60 to 65°C for 15 min to kill microorganisms and destroy enzymes. This pasteurization process produces a slight turbidity. The precursors of the turbid material are enzyme proteins, especially the saccharifying amylase, derived from rice koji and dissolved in sake during fermentation (1). The soluble proteins form insoluble materials that are denatured by heat (pasteurization) and gradually coagulate into large particles, giving the sake a slight turbidity. These particles are too small to remove by filtration only. Therefore, sake is fined by the following method.

a. Physical Fining Method. Protein precipitants such as egg white, wheat flour, and gelatin are used in combination with persimmon tannin. This method cannot be adapted to fresh sake (2,11,16).

b. Enzymatic Fining Method. Bacterial or fungus protease preparations are employed. This method also cannot be adapted to fresh sake (53).

c. Ultrafiltration Method. The precursor of the turbid material in sake and the turbid material in pasteurized sake may be removed by ultrafiltration. The protein fraction containing the precursor of turbid material is isolated from fresh sake by ultrafiltration. The precursors of the turbid material are found to be glucamylase GI, GII, and α-amylase. From the results of gel filtration on Sephadex G-150, the molecular weights of GI, GII, and α-amylase have been estimated at 190,000, 69,000, and 37,000, respectively (23,34).

d. Immobilized Tannin Method. The protein precursors of turbidity in fresh sake may be absorbed through a column of an immobilized tannin, prepared by coupling the BrCN-activated gallotannin to aminohexylcellulose (21).

e. Silica Sol Method. Silica sol is capable of absorbing the protein and coagulate. The turbid materials in pasteurized sake may be removed with silica sol in combination with gelatin (25).

E. Activated Carbon

The amount of activated carbon used for sake filtration is usually 0.3-1.5 kg/kiloliter of sake. Treatment with activated carbon improves the flavors and adjusts the maturity of sake. Activated carbon absorbs not only the bitter compounds and colorants but also their precursors, and it stabilizes the quality of stored and bottled sake. Moreover, since it absorbs the growth factors of hiochi bacteria such as certain peptides, amino acids, vitamins, and mevalonic acid, the growth of hiochi bacteria is prevented (13). However, it is unable to absorb the colorant ferrichrysin or the smell caused by using old rice fumigated with methyl bromide. The recently developed molecular sieving carbon effectively absorbs the odor caused by aging during storage.

III. CONCLUSION

The flavors and tastes of fresh sake improve gradually with aging during storage. The optimum aging term is usually 6-12 months. Overaging not only deteriorates the flavors and tastes but also causes the coloration of sake. Although half the quality of sake is attributed to the raw materials and fermentation techniques, the techniques that follow filtration of the main mash (e.g., pasteurization, storage, filtration, blending, fining, and bottling) are also important.

Unpasteurized fresh sake with the characteristic fragrant flavor, which is stored at temperatures $<5^\circ C$ and filtrated by superfine filtration with a membrane filter, has come onto the market. Sake is usually sold in a bottle of 1800 ml capacity, but bottles of various capacities and paper containers have also been used.

In conclusion, bottled sake should be consumed as soon as possible--at the latest, within 10 months after bottling.

TABLE I

Export of Sake in Recent Years (kiloliters)

Country	1981	1983
U.S.A.	1,657 (50.6 %)	2,287 (51.0 %)
Formosa	411 (12.6)	591 (13.2)
Canada	284 (8.7)	427 (9.5)
W.Germany	181 (5.5)	219 (4.9)
Hong Kong	133 (4.1)	176 (3.9)
England	103 (3.1)	163 (3.6)
Others	502 (15.3)	623 (13.9)
Total	3,272 (100.0)	4,486 (100.0)

TABLE II

Composition of an Average Sake

Alcohol, %(v/v)	16.0
Total sugar as glucose, %(w/v)	3.7
Acidity,meq/100 ml	1.4
Formol-N (ml of 0.1 N NaOH/10 ml)	1.8
pH	4.4

TABLE III

Ferrichrysin and Deferriferrichrysin Contents in Sake and Rice-Koji

	Sake(ppm as Ferrichrysin)	Rice-koji(mg/Kg as Ferrichrysin)
Ferrichrysin	0- 7	0-80
Deferriferrichrysin	4-14	0-400

TABLE IV

Odor Units of HDMF

	Odor threshold value (a) (ppb) (A)	Concentration in aged sake (ppb) (B)	Odor unit (B/A)
I (b)	5,804	140-430	0.02-0.07
II(c)	76.3		1.84-5.64

(a). Sample was dissolved in distilled water.
(b). Odor threshold value obtained by sniffing the sample solution.
(c). Odor threshold value obtained by tasting the sample solution.

TABLE V

Threshold Values of Volatile Sulfur Compound in Sake

Compound	Threshold value (ppb) (A)	Content (ppb) (B)	Odor unit (B/A)
CH_3SH	4	0–14	~3.5
DMS	10	0–10	1.0
DMDS	6.5	0– 7	1.1

TABLE VII

DMS Precursor Content of Unpolished Rice Fumigated with Methyl Bromide and Phostoxine

Fumigating agent	Sample	DMS pre. content (μg DMS/Kg dry wt.)	
		Before fumigation	After fumigation
Methyl bromide	A	250	41,000
	B	190	38,000
	C	170	33,000
Phostoxine	A	250	270
	B	190	190
	C	170	170

A; Rice sample cropped in 1978.
B and C; Rice sample cropped in 1979.
A and B; Fumigated as unpolished rice.
C; Fumigated as unhulled rice.

Fumigation was carried out in a desiccator for 5 days at 25°C with 100 ppm of fumigating agent.

TABLE VI

Flavor Compounds Detected from Sake

Alcohol	methanol, ethanol, n-propanol, iso-butanol, n-butanol, iso-amylalcohol, n-amylalcohol, active-amylalcohol, n-hexanol, β-phenetyl alcohol, tryptohol
Ester	ethyl ester of C_1-C_{14} straight-chain aliphatic acid, propyl acetate, butyl acetate, iso-butyl acetate, n-amyl acetate, iso-amyl acetate, β-phenethyl acetate, ethyl iso-valerate, ethyl iso-caproate, ethyl pyruvate, ethyl keto-iso-valerate, ethyl keto-iso-caproate, ethyl oxy-iso-valerate, ethyl oxy-iso-caproate, ethyl p-oxycinnamic acid ethyl ester, ethyl lactate, * ethyl phenylacetate, * ethyl vanillate, * ferulic acid ethyl ester, * ethyl p-oxybenzoate
Acid	C_1-C_{16} straight-chain aliphatic acid, iso-butyric acid, iso-valeric acid, iso-caproic acid, iso-capric acid, iso-caprylic acid, oleic acid, pyruvic acid, keto-valeric acid, phenyl pyruvic acid, benzoic acid, ferulic acid, indole acetic acid, protocatechuic acid, gallic acid, * coumalic acid, * p-oxy-cinnamic acid, * p-hydroxy-phenyl acetic acid, * phenyl acetic acid, * vanillic acid
Carbonyl compound	formaldehyde, acetaldehyde, propionaldehyde, iso-valeraldehyde, caproic aldehyde, diacetyl, acetoin, acetone, * benzaldehyde, *cinnamic aldehyde, * fulfural, * p-oxybenzaldehyde, * vanillin, * phenyl acetaldehyde
Amine	ethanol amine, iso-butyl amine, putrescine, cadaverine, phenethyl amine
Sulfur compound	hydrogen sulfide, methyl sulfide, dimethyl sulfide, * dimethyl disulfide
Others	* 3-hydroxy-4,5-dimethyl-2(5H)-furanone, * maltol

* : Detected only from aged sake.

TABLE VIII

Content and Threshold Value of Bitter Compound in Sake

<table>
<tr><th rowspan="2">Compound</th><th rowspan="2">Threshold value(ppm)</th><th colspan="2">Content (ppm)</th></tr>
<tr><th>Fresh sake</th><th>Old sake</th></tr>
<tr><td>Prolyl-leucine anhydride</td><td>259</td><td>1- 2</td><td>5-15</td></tr>
<tr><td>Methyl thioadenosine</td><td>215</td><td>10-14</td><td>20-30</td></tr>
<tr><td>Kynurenic acid</td><td>78</td><td colspan="2">1-3</td></tr>
<tr><td>Tyrosol</td><td>346</td><td colspan="2">80-140</td></tr>
<tr><td>Tetrahydro-harman-3-carboxylic acid</td><td>220</td><td></td><td>3</td></tr>
<tr><td>Harman</td><td>57</td><td></td><td>0.4</td></tr>
<tr><td>L-Leucine</td><td>4,099</td><td colspan="2">100-300</td></tr>
<tr><td>L-Tryptophan</td><td>798</td><td>53</td><td>5</td></tr>
</table>

TABLE IX

Taste Compounds in Sake

Taste	Compound
Sweet	glucose, oligo-saccharide, glycerol 2,3-butylene glycol, ethanol, ethyl α-D-glucoside, glycine, alanine, proline
Sour	saturated mono-carboxylic acid, saturated di-carboxylic acid, unsaturated di-carboxylic acid, hydroxy carboxylic acid, keto-carboxylic acid, pyrrolidone carboxylic acid
Pungent	ethanol, aldehyde, sour taste compounds, bitter taste compounds
Bitter	choline, tyramine, histidine, arginine, methionine, valine, leucine, isoleucine, phenylalanine, tryptophan, tyrosol, harman, THCA, MTA, kynurenic acid, 5'-methyl thioadenosine, L-prolyl-L-leucine anhydride, hypoxanthine, succinic acid, inorganic salt
Astringent	tyrosine, lactic acid, succinic acid, inorganic salt

TABLE X

Change of Content of Amino Acid in Sake during Storage

Amino acid	Content(mM) Storage(day)		Decrease rate(%)	Amino acid	Content(mM) Storage(day)		Decrease rate(%)
	0	60			0	60	
Thr	0.96	0.25	-74.0	Val	1.54	1.32	-14.3
Trp	0.31	0.20	-35.5	Leu	2.13	1.88	-11.7
Asp	0.99	0.79	-20.2	Gly	2.71	2.44	-10.0
Phe	0.97	0.78	-19.6	Lys	1.18	1.08	- 8.5
Met	0.39	0.32	-18.0	Arg	2.44	2.25	- 7.8
Ile	0.84	0.69	-17.9	Ala	3.89	3.65	- 6.2
Pro	2.16	1.80	-16.7	His	0.97	0.77	- 2.5
Glu	2.19	1.83	-16.4	Ser	1.29	1.57	+21.7
Tyr	1.22	1.03	-15.6	NH_3	6.35	6.89	+ 8.5

TABLE XI

System of Classification of Hiochi-bacteria

	Group name (Species)	Growth at pH 7	Fermentable sugars	Effect of alcohols on growth	Highest alcohol concent-ration on growth	Requirement of mevalonic acid
Hetero-fermenters	True hiochi-bacilli *(L.heterohiochii)*	No growth	Glucose Fructose	Promotive	20–21%	Essential requirement
	Hiochi-lactobacilli	Growth	Many sugars	Inhibitory	17–18%	No requirement
Homo-fermenters	True hiochi-bacilli *(L.homohiochii)*	No growth	Glucose Mannose	Promotive	21–25%	No-essential requirement
	Hiochi-lactobacilli	Growth	Many sugars	Inhibitory	17–18%	No requirement

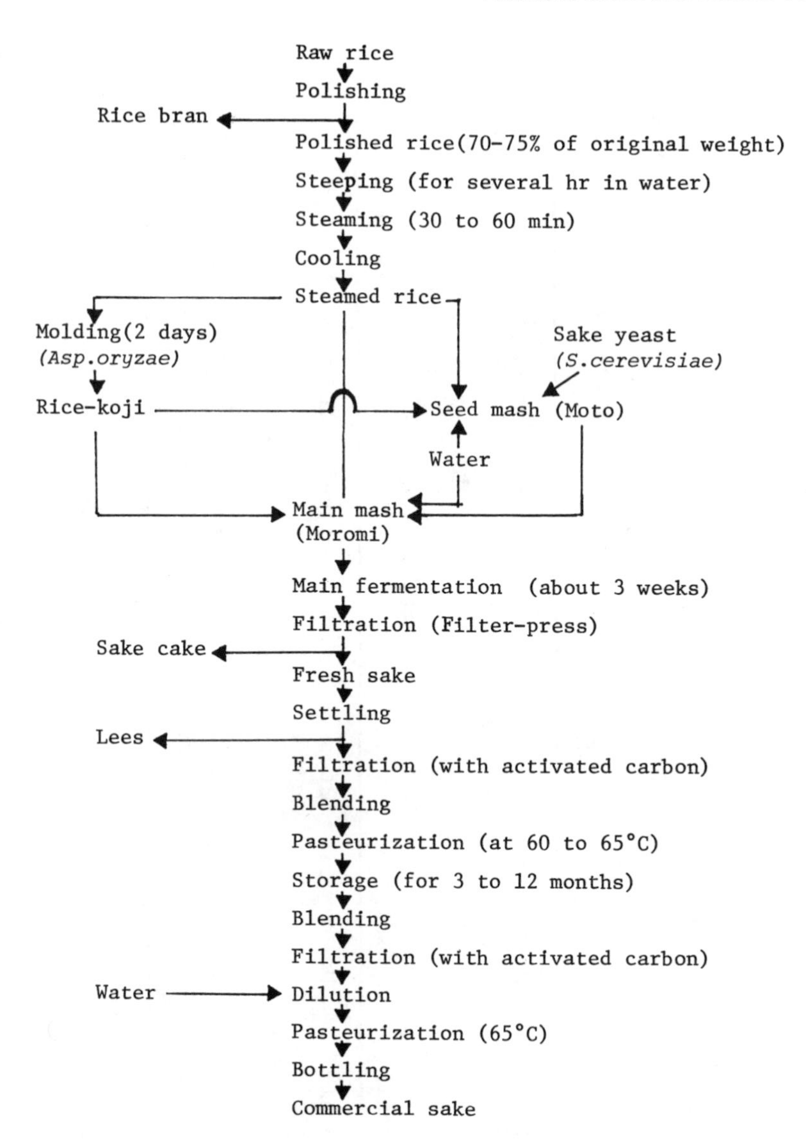

Fig. 1. Process flowchart of sake brewing.

(Reaction I)

Deferriferrichrysin
Kynurenic acid or flavin
Tyrosine or tryptophan
Hydroxy or keto acids
$\xrightarrow[O_2]{\text{Mn, Light}}$ Colorant

(Reaction II)

Tyrosine or tryptophan
Kynurenic acid or flavin
$\xrightarrow[O_2]{\text{Light}}$ Colorant

(Reaction III)

Indole acetate
or Protocatechuic acid
$\xrightarrow[O_2]{\text{Light}}$ Colorant

Fig. 2. Photochemical coloring reaction of sake.

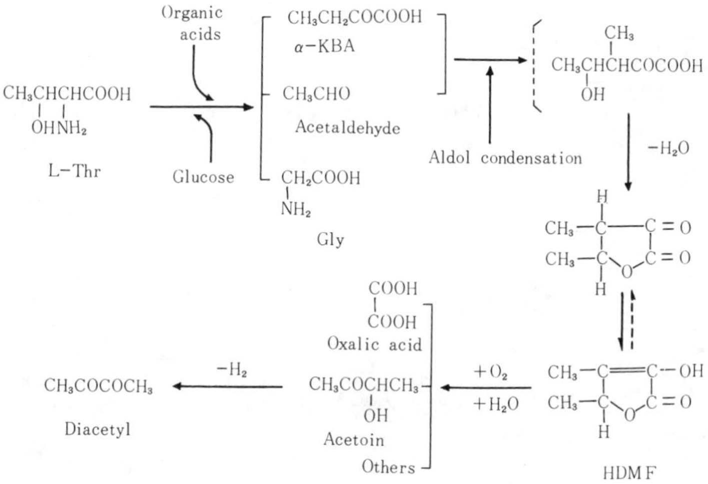

$CH_3CH(OH)CH(NH_2)COOH$ (L-Thr) + Organic acids, Glucose → [$CH_3CH_2COCOOH$ (α-KBA); CH_3CHO (Acetaldehyde); $CH_2(NH_2)COOH$ (Gly)] → (Aldol condensation) [$CH_3CH(OH)CH(CH_3)COCOOH$] → ($-H_2O$) → $CH_3-CH-C{=}O$ / $CH_3-CH-O-C{=}O$ (ring) ⇌ HDMF ($CH_3-C{=}C-OH$ / $CH_3-CH-O-C{=}O$, ring)

HDMF → ($+O_2$, $+H_2O$) → [$COOH-COOH$ (Oxalic acid); $CH_3COCH(OH)CH_3$ (Acetoin); Others] → ($-H_2$) → $CH_3COCOCH_3$ (Diacetyl)

Fig. 3. Formation and degradation of 3-hydroxy-4,5-dimethyl-2(5*H*)-furanone (HDMF).

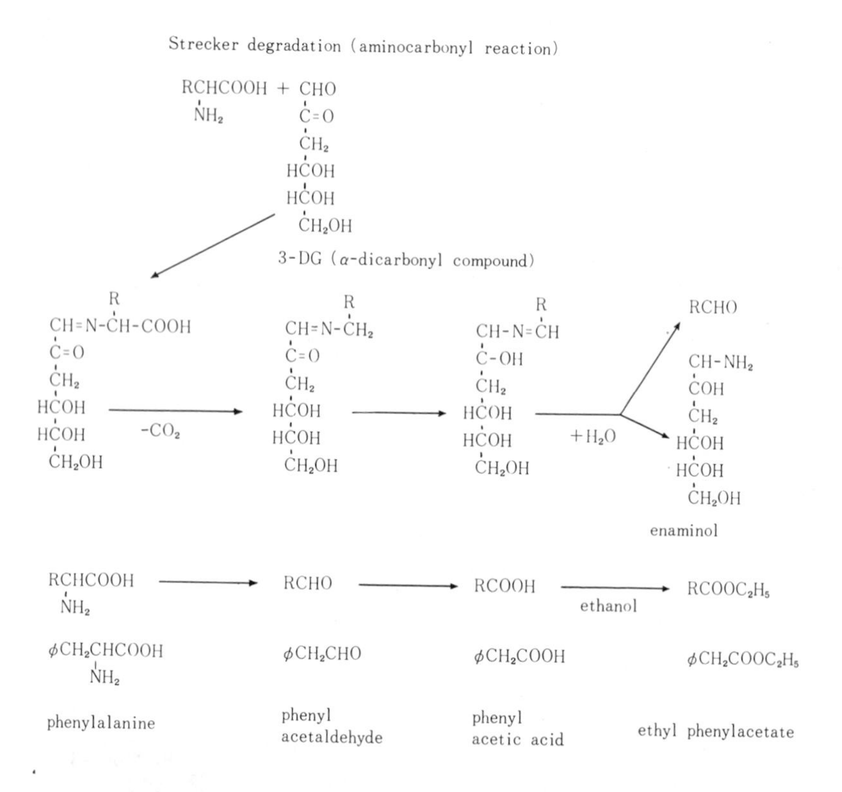

Fig. 4. Mechanism of formation of phenyl acetic acid.

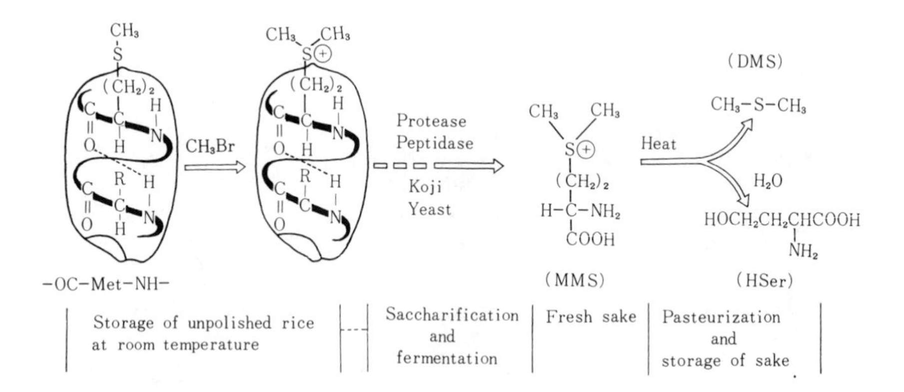

Fig. 5. Mechanism of evolution of DMS in sake brewed from old rice.

REFERENCES

1. Akiyama, H. (1962). *J. Agric. Chem. Soc. Jpn. 36*, 825-829, 903-907.
2. Akiyama, H., Uchiyama, K., Koide, Y., Himeno, K., and Noda, Y. (1969). *J. Brew. Soc. Jpn. 64*, 889-893.
3. Dubois, P., Rigaud, J., and Dekimpe, J. (1976). *Lebensm.-Wiss. Technol. 9*, 366.
4. Iwano, K., and Nunokawa, Y. (1978). *J. Brew. Soc. Jpn. 73*, 968-970.
5. Keller-Schierlein, W., and Deer, A. (1963). *Helv. Chim. Acta 46*, 1907.
6. Keller-Schierlein, W. (1963). *Helv. Chim. Acta 46*, 1920.
7. Kitamoto, K., Ohba, T., and Namba, Y. (1982). *J. Ferment. Technol. 60*, 417-422.
8. Kitamoto, K., Ohba, T., and Namba, Y. (1981). *J. Brew. Soc. Jpn. 76*, 491-494.
9. Kodama, K., and Yoshizawa, K. (1977). *Econ. Microbiol. 1*, 423-475.
10. Maegawa, K., Yokoyama, T., Tabata, N., Shinke, R., and Nishina, H. (1981). *J. Ferment. Technol. 59*, 483-489.
11. Miyai, K. (1957). *J. Brew. Soc. Jpn. 52*, 39-41.
12. Momose, H., Ando, K., Sato, Y., and Akiyama, H. (1971). *J. Brew. Soc. Jpn. 66*, 1190-1193.
13. Momose, H., Morinaga, K., Tukioka, H., and Akiyama, H. (1972). *J. Brew. Soc. Jpn. 67*, 723-725.
14. Momose, H., Yamanaka, E., Akiyama, H., and Nosiro, K. (1974). *J. Gen. Appl. Microbiol. 20*, 179-185.
15. Nagatani, M., Kikuchi, K., and Sugama, S. (1970). *J. Brew. Soc. Jpn. 65*, 159-165.
16. Nakabayashi, T. (1968). *J. Food Sci. Technol. (Tokyo) 15*, 502-506.
17. Nakamura, K. (1971). *J. Brew. Soc. Jpn. 66*, 13-18.
18. Namba, Y., Ohba, T., Kitamoto, K., and Hirasawa, A. (1982). *J. Ferment. Technol. 60*, 35-39.
19. Neiland, J. B. (1966). *Struct. Bonding (Berlin) 1*, 70.
20. Neiland, J. B. (1952). *J. Am. Chem. Soc. 74*, 48-46.
21. Nunokawa, Y., Shiinoki, T., and Watanabe, T. (1978). *J. Ferment. Technol. 56*, 776-781.
22. Ohba, T., and Sato, M. (1984). *J. Brew. Soc. Jpn. 79*, 271-273.
23. Ohba, T., Takahashi, K., Hayakawa, S., Namba, Y., and Sato, S. (1978). *J. Brew. Soc. Jpn. 73*, 951-954.
24. Oka, T., Ido, K., Shimizu, T., and Sakai, M. (1965). *J. Agric. Chem. Soc. Jpn. 39*, 415-419.
25. Sato, M., Ohba, T., Mori, T., Obitsu, M., Watanabe, Y., and Tasaki, K. (1982). *J. Brew. Soc. Jpn. 77*, 553-556, 822-824.

26. Sato, S., Ohba, T., Takahashi, K., and Kokubu, S. (1978). *J. Brew. Soc. Jpn. 73*, 945-950.
27. Sato, S., and Tadenuma, M. (1967). *J. Brew. Soc. Jpn. 62*, 1279-1287.
28. Sato, S., Nakamura, K., and Tadenuma, M. (1971). *Agric. Biol. Chem. 35*, 308-313.
29. Sato, S., Tadenuma, M., Takahashi, K., and Shimizu, T. (1971). *J. Brew. Soc. Jpn. 66*, 508-511.
30. Sato, S., Tadenuma, M., Takahashi, K., and Koike, K. (1975). *J. Brew. Soc. Jpn. 70*, 588-591.
31. Sato, S., Tadenuma, M., Takahashi, K., and Koike, K. (1975). *J. Brew. Soc. Jpn. 70*, 592-594.
32. Sato, S., Tadenuma, M., Takahashi, K., and Nakamura, K. (1975). *J. Brew. Soc. Jpn. 70*, 821-824.
33. Sato, S., Ohba, T., Takahashi, K., and Sugitani, M. (1978). *J. Brew. Soc. Jpn. 73*, 473-478.
34. Sato, S., Ohba, T., Takahashi, K., and Umehara, Y. (1977). *J. Ferment. Technol. 55*, 84-94.
35. Schinz, H., and Hinder, M. (1947). *Helv. Chim. Acta 30*, 3149.
36. Schinz, H., and Rossi, A. (1948). *Helv. Chim. Acta 31*, 1953.
37. Sugama, S., Nishiya, S., Ohba, T., Murai, S., Egashira, S., and Hara, S. (1975). *J. Brew. Soc. Jpn. 70*, 666-670.
38. Sugama, S., and Iguchi, T. (1970). *J. Brew. Soc. Jpn. 65*, 720-725.
39. Tadenuma, M., and Sato, S. (1967). *Agric. Biol. Chem. 31*, 1482-1489.
40. Tadenuma, M., Takahashi, K., Hayashi, S., and Sato, S. (1975). *J. Brew. Soc. Jpn. 70*, 585-587.
41. Takahashi, K., Tadenuma, M., Kitamoto, K., and Sato, S. (1974). *Agric. Biol. Chem. 38*, 927-932.
42. Takahashi, K., Sato, S., Tadenuma, M., and Arai, T. (1972). *J. Brew. Soc. Jpn. 67*, 726-730.
43. Takahashi, K., Kokubu, S., Ohji, S., Ohba, T., and Sato, S. (1978). *J. Brew. Soc. Jpn. 73*, 886-890.
44. Takahashi, K., Tadenuma, M., and Sato, S. (1976). *Agric. Biol. Chem. 40*, 325-330.
45. Takahashi, K., Ohba, T., Takagi, M., Sato, S., and Namba, Y. (1979). *J. Ferment. Technol. 57*, 148-157.
46. Takahashi, M. (1963). *J. Brew. Soc. Jpn. 58*, 1200-1204.
47. Takase, S., and Murakami, H. (1967). *Agric. Biol. Chem. 31*, 142-149.
48. Takimoto, T., Kobayashi, A., and Yamanishi, T. (1980). *Proc. Jpn. Acad., Ser. B 56*, 457-462.
49. Tamura, G. (1958). *J. Agric. Chem. Soc. Jpn. 32*, 701-713, 778-790.

50. Totsuka, A., and Namba, Y. (1978). *J. Brew. Soc. Jpn. 73*, 789-792.
51. Totsuka, A., Sumikawa, T., Ogino, H., Namba, Y., and Kobuyama, Y. (1971). *J. Brew. Soc. Jpn. 66*, 495-499.
52. Ueno, H., Yao, T., Tadenuma, M., and Sato, S. (1966). *J. Brew. Soc. Jpn. 61*, 1169-1173.
53. Yamada, M., Akiyama, H., Shimizu, A., Fujii, R., and Harada, M. (1957). *J. Agric. Chem. Soc. Jpn. 31*, 127-132.
54. Yoshizawa, K., Ishikawa, T., Unemoto, F., and Noshiro, K. (1973). *J. Brew. Soc. Jpn. 68*, 705-707.
55. Yoshizawa, K., Komatsu, S., Takahashi, I., and Ohtsuka, K. (1970). *Agric. Biol. Chem. 34*, 170-180.

CHAPTER 17

MATURATION OF POTABLE SPIRITS

JAMES S. SWAN
Pentlands Scotch Whisky Research Ltd.
Edinburgh, Scotland

I. INTRODUCTION

Many types of distilled spirits undergo prolonged storage in wooden containers prior to sale. During this time major changes take place in the volume and chemical composition, and the product becomes mellow and palatable. Examples of such products are whiskies, including Scotch, Irish, bourbon, rye, Canadian, and Japanese, brandies including cognac, armagnac, Spanish, Californian, and Australian, and rum, which is derived from sugar cane.

The production and maintenance of wooden containers for maturation is undertaken by labor-intensive industries that are often major employers in their localities, and the cost to beverage producers of investment in wood and warehoused spirits represents a significant proportion of production costs.

The metamorphosis that takes place during maturation is essential to product character and is the result of complex chemical changes. Selection of wood for maturation, pretreatment of the wood (particularly on the internal surface of the container), and variability in warehousing conditions can all influence the quality of the final product; hence an understanding of their chemistry is essential to continued quality in an ever-changing world.

II. TIGHT COOPERAGE FOR SPIRITS MATURATION

A. The Maturation Container

Casks are produced for many storage and transportation purposes. Sharp (1983) noted their strength and durability, and how they may be pivoted on a small contact area and thus easily moved by one man, although they may substantially exceed his weight. Casks used for storage of liquids are known as tight cooperage, and those used for maturation are produced from the highest grade of oak timber (Hankerson, 1947). Figure 1 shows details of a typical maturation cask.

The size, shape, and methods of preparation of the internal surfaces of casks vary somewhat among locations. In the Western World the major cask manufacturing areas are in the United States, Spain, and France. Casks produced in the United States are referred to as barrels and have a capacity of ∿180 liters[1]; they are generally charred inside. In Spain the most common casks are butts with capacities ∿500 liters. The internal surfaces are less severely treated, being flamed or toasted. Casks in France are commonly ∿300 liters.

In Scotland use is made of American barrels and Spanish (sherry) butts. In addition, barrels are rebuilt by adding extra staves and new heading to produce an increased volume in the region of 250 liters and are known as remade hogsheads.

The production, usage, and export of American barrels has been discussed by Williams (1983) and is summarized in Fig. 2.

Rickards (1983) estimated the cask requirement for Scotch whisky production in 1978 as 0.33 million (butts equivalent). The total production in that year was 1.5 million butts, that is, 750 million liters at the traditional filling strength of 63% (v/v).

The preparation of staves for cask production is shown in Fig. 3. Selected logs are cut into stave lengths, then cleft in line with the hearts. Staves are then split from these "bolts" and the sapwood removed. Blemishes are cut out, if present; then the wood is used for cask headings.

The production process involves considerable loss, with less than half the volume of logs finishing up as staves or headings (Williams, 1983). Before using, staves are air-dried for up to 18 months (usually less in the United States); shrinkage is 13% in length, 4% radial, and 13% volume (Rickards, 1983). Staves are worked at 14 to 16% moisture content; they must be odor free and straight in grain for not less than 45% of the stave. There should be no evidence of fungal disease.

B. Selection of Wood for Spirits Maturation

Many types of wood have been used for tight cooperage. Singleton (1974) listed 16 types of wood known to have been used in the United States, eight types used in Europe for wine or brandy, and seven more types in Australia and the Far East. The list dwindles rapidly if woods with unsatisfactory characteristics are rejected. The choice narrows down to a few species of oak because of both chemical and physiological characteristics.

[1]For convenience, volumes are quoted throughout in liters. The following conversion factors may be useful: gallons U.S. to liters × 3.785; gallons imperial to liters × 4.545; gallons U.S. to gallons imperial × 0.833.

C. Physiology of Oak Wood

In the United States casks are constructed from commercial white oak, of which *Quercus alba* constitutes some 45% of standing reserves. Other lesser species include *Q. stellata, Q. prinus, Q. lyrata, Q. durandii,* and *Q. bicolor* (Singleton, 1974). *Quercus alba* grows widely throughout most of the United States, with the main centers of production being in Kentucky, Missouri, and Arkansas. Singleton (1974) has commented that use of the lesser species in commercial cooperage does not appear to cause differences in stored beverages.

In Europe the most common species are *Quercus robur* and *Q. sessilis* (Singleton, 1974; Rickards, 1983). Both are widespread throughout the continent.

The particular features of oak wood species that make them suitable for tight cooperage are their rays and tyloses. Rays (see Fig. 3) form diffusion channels from the pith to bark; rays, one cell wide (uniseriate) and two or more cells wide (multiseriate), occur in hardwoods (Parham and Gray, 1984). In oak, however, the multiseriate rays are so wide and significant that they are often considered as a separate type, that is, compound rays that form the pattern observed on the sawn timber (Brown *et al.*, 1949; Singleton, 1974). Rays contribute to the strength and flexibility of oak as well as its dimensional stability and relative impermeability. In *Quercus alba*, rays represent ∿28% of the wood volume.

Hardwoods such as oak are porous compared to softwoods, because they contain tubes or vessels (Parham and Gray, 1984), that is, porous vertical elements. In oak the vessels in spring wood are much larger and the transition to summer wood is abrupt; such timbers are described as ring porous (Brown *et al.*, 1949). In *Quercus* used for cooperage, spring wood pores can frequently attain a size of 300 μm. Staves made from such woods would leak from the ends if not plugged in their vessels. Oaks used in cooperage are rich in tyloses, which plug the vessels by ballooning of adjacent cell walls into the channel of the vessels (Fig. 4).

Tylose formation usually coincides with the conversion of sapwood into heartwood, although they can be found in sapwood where the water content falls below normal. Vessels may be blocked or contain gummy deposits, and tyloses may be present at the same time (Gerry, 1941).

D. Chemistry of Oak Wood Used in Maturation

It is convenient to divide the chemical constituents of oak wood into two major groups: (1) cell wall components and (2) extractives and other infiltration substances.

1. Cell Wall Components

Wood cells consist mainly of cellulose, hemicelluloses, and lignin. A simplified explanation is that cellulose forms the basic skeletal structure, with hemicelluloses serving as a matrix. Lignin acts as an encrusting material, permeating the cell walls and intercellular regions (middle lamella), rendering the wood rigid and able to withstand mechanical stress (Parham and Gray, 1984; Sjöström, 1981).

Cellulose is a linear polymer with molecules held together by hydrogen bonding to form bundles. The fibrous structure renders cellulose resistant to solvents such as newly distilled spirit. Hemicelluloses are formed in wood by different biosynthetic routes than those of cellulose (Sjöström, 1981) and are relatively easily hydrolyzed. They are heteropolysaccharides and in oak wood, are predominantly xylose based (i.e., xylans). Most of the xylose residues contain an acetyl group; these are slowly hydrolyzed in the living tree to acetic acid. Evidence has been presented that demonstrates the existence of chemical bonds between hemicelluloses and lignin (Eriksson and Lindgren, 1977).

Lignins are complex branched-chain polymers of both hydroxy- and methoxy-substituted phenylpropane units (Pettersen, 1984). The former produces "guaiacol lignin" and the latter "syringyl lignin." "Guaiacyl-syringyl lignin" is a copolymer of the two precursors coniferyl and sinapyl alcohol, and typically occurs in hardwoods (Fig. 5).

Many aspects of the structural features of lignin in various morphological regions of the wood still remain unclear; however, studies on oak fibers indicate a rather uniform syringyl:guaiacyl ratio.

Because of separation problems, lignin is often classified according to its method of extraction (Bolker, 1974). Two types of particular interest in maturation chemistry are dioxan lignin and ethanol lignin, the latter often referred to as Brauns or native lignin. Puech (1978) has described the preparation of these fractions from oak wood. The amount of lignin in oak wood may be determined by estimating the methoxyl content in dioxan lignin and applying a multiplication factor (×5.37).

2. Wood Extractives and Infiltration Substances

These substances are not integral parts of the cell structure and therefore may be easily removed by solvents without affecting the physical structure and strength of the wood (Brown *et al.*, 1949). Wood, in general, contains a wide range of extractable substances including volatile oils, fats, resins, tannins, carbohydrates, sterols, and inorganic salts, particularly calcium oxalate. It is these compounds that impart characteristic odors, thus forming the main impediment to using readily available wood for spirits maturation. The group of extractable compounds of particular importance in oak wood are the tannins, and their importance to matured flavor cannot be underestimated.

Tannins are polyphenols and are classified as either condensed or hydrolyzable (Hathway and Jurd, 1962). The hydrolyzable tannins are esters of a sugar, usually glucose, with one or more polyphenolic moieties. Hydrolyzable tannins are usually subdivided into gallotannins and ellagotannins, yielding these acids on hydrolysis (Fig. 6).

Both types of hydrolyzable tannins are present in cooperage oak (Chen, 1970). Condensed tannins are flavonoid phenolics and have a C_6-C_3-C_6 skeleton.

The distribution of tannins within the tree is extremely variable; hence extractable tannins may vary greatly in casks from apparently similar sources (Puech, 1984). Singleton (1974) has cited variations in tannin content within one example of *Quercus robur*.

Tannins are present in the wood rays and are therefore fairly accessible to maturing spirit; they have germicidal properties, contribute to the coloring of the wood, and are considered to protect the tree from attack and decay.

Heartwood American oak contains a number of sterols that can cause problems related to whisky stability (see Section III,A,2). Chen (1970) has identified β-sitosterol, stigmasterol, camposterol, and traces of β-dihydrosterol.

3. Composition of Heartwood Oak

Clearly the composition of oak is somewhat variable. Singleton (1974) has given the following data for the composition of heartwood American white oak: 50% cellulose, 22% hemicellulose, 32% lignin, 2.8% acetyl groups, and 5-10% hot-water extractables. Puech (1978) gave the following composition of European oak: 23-50% cellulose, 17-30% hemicelluloses, 17-30% lignin, 2-10% tannins, and 0.3-0.6% resins.

III. CHEMISTRY OF MATURATION

The earliest detailed studies of chemical changes taking place in maturing spirits were reported for Scotch whisky (Schidrowitz and Kaye, 1905) and American bourbon and rye whiskies (Crampton and Tolman, 1908).

In the former study 77 samples from plain, sherry, brandy, and refill casks were studied, but the samples representing new distillate and matured product were not drawn from the same batch of distillate. Samples in the latter study did meet this criterion but were distilled and matured by techniques that are no longer used. Both studies, together with another carried out in the United States on whisky produced during the years of prohibition (Valaer and Frazier, 1936), measured changes in alcoholic strength, color, total solids, acidity, fusel alcohols, aldehydes, and furfural. In a later study of American whiskies (Liebmann and Scherl, 1949), tannin content and pH were also determined; the variety and number of samples studied over an 8-year period were included in a design that permitted statistically sound conclusions to be drawn. In this study Liebmann and Scherl showed that an increase took place in all groups of compounds. These increases could be modeled by hyperbolic equations, while pH decreased in a similar manner.

Baldwin and Andreasen (1974) studied similar changes in bourbon whiskies matured at various filling strengths, namely 55% (v/v, control), 58%, 63%, 73%, and 78% over a 12-year period. Equations derived for the original strengths of 55 and 63% are reproduced in Table I. Graphs of these component changes are shown in Fig. 7.

Changes that take place during maturation are produced by the following mechanisms (Nishimura 1983):

1. Extraction of wood components from the cask
2. Interaction of ethanol and the cask wood
3. Reactions that involve only the unaged distillate

In the case of products like Scotch whisky, an additional mechanism accounting for the presence of materials left behind, or modified, by previous cask contents should be added (Shortreed *et al.*, 1979).

Baldwin and Andreasen (1974) proposed the parameter "total barrel-derived material" (TBDM), which they defined as the sum of the increase in volatiles and nonvolatiles in an aged product, and suggested that this parameter could be used for direct comparison of whiskies stored under different warehousing conditions and that it could provide a better understanding of the course of maturation.

A. Flavor Components Produced by Extraction of Oak Wood

The studies of Liebmann and Scherl (1949) and Baldwin and Andreasen (1974) indicated that maturation in American oak wood produces an extract of ∿4000 mg/liter of pure alcohol, whereas Puech (1984) showed that concentrations up to 9000 mg/liter could be extracted from European oak. (Values are expressed as milligrams per liter of absolute alcohol.) Puech also showed that the optimum extraction strength is 55% v/v and that tannins accounted for around two-thirds of the extract, with lignin making up most of the remainder.

1. Extraction of Lignin and Tannins

Studies by Puech (1984) measured the lignin and tannin content of oak woods from various sources and the proportions extracted into spirit. Table II shows the lignin and tannin content in the woods and Table III the percentage extracted into Armagnac brandy.

The results shown in Tables II and III indicate that lignin content shows some consistency between woods, while tannin results show heterogeneity. Consequently tannins extracted into spirit may very substantially from one apparently similar cask to another. Oak woods contain ∿30% lignin, of which only 4% is extractable, while >50% of tannins are extractable. Puech (1984) noted the low tannin content of American white oak compared to European oak, confirming earlier studies of Guymon and Crowell (1970).

Rous and Alderson (1983) studied the extraction of total and nonflavonoid phenolics from American and European oak into wine at 13% (v/v). No similar study appears to have been reported at the alcoholic strength of spirits, but the essential conclusions are likely to be similar. The results of this study are summarized in Fig. 8.

These results indicate that the total of phenolics available is very much greater in European oak than in American oak and that initial extraction follows a power curve, indicating diffusion kinetics. While the amount of phenols available for the second filling in European oak was only ∿25% of the first fill, the second fill into American oak gave almost as much as the first fill. The third filling with wine produced extraction curves that were linear, indicating that the easily extractable phenols were depleted and only the hydrolyzable tannins remained. Also, by the third fill there were no quantitative differences between total and nonflavonoid phenols. Rous and Alderson also noted that there were strong sensory differences between the two types of oak.

2. Steroids in Aged Whisky

Section II,D,2 mentions the presence of steroids in oak wood. These appear to be readily extractable into maturing spirit, and reports have been presented of their presence in American whisky (Black and Andreasen, 1974; Byrne *et al.*, 1981). Black and Andreasen identified β-sitosterol-D-glucoside, β-sitosterol, stigmasterol, and camposterol. Despite their low levels--Byrne *et al.* (1981) quoted average figures of 0.365 mg/liter of β-sitosterol and 0.533 mg/liter of the glucoside--their presence can cause a permanent floc in bottled whiskies. This problem is uncommon in Scotch whiskies and does not appear to have been reported in other spirits.

3. Extraction of Volatile Aroma Compounds

Nishimura and co-workers (1983) listed 118 volatile compounds identified in the steam distillate of methanol extracts of white oak. These include hydrocarbons, acids, esters, phenols, alcohols, and terpenes, and it was noted that so far 51 of these have been found in matured spirits. These workers commented that while the odor of oak wood seems to be important in matured spirits, most studies have concentrated on the nonvolatile components.

B. Flavor Components Produced by Interaction of Distillate with Oak Wood

1. Aromatic Aldehydes

During maturation in oak wood a series of aromatic aldehydes are produced that are believed to be formed by breakdown of lignin. Many of the compounds produced are similar to those involved in biosynthesis in the living tree. Reazin and co-workers (1976) included investigation of aromatic aldehyde formation in a study in which a small amount of radioactive ethanol was added to a cask and the uptake of radioactivity by group of compounds was monitored over a 56-month period. The reactions leading to aromatic aldehyde formation are shown in Fig. 9.

Reazin states:

Reaction la shows a portion of the wood lignin reacting with ethanol to form ethanol lignin. In addition, some of the ethanol lignin breaks down to form ethyl, coniferyl and sinapic alcohols. The coniferyl and sinapic alcohols can then be oxidized into sinapic- and coniferaldehydes. Finally they are transformed into syringaldehyde and vanillin, respectively.

The process of aromatic aldehyde formation is often referred to as ethanolysis of lignin; however, Puech (1984) has shown that the process is more correctly one of hydroalcoholysis at room temperature. Thus, the aromatic aldehydes of Armagnac could be produced by heating oak wood in an aqueous medium (hydrolysis) or heating with Armagnac (alcoholysis). On the other hand, treatment with absolute alcohol (ethanolysis) yields a mixture of phenyl propanoid ketones known as Hibbert's ketones (Hibbert, 1942; Hibbert *et al.*, 1939), which are not found in matured spirits. Figure 10 shows an HPLC trace of aromatic aldehydes in Armagnac.

Lehtonen (1984) analyzed the aromatic aldehydes in whisky, cognac, and rum. Vanillin and syringaldehyde were found in each of them; coniferaldehyde and sinapaldehyde were found in whisky and cognac, and salicylaldehyde only in whisky. The structures of these compounds are shown in Fig. 11.

Two coumarin derivatives believed to be formed from related sources, scopoletin and escutelin, have been reported in cognac (Joseph and Marche, 1972). Their structures are shown in Fig. 12.

Nishimura and co-workers (1983) studied the effect of charring or toasting casks during construction. Charring is commonly practiced in the United States, while treatments in France and Spain are milder (i.e., toasting). Casks used in Scotch whisky production are occasionally decharred before use. Nishimura and co-workers found that charred-oak casks contained much higher amounts of aromatic aldehydes than uncharred ones; moreover, quantities increased with toasting temperature up to 200°C, whereas charred casks contained lesser amounts, possibly because of carbonization of aromatics. Nevertheless, heating wood appears to break down lignin and facilitate aromatic aldehyde formation. It was also shown that ethanolysis of native lignin (i.e., the portion that is soluble in alcohol) is not the major source of lignin-related compounds in spirits. The proposed pathways for lignin-derived flavor component formation are shown in Fig. 13.

It was suggested that for bourbon whisky pathway A → B → D is more relevant, while pathway C → D is the main route for Scotch whisky and cognac.

2. Sugars

Matured spirit contains a number of pentose and hexose sugars (Otsuka *et al.*, 1963; Black and Andreasen, 1974). Although their concentrations are too low to impart any significant sweetening effect, it has been speculated that their contribution may be of the same order as that of the aromatic aldehydes (Salo *et al.*, 1976). Arabinose, glucose, xylose, galactose, and rhamnose have been shown to increase with time in a hyperbolic manner and probably result from hemicellulose breakdown. Fructose and glycerol have also been detected, but in these cases the rate of formation is linear, suggesting an alternative mechanism (Reazin, 1983). Glucose and arabinose together account for around two-thirds of the total sugars in matured spirits.

3. Lactones

One group of compounds that characterize matured spirit and have attracted the attention of many research workers are lactones. "Oak lactone" (β-methyl-γ-octalactone) was first reported in 1970 (Suomalainen and Nykanen, 1970). Later Nishimura and Masuda (1971) identified both cis and trans isomers in Scotch and Japanese whiskies. Onishi *et al.* (1977) found oak lactone more abundant in American oak than European oak, and lesser amounts were obtained in brandies matured in reused casks. Swan *et al.* (1981), found both isomers, together with γ-nonalactone and β-nonalactone in matured Scotch whisky. The configuration of the two isomers of oak lactone are shown in Fig. 14.

Oak lactone has a coconutlike aroma and together with γ-nonalactone and eugenol can contribute a matured flavor (Nishimura *et al.*, 1983). Otsuka and co-workers (1974) studied the sensory characteristics of both isomers; they reported the odor threshold of the trans isomer to be 0.067 mg/liter and that of the cis isomer about 10 times higher. They also found the cis isomer to be more fragrant.

Otsuka *et al.* (1980) has suggested that oak lactone may be formed from a precursor containing both oak lactone and gallic acid-type moieties.

C. Changes in Volatile Aroma Compounds

In the radioactivity study Reazin and co-workers (1976) also studied the production of volatile aroma compounds. They showed that in all ∿0.3% of the original ethanol is transformed into components during maturation. Acetaldehyde, acetic acid, and ethyl acetate were all formed from ethanol, but they reasoned that more acetic acid was formed from other sources. Nishimura and co-workers (1983) confirmed this finding and suggested that acetic acid is formed from the acetyl group of hemicellulose by degradation. They also confirmed that acetaldehyde increase requires the presence of oak wood and air. The study by Reazin and coworkers concluded that the law of mass action governs the formation of ethyl acetate.

Puech and co-workers (1984) have studied the changes in flavor compounds in cognac from the Grand Champagne region produced over a period of 50 years. Table IV shows the changes in ethyl esters over this period, and acetaldehyde and miscellaneous compounds are shown in Table V.

Organosulfur compounds are often noticeable in new spirit aroma but disappear (or become masked) during maturation. The presence of organosulfur compounds may necessitate extended maturation time (and cost) in order to mask their undesirable influence. Nishimura *et al.* (1983) showed that dimethyl sulfide and dimethyl disulfide decrease during maturation but require the presence of oak wood. Similarly, methionyl acetate and ethyl methionate decrease but require the presence of both oak wood and air.

D. Miscellaneous Compounds Produced by Cask Pretreatment

Section III,B,1 refers to the role of toasting or charring casks in increasing the availability of aromatic aldehydes. Nishimura and co-workers (1983) also found an increase in furfural, guaiacol, and 4-methyl guaiacol with roasting, and in addition they detected 2-hydroxy-3-methyl-2-cyclopentenone and maltol, which they believe to be principal contributors to a sweet, burnt aroma. They were able to identify the presence of these compounds in bourbon whisky but not in Scotch.

E. Role of Sherry Casks

Traditionally Scotch whisky was matured in sherry casks shipped to the United Kingdom. Apart from imparting a distinct sherrylike aroma to matured Scotch, sherry casks have an effect that cannot be imitated by simply adding sherry. Although the chemistry of Spanish sherry production is well documented (Amerine *et al.*, 1972; Criddle *et al.*, 1983), little has been published on the effect on spirit matured in sherry casks. Sharp (1983) noted that high tannin levels, if present with high hexose levels, may indicate the use of sherry casks.

F. Studies on the Causes of Mellowing

When rapidly frozen samples of matured whisky are examined by differential scanning colorimetry (DSC), an interaction is observed between ethanol and water that is not present in aqueous alcohol solutions or unmatured whisky (Akahoshi, 1963; Koga and Yoshizumi, 1977, 1979). In Fig. 15 peak 3 demonstrates the strong interaction for the melting of water and ethanol, while peaks 1 and 4 are attributable to the melting of free water and ethanol. Studies by Nishimura and co-workers (1983) showed that this interaction is lost if the matured spirit is distilled, but is regained if the residue from the distillation is recombined with the distillate.

Further study showed that inorganic salts contribute 65% of the alcohol-water interactions, with the remainder coming largely from wood components such as lignins and tannins.

G. Influence of Warehousing Conditions on Alcohol Losses and Flavor Production

1. Ethanol and Water Losses

During the initial weeks after filling into casks there is a settling-in period during which alcohol and water may be lost from the bulk within the cask in variable amounts while the inner surface of the wood is permeated and leaks sealed. Thereafter alcohol and water continue to diffuse through the staves and evaporate from the wood surface. Environmental features within the warehouse--namely, temperature, humidity, and ventilation--are then intimately related to the relative losses of alcohol and water and to flavor development (Liebmann and Scherl, 1949).

If warehouse humidity is low the rate of loss of water is greater than that of alcohol, whereas diffusion of water is depressed when humidity is high and the alcoholic strength of the contents of the cask steadily declines with age. Guymon and Crowell (1970) have shown that casks stored at 20°C and 40% RH may gain 6% in alcoholic strength over 32 months, compared to losses of up to 1.2% at 80% RH. The influence of RH and temperature on alcoholic strength have been shown to be linear. Nevertheless, there is an overall loss of alcohol in all locations during maturation. Baldwin and Andreasen (1974) quote figures ∿3% per annum in barrels and average temperatures of 65°F (18°C) in the United States. Losses in Scotland average 2.2% per annum in hogsheads (J. D. Gray and J. S. Swan, unpublished data, 1984) and average temperatures of ∿18°F (9°C). In France losses of 2 to 4%, averaging 2.8% in 350-liter casks, have been reported (Lafon *et al.*, 1973). Kluezko (1978) quoted figures of 3 to 5% per annum for Australian brandy maturation.

2. Flavor Production

Clearly the temperature of maturation would be expected to influence the chemical reactions and extractions already discussed. Reazin (1983) studied the effect of average temperatures ranging from 18.1 to 23°C upon flavor component development; his results are shown in Table VI.

In this study acetaldehyde and nonvolatile acids increased by up to 48%, being the greatest difference found, while oak lactone and furfural did not appear to change at all. The majority of components increased at 4% per degree Celsius. Reazin pointed out, however, that although the reactions could be increased with temperature there does exist an optimum temperature to produce the desired product quality. It should also be noted that the changing alcoholic strength will influence the time in the maturation cycle at which the optimum reaction rates relative to the alcoholic strength will occur.

Hasuo and co-workers (1983) studied the effect of environmental humidity on flavor component development in experimental white oak casks of only 2.5-liter capacity. Acetic acid, acetaldehyde, and ethyl acetate increased with time but did not appear to be dependent on humidity. By contrast, the increase in phenolic compounds was lowest at low humidity and that in vanillin was highest. It was claimed that the sample stored at the lowest of the three humidities tested had the best sensory characteristics.

IV. CONCLUSION

The maturation of potable spirits in oak wood is extremely complex, and the practice varies not only from one spirit-producing region to another but to a lesser extent among different distillers. Since the turn of the century, however, the chemical mechanisms taking place and the role of oak wood have been gradually elucidated, and the principles are now understood.

To date few attempts have been made to relate the process changes to sensory quality or commercial optimization. With the advent of recent instrumental and computing techniques, future developments may be concentrated in this area.

Table I

Hyperbolic Equation Constants[a] for Selected Components in Bourbon Matured for 12 years.

Original Cask Strength % Vol	A	B	D[b]	Std Error
	Colour			
55	0.490	0.218	0.0637	0.0642
63	0.489	0.271	0.108	0.0453
	Solids			
55	56.6	49.1	0.124	8.15
63	61.3	38.7	0.111	4.00
	Total Acids			
55	29.8	3.20	-0.0122	4.80
63	24.0	5.92	0.0526	3.08
	Esters			
55	2.08	3.81	-0.0182	3.81
63	2.08	3.87	-0.00849	3.87
	Aldehydes			
55	2.03	0.870	0.0348	0.526
63	2.28	0.533	0.00481	0.483
	Furfural			
55	0.945	0.171	0.0939	0.0519
63	0.775	0.531	0.375	0.0512
	Tannins			
55	29.8	3.20	-0.0122	4.80
63	24.0	5.92	0.0526	3.08

Source : Baldwin and Andreasen, 1974.

[a] Hyperbolic equation $y = (A + Bt)/(1 + Dt)$ where y = individual component concentration, t = time in years, A, B & D are constants.

[b] When D is 0 or very small (c ± 0.02) the hyperbolic curve reduces to a straight line.

Table II

Lignin and Tannin Contents in Various Oak Woods

Type of Oak	Lignin (mg/g wood)	Tannin (mg/g wood)
French Oaks		
Troncais	289	135
	315	96
	307	84
Limousin	322	73
	288	154
	334	89
Gascony	300	111
	299	153
	322	120
	291	105
	286	150
	294	80
	299	82
	328	81
Bulgarian Oak	300	79
Russian Oak	315	105
American White Oak	336	39

Source : Puech, 1984.

Table III

Extraction of Methoxyl Compounds and Tannin from various Oak Woods

Type of Oak	OCH_3 extract (mg/g dry wood)	% methoxyl compounds extracted	Tannin extract (mg/g dry wood)	% tannins extracted
French Oaks				
Tronçais	2.25	4.2	86	63.7
	2.84	4.8	62.6	65.2
	2.93	5.1	52.6	62.4
Limousin	1.52	2.5	29.9	41
	2.42	4.5	102	66.1
	1.97	3.2	59	66.5
Gascony	2.29	4.1	64.6	58.3
	3.25	5.8	82.3	54
	1.87	3.1	56.9	47.4
	1.81	3.3	60.8	57.7
	2.17	4.1	90.3	60.2
	2.27	4.1	48.3	60.7
	1.89	3.4	40	48.7
Bulgarian Oak	1.86	3.3	39.5	49.9
Russian Oak	1.91	4.7	58.3	55.5
American White	1.91	3.0	19.7	51

Table IV

Ethyl esters in different cognacs from 0 to 50 years in wood. Source : Peuch *et al* (1984)

[a] Results in mg/litre.

Age (Years)	0	1	3	5
Ethyl acetate	348,0	365,0	419,0	484,0
Ethyl propionate	2,4		1,7	2,0
Ethyl caproate	3,9	2,4	5,0	T
Ethyl caprylate	23,4	19,7	38,9	26,9
Ethyl caprate	64,2	56,3	85,4	56,5
Ethyl laurate	34,4	29,3	39,3	24,9
Ethyl myristate	8,8	6,7	6,3	4,3
Ethyl palmitate	13,0	11,4	13,0	8,2
Ethyl palmitoleate	2,0	1,3	1,7	1,2
Ethyl stearate	0,8	0,9	0,7	0,4
Ethyl oleate	2,0	2,1	1,7	1,3
Ethyl linoleate	10,8	9,6	10,1	6,1
Ethyl linolenate	3,0	2,5	2,6	1,6

Age (Years)	10	15	25	50
Ethyl acetate	668,0	622,0	971,0	1458,0
Ethyl propionate	1,3	1,8	3,7	0,7
Ethyl caproate	5,9	3,7	8,4	9,4
Ethyl caprylate	51,7	42,0	63,0	116,0
Ethyl caprate	82,2	64,9	95,4	128,0
Ethyl laurate	31,0	24,9	40,8	51,7
Ethyl myristate	6,6	6,6	10,5	13,8
Ethyl palmitate	13,7	18,8	18,5	36,5
Ethyl palmitoleate	2,3	2,7	2,2	3,8
Ethyl stearate	0,9	1,5	1,3	3,0
Ethyl oleate	2,8	4,2	4,0	15,6
Ethyl linoleate	12,0	15,3	12,7	20,6
Ethyl linolenate	2,6	3,1	2,2	3,3

Table V
Miscellaneous Compounds in Different Cognacs from 0 to 50 years in wood
Source : Puech *et al* (1984)

Age (Years)	0	1	3	5
Acetaldehyde	22,0	16,5	27,5	24,9
1,1-diethoxymethane	T	T	T	T
1,1-diethoxyethane	42,0	29,0	57,0	53,0
Isobutyl acetate	2,8	3,1	2,9	3,2
Isoamyl acetate	5,7	4,0	6,4	2,8
2-Nonanone	0,25	0,15	0,43	0,24
1,1,3-triethoxypropane	1,1	1,7	1,4	1,9
Ethyl-3-ethoxypropionate	0	0,4	T	0,5
Ethyl lactate	214,0	268,0	200,0	281,0
Diethyl succinate	6,2	5,1	3,9	3,0
Furfural	20,1	16,4	22,8	22,8
1,1-diethoxyacetone	0	0,04	0,56	0,80
β-methyl-γ-octalactone	0	0,4	0,44	0,34

Age (Years)	10	15	25	50
Acetaldehyde	59,0	79,0	105,0	158,0
1,1-diethoxymethane	0,6	1,0	3,8	9,0
1,1-diethoxyethane	64,0	92,0	120,0	153,0
Isobutyl acetate	3,0	4,1	3,2	3,2
Isoamyl acetate	4,8	1,1	2,6	8,8
2-Nonanone	0,38	0,27	0,29	T
1,1,3-triethoxypropane	2,4	2,4	1,0	2,3
Ethyl-3-ethoxypropionate	0,7	0,65	0,6	2,1
Ethyl lactate	231,0	215,0	258,0	483,0
Diethyl succinate	3,5	5,1	5,7	33,1
Furfural	20,9	20,2	21,2	22,2
1,1-diethoxyacetone	0,90	1,6	1,5	6,9
β-methyl-γ-octalactone	0,80	0,9	1,3	2,9

[a] Results in mg/litre

T = Trace amounts

Table VI

Influence of temperature upon Flavour Component Development

	Tier Level		
	Bottom	Middle	Top
Average temperature, °C	18.1	19.2	23.0
Proof, °P	101.5	103.9	105.5
Congeners, g/100PL			
Acetaldehyde	1.9	2.2	2.8
Acids: Fixed	6.0	6.3	8.9
Volatile	43.0	46.7	52.1
Ethyl acetate	24.0	27.0	31.0
Furfural	1.0	1.2	0.9
Oak lactone	0.3	0.3	0.3
Solids	131.0	138.0	173.0
Sugars: Arabinose	7.3	7.7	8.9
Fructose	0.6	0.6	0.8
Glucose	4.5	4.8	5.8
Tannins	39.0	40.0	47.0

Source : Reazin, 1983.

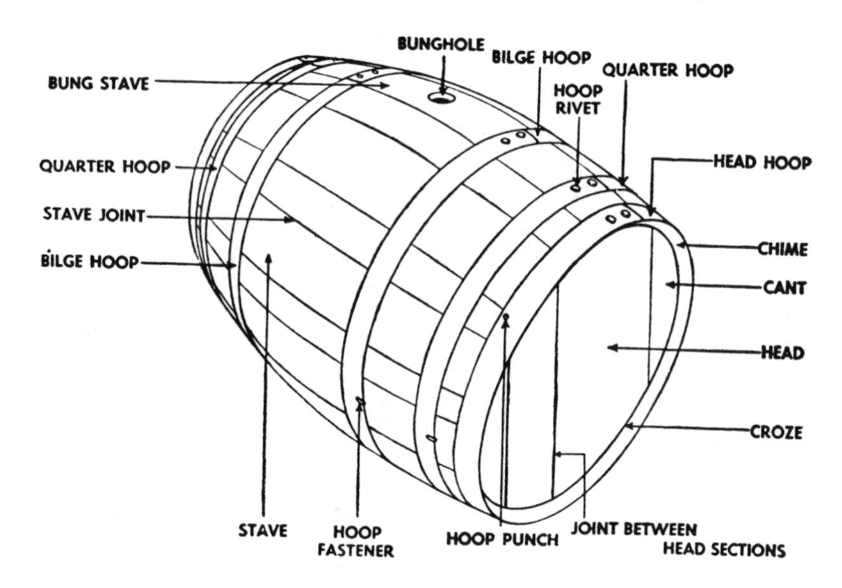

Fig. 1. Construction details of a maturation cask (from Hankerson, 1947).

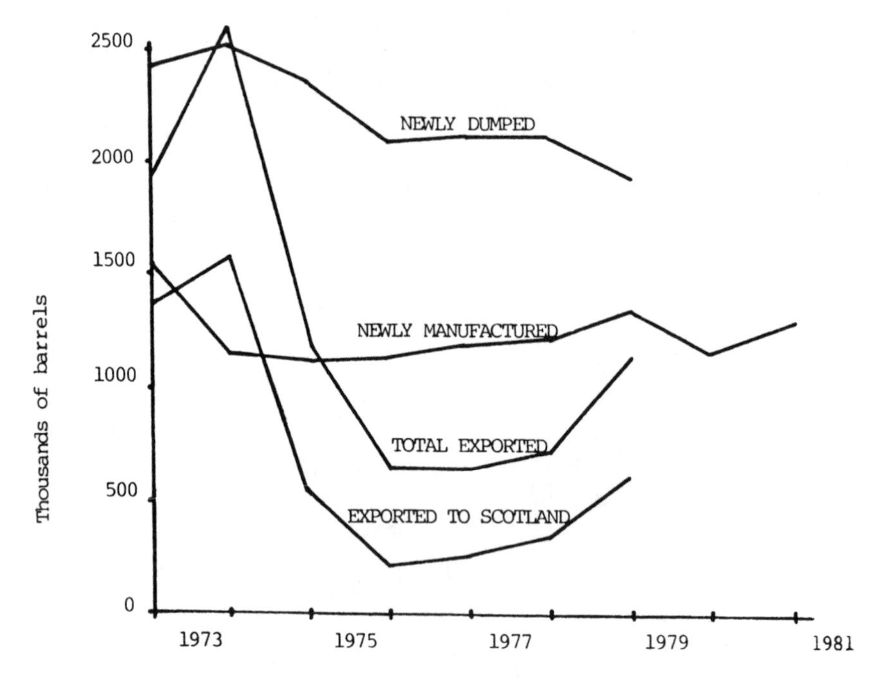

Fig. 2. Production, usage, and export of U.S. whisky barrels (from Williams, 1983).

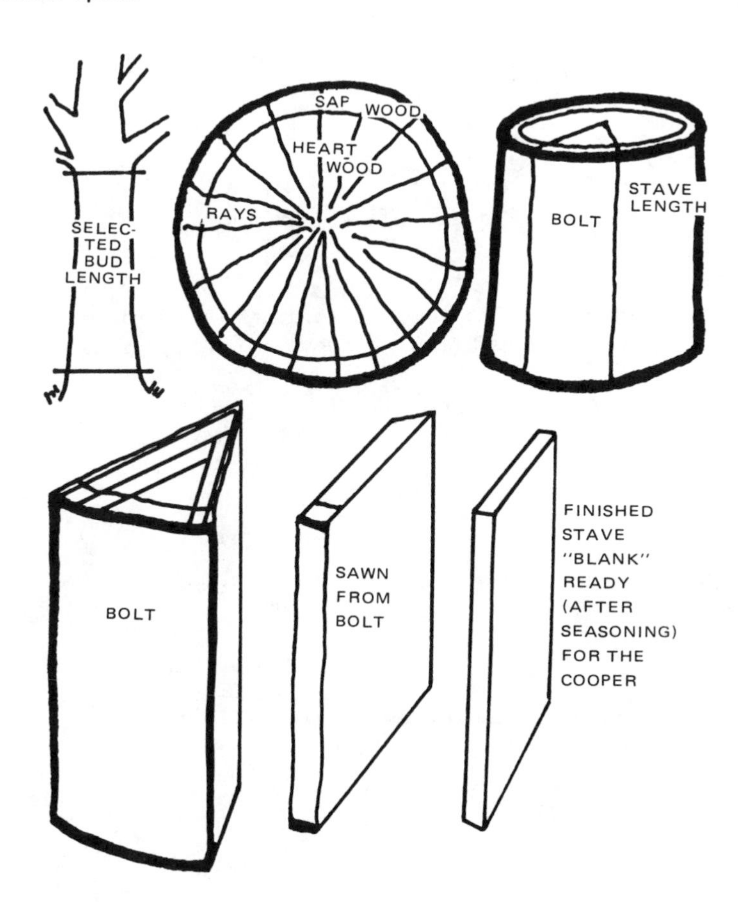

Fig. 3. Preparation of wood for cask production (from Williams, 1983).

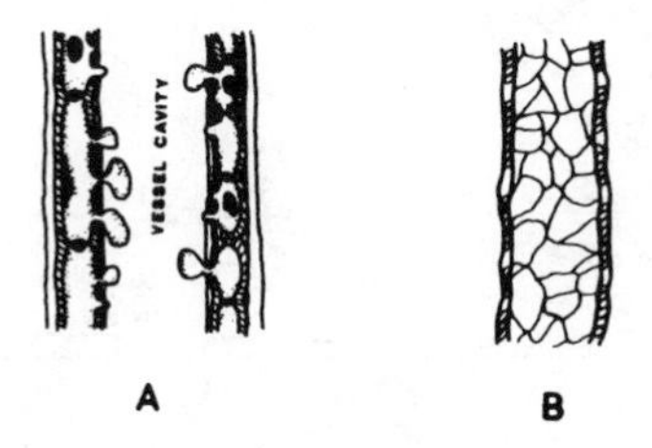

Fig. 4. Tyloses at the bud stage (A) and at a later stage filling the vessel cavity (B), both in lateral view (from Sjöström, 1981).

Fig. 5. Precursors of hardwood lignin. (A) Coniferyl alcohol, "guaiacol type"; (B) sinapyl alcohol, "syringyl type".

Fig. 6. Polyphenolic moieties: (A) gallic acid, (B) ellagic acid, and (C) gallocatechin.

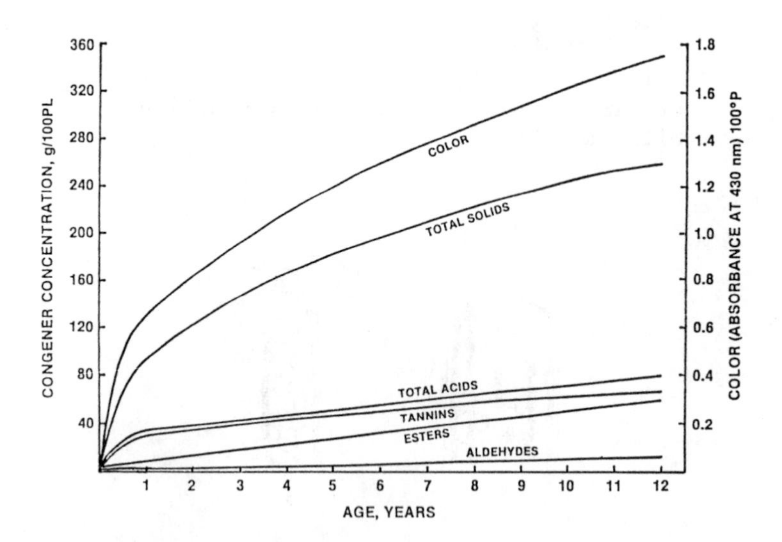

Fig. 7. Component changes during maturation of bourbon whisky (from Reazin, 1983).

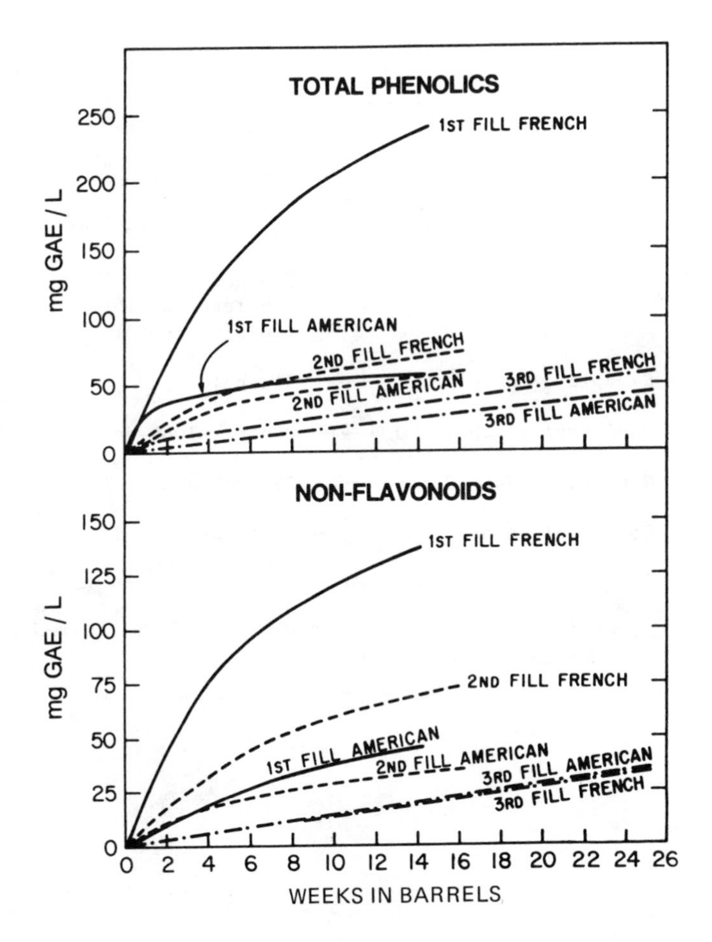

Fig. 8. Changes in phenolics over time for French and American oak casks (from Rous and Alderson, 1983).

Wood-lignin$_x$ + Ethanol	⟶	Wood-lignin$_{(x-n)}$ + Ethanol-lignin
Wood-lignin$_x$ + Ethanol	⟶	Coniferyl alcohol + Sinapic alcohol
Ethanol-lignin	⟶	Ethanol + Coniferyl alcohol + Sinapic alcohol
Sinapic alcohol + O_2	⟶	Sinapaldehyde
Coniferyl alcohol + O_2	⟶	Coniferaldehyde
Sinapaldehyde + O_2	⟶	Syringaldehyde
Coniferaldehyde + O_2	⟶	Vanillin

Fig. 9. Proposed mechanism for aromatic aldehyde formation (from Reazin *et al.*, 1976).

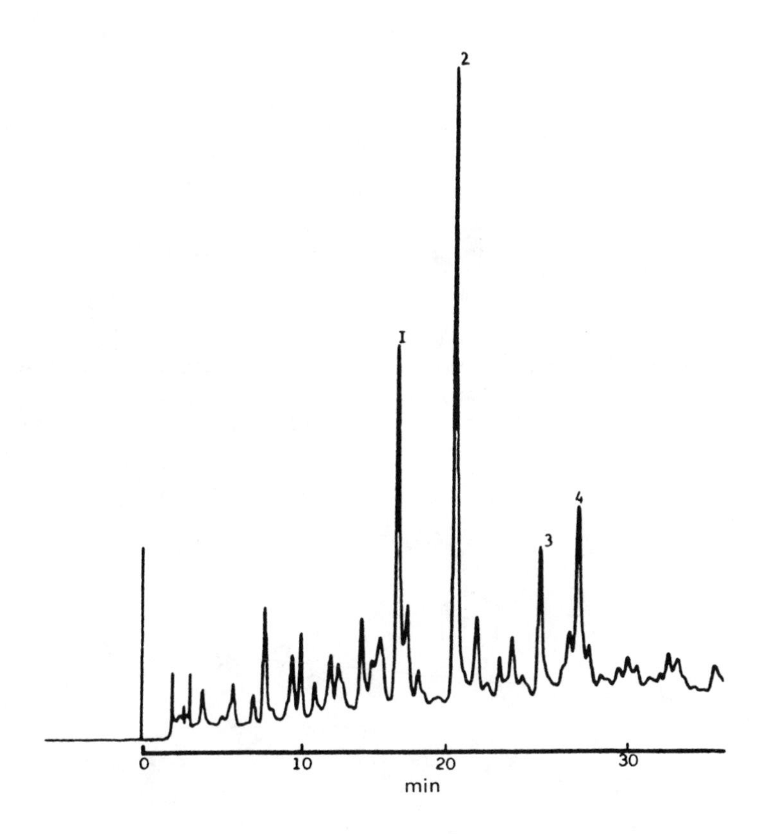

Fig. 10. Oak wood hydroalcoholysis by Armagnac. Peaks: 1, vanillin; 2, syringaldehyde; 3, coniferaldehyde; 4, sinpaldehyde (from Puech, 1984).

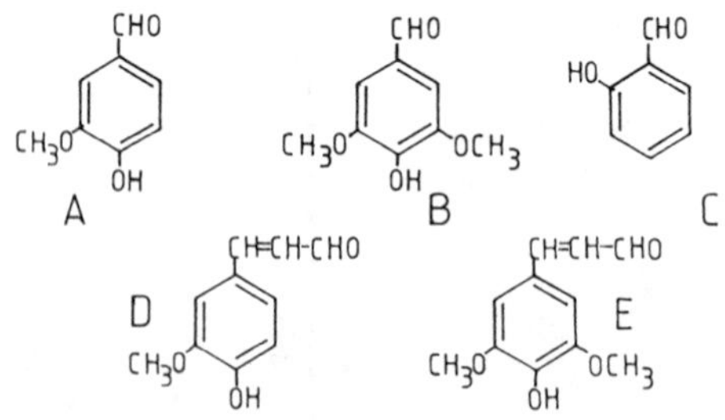

Fig. 11. Structures of aromatic aldehydes found in matured spirits. (A) Vanillin; (B) syringaldehyde; (C) salicylaldehyde; (D) coniferaldehyde; (E) sinapaldehyde.

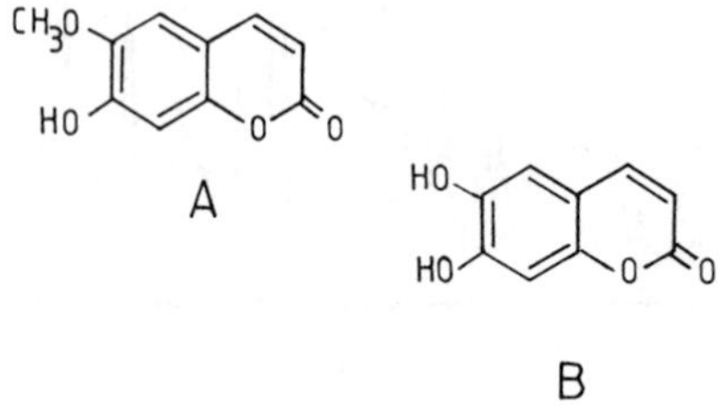

Fig. 12. Structural formulas of (A) scopoletin and (B) escutelin.

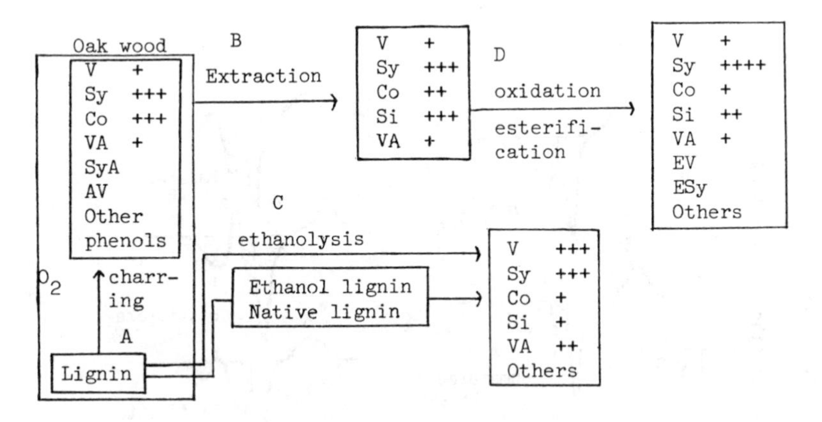

Fig. 13. Proposed pathways for lignin-derived compound formation. V, Vanillin; Sy, syringaldehyde; Co, coniferylaldehyde; Si, sinapaldehyde; VA, vanillic acid; AV, acetovanillone; Asy, acetosyringone; PV, propiovanillone; EV, ethyl vanillate; ESy, ethyl syringate (from Nishimura *et al.*, 1983).

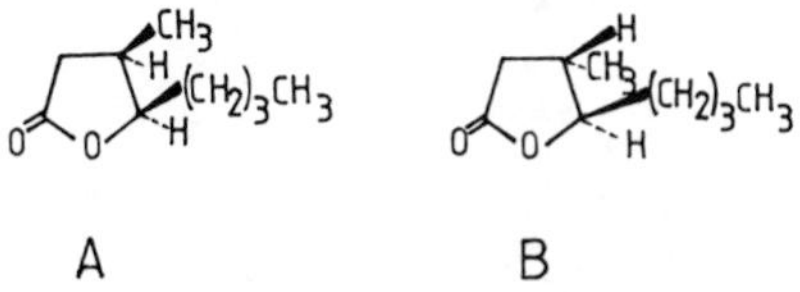

A B

Fig. 14. Configuration of (A) cis and (B) trans isomers of oak lactone.

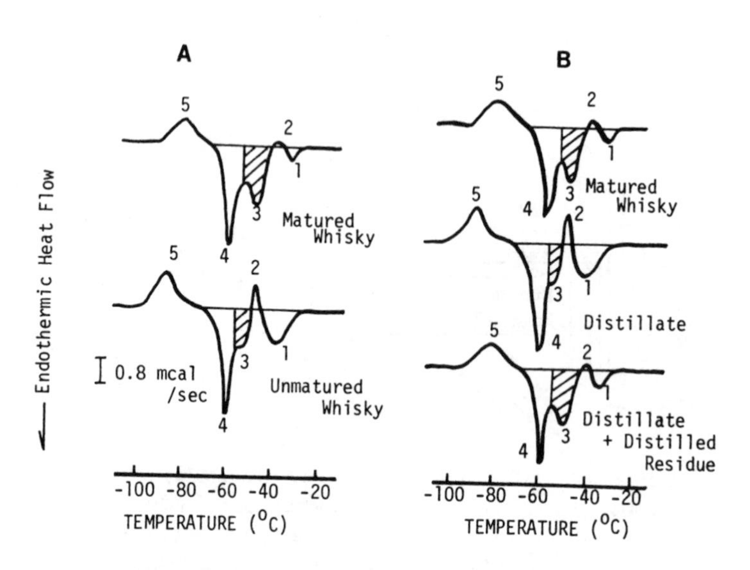

Fig. 15. DSC thermograms of the melting of rapidly frozen matured and unmatured whisky. (B) DSC thermograms of the melting of rapidly frozen matured whisky, distillate of matured whisky, and restored whisky (from Nishimura *et al.*, 1983).

REFERENCES

Akahoshi, R. (1963). *Nippon Nogei Kagaku Kaishi 37* (8), 433-438.

Amerine, M. A., Berg, H. W., and Cruess, W. V. (1972). "The Technology of Wine Making." AVI Publ. Co., Westport, Connecticut.

Baldwin, S., and Andreasen, A. A. (1974). *J. Assoc. Off. Anal. Chem. 57* (4), 940-950.

Black, R. A., and Andreasen, A. A. (1974). *J. Assoc. Off. Anal. Chem. 57*, 111-117.

Bolker, H. I. (1974). "Natural and Synthetic Polymers." Dekker, New York.

Brown, H. P., Banshin, A. J., and Forsaith, C. C. (1949). "Textbook of Wood Technology," Vol. I. McGraw-Hill, New York.

Byrne, K. J., Reazin, G. H., Andreasen, A. A. (1981). *J. Assoc. Off. Anal. Chem. 64*, 181-185.

Chen, C.-L. (1970). *Phytochemistry 9*, 1149.

Crampton, C. A., and Tolman, L. M. (1908). *J. Amer. Chem. Soc. 30*, 98-136.

Criddle, W. J., Goswell, R. W., and Williams, M. A. (1983). *Am. J. Enol. Vitic. 34* (2), 61-71.

Eriksson, O., and Lindgren, B. S. (1977). *Sven Papperstidn. 80*, 59-63.

Gerry, E. (1941). *J. Agric. Res. 1*, 445-469.

Guymon, J. F., and Crowell, E. A. (1970). *Wines & Vines* January, pp. 23-25.

Hankerson, F. P. (1947). "The Cooperage Handbook." Chem. Publ. Co., New York.

Hasuo, T., Saito, K., Terauchi, T., Tadenuma, M., and Sato, S. (1983). *Nippon Jozo Kyokai Zasshi 78* (12), 966-969.

Hathway, D. E., and Jurd, L. (1962). *In* "Wood Extractives" (W. E. Hillis, ed.), pp. 191-258. Academic Press, New York.

Hibbert, H. (1942). *Annu. Rev. Biochem. 11*, 183-202.

Hibbert, H., Cramer, A. B., and Hunter, J. (1939). *J. Am. Chem. Soc. 61*, 509-516.

Joseph, E., and Marche, M. (1972). *Connaiss. Vigne Vin 6* (3), 1-58.

Kluezko, A. (1978). *Aust. Wine, Brew. Spirit Rev.* August, pp. 35-38.

Koga, K., and Yoshizumi, H. (1977). *J. Food Sci. 42*, 1213-1217.

Koga, K., and Yoshizumi, H. (1979). *J. Food Sci. 44*, 1386-1389.

Lafon, J., Couilland, P., and Gaybelile, F. (1973). "Le Cognac." Baillière, Paris.

Lehtonen, P. (1984). *In* "Flavour Research of Alcoholic Beverages" (L. Nykenen and P. Lehtonen, eds.), pp. 121-130. Foundation for Biotechnical and Industrial Fermentation Research, Helsinki.

Liebmann, A. J., and Scherl, B. (1949). *Ind. Eng. Chem. 41* (3), 534-543.

Nishimura, K., and Masuda, M. (1971). *Phytochemistry 10*, 1401-1402.

Nishimura, K., Ohnishi, M., Masuda, M., Koga, K., and Matsuzama, . (1983). *In* "Flavour of Distilled Beverages" (J. R. Piggot, ed.), pp. 241-255. Horwood Ellis/SCI, London.

Onishi, M., Guymon, J. F., and Crowell, E. A. (1977). *Am. J. Enol. Vitic. 28* (3), 152-158.

Otsuka, K., Morinaga, K., and Imai, S. (1963). *Nippon Jozo Kyokai Zasshi 59*, 448; *Chem. Abstr. 63*, 14005e (1965).

Otsuka, K., Zenibayashi, Y., ItoH, M., and Totsuka, A. (1974). *Agric. Biol. Chem. 38* (3), 485-490.

Otsuka, K., Sato, K., and Yamashita, T. (1980). *J. Ferment. Technol. 58*, 395-398.

Parham, R. A., and Gray, R. L. (1984). *In* "The Chemistry of Solid Wood (R. Rowell, ed.), pp. 3-57. Am. Chem. Soc., Washington, D.C.

Pettersen, R. C. (1984). *In* "The Chemistry of Solid Wood" (R. Rowell, ed.), pp. 57-127. Am. Chem. Soc., Washington, D.C.

Puech, J.-L. (1978). Ph.D. Thesis, University of Paul Sabatier de Toulouse.

Puech, J.-L. (1984). *Am. J. Enol. Vitic. 35* (2), 77-81.

Puech, J.-L., Leaute, R., Clot, G., and Nomdedeu, L. (1984). *Sci. Aliment. 4*, 65-80.

Reazin, G. H. (1983). *In* "Flavour of Distilled Beverages" (J. R. Piggot, ed.), pp. 225-240. Horwood Ellis/SCI, London.

Reazin, G. H., Baldwin, S., Scales, H. S., Washington, H. W., and Andreasen, A. A. (1976). *J. Assoc. Off. Anal. Chem. 59* (4), 770-776.

Rickards, P. (1983). *In* "Current Developments in Malting, Brewing and Distilling" (F. G. Priest and I. Campbell, eds.), pp. 199-203. Institute of Brewing, London.

Rous, C., and Alderson, B. (1983). *Am. J. Enol. Vitic. 34* (4), 211-215.

Salo, P., Lehtonen, M., and Suomalainen, H. (1976). *Proc.--Nord. Symp. Sens. Prop. Foods, 4th, 1976*, pp. 87-108.

Schidrowitz, P., and Kaye, P. (1905). *J. Soc. Chem. Ind., London* June, pp. 585-589.

Sharp, R. (1983). *In* "Current Developments in Malting, Brewing and Distilling" (F. G. Priest and I. Campbell, eds.), pp. 143-157. Institute of Brewing, London.

Shortreed, G. W., Rickards, P., Swan, J. S., and Burtles, S. M. (1979). *Brew. Guardian 109*, 1-10.

Singleton, V. L. (1974). *In* "Chemistry of Winemaking" (A. Dinsmoor Webb, ed.), pp. 254-278. Am. Chem. Soc., Washington, D.C.

Sjöström, E. (1981). "Wood Chemistry: Fundamentals and Applications." Academic Press, New York

Suomalainen, H., and Nykanen, L. (1970). *Haringsmiddel-industrien 23*, 15.

Swan, J. S., Howie, D., Burtles, S. M., Williams, A. A., and Lewis, M. J. (1981). *In* "Quality of Foods and Beverages: Chemistry and Technology" (G. Charalambous and G. E. Inglett, eds.), Vol. 1, pp. 201-225. Academic Press, New York.

Valaer, P., and Frazier, W. H. (1936). *Ind. Eng. Chem. 28* (1), 92-105.

Williams, L. I. (1983). *In* "Current Developments in Malting, Brewing and Distilling" (F. G. Priest and I. Campbell, eds.), pp. 193-197. Institute of Brewing, London.

Index*

*For index to Chapter 4, Shelf-Life of Fish and Shellfish, see pages 340–342.

D